CONTEMPORARY MATHEMATICS

483

Algebras, Representations and Applications

Conference in Honour of Ivan Shestakov's 60th Birthday
August 26–September 1, 2007
Maresias, Brazil

Vyacheslav Futorny
Victor Kac
Iryna Kashuba
Efim Zelmanov
Editors

American Mathematical Society
Providence, Rhode Island

2000 *Mathematics Subject Classification.* Primary 17A70, 17B10, 17B65, 17C10, 17D05, 16G60, 16W30, 18D10.

Library of Congress Cataloging-in-Publication Data

Algebras, representations and applications : a conference in honour of Ivan Shestakov's 60th birthday, August 26–September 1, 2007, Maresias, Brazil / Vyacheslav Futorny ... [et al.], editors.
p. cm. — (Contemporary mathematics ; v. 483)
Includes bibliographical references and index.
ISBN 978-0-8218-4652-0 (alk. paper)
1. Representations of algebras—Congresses. I. Shestakov, I. P. II. Futorny, V.

QA150.A467 2009
512—dc22 2008044490

∞ The paper used in this book is acid-free and falls within the guidelines
established to ensure permanence and durability.
Visit the AMS home page at http://www.ams.org/

10 9 8 7 6 5 4 3 2 1 14 13 12 11 10 09

Contents

Preface v

List of Participants vii

Classification of complex naturally graded quasi-filiform Zinbiel algebras
J. Q. Adashev, A. Kh. Khuhoyberdiyev, and B. A. Omirov 1

Akivis superalgebras and speciality
Helena Albuquerque and Ana Paula Santana 13

Properties of some semisimple Hopf algebras
V.A. Artamonov and I.A. Chubarov 23

Classifying simple color Lie superalgebras
Yuri Bahturin and Dušan Pagon 37

Derived categories for algebras with radical square zero
Viktor Bekkert and Yuriy Drozd 55

Tits construction, triple systems and pairs
Pilar Benito and Fabián Martín-Herce 63

Universal enveloping algebras of the four-dimensional Malcev algebras
Murray R. Bremner, Irvin R. Hentzel, Luiz A. Peresi, and Hanid Usefi 73

The supermagic square in characteristic 3 and Jordan superalgebras
Isabel Cunha and Alberto Elduque 91

Monoidal categories of comodules for coquasi Hopf algebras and Radford's formula
Walter Ferrer Santos and Ignacio López Franco 107

Group identities on symmetric units in alternative loop algebras
Edgar G. Goodaire and César Polcino Milies 137

On multiplicity problems for finite-dimensional representations of hyper loop algebras
Dijana Jakelić and Adriano Moura 147

Maximal subalgebras of simple alternative superalgebras
Jesús Laliena and Sara Sacristán 161

Automorphisms of elliptic Poisson algebras
LEONID MAKAR-LIMANOV, UMUT TURUSBEKOVA, and UALBAI UMIRBAEV 169

Jordan superalgebras and their representations
CONSUELO MARTÍNEZ and EFIM ZELMANOV 179

A new proof of Itô's theorem
KURT MEYBERG 195

The ideal of the Lesieur-Croisot elements of a Jordan algebra. II
FERNANDO MONTANER and MARIBEL TOCÓN 199

Unital algebras, Ternary derivations, and local triality
JOSÉ M. PÉREZ-IZQUIERDO 205

A decomposition of LF-quasigroups
PETER PLAUMANN, LIUDMILA SABININA, and LARISSA SBITNEVA 221

Braided and coboundary monoidal categories
ALISTAIR SAVAGE 229

Bases for direct powers
PHILL SCHULTZ 253

Structure and representations of Jordan algebras arising from intermolecular recombination
SERGEI R. SVERCHKOV 261

Preface

The conference "Algebras, Representations and Applications" was held in Maresias, Brazil on August 26 - September 1, 2007. It had 124 participants from 23 countries. The conference was organized by the Department of Mathematics of the University of São Paulo to celebrate the 60th birthday of Ivan Shestakov.

Ivan Pavlovich Shestakov was born in a small village in Eastern Siberia. He distinguished himself early in the mathematical olympiads and as a teenager was invited to the special physical and mathematical boarding school at the Academic Town of Novosibirsk. His masters thesis "On a Class of Non-commutative Jordan Ring" was awarded the Medal of the Academy of Science of USSR for students. Soon after having obtained the BC and PhD degrees at the Novosibirsk State University under the supervision of A.I. Shirshov and K.A. Zhevlakov he became a leading force in the area of nonassociative algebra and carried the torch of the Novosibirsk Nonassociative School.

From 1993 to 1998 Ivan held visiting positions at the Universities of Oviedo and Logroño in Spain. In 1999 he accepted a position of Professor Titular at the University of São Paulo and moved to Brazil.

In 2002 Ivan Shestakov and his former student Ualbai Umirbaev proved the long standing Nagata Conjecture on the existence of wild automorphisms in polynomial algebras. For this remarkable work the authors were awarded the Moore prize of the AMS.

This volume contains selected papers contributed by the participants in the conference. We gratefully acknowledge that the conference would not be possible without the generous financial support from CAPES, CNPq, FAPESP, Instituto Millenium and the Institute of Mathematics and Statistics of the University of São Paulo.

Our special thanks are to Christine Thivierge of the AMS for her invaluable help during the preparation of this volume and to many anonymous referees whose deep remarks greatly improved the manuscripts.

V. Futorny
V. Kac
I. Kashuba
E. Zelmanov

List of Participants

Nicolás Andruskiewitsch
Universidad Nacional de Córdoba, Argentina

Fernando Martins Antoneli Junior
Universidade do Porto, Portugal

Gonzalo Aranda-Pino
Universidad Complutense de Madrid, Spain

Manuel Arenas
Universidad de Chile, Chile

Viatcheslav Artamonov
Moscow State University, Russia

Jesus Antonio Avila-Guzman
UFRGS, Brazil

Yuri Bahturin
Memorial University of Newfoundland, Canada

Viktor Bekkert
UFMG, Brazil

Pilar Benito
Universidad de La Rioja, Spain

Vagner Rodrigues de Bessa
UnB, Brazil

Maria De Nazaré C. Bezerra
UFPA, Brazil

Brian Boe
University of Georgia, USA

Victor Bovdi
Univesity of Debrecen, Hungary

Murray R. Bremner
University of Saskatchewan, Canada

Thomas Bunke
IME-USP, Brazil

André Gimenez Bueno
IME-USP, Brazil

Paula Andrea Cadavid
IME-USP, Brazil

Sheila Campos Chagas
Universidade Federal do Amazonas, Brazil

Flávio Ulhoa Coelho
IME-USP, Brazil

Jones Colombo
UFMT, Brazil

Ben Cox
College of Charleston, USA

José Antonio Cuenca-Mira
Universidad de Malaga, Spain

Onofrio Mario Di Vincenzo
Università di Bari, Italy

Ivan Dimitrov
Queen's University, Canada

Vlastimil Dlab
Carleton University, Canada

Michael Dokuchaev
IME-USP, Brazil

Askar Dzumadil'daev
Kazakh-British University Almaty, Kazakhstan

Alberto Elduque
Universidad de Zaragoza, Spain

Raul Antonio Ferraz
IME-USP, Brazil

Vitor de Oliveira Ferreira
IME-USP, Brazil

Octávio Bernardes Ferreira Neto
IME-USP, Brazil

Walter Ferrer Santos
Universidad de la Republica, Uruguay

Miguel Ferrero
UFRGS, Brazil

Frank Michael Forger
IME-USP, Brazil

Vyacheslav Futorny
IME-USP, Brazil

Paola Andrea Gaviria
IME-USP, Brazil

Dimitry Gitman
IF-USP, Brazil

Nataliya Golovashchuk
Kyiv T.Shevchenko University, Ukraine

Faber Gómez
Universidad de Antioquia, Colombia

Jairo Zacarias Gonçalves
IME-USP, Brazil

Alexander Grishkov
IME-USP, Brazil

Marinês Guerreiro
Universidade Federal de Viçosa, Brazil

Juan Carlos Gutierrez-Fernandez
IME-USP, Brazil

Henrique Guzzo Junior
IME-USP, Brazil

Edson Ryoji Okamoto Iwaki
IME-USP, Brazil

Sandra Mara Alves Jorge
UFMG, Brazil

Orlando Stanley Juriaans
IME-USP, Brazil

Victor Kac
MIT, USA

Seok-Jin Kang
Seoul National University, South Korea

Laiachi El Kaoutit
Universidad de Granada, Spain

Iryna Kashuba
IME-USP, Brazil

Vladislav Khartchenko
Universidad Autonoma de Mexico, Mexico

Plamen Koshlukov
IMECC-UNICAMP, Brazil

Dessislava Koshlukova
IMECC-UNICAMP, Brazil

Mikhail Kotchetov
Memorial University of Newfoundland, Canada

Alexei Krasilnikov
UnB, Brazil

Vladislav Kupriyanov
IF-USP, Brazil

Alicia Labra
Universidad de Chile, Chile

Piroska Lakatos
University of Debrecen, Hungary

Jesus Laliena
Universidad de La Rioja, Spain

Yves Lequain
IMPA, Brazil

Fernando Levstein
Universidad Nacional de Córdoba, Argentina

Fang Li
Zhejiang University, China

Jose Ignacio Liberati
Universidad Nacional de Córdoba, Argentina

Aline de Souza Lima
UnB, Brazil

Tiago Rodrigues Macedo
IMECC-UNICAMP, Brazil

Eduardo do Nascimento Marcos
IME-USP, Brazil

Fabián Martín-Herece
Universidad de La Rioja, Spain

Consuelo Martínez-López
University of Oviedo, Spain

Renato Alessandro Martins
IME-USP, Brazil

Olivier Mathieu
Université Louis Pasteur, France

Kurt Meyberg
Technische Universitaet Muenchen, Germany

Edilson Soares Miranda
UFRGS, Brazil

Daniel Mondoc
Royal Institute of Technology, Sweden

Fernando Montaner
Universidad de Zaragoza, Spain

Pablo Salermo Monteiro
IME-USP, Brazil

Adriano Adrega de Moura
IMECC-UNICAMP, Brazil

Lucia Satie Ikemoto Murakami
IME-USP, Brazil

Daniel K. Nakano
University of Georgia, USA

Sonia Natale
Universidad Nacional de Córdoba, Argentina

Karise Goncalves Oliveira
UnB, Brazil

Sergiy Ovsienko
Kyiv T.Shevchenko University, Ukraine

Vehbi Emrah Paksoy
Pomona College, USA

Irene Paniello-Alastruey
Universidad Pública de Navarra, Spain

Hyungju Park
Korea Institute for Advanced Study (KIAS), South Korea

Jiri Patera
Universite de Montreal, Canada

Luiz Antonio Peresi
IME-USP, Brazil

José M. Pérez-Izquierdo
Universidad de La Rioja, Spain

Victor Petrogradsky
Ulyanovsk State University, Russia

Aline Gomes da Silva Pinto
UnB, Brazil

Bárbara Seelig Pogorelsky
UFRGS, Brazil

Francisco César Polcino-Milies
IME-USP, Brazil

Anderson Luiz Pedrosa Porto
UnB, Brazil

Daniela Moura Prata
IMECC-UNICAMP, Brazil

Michel L. Racine
University of Ottawa, Canada

Flávia Ferreira Ramos
UnB, Brazil

Evander Pereira de Rezende
UnB, Brazil

Roldão da Rocha
Universidade Estadual de Campinas/IFT-UNESP, Brazil

Carlos Rojas-Bruna
Universidad de Chile, Chile

Nikolay Romanovskii
Russian Academy of Sciences, Russia

Liudmila Sabinina
UAEM, Mexico

Ednei Aparecido Santulo Junior
IMECC-UNICAMP, Brazil

Manuel Saorin
Universidad de Murcia, Spain

Anliy Natsuyo Nashimoto Sargeant
IME-USP, Brazil

Alistair Savage
University of Ottawa, Canada

Phill Schultz
University of Western Australia, Australia

Ivan Shestakov
IME-USP, Brazil

Pavel Shumyatsky
UnB, Brazil

Said Najati Sidki
UnB, Brazil

Viviane Ribeiro T. da Silva
UFMG, Brazil

Arkadii Slinko
University of Auckland, New Zealand

Oleg Smirnov
College of Charleston, USA

Antônio Calixto de Souza Filho
IME-USP, Brazil

Alexander Stolin
University of Göteborg, Sweden

Eduardo Tengan
IME-USP, Brazil

Maribel Tocon
Universidad de Cordoba, Spain

Ualbai Umirbaev
Eurasian National University, Kazakhstan

Paula Murgel Veloso
IME-USP, Brazil

Ana Cristina Vieira
UFMG, Brazil

Benjamin Wilson
IME-USP, Brazil/University of Sydney, Australia

Pavel Zalesski
UnB, Brazil

Andrei Zavarnitsine
Sobolev Institute of Mathematics, Russia

Efim Zelmanov
University of California - San Diego, USA

Natalia Zhukavets
Czech Technical University in Prague, Czech Republic

Contemporary Mathematics
Volume **483**, 2009

Classification of complex naturally graded quasi-filiform Zinbiel algebras

J.Q. Adashev, A.Kh. Khuhoyberdiyev, and B.A. Omirov

This work is dedicated to the 60-th anniversary of Professor Shestakov I.P.

ABSTRACT. In this work the description up to isomorphism of complex naturally graded quasi-filiform Zinbiel algebras is obtained.

1. Introduction

In the present paper we investigate algebras which are Koszul dual to Leibniz algebras. The Leibniz algebras were introduced in the work [**L2**] and they present a "non commutative" (to be more precise, a "non antisymmetric") analogue of Lie algebras. Many works, including [**L2**]-[**O1**], were devoted to the investigation of cohomological and structural properties of Leibniz algebras. Ginzburg and Kapranov introduced and studied the concept of Koszul dual operads [**G**]. Following this concept, it was shown in [**L1**] that the category of dual algebras to the category of Leibniz algebras is defined by the identity:

$$(x \circ y) \circ z = x \circ (y \circ z) + x \circ (z \circ y).$$

In this paper, dual Leibniz algebras will called Zinbiel algebras (Zinbiel is obtained from Leibniz written in inverse order). Some interesting properties of Zinbiel algebras were obtained in [**A**], [**D1**], and [**D2**]. In particular, the nilpotency of an arbitrary finite-dimensional complex Zinbiel algebra was proved in [**D2**], and zero-filiform and filiform Zinbiel algebras were classified in [**A**]. The classification of complex Zinbiel algebras up to dimension 4 is obtained in works [**D2**] and [**O2**]. In this work, we present the classification of complex naturally graded quasi-filiform Zinbiel algebras.

Examples of Zinbiel algebras can be found in [**A**], [**D2**] and [**L1**].

We consider below only complex algebras and, for convenience, we will omit zero products the algebra's multiplication table.

1991 *Mathematics Subject Classification.* Primary 17A32.

Key words and phrases. Zinbiel algebra, Leibniz algebra, nilpotency, nul-filiform algebra, filiform and quasi-filiform algebras.

The third author was supported by DFG project 436 USB 113/10/0-1 project (Germany).

2. Preliminaries

DEFINITION 2.1. An algebra A over a field F is called a Zinbiel algebra if for any $x, y, z \in A$ the identity

$$(x \circ y) \circ z = x \circ (y \circ z) + x \circ (z \circ y) \tag{2.1}$$

holds.

For an arbitrary Zinbiel algebras define the lower central series

$$A^1 = A, \quad A^{k+1} = A \circ A^k, \quad k \geq 1.$$

DEFINITION 2.2. A Zinbiel algebra A is called nilpotent if there exists an $s \in N$ such that $A^s = 0$. The minimal number s satisfying this property (i.e. $A^{s-1} \neq 0$ and $A^s = 0$) is called nilindex of the algebra A.

It is not difficult to see that nilindex of an arbitrary n-dimensional nilpotent algebra does not exceed the number $n + 1$.

DEFINITION 2.3. An n-dimensional Zinbiel algebra A is called zero-filiform if $dimA^i = (n + 1) - i$, $1 \leq i \leq n + 1$.

Clearly, the definition of a zero-filiform algebra A amounts to requiring that A has a maximal nilindex.

THEOREM 2.4. **[A]** *An arbitrary n-dimensional zero-filiform Zinbiel algebra is isomorphic to the algebra*

$$e_i \circ e_j = C^j_{i+j-1} e_{i+j}, \quad \text{for } 2 \leq i + j \leq n \tag{2.2}$$

where symbol C^t_s *is a binomial coefficient defined as* $C^t_s = \frac{s!}{t!(s-t)!}$, *and* $\{e_1, e_2, \ldots, e_n\}$ *is a basis of the algebra.*

We denote the algebra from Theorem 2.4 as NF_n.

It is easy to see that an n-dimensional Zinbiel algebra is one-generated if and only if it is isomorphic to NF_n.

DEFINITION 2.5. An n-dimensional Zinbiel algebra A is called filiform if $dimA^i = n - i$, $2 \leq i \leq n$.

The following theorem gives classification of filiform Zinbiel algebras.

THEOREM 2.6. **[A]** *Any n-dimensional* $(n \geq 5)$ *filiform Zinbiel algebra is isomorphic to one of the following three pairwise non isomorphic algebras:*

$$F^1_n : e_i \circ e_j = C^j_{i+j-1} e_{i+j}, \quad 2 \leq i + j \leq n - 1;$$
$$F^2_n : e_i \circ e_j = C^j_{i+j-1} e_{i+j}, \quad 2 \leq i + j \leq n - 1, \quad e_n \circ e_1 = e_{n-1};$$
$$F^3_n : e_i \circ e_j = C^j_{i+j-1} e_{i+j}, \quad 2 \leq i + j \leq n - 1, \quad e_n \circ e_n = e_{n-1}.$$

Since the direct sum of nilpotent Zinbiel algebras is nilpotent one, we shall consider only non split algebras.

Summarizing the results of **[A]**, **[D2]**, and **[O2]**, we give the classification of complex Zinbiel algebras up to dimension ≤ 4.

THEOREM 2.7. *An arbitrary non split Zinbiel algebra is isomorphic to the following pairwise non isomorphic algebras:*

$DimA = 1 :$ *Abelian*

$DimA = 2 : e_1 \circ e_1 = e_2$

$DimA = 3 :$

Z_3^1: $e_1 \circ e_1 = e_2$, $e_1 \circ e_2 = \frac{1}{2}e_3$, $e_2 \circ e_1 = e_3$;
Z_3^2: $e_1 \circ e_2 = e_3$, $e_2 \circ e_1 = -e_3$;
Z_3^3: $e_1 \circ e_1 = e_3$, $e_1 \circ e_2 = e_3$, $e_2 \circ e_2 = \alpha e_3$, $\alpha \in C$;
Z_3^4: $e_1 \circ e_1 = e_3$, $e_1 \circ e_2 = e_3$, $e_2 \circ e_1 = e_3$.

$DimA = 4$:

Z_4^1: $e_1 \circ e_1 = e_2$, $e_1 \circ e_2 = e_3$, $e_2 \circ e_1 = 2e_3$, $e_1 \circ e_3 = e_4$, $e_2 \circ e_2 = 3e_4$, $e_3 \circ e_1 = 3e_4$;
Z_4^2: $e_1 \circ e_1 = e_3$, $e_1 \circ e_2 = e_4$, $e_1 \circ e_3 = e_4$, $e_3 \circ e_1 = 2e_4$;
Z_4^3: $e_1 \circ e_1 = e_3$, $e_1 \circ e_3 = e_4$, $e_2 \circ e_2 = e_4$, $e_3 \circ e_1 = 2e_4$;
Z_4^4: $e_1 \circ e_2 = e_3$, $e_1 \circ e_3 = e_4$, $e_2 \circ e_1 = -e_3$;
Z_4^5: $e_1 \circ e_2 = e_3$, $e_1 \circ e_3 = e_4$, $e_2 \circ e_1 = -e_3$, $e_2 \circ e_2 = e_4$;
Z_4^6: $e_1 \circ e_1 = e_4$, $e_1 \circ e_2 = e_3$, $e_2 \circ e_1 = -e_3$, $e_2 \circ e_2 = -2e_3 + e_4$;
Z_4^7: $e_1 \circ e_2 = e_3$, $e_2 \circ e_1 = e_4$, $e_2 \circ e_2 = -e_3$;
$Z_4^8(\alpha)$: $e_1 \circ e_1 = e_3$, $e_1 \circ e_2 = e_4$, $e_2 \circ e_1 = -\alpha e_3$, $e_2 \circ e_2 = -e_4$, $\alpha \in C$
$Z_4^9(\alpha)$: $e_1 \circ e_1 = e_4$, $e_1 \circ e_2 = \alpha e_4$, $e_2 \circ e_1 = -\alpha e_4$, $e_2 \circ e_2 = e_4$, $e_3 \circ e_3 = e_4$, $\alpha \in C$
Z_4^{10}: $e_1 \circ e_2 = e_4$, $e_1 \circ e_3 = e_4$, $e_2 \circ e_1 = -e_4$, $e_2 \circ e_2 = e_4$, $e_3 \circ e_1 = e_4$;
Z_4^{11}: $e_1 \circ e_1 = e_4$, $e_1 \circ e_2 = e_4$, $e_2 \circ e_1 = -e_4$, $e_3 \circ e_3 = e_4$;
Z_4^{12}: $e_1 \circ e_2 = e_3$, $e_2 \circ e_1 = e_4$;
Z_4^{13}: $e_1 \circ e_2 = e_3$, $e_2 \circ e_1 = -e_3$, $e_2 \circ e_2 = e_4$;
Z_4^{14}: $e_2 \circ e_1 = e_4$, $e_2 \circ e_2 = e_3$;
$Z_4^{15}(\alpha)$: $e_1 \circ e_2 = e_4$, $e_2 \circ e_2 = e_3$, $e_2 \circ e_1 = \frac{1+\alpha}{1-\alpha}e_4$, $\alpha \in C \setminus \{1\}$;
Z_4^{16}: $e_1 \circ e_2 = e_4$, $e_2 \circ e_1 = -e_4$, $e_3 \circ e_3 = e_4$;

Let us introduce some definitions and notations.

The set $R(A) = \{a \in A|\ \ b \circ a = 0 \text{ for any } b \in A\}$ is called the right annihilator of the Zinbiel algebra A.

The set $L(A) = \{a \in A|\ \ a \circ b = 0 \text{ for any } b \in A\}$ is called the left annihilator of the Zinbiel algebra A.

$Z(a, b, c)$ denotes the following polynomial:

$$Z(a, b, c) = (a \circ b) \circ c - a \circ (b \circ c) - a \circ (c \circ b).$$

It is obvious that Zinbiel algebras are determined by the identity $Z(a, b, c) = 0$.

3. Classification of naturally graded quasi-filiform Zinbiel algebras

DEFINITION 3.1. A Zinbiel algebra A is called quasi-filiform if $A^{n-2} \neq 0$ and $A^{n-1} = 0$, where $dimA = n$.

Let A be a quasi-filiform Zinbiel algebra. Putting $A_i = A^i/A^{i+1}$, $1 \leq i \leq n-2$, we obtain the graded Zinbiel algebra

$$GrA = A_1 \oplus A_2 \oplus \ldots \oplus A_{n-2}, \quad \text{where} \quad A_i \circ A_j \subseteq A_{i+j}.$$

An algebra A is called naturally graded if $A \cong GrA$. It is not difficult to see that $A_1 \circ A_j = A_{j+1}$ in the naturally graded algebra A. Let A be an n-dimensional graded quasi-filiform algebra. Then there exists a basis $\{e_1, e_2, \ldots, e_n\}$ of the algebra A such that $e_i \in A_i$, $1 \leq i \leq n-2$. It is evident that $dimA_1 > 1$. In fact, if $dimA_1 = 1$, then the algebra A is one-degenerated and therefore it is the zero-filiform algebra, but it is not quasi-filiform. Without loss of generality, one can assume $e_{n-1} \in A_1$. If for a Zinbiel algebra A, the condition $e_n \in A_r$ holds, the algebra is said to be of type $A_{(r)}$.

3.1. The case $r = 1$.

THEOREM 3.2. *Any n-dimensional ($n \geq 6$) naturally graded quasi-filiform Zinbiel algebra of the type $A_{(1)}$ is isomorphic to the algebra:*

$$KF_n^1 : e_i \circ e_j = C_{i+j-1}^j e_{i+j} \quad \text{for} \quad 2 \leq i+j \leq n-2. \tag{3.1}$$

PROOF. Let an algebra A satisfy the conditions of the theorem. Then there exists a basis $\{e_1, e_2, \dots, e_n\}$ such that

$$A_1 = \langle e_1, e_{n-1}, e_n \rangle, \quad A_2 = \langle e_2 \rangle, \quad A_3 = \langle e_3 \rangle, \quad \dots, \quad A_{n-2} = \langle e_{n-2} \rangle.$$

Using arguments similar to the ones from [**V**], we obtain

$$e_1 \circ e_i = e_{i+1} \quad \text{for} \quad 2 \leq i \leq n-3.$$

Now introduce the notations

$$\begin{array}{lll} e_1 \circ e_1 = \alpha_{1,1} e_2, & e_1 \circ e_{n-1} = \alpha_{1,2} e_2, & e_1 \circ e_n = \alpha_{1,3} e_2, \\ e_{n-1} \circ e_1 = \alpha_{2,1} e_2, & e_{n-1} \circ e_{n-1} = \alpha_{2,2} e_2, & e_{n-1} \circ e_n = \alpha_{2,3} e_2, \\ e_n \circ e_1 = \alpha_{3,1} e_2, & e_n \circ e_{n-1} = \alpha_{3,2} e_2, & e_n \circ e_n = \alpha_{3,3} e_2. \end{array}$$

We consider the following cases:

Case 1. Let $(\alpha_{1,1}, \alpha_{2,2}, \alpha_{3,3}) \neq (0,0,0)$. Then without loss of generality, one can take $\alpha_{1,1} \neq 0$. Moreover, making the change $e_2' = \alpha_{1,1} e_2, \ e_3' = \alpha_{1,1} e_3, \ \dots, \ e_{n-2}' = \alpha_{1,1} e_{n-2}$, we can assume $\alpha_{1,1} = 1$.

Obviously, the linear span $lin\langle e_1, e_2, \dots, e_{n-2} \rangle$ forms zero-filiform Zinbiel algebra. Hence

$$e_i \circ e_j = C_{i+j-1}^j e_{i+j} \quad \text{at} \quad 2 \leq i+j \leq n-2$$

and omitted products of the basic elements $\{e_1, e_2, \dots, e_{n-2}\}$ are equal to zero. Taking into account equalities

$$Z(e_1, e_{n-1}, e_1) = Z(e_1, e_1, e_{n-1}) = Z(e_1, e_{n-1}, e_{n-1}) = Z(e_{n-1}, e_1, e_1) = 0$$

we obtain

$$\alpha_{1,2} = \alpha_{2,1}, \quad e_2 \circ e_{n-1} = 2\alpha_{1,2} e_3, \quad \alpha_{1,2}^2 = \alpha_{2,2}, \quad e_{n-1} \circ e_2 = \alpha_{1,2} e_3.$$

Substituting $e_{n-1}' = -\alpha_{1,2} e_1 + e_{n-1}, \ e_i' = e_i$ for $i \neq n-1$, one can suppose

$$\alpha_{1,2} = \alpha_{2,1} = \alpha_{2,2} = 0.$$

Analogously, we have

$$e_1 \circ e_n = e_n \circ e_1 = e_n \circ e_n = e_2 \circ e_n = e_n \circ e_2 = 0 \quad \text{and} \quad e_n \circ e_{n-1} = e_{n-1} \circ e_n = 0.$$

Hence, using induction and the following chain of equalities

$$e_{n-1} \circ e_i = e_{n-1} \circ (e_1 \circ e_{i-1}) = (e_{n-1} \circ e_1) \circ e_{i-1} - e_{n-1} \circ (e_{i-1} \circ e_1) = -(i-1) e_{n-1} \circ e_i$$

we prove validity of the equality $e_{n-1} \circ e_i = 0$ for $1 \leq i \leq n$.
In the same way, equalities

$$e_{i+1} \circ e_{n-1} = (e_1 \circ e_i) \circ e_{n-1} = e_1 \circ (e_i \circ e_{n-1}) + e_1 \circ (e_{n-1} \circ e_i) = 0$$

lead to $e_{i+1} \circ e_{n-1} = 0, \ 1 \leq i \leq n$. But it means that the algebra A is split, i.e. $A = F_{n-1}^1 \oplus C$, where $F_{n-1}^1 = NF_{n-2} \oplus C$. Hence, $A = KF_n^1$.
Case 2. Let $(\alpha_{1,1}, \alpha_{2,2}, \alpha_{3,3}) = (0,0,0)$. Then

$$(\alpha_{1,2}, \alpha_{2,1}, \alpha_{1,3}, \alpha_{3,1}, \alpha_{2,3}, \alpha_{3,2}) \neq (0,0,0,0,0,0).$$

Put $e_1' = ae_1 + be_{n-1} + ce_n$. Then

$$e_1' \circ e_1' = [ab(\alpha_{1,2} + \alpha_{2,1}) + ac(\alpha_{1,3} + \alpha_{3,1}) + bc(\alpha_{2,3} + \alpha_{3,2})] e_2.$$

From this it follows $\alpha_{1,2}+\alpha_{2,1}=0$, $\alpha_{1,3}+\alpha_{3,1}=0$, and $\alpha_{2,3}+\alpha_{3,2}=0$. In fact, otherwise revert to the conditions of the case 1.
Without loss of generality, we can assume $\alpha_{1,2}=1$. The equality $Z(e_1,e_{n-1},e_1)=0$ implies $e_2\circ e_1=0$. In addition,

$$0=(e_1\circ e_1)\circ e_2=e_1\circ(e_1\circ e_2)+e_1\circ(e_2\circ e_1)=e_1\circ e_3=e_4.$$

Thus we obtain a contradiction with existence of an algebra in case 2. □

PROPOSITION 3.3. *Let A be a five-dimensional naturally graded quasi-filiform Zinbiel algebra of type $A_{(1)}$. Then it is isomorphic to one of the following three pairwise non isomorphic algebras:*

$$KF_5^1: e_1\circ e_1=e_2,\quad e_1\circ e_2=e_3,\quad e_2\circ e_1=e_3;$$
$$KF_5^2: e_1\circ e_4=e_2,\quad e_1\circ e_2=e_3,\quad e_4\circ e_1=-e_3;$$
$$KF_5^3: e_1\circ e_4=e_2,\quad e_1\circ e_2=e_3,\quad e_4\circ e_1=-e_3,\quad e_4\circ e_1=e_3.$$

PROOF. Let an algebra A satisfy the conditions of the proposition. If the conditions of the case 1 of Theorem 3.2 hold, then A is isomorphic to the algebra KF_5^1. if the conditions of the case 2 of Theorem 3.2 are valid for A, we obtain the following multiplication in A :

$$\begin{array}{lll} e_1\circ e_4=e_2, & e_1\circ e_5=\alpha_{1,3}e_2, & e_1\circ e_2=e_3,\\ e_4\circ e_1=-e_2, & e_4\circ e_5=\alpha_{2,3}e_2, & e_4\circ e_2=\beta_1e_3,\\ e_5\circ e_1=-\alpha_{1,3}e_2, & e_5\circ e_4=-\alpha_{2,3}e_2, & e_5\circ e_2=\beta_2e_3. \end{array}$$

Changing basic elements by the rules

$$e_1'=e_1,\quad e_2'=e_2,\quad e_3'=e_3,\quad e_4'=-\beta_1e_1+e_4,\quad e_5'=\alpha_{2,3}e_1-\alpha_{1,3}e_4+e_5,$$

we can assume $\alpha_{1,3}=\alpha_{2,3}=\beta_1=0$. If $\beta_2=0$, then we obtain the algebra KF_5^2. But if $\beta_2\neq 0$, then putting $e_5'=\frac{1}{\beta_2}e_5$, we have $\beta_2=1$ and obtain isomorphism to algebra KF_5^3. By virtue of $dimL(KF_5^2)=3$, $dimL(KF_5^3)=2$, and taking into account that generating elements of algebras KF_5^2 and KF_5^3 satisfy the identity $x\circ x=0$, but this identity does not hold for generating elements of the algebra KF_5^1 (in particular, $e_1\circ e_1=e_2$), we obtain pairwise non isomorphic of obtained algebras. □

3.2. The case $r=2$. Classification of naturally graded quasi-filiform Zinbiel algebras of type $A_{(2)}$ is given in the following theorem.

THEOREM 3.4. *Any n-dimensional ($n\geq 8$) naturally graded quasi-filiform Zinbiel algebra of type $A_{(2)}$ is isomorphic to one of the following pairwise non isomorphic algebras:*

$$KF_n^1:\begin{cases} e_i\circ e_j=C_{i+j-1}^j e_{i+j}, & 2\leq i+j\leq n-2,\\ e_1\circ e_{n-1}=e_n, & e_{n-1}\circ e_1=\alpha e_n; \end{cases}$$

$$KF_n^2:\begin{cases} e_i\circ e_j=C_{i+j-1}^j e_{i+j}, \quad 2\leq i+j\leq n-2,\\ e_1\circ e_{n-1}=e_n,\quad e_{n-1}\circ e_1=e_n,\quad e_{n-1}\circ e_{n-1}=e_n; \end{cases}$$

$$KF_n^3:\begin{cases} e_i\circ e_j=C_{i+j-1}^j e_{i+j}, & 2\leq i+j\leq n-2,\\ e_1\circ e_{n-1}=e_n, & e_{n-1}\circ e_{n-1}=e_n; \end{cases}$$

$$KF_n^4:\begin{cases} e_i\circ e_j=C_{i+j-1}^j e_{i+j}, & 2\leq i+j\leq n-2,\\ e_{n-1}\circ e_1=e_n. \end{cases}$$

PROOF. Let an algebra A satisfy the conditions of theorem and $\{e_1, e_2, \ldots, e_n\}$ be a basis of A such that $A_1 = \langle e_1, e_{n-1}\rangle$, $A_2 = \langle e_2, e_n\rangle$, $A_i = \langle e_i\rangle$ for $3 \le i \le n-2$. Applying the arguments similar to those used in the proof of Theorem 3.2, we obtain the following multiplications of basic elements, which are generators of the algebra:

$$e_1 \circ e_1 = e_2, \qquad e_1 \circ e_{n-1} = \alpha_3 e_2 + \alpha_4 e_n,$$
$$e_{n-1} \circ e_1 = \alpha_5 e_2 + \alpha_6 e_n, \quad e_{n-1} \circ e_{n-1} = \alpha_7 e_2 + \alpha_8 e_n.$$

The proof of the theorem is complete due to the consideration of possible cases for structural constants and taking appropriate transformations of basis. □

THEOREM 3.5. *Any five-dimensional naturally graded quasi-filiform Zinbiel algebra of type $A_{(2)}$ is isomorphic to one of the following pairwise non isomorphic algebras:*

$$KF_5^1: \begin{cases} e_1 \circ e_1 = e_2, \quad e_4 \circ e_4 = e_5, \quad e_1 \circ e_2 = e_3, \quad e_4 \circ e_5 = e_3, \\ e_2 \circ e_1 = 2e_3, \quad e_5 \circ e_4 = 2e_3; \end{cases}$$

$$KF_5^2: \begin{cases} e_1 \circ e_1 = e_2, \quad e_1 \circ e_2 = e_3, \quad e_1 \circ e_4 = e_5, \quad e_1 \circ e_5 = e_3, \\ e_2 \circ e_1 = 2e_3, \quad e_4 \circ e_1 = -e_2, \quad e_4 \circ e_2 = -e_3, \quad e_4 \circ e_5 = -e_3; \end{cases}$$

$$KF_5^3(\beta): \begin{cases} e_1 \circ e_1 = e_2, \quad e_1 \circ e_2 = e_3, \quad e_1 \circ e_4 = e_5, \quad e_1 \circ e_5 = \beta e_3, \\ e_2 \circ e_1 = 2e_3, \quad e_2 \circ e_4 = (\beta-1)e_3, \quad e_4 \circ e_1 = -e_2, \quad e_4 \circ e_4 = -e_5, \\ e_4 \circ e_2 = -e_3, \quad e_4 \circ e_5 = -\beta e_3, \quad e_5 \circ e_1 = (\beta-1)e_3, \quad e_5 \circ e_4 = -2\beta e_3, \quad \beta \in C; \end{cases}$$

$$KF_5^4: \begin{cases} e_1 \circ e_1 = e_2, \quad e_1 \circ e_2 = e_3, \quad e_1 \circ e_4 = e_5, \quad e_2 \circ e_1 = 2e_3, \\ e_4 \circ e_4 = -e_5; \end{cases}$$

$$KF_5^5: \begin{cases} e_1 \circ e_4 = e_2, \quad e_1 \circ e_2 = e_3, \quad e_1 \circ e_5 = -e_3, \quad e_4 \circ e_1 = e_5, \\ e_4 \circ e_2 = e_3, \quad e_4 \circ e_5 = -e_3; \end{cases}$$

$$KF_5^6: e_1 \circ e_4 = e_2, \quad e_1 \circ e_2 = e_3, \quad e_1 \circ e_5 = -e_3, \quad e_4 \circ e_1 = e_5;$$

$$KF_5^7: \begin{cases} e_1 \circ e_1 = e_2, \quad e_1 \circ e_2 = e_3, \quad e_1 \circ e_4 = e_5, \quad e_1 \circ e_5 = e_3, \\ e_2 \circ e_1 = 2e_3, \quad e_2 \circ e_4 = e_3, \quad e_5 \circ e_1 = e_3; \end{cases}$$

$$KF_5^8: e_1 \circ e_1 = e_2, \quad e_1 \circ e_2 = e_3, \quad e_1 \circ e_4 = e_5, \quad e_2 \circ e_1 = 2e_3;$$

$$KF_5^9: \begin{cases} e_1 \circ e_1 = e_2, \quad e_1 \circ e_2 = e_3, \quad e_1 \circ e_4 = -e_5, \quad e_2 \circ e_1 = 2e_3, \\ e_4 \circ e_1 = e_5, \quad e_4 \circ e_5 = e_3; \end{cases}$$

$$KF_5^{10}: e_1 \circ e_1 = e_2, \quad e_1 \circ e_4 = -e_5, \quad e_4 \circ e_1 = e_5, \quad e_4 \circ e_5 = e_3;$$

$$KF_5^{11}: \begin{cases} e_1 \circ e_1 = e_2, \quad e_1 \circ e_2 = e_3, \quad e_1 \circ e_4 = -e_5, \quad e_1 \circ e_5 = e_3, \\ e_2 \circ e_1 = 2e_3, \quad e_4 \circ e_1 = e_5; \end{cases}$$

$$KF_5^{12}: \begin{cases} e_1 \circ e_1 = e_2, \quad e_1 \circ e_2 = e_3, \quad e_1 \circ e_4 = -e_5, \quad e_2 \circ e_1 = 2e_3, \\ e_4 \circ e_1 = e_5; \end{cases}$$

$$KF_5^{13}: e_1 \circ e_1 = e_2, \quad e_1 \circ e_4 = -e_5, \quad e_1 \circ e_5 = e_3, \quad e_4 \circ e_1 = e_5;$$

$$KF_5^{14}(\alpha): \begin{cases} e_1 \circ e_1 = e_2, \quad e_1 \circ e_2 = e_3, \quad e_1 \circ e_4 = \alpha e_5, \quad e_2 \circ e_1 = 2e_3, \\ e_4 \circ e_1 = e_5, \quad \alpha \in C \backslash \{-1\}; \end{cases}$$

$$KF_5^{15}(\alpha): \begin{cases} e_1 \circ e_1 = e_2, \quad e_1 \circ e_4 = \alpha e_5, \quad e_1 \circ e_5 = \frac{2\alpha}{1+\alpha} e_3, \quad e_2 \circ e_4 = 2\alpha e_3, \\ e_4 \circ e_1 = e_5, \quad e_4 \circ e_2 = e_3, \quad e_5 \circ e_1 = 2e_3, \quad \alpha \in C \backslash \{-1, -\frac{1}{2}\}; \end{cases}$$

$$KF_5^{16}: \begin{cases} e_1 \circ e_1 = e_2, \quad e_1 \circ e_2 = e_3, \quad e_1 \circ e_4 = -\frac{1}{2}e_5, \quad e_1 \circ e_5 = -2e_3, \\ e_2 \circ e_1 = 2e_3, \quad e_2 \circ e_4 = -e_3, \quad e_4 \circ e_1 = e_5, \quad e_4 \circ e_2 = e_3, \\ e_5 \circ e_1 = 2e_3; \end{cases}$$

$$KF_5^{17}: \begin{cases} e_1 \circ e_1 = e_2, \quad e_1 \circ e_4 = -\frac{1}{2}e_5, \quad e_1 \circ e_5 = -2e_3, \quad e_2 \circ e_4 = -e_3, \\ e_4 \circ e_1 = e_5, \quad e_4 \circ e_2 = e_3, \quad e_5 \circ e_1 = 2e_3. \end{cases}$$

PROOF. Let A be an algebra satisfying the conditions of the theorem and let $\{e_1, e_2, e_3, e_4, e_5\}$ be a basis of the algebra satisfying the natural gradating,

$$A_1 = \langle e_1, e_4 \rangle, \quad A_2 = \langle e_2, e_5 \rangle, \quad A_3 = \langle e_3 \rangle.$$

Write the multiplication of the basic elements in the form

$$\begin{aligned} &e_1 \circ e_1 = \alpha_1 e_2 + \alpha_2 e_5, \quad e_1 \circ e_4 = \alpha_3 e_2 + \alpha_4 e_5, \\ &e_4 \circ e_1 = \alpha_5 e_2 + \alpha_6 e_5, \quad e_4 \circ e_4 = \alpha_7 e_2 + \alpha_8 e_5, \\ &e_1 \circ e_2 = \beta_1 e_3, \quad e_1 \circ e_5 = \beta_2 e_3, \quad e_4 \circ e_2 = \beta_3 e_3, \quad e_4 \circ e_5 = \beta_4 e_3, \end{aligned}$$

where $(\beta_1, \beta_2, \beta_3, \beta_4) \neq (0, 0, 0, 0)$.
It is easy to see that the linear span $\langle e_3 \rangle$ is an ideal of A. Consider now the quotient algebra $A/I = \{\bar{e}_1, \bar{e}_4, \bar{e}_2, \bar{e}_5\}$. It is a four-dimensional Zinbiel algebra for which conditions $dim(A/I)^2 = 2$ and $dim(A/I)^3 = 0$ hold. Using classification of four-dimensional Zinbiel algebras according to Theorem 2.7, we conclude that A/I is isomorphic to the following pairwise non isomorphic algebras:

M_1: $e_1 \circ e_1 = e_2, \quad e_4 \circ e_4 = e_5;$
M_2: $e_1 \circ e_1 = e_2, \quad e_1 \circ e_4 = e_5, \quad e_4 \circ e_1 = -e_5, \quad e_4 \circ e_4 = e_2 - 2e_5;$
M_3: $e_1 \circ e_1 = e_2, \quad e_1 \circ e_4 = e_5, \quad e_4 \circ e_1 = -e_2;$
$M_4(\alpha)$: $e_1 \circ e_1 = e_2, \quad e_1 \circ e_4 = e_5, \quad e_4 \circ e_1 = \alpha e_2, \quad e_4 \circ e_4 = -e_5, \quad \alpha \in C;$
M_5: $e_1 \circ e_4 = e_2, \quad e_4 \circ e_1 = e_5;$
M_6: $e_1 \circ e_1 = e_2, \quad e_1 \circ e_4 = e_5;$
$M_7(\alpha)$: $e_1 \circ e_1 = e_2, \quad e_1 \circ e_4 = \alpha e_5, \quad e_4 \circ e_1 = e_5 \quad \alpha \in C.$

Hence we can get the values for structural constants α_i of the algebra A, namely by equating the values of α_i to the corresponding ones in the algebra from the above list. Applying standard classification methods in each of the seven cases complete the proof of the theorem. □

THEOREM 3.6. *Any six-dimensional naturally graded quasi-filiform Zinbiel algebra of type $A_{(2)}$ is isomorphic to one of the following pairwise non isomorphic algebras:*

$$KF_6^1: \begin{cases} e_i \circ e_j = C_{i+j-1}^j e_{i+j}, \quad 2 \leq i+j \leq 4, \\ e_1 \circ e_5 = e_6, \quad e_5 \circ e_1 = e_6; \end{cases}$$

$$KF_6^2: \begin{cases} e_i \circ e_j = C_{i+j-1}^j e_{i+j}, \quad 2 \leq i+j \leq 4, \\ e_1 \circ e_5 = e_6, \quad e_5 \circ e_1 = e_6, \quad e_5 \circ e_5 = e_6; \end{cases}$$

$$KF_6^3: \begin{cases} e_i \circ e_j = C_{i+j-1}^j e_{i+j}, \quad 2 \leq i+j \leq 4, \\ e_1 \circ e_5 = e_6, \quad e_5 \circ e_5 = e_6; \end{cases}$$

$$KF_6^4: \begin{cases} e_i \circ e_j = C_{i+j-1}^j e_{i+j}, \quad 2 \leq i+j \leq 4, \\ e_5 \circ e_1 = e_6; \end{cases}$$

$$KF_6^5: \begin{cases} e_1\circ e_1=e_2, \quad e_1\circ e_3=e_4, \quad e_1\circ e_5=e_6, \quad e_1\circ e_6=e_3, \\ e_2\circ e_5=-e_3, \quad e_5\circ e_1=-3e_2-2e_6, \quad e_5\circ e_2=e_3, \quad e_5\circ e_5=2e_2+e_6, \\ e_5\circ e_6=-2e_3, \quad e_6\circ e_1=-e_3, \quad e_6\circ e_5=e_3; \end{cases}$$

$$KF_6^6: \begin{cases} e_1\circ e_1=e_2, \quad e_1\circ e_3=e_4, \quad e_1\circ e_5=e_6, \quad e_1\circ e_6=e_3, \\ e_2\circ e_5=-e_3, \quad e_5\circ e_1=-2e_6, \quad e_5\circ e_2=e_3, \quad e_6\circ e_1=-e_3; \end{cases}$$

$$KF_6^7: \begin{cases} e_1\circ e_1=e_2, \quad e_1\circ e_3=e_4, \quad e_1\circ e_5=e_6, \quad e_1\circ e_6=e_3, \\ e_3\circ e_1=e_4, \quad e_5\circ e_1=-e_6. \end{cases}$$

PROOF. The algebras

$$KF_6^1, \ KF_6^2, \ KF_6^3, \ KF_6^4$$

are obtained as in Theorem 3.4. To study the remaining possible cases, we should consider an algebra with the following table of multiplication:

$$\begin{gathered} e_1\circ e_1=e_2, \quad e_1\circ e_5=e_6, \quad e_1\circ e_2=0, \quad e_1\circ e_6=e_3, \quad e_1\circ e_3=e_4, \\ e_5\circ e_1=\alpha_5e_2+\alpha_6e_6, \quad e_5\circ e_5=\alpha_7e_2+\alpha_8e_6, \quad e_5\circ e_2=-(1+\alpha_6)e_3, \ . \\ e_5\circ e_6=\beta_4e_3, \quad e_5\circ e_3=\gamma e_4. \end{gathered}$$

The rest of the algebras in the list of the theorem are obtained by applying the fundamental identity for the above table of multiplication of algebra and consideration of possible cases for structural constants. □

The following theorem can be proved in the same manner.

THEOREM 3.7. *Any seven-dimensional naturally graded quasi-filiform Zinbiel algebra of type $A_{(2)}$ is isomorphic to one of the following pairwise non isomorphic algebras:*

$$KF_7^1: \begin{cases} e_i\circ e_j=C_{i+j-1}^j e_{i+j}, \quad 2\le i+j\le 5, \\ e_1\circ e_6=e_7, \quad e_6\circ e_1=\alpha e_6; \end{cases}$$

$$KF_7^2: \begin{cases} e_i\circ e_j=C_{i+j-1}^j e_{i+j}, \quad 2\le i+j\le 5, \\ e_1\circ e_6=e_7, \quad e_6\circ e_1=e_7, \quad e_6\circ e_6=e_7; \end{cases}$$

$$KF_7^3: \begin{cases} e_i\circ e_j=C_{i+j-1}^j e_{i+j}, \quad 2\le i+j\le 5, \\ e_1\circ e_6=e_7, \quad e_6\circ e_6=e_7; \end{cases}$$

$$KF_7^4: \begin{cases} e_i\circ e_j=C_{i+j-1}^j e_{i+j}, \quad 2\le i+j\le 5, \\ e_6\circ e_1=e_7; \end{cases}$$

$$KF_7^5: \begin{cases} e_1\circ e_1=e_2, \quad e_1\circ e_3=e_4, \quad e_1\circ e_4=e_5, \quad e_1\circ e_6=e_7, \\ e_1\circ e_7=e_3, \quad e_2\circ e_6=-e_3, \quad e_3\circ e_1=-e_4, \quad e_6\circ e_1=-2e_7, \\ e_6\circ e_2=e_3, \quad e_7\circ e_1=-e_3; \end{cases}$$

$$KF_7^6: \begin{cases} e_1\circ e_1=e_2, \quad e_1\circ e_3=e_4, \quad e_1\circ e_4=e_5, \quad e_1\circ e_6=e_7, \\ e_1\circ e_7=e_3, \quad e_2\circ e_6=-e_3, \quad e_3\circ e_1=-e_4, \quad e_6\circ e_1=-2e_7, \\ e_6\circ e_2=e_3, \quad e_6\circ e_4=e_5, \quad e_7\circ e_1=-e_3. \end{cases}$$

3.3. The case $r>2$. The proving the remaining cases, we need the following lemmas.

LEMMA 3.8. *Let A be a naturally graded quasi-filiform Zinbiel algebra of type $A_{(r)}$ $(r>2)$. Then $x\circ x=0$ for any $x\in A_1$.*

PROOF. Assume to the contrary, that is there exists $x \in A_1$ such that $x \circ x \neq 0$. Then we choose a basis $\{e_1, e_2, \dots, e_n\}$ of A such that $e_1 = x$, $e_2 = x \circ x$, and $A_1 = \langle e_1, e_{n-1} \rangle$, $A_2 = \langle e_2 \rangle$, $\dots$, $A_r = \langle e_r, e_n \rangle$, $A_{r+1} = \langle e_{r+1} \rangle$, $\dots$, $A_{n-2} = \langle e_{n-2} \rangle$. Thus, we can assume

$$e_1 \circ e_i = e_{i+1} \quad \text{for} \quad 2 \leq i \leq r-1,$$
$$e_{n-1} \circ e_{r-1} = e_n.$$

On the other hand, similar to the case of a filiform Zinbiel algebra in [**A**], we obtain $e_{n-1} \circ e_i = 0$ for $2 \leq i \leq r-1$, which contradicts the existence of an element x such that $x \circ x \neq 0$. □

LEMMA 3.9. *Let A be a naturally graded quasi-filiform Zinbiel algebra of type $A_{(r)}$. Then $r \leq 3$.*

PROOF. Assume to the contrary, i.e. $r > 3$. By Lemma 3.8, $x \circ x = 0$ for any $x \in A_1$.
Choose a basis $\{e_1, e_2, \dots, e_n\}$ of the algebra A such that

$$e_1 \circ e_i = e_{i+1} \quad \text{for} \quad 2 \leq i \leq r-1$$
$$e_1 \circ e_1 = e_{n-1} \circ e_{n-1} = 0, \quad e_1 \circ e_{n-1} = -e_{n-1} \circ e_1 = e_2.$$

We get a contradiction from the equalities

$$e_2 \circ e_1 = (e_1 \circ e_{n-1}) \circ e_1 = e_1 \circ (e_{n-1} \circ e_1) + e_1 \circ (e_1 \circ e_{n-1}) = e_1 \circ (-e_2 + e_2) = 0,$$
$$0 = (e_1 \circ e_1) \circ e_2 = e_1 \circ (e_1 \circ e_2) + e_1 \circ (e_2 \circ e_1) = e_1 \circ e_3 = e_4,$$

thus completing the proof of the lemma. □

LEMMA 3.10. *Let A be a naturally graded quasi-filiform Zinbiel algebra of type $A_{(3)}$. Then $dimA \leq 7$.*

PROOF. Suppose that $dimA > 7$ and let $\{e_1, e_2, \dots, e_n\}$ be a basis satisfying the conditions $A_1 = \langle e_1, e_{n-1} \rangle$, $A_2 = \langle e_2 \rangle$, $A_3 = \langle e_3, e_n \rangle$, $A_4 = \langle e_4 \rangle$, $\dots$, $A_{n-2} = \langle e_{n-2} \rangle$,

$$e_1 \circ e_1 = e_{n-1} \circ e_{n-1} = 0,$$
$$e_1 \circ e_{n-1} = -e_{n-1} \circ e_1 = e_2,$$
$$e_1 \circ e_2 = e_3, \quad e_{n-1} \circ e_2 = e_n.$$

The equalities

$$Z(e_1, e_{n-1}, e_1) = Z(e_1, e_{n-1}, e_{n-1}) = Z(e_1, e_1, e_2) = Z(e_{n-1}, e_{n-1}, e_2) =$$
$$Z(e_1, e_{n-1}, e_2) = Z(e_{n-1}, e_1, e_2) = Z(e_1, e_2, e_{n-1}) = Z(e_{n-1}, e_2, e_1) =$$
$$Z(e_1, e_2, e_1) = Z(e_{n-1}, e_2, e_{n-1}) = 0,$$

lead to

$$e_2 \circ e_1 = e_2 \circ e_{n-1} = e_1 \circ e_3 = e_{n-1} \circ e_n = 0$$
$$e_3 \circ e_1 = e_1 \circ e_3 = e_n \circ e_{n-1} = e_{n-1} \circ e_n = 0,$$
$$e_2 \circ e_2 = e_1 \circ e_n = e_3 \circ e_{n-1} = -e_{n-1} \circ e_3 = -e_n \circ e_1 = \gamma e_4.$$

Note that $\gamma \neq 0$. Otherwise $A_4 = 0$, i.e. $dimA \leq 5$.
Without loss of generality we can assume that $\gamma = 1$ and

$$e_1 \circ e_i = e_{i+1} \quad \text{for} \quad 4 \leq i \leq n-2,$$
$$e_1 \circ e_n = e_4.$$

Using the identity (2.1) for elements $\{e_1, e_n, e_1\}$, we obtain $e_4 \circ e_1 = 0$.
On the other hand, the equalities

$$(e_1 \circ e_1) \circ e_4 = e_1 \circ (e_1 \circ e_4) + e_1 \circ (e_4 \circ e_1)$$

imply $0 = e_1 \circ e_5 = e_6$, i.e. we arrive at a contradiction, which completes the proof of the lemma. □

From Lemma 3.10 one can easily derive the following corollaries.

COROLLARY 3.11. *Any five-dimensional naturally graded quasi-filiform Zinbiel algebra of type $A_{(3)}$ is isomorphic to the algebra*

$$e_1 \circ e_2 = e_3, \quad e_2 \circ e_1 = -e_3, \quad e_1 \circ e_3 = e_4, \quad e_2 \circ e_3 = e_5.$$

COROLLARY 3.12. *Any six-dimensional naturally graded quasi-filiform Zinbiel algebra of type $A_{(3)}$ is isomorphic to the algebra*

$$\begin{array}{l} e_1 \circ e_2 = e_3, \quad e_1 \circ e_3 = e_4, \quad e_1 \circ e_5 = e_6, \quad e_2 \circ e_1 = -e_3, \quad e_2 \circ e_3 = e_5, \\ e_2 \circ e_4 = -e_6, \quad e_3 \circ e_3 = e_6, \quad e_4 \circ e_2 = e_6, \quad e_5 \circ e_1 = -e_6. \end{array}$$

COROLLARY 3.13. *Any seven-dimensional naturally graded quasi-filiform Zinbiel algebra of type $A_{(3)}$ is isomorphic to the algebra*

$$\begin{array}{llll} e_1 \circ e_2 = e_3, & e_1 \circ e_3 = e_4, & e_1 \circ e_5 = e_6, & e_1 \circ e_6 = e_7, \\ e_2 \circ e_1 = -e_3, & e_2 \circ e_3 = e_5, & e_2 \circ e_4 = -e_6, & e_3 \circ e_3 = e_6, \\ e_4 \circ e_2 = e_6, & e_4 \circ e_3 = 2e_7, & e_5 \circ e_1 = -e_6. & \end{array}$$

Thus, we obtain the classification of complex naturally graded quasi-filiform Zinbiel algebras of an arbitrary dimension. In fact, the results of this paper complete the classification of n-dimensional nilpotent naturally graded algebras A satisfying the condition $A^{n-2} \neq 0$.

References

[A] Adashev J.Q., Omirov B.A. and Khudoyberdiyev A.Kh. On some nilpotent classes of Zinbiel algebras and their applications. Third International Conference on Research and Education in Mathematics. 2007. Malaysia. pp. 45-47.

[D1] Dzhumadil'daev A.S., Identities for multiplications derived by Leibniz and Zinbiel multiplications. Abstracts of short communications of International conference "Operator algebras and quantum theory of probability" (2005), Tashkent, pp. 76-77.

[D2] Dzhumadil'daev A.S. and Tulenbaev K.M., Nilpotency of Zinbiel algebras. J. Dyn. Control. Syst., vol. 11(2), 2005, pp. 195-213.

[G] Gunzburg V. and Kapranov M. Koszul duality for operads, Duke Math. J. vol. 76, 1994, pp. 203-273.

[L1] Loday J.-L., Cup product for Leibniz cohomology and dual Leibniz algebras. Math Scand., vol. 77, 1995, pp. 189-196.

[L2] Loday J.-L. and Pirashvili T., Universal enveloping algebras of Leibniz algebras and (co)homology. Math.Ann. vol. 296, 1993, pp. 139-158.

[O1] Omirov B.A., On derivations of filiform Leibniz algebras. Math. Notes, v. 77(5), 2005, pp. 733-742.

[O2] Omirov B.A., Classification of two-dimensional complex Zinbiel algebras, Uzbek. Mat. Zh., vol. 2, 2002, pp. 55-59.

[V] Vergne M. Cohomologie des algèbres de Lie nilpotentes. Application à l'étude de la variété des algèbres de Lie nilpotentes. Bull. Soc. Math. France, v. 98, 1970, pp. 81 - 116.

Institute of Mathematics and Information Technologies, Uzbek Academy of Sciences, Tashkent, Uzbekistan
E-mail address: adashevjq@mail.ru

Institute of Mathematics and Information Technologies, Uzbek Academy of Sciences, Tashkent, Uzbekistan
E-mail address: khabror@mail.ru

Institute of Mathematics and Information Technologies, Uzbek Academy of Sciences, Tashkent, Uzbekistan
E-mail address: omirovb@mail.ru

Contemporary Mathematics
Volume **483**, 2009

Akivis superalgebras and speciality

Helena Albuquerque and Ana Paula Santana

ABSTRACT. In this paper we define Akivis superalgebras and study enveloping superalgebras for this class of algebras, proving an analogous of the Poincaré-Birkhoff-Witt Theorem.

Lie and Malcev superalgebras are examples of Akivis superalgebras. For these particular superalgebras, we describe the connection between the classical enveloping superalgebras and the corresponding generalized concept defined in this work.

In honour of Ivan Shestakov on the occasion of his 60th birthday

1. Introduction

DEFINITION 1.1. The supervector space $M = M_0 \oplus M_1$ is called an *Akivis superalgebra* if it is endowed with two operations:

- a bilinear superanticommutative map $[,]$ that induces on M a structure of superalgebra;
- a trilinear map A, compatible with the gradation (i.e, $A(M_\alpha, M_\beta, M_\gamma) \subseteq M_{\alpha+\beta+\gamma}$, all $\alpha, \beta, \gamma \in Z_2$), satisfying the following identity:

$$[[x,y],z] + (-1)^{\alpha(\beta+\gamma)}[[y,z],x] + (-1)^{\gamma(\beta+\alpha)}[[z,x],y] =$$
$$A(x,y,z) + (-1)^{\alpha(\beta+\gamma)}A(y,z,x) + (-1)^{\gamma(\beta+\alpha)}A(z,x,y) -$$
$$-(-1)^{\alpha\beta}A(y,x,z) - (-1)^{\alpha(\beta+\gamma)+\beta\gamma}A(z,y,x) - (-1)^{\gamma\beta}A(x,z,y),$$

for homogeneous elements $x \in M_\alpha, y \in M_\beta, z \in M_\gamma$, all $\alpha, \beta, \gamma \in Z_2$.

This superalgebra will be denoted in this work by $(M, [,], A)$, or simply M, if no confusion arises.

This definition is a generalization of the notion of Akivis algebra presented by I. Shestakov in [6]. In fact, the even part of an Akivis superalgebra is an Akivis algebra. Akivis algebras were introduced by M. A. Akivis in [1] as local algebras of local analytic loops.

In this paper, we consider Akivis superalgebras over a field K of characteristic different from 2 and 3. It is our aim to study the enveloping superalgebra of an Akivis superalgebra and to prove an analogous of the PBW Theorem. Our approach

2000 *Mathematics Subject Classification.* 15A63, 17A70.

Financial support from CMUC-FCT is gratefully acknowledged.

is similar to the one used in [6] but, as it is expected, the superization of the results imply more elaborated calculations and arguments. This is particularly evident in the definition of the superalgebra $\tilde{V}(M)$ studied in Section 5.

Given a superalgebra W with multiplication $(,)$, we will denote by W^- the superalgebra with underlying supervector space W and multiplication $[,]$ given by $[x,y] = (x,y) - (-1)^{\alpha\beta}(y,x)$ for homogeneous elements $x \in W_\alpha$, $y \in W_\beta$ (and extended by linearity to every element of W). It is known that if W is an associative superalgebra then the superalgebra W^- is a Lie superalgebra, and if W is an alternative superalgebra then W^- is a Malcev superalgebra. Standard calculations show that if W is any superalgebra, the superalgebra W^- is an Akivis superalgebra for the trilinear map $A(x,y,z) = (xy)z - x(yz)$. This superalgebra will be denoted in this work by W^A.

We recall that a superalgebra S is said to be special if it is isomorphic to U^- for some superalgebra U.

It is well known that every Lie superalgebra is isomorphic to a superalgebra S^-, where S is an associative superalgebra. In fact, for any Lie superalgebra L, let $T(L)$ denote the associative tensor superalgebra of the vector space L, and consider its bilateral ideal I generated by the homogeneous elements $x \otimes y - (-1)^{\alpha\beta} y \otimes x - [x,y]$, all $x \in M_\alpha, y \in M_\beta, \alpha, \beta \in Z_2$. Then the associative superalgebra $T(L)/I$ is the universal enveloping superalgebra of L and L is isomorphic to a subsuperalgebra of $(T(L)/I)^-$.

It is an open problem to know if a Malcev superalgebra is isomorphic to S^- for some alternative superalgebra S. This problem was solved only partially in [4] and [5]. There it was shown that, in some cases, a Malcev algebra is isomorphic to a subalgebra of $Nat(T)^-$ for an algebra T, where $Nat(T)$ denotes the generalized alternative nucleous of T.

In this work, we prove that an Akivis superalgebra M, defined over a field K of characteristic different from 2 and 3, is isomorphically embedded in $\tilde{U}(M)^A$, where $\tilde{U}(M)$ is its enveloping superalgebra. So M is special.

2. Examples of Akivis superalgebras

Lie superalgebras and more generally Malcev superalgebras are Akivis superalgebras. For the first class, we consider the trilinear map A to be the zero map, and for the second class, we take $A(x,y,z) = 1/6\, SJ(x,y,z)$. Here $SJ(x,y,z)$ denotes the superjacobian

$$SJ(x,y,z) = [[x,y],z] + (-1)^{\alpha(\beta+\gamma)}[[y,z],x] + (-1)^{\gamma(\beta+\alpha)}[[z,x],y],$$

of the homogeneous elements $x \in M_\alpha, y \in M_\beta, z \in M_\gamma$, $(\alpha, \beta, \gamma \in Z_2)$.

Next, we give two examples of Akivis superalgebras which are not included in these classes.

Consider the algebra of octonions O as the algebra obtained by Cayley-Dickson Process from the quaternions Q, with the Z_2 gradation

$$O_0 = Q = < 1, e_1, e_2, e_3 > \quad \text{and} \quad O_1 = e_4 Q = < e_4, e_5, e_6, e_7 > .$$

The multiplication table of the Akivis superalgebra O^A is shown below. Note that the even part of this superalgebra is the simple Lie algebra $sl(2, K) =< e_1, e_2, e_3 >$ together with $1 \in Z(O)$.

	1	e_1	e_2	e_3	e_4	e_5	e_6	e_7
1	0	0	0	0	0	0	0	0
e_1	0	0	$-2e_3$	$2e_2$	$-2e_5$	$2e_4$	$2e_7$	$-2e_6$
e_2	0	$2e_3$	0	$-2e_1$	$-2e_6$	$-2e_7$	$2e_4$	$2e_5$
e_3	0	$-2e_2$	$2e_1$	0	$-2e_7$	$2e_6$	$-2e_5$	$2e_4$
e_4	0	$2e_5$	$2e_6$	$2e_7$	-2	0	0	0
e_5	0	$-2e_4$	$2e_7$	$-2e_6$	0	-2	0	0
e_6	0	$-2e_7$	$-2e_4$	$2e_5$	0	0	-2	0
e_7	0	$2e_6$	$-2e_5$	$-2e_4$	0	0	0	-2

This superalgebra is an Akivis superalgebra that is neither a Lie nor a Malcev superalgebra. Indeed, we have that

$$SJ(e_3, e_7, e_2) \neq 0 \quad \text{and} \quad ((e_4e_2)e_3)e_5 - ((e_2e_3)e_5)e_4 \neq (e_4e_3)(e_2e_5).$$

The second class of examples can be obtained using antiassociative superalgebras (nonassociative Z_2-graded quasialgebras). Consider D a division algebra and n, m natural numbers. In [3] the authors studied the superalgebra $\widetilde{Mat}_{n,m}(D)$ of the $(n+m) \times (n+m)$ matrices over D, with the chess-board Z_2-grading

$$\widetilde{Mat}_{n,m}(D)_0 = \left\{ \begin{pmatrix} a & 0 \\ 0 & b \end{pmatrix} : a \in Mat_n(D),\ b \in Mat_m(D) \right\},$$

$$\widetilde{Mat}_{n,m}(D)_1 = \left\{ \begin{pmatrix} 0 & v \\ w & 0 \end{pmatrix} : v \in Mat_{n\times m}(D),\ w \in Mat_{m\times n}(D) \right\}$$

and with multiplication given by

$$\begin{pmatrix} a_1 & v_1 \\ w_1 & b_1 \end{pmatrix} \cdot \begin{pmatrix} a_2 & v_2 \\ w_2 & b_2 \end{pmatrix} = \begin{pmatrix} a_1a_2 + v_1w_2 & a_1v_2 + v_1b_2 \\ w_1a_2 + b_1w_2 & -w_1v_2 + b_1b_2 \end{pmatrix}.$$

This superalgebra is antiassociative, with even part isomorphic to $Mat_n(D) \times Mat_m(D)$.

$\widetilde{Mat}_{n,m}(D)^A$ is an Akivis superalgebra that is neither a Lie nor a Malcev superalgebra. In fact, if $E_{i,j}$ denotes the $(n+m) \times (n+m)$ matrix with (ij) entry equal to 1 and all the other entries equal to 0, then $SJ(E_{1,n+1}, E_{n+1,1}, E_{1,n+1}) \neq 0$ and $2((E_{1,n+1}E_{n+1,1})E_{1,n+1})E_{n+1,1} - ((E_{n+1,1}E_{1,n+1})E_{n+1,1})E_{1,n+1} \neq E^2_{1,n+1}E^2_{n+1,1}$.

In the case $m = n = 1$, $\tilde{Mat}_{1,1}(D)^A$ has abelian even part and multiplication given by the following table, where $a = E_{11}, b = E_{11} - E_{22}, x = E_{12}, y = E_{21}$:

	a	b	x	y
a	0	0	x	$-y$
b	0	0	$2x$	$-2y$
x	$-x$	$-2x$	0	b
y	y	$2y$	b	0

3. Enveloping superalgebra of an Akivis superalgebra

In this section we construct and study the universal enveloping superalgebra of an Akivis superalgebra.

Given the Akivis superalgebras $(M, [,], A)$ and $(N, [,]', A')$, by an *Akivis homomorphism* we mean a superalgebra homomorphism of degree 0, $f : M \to N$, such that, for all $x, y, z \in M$, $f(A(x, y, z)) = A'(f(x), f(y), f(z))$.

DEFINITION 3.1. Let M be an Akivis superalgebra. A pair $(\tilde{U}, \iota)$ is a *universal enveloping superalgebra* of M if $\tilde{U}$ is a superalgebra and $\iota : M \to \tilde{U}^A$ is an Akivis homomorphism satisfying the following condition: given any superalgebra W and any Akivis homomorphism $\theta : M \to W^A$, there is a unique superalgebra homomorphism of degree 0, $\tilde{\theta} : \tilde{U} \to W$, such that $\theta = \tilde{\theta}\iota$.

In a similar way to the one used in the classical case for Lie superalgebras we can prove the following:

THEOREM 3.2. *1) The universal enveloping superalgebra of an Akivis superalgebra is unique up to isomorphism;*

2) Let M be an Akivis superalgebra and $(\tilde{U}, \iota)$ its universal enveloping superalgebra. Then the superalgebra $\tilde{U}$ is generated by $\iota(M)$ and K;

3) Consider two Akivis superalgebras M_1 and M_2 with universal enveloping superalgebras $(\tilde{U}_1, \iota_1)$ and $(\tilde{U}_2, \iota_2)$, respectively. If there is an Akivis homomorphism $\phi : M_1 \to M_2$ then there is a superalgebra homomorphism $\tilde{\phi} : \tilde{U}_1 \to \tilde{U}_2$ such that $\tilde{\phi}\iota_1 = \iota_2\phi$.

4) Let M be an Akivis superalgebra with universal enveloping superalgebra $(\tilde{U}, \iota)$. Let I be a graded ideal in M and let J be the graded ideal of $\tilde{U}$ generated by $\iota(I)$. If $m \in M$, the map $\lambda : m + I \to \iota(m) + J$ is an Akivis homomorphism from M/I to $(\tilde{U}/J)^A$ and $(\tilde{U}/J, \lambda)$ is the universal enveloping superalgebra of M/I.

Next we will construct the universal enveloping superalgebra of the Akivis superalgebra $(M, [,], A)$. We start by considering the nonassociative Z-graded tensor algebra of M

$$\tilde{T}(M) = \oplus_{n \in Z} \tilde{T}^n(M),$$

where $\tilde{T}^n(M) = 0$, if $n < 0$, $\tilde{T}^0(M) = K$, $\tilde{T}^1(M) = M$ and $\tilde{T}^n(M) = \oplus_{i=1}^{n-1} \tilde{T}^i(M) \otimes \tilde{T}^{n-i}(M)$, for $n \geq 2$, with multiplication defined by $xy = x \otimes y$. $\tilde{T}(M)$ is a superalgebra with the Z_2-gradation

$$\tilde{T}(M) = \tilde{T}(M)_0 \oplus \tilde{T}(M)_1 = (\oplus_{n \geq 0} \tilde{T}^n(M)_0) \oplus (\oplus_{n \geq 1} \tilde{T}^n(M)_1),$$

where

$$\tilde{T}^n(M)_\gamma = \oplus_{i=1}^{n-1} (\oplus_{\alpha+\beta=\gamma} (\tilde{T}^i(M)_\alpha \otimes \tilde{T}^{n-i}(M)_\beta)), \; \gamma \in Z_2.$$

Let I be the Z_2-graded ideal of $\tilde{T}(M)$ generated by the homogeneous elements

$$x \otimes y - (-1)^{\alpha\beta} y \otimes x - [x, y] \quad \text{and} \quad (x \otimes y) \otimes z - x \otimes (y \otimes z) - A(x, y, z),$$

for $x \in M_\alpha, y \in M_\beta, z \in M_\gamma$. The quocient algebra $\tilde{U}(M) = \tilde{T}(M)/I$ is a superalgebra with the natural Z_2-gradation induced by the graded ideal I. Consider the map $\iota : M \to \tilde{T}(M) \to \tilde{U}(M)$ obtained by the composition of the canonical injection with the quocient map. It is obvious that ι is an Akivis homomorphism from M to $\tilde{U}(M)^A$.

THEOREM 3.3. *The superalgebra $(\tilde{T}(M)/I, \iota)$ is the universal enveloping superalgebra of the Akivis superalgebra M.*

Proof: Given any superalgebra W and an Akivis homomorphism $f : M \to W^A$, we need to prove that there is a unique homomorphism $\tilde{f} : \tilde{U}(M) \to W$ such that $f = \tilde{f}\iota$.

Using the universal property of tensor products it is easy to see that there is a unique superalgebra homomorphism of degree 0, $f^* : \tilde{T}(M) \to W$, such that $f^*(m) = f(m)$, for all $m \in M$. The fact that f is an Akivis homomorphism implies that $I \subseteq \ker f^*$. Hence, there is an homomorphism of superalgebras of degree 0, $\tilde{f} : \tilde{U}(M) \to W$ such that $\tilde{f}(\iota(m)) = f^*(m) = f(m)$, all $m \in M$. As $\tilde{U}(M)$ is generated by K and $\iota(M)$, the unicity of $\tilde{f}$ follows.■

4. Enveloping superalgebras of Lie and Malcev superalgebras

In [4], Pérez-Izquierdo studied enveloping algebras of Sabinin algebras, showing that these generalize the classical notions of enveloping algebras for the particular cases of Lie algebras and Malcev algebras. In this section, we will study the connections between the classical definitions and the definition of enveloping superalgebras of Lie and Malcev superalgebras considered as Akivis superalgebras.

Given a Lie superalgebra L, denote by $(\tilde{U}(L), \iota)$ its enveloping superalgebra as an Akivis superalgebra and by $(U(L), \sigma)$ its classical universal enveloping superalgebra. Clearly σ is an Akivis homomorphism from L to $U(L)^A$. So there is a unique homomorphism of superalgebras $\tilde{\sigma}$ such that $\tilde{\sigma}\iota = \sigma$. As $U(L)$ is generated by K and $\sigma(L)$, $\tilde{\sigma}$ is surjective. So $U(L)$ is an epimorphic image of $\tilde{U}(L)$ by $\tilde{\sigma}$, i.e.,

$$U(L) \simeq \tilde{U}(L)/\ker\tilde{\sigma}\,.$$

Suppose now that N is a Malcev superalgebra. Superizing the theory exposed in [5], we can naturally define the classical enveloping superalgebra of N as the superalgebra $(\tilde{T}(N)/\tilde{I}, \tilde{\iota})$, where $\tilde{I}$ is the graded ideal of $\tilde{T}(N)$ generated by the homogeneous elements

$$a \otimes b - (-1)^{\alpha\beta} b \otimes a - [a,b],\ \ (a,x,y) + (-1)^{\alpha\gamma}(x,a,y),\ \ (x,a,y) + (-1)^{\alpha\xi}(x,y,a),$$

all $a \in N_\alpha$, $b \in N_\beta$, $x \in \tilde{T}(N)_\gamma$, $y \in \tilde{T}(N)_\xi$, and where $(x,y,z) = (xy)z - x(yz)$ denotes the usual associator. The map $\tilde{\iota}$ is the composition of the canonical injection from N to $\tilde{T}(N)$ with the canonical epimorphism $\mu : \tilde{T}(N) \to \tilde{T}(N)/\tilde{I}$.

We will prove that $\tilde{T}(N)/\tilde{I}$ is an epimorphic image of $\tilde{U}(N)$. For this, we show that $I \subseteq \tilde{I}$ and so μ gives rise to the epimorphism $\tilde{\mu} : \tilde{T}(N)/I \to \tilde{T}(N)/\tilde{I}$ defined by $\tilde{\mu}(n + I) = n + \tilde{I}$, all $n \in \tilde{T}(N)$. To see that $I \subseteq \tilde{I}$ notice that, for all homogeneous elements $x \in N_\gamma, y \in N_\xi, z \in N_\alpha$, we have that

$$\begin{aligned}&SJ(x,y,z) - 3(xy)z = \\ &(-1)^{\alpha(\gamma+\xi)}[(z,x,y) + (-1)^{\alpha\gamma}(x,z,y)] - (-1)^{\alpha\xi}[(x,z,y) + (-1)^{\alpha\xi}(x,y,z)]\end{aligned}$$

is an element of $\tilde{I}$ and so $SJ(x,y,z) - 3(xy)z + SJ(y,z,x) - 3(yz)x \in \tilde{I}$. Therefore, $SJ(x,y,z) \in \tilde{I}$ and $(x,y,z) \in \tilde{I}$. As the generators of I are in $\tilde{I}$, we can conclude that $I \subseteq \tilde{I}$.

5. Speciality of Akivis superalgebras

The canonical filtration of $\tilde{T}(M)$, $\tilde{T}_0(M) \subseteq \tilde{T}_1(M) \subseteq ...$, where $\tilde{T}_0(M) = K$, and $\tilde{T}_n(M) = \oplus_{i=0}^{n} \tilde{T}^i(M)$, $n > 0$, gives rise to the canonical filtration of $\tilde{U}(M)$, $\tilde{U}_0(M) \subseteq \tilde{U}_1(M) \subseteq ...$, where $\tilde{U}_n(M) = \tilde{T}_n(M) + I$.

Associated with this filtration there is the Z-graded superalgebra,

$$gr\tilde{U}(M) = \oplus_{n\in Z}(gr\tilde{U}(M))_n,$$

where $(gr\tilde{U}(M))_n = 0$, if $n < 0$, $(gr\tilde{U}(M))_0 = K$, $(gr\tilde{U}(M))_n = \tilde{U}_n(M)/\tilde{U}_{n-1}(M)$, for $n \geq 1$, with multiplication given by

$$(a + \tilde{U}_{i-1}(M))(b + \tilde{U}_{j-1}(M)) = ab + \tilde{U}_{i+j-1}(M)$$

for $a \in \tilde{U}_i(M)$ and $b \in \tilde{U}_j(M)$. For simplicity we identify $(gr\tilde{U}(M))_1$ with $\iota(M)$.

Now consider the classical tensor algebra $T(M)$ of M, that is naturally a Z-graded associative superalgebra,

$$T(M) = \oplus_{n\in Z} T^n(M)$$

where $T^n(M) = 0$ if $n < 0$, $T^0(M) = K$, $T^n(M) = M \otimes M \otimes ... \otimes M$ (n times) if $n > 0$. Let J be the ideal of $T(M)$ generated by the homogeneous elements $x \otimes y - (-1)^{\alpha\beta} y \otimes x$, all $x \in M_\alpha, y \in M_\beta$, $\alpha, \beta \in Z_2$. The associative Z-graded quotient superalgebra $S(M) = T(M)/J$ is called the *supersymmetric superalgebra* of M. Note that as an associative Z-graded algebra the homogeneous spaces of $S(M)$ are $S^n(M) = T^n(M) + J$ and as a superalgebra we have $S(M)_\alpha = \oplus_{n\in Z}(\oplus_{\alpha_1+...+\alpha_n=\alpha}(M_{\alpha_1} \otimes ... \otimes M_{\alpha_n} + J))$. Since the generators of J lie in $T^2(M)$, we identify $S^0(M)$ with K and $S^1(M)$ with M.

We will now construct from $S(M)$ a nonassociative superalgebra $V(M)$ which will play in this work the role that the symmetric algebra plays in the classical case of the PBW Theorem.

We define the Z-graded supervector space $V(M) = \oplus_{n\in Z} V^n(M)$, where the subspaces $V^n(M)$ are defined by

$$V^n(M) = S^n(M), \text{ if } n \leq 3, \text{ and } V^n(M) = \oplus_{i=1}^{n-1}(V^i(M) \otimes V^{n-i}(M)), \text{ if } n > 3.$$

We turn $V(M)$ into a superalgebra by defining the multiplication for homogeneous elements $v_i \in V^i(M), v_j \in V^j(M)$ by $v_i.v_j = v_i v_j$ if $i + j \leq 3$, and $v_i.v_j = v_i \otimes v_j$ if $i + j > 3$ (here juxtaposition of elements means the product of these elements in $S(M)$).

LEMMA 5.1. *The superalgebra $V(M)$ is the enveloping superalgebra of the trivial Akivis superalgebra with underlying vector space M, i.e., $V(M) \cong \tilde{T}(M)/I^*$, where I^* is the ideal of $\tilde{T}(M)$ generated by the homogeneous elements $x \otimes y - (-1)^{\alpha\beta} y \otimes x$, $(x \otimes y) \otimes z - x \otimes (y \otimes z)$, all $x \in M_\alpha, y \in M_\beta, z \in M_\gamma$.*

Proof: The inclusion map $\pi : M \to V(M)^A$ is an Akivis homomorphism. Therefore, as $\tilde{T}(M)/I^*$ is the enveloping superalgebra of the trivial Akivis superalgebra obtained from M, there is a superalgebra map of degree 0, $\tilde{\pi} : \tilde{T}(M)/I^* \to V(M)$ such that $\tilde{\pi}(m + I^*) = m$, all $m \in M$.

Notice that in $\tilde{T}(M)/I^*$ we have

$$x \otimes y - (-1)^{\alpha\beta} y \otimes x + I^* = 0 \quad \text{and} \quad (x \otimes y) \otimes z - x \otimes (y \otimes z) + I^* = 0,$$

for homogeneous elements x, y, z. Hence, using the universal property of the tensor products, we may define linear maps $\rho_1 : M \to \tilde{T}(M)/I^*$, $\rho_2 : M \otimes M \to \tilde{T}(M)/I^*$, $\rho_3 : M \otimes M \otimes M \to \tilde{T}(M)/I^*$ by $\rho_i(a) = a + I^*$, $i = 1, 2, 3$.

Clearly, $J \cap M \otimes M \subseteq \ker \rho_2$ and $J \cap M \otimes M \otimes M \subseteq \ker \rho_3$. Therefore, there are linear maps $\tilde{\rho}_i : V^i(M) \to \tilde{T}(M)/I^*$ defined by $\tilde{\rho}_i(a + J) = a + I^*$, i=1, 2, 3. Using once more the universal property of tensor products and induction, one can extend these maps to a homomorphism of superalgebras of degree 0, $\tilde{\rho} : V(M) \to \tilde{T}(M)/I^*$ satisfying $\tilde{\rho}(m) = m + I^*$ for all $m \in M$. The maps $\tilde{\rho}$ and $\tilde{\pi}$ are inverse of each other. So $\tilde{\pi}$ is an isomorphism. ■

LEMMA 5.2. *There exists an epimorphism of Z-graded superalgebras $\tilde{\tau} : \tilde{T}(M)/I^* \to gr\tilde{U}(M)$, of degree 0, such that $\tilde{\tau}(m + I^*) = \iota(m)$, all $m \in M$.*

Proof: Consider the natural epimorphism of Z-graded algebras $\tau : \tilde{T}(M) \to gr\tilde{U}(M)$ given by $\tau(a) = (a + I) + \tilde{U}_{n-1}(M)$, for each $a \in \tilde{T}^n(M)$. Notice that, since we identify $(gr\tilde{U}(M))_1$ with $\iota(M) = M + I$, then $\tau(m) = \iota(m)$, for all $m \in M$. To see that τ preserves the gradation, recall that for $\alpha \in Z_2$, $\tilde{T}(M)_\alpha = \oplus_{n \geq 0} \tilde{T}^n(M)_\alpha$ and

$$(gr\tilde{U}(M))_\alpha = \oplus_{n \geq 0} (\tilde{U}_n(M)/\tilde{U}_{n-1}(M))_\alpha = \oplus_{n \geq 0} (\oplus_{i=0}^n (\tilde{T}^i(M)_\alpha + I) + \tilde{U}_{n-1}(M)).$$

So if $a \in \tilde{T}^n(M)_\alpha$, then $\tau(a) = (a + I) + \tilde{U}_{n-1}(M) \in (\tilde{T}^n(M)_\alpha + I) + \tilde{U}_{n-1}(M)$. Hence

$$\tau(\tilde{T}(M)_\alpha) = \sum_{n \geq 0} \tau(\tilde{T}^n(M)_\alpha) \subseteq \oplus_{n \geq 0} (\tilde{T}^n(M)_\alpha + I) + \tilde{U}_{n-1}(M)) \subseteq (gr\tilde{U}(M))_\alpha,$$

as desired. Now for any $x \in M_\alpha, y \in M_\beta, z \in M_\gamma$, consider $\bar{x} = \iota(x)$, $\bar{y} = \iota(y)$, $\bar{z} = \iota(z) \in (gr\tilde{U}(M))_1$. Then, in $gr\tilde{U}(M)$ there holds

$$\bar{x}\bar{y} - (-1)^{\alpha\beta}\bar{y}\bar{x} = (\iota(x)\iota(y) - (-1)^{\alpha\beta}\iota(y)\iota(x)) + \tilde{U}_1(M) = \iota([x, y]) + \tilde{U}_1(M) = 0;$$

$$(\bar{x}\bar{y})\bar{z} - \bar{x}(\bar{y}\bar{z}) = (\iota(x)\iota(y))\iota(z) - \iota(x)(\iota(y)\iota(z)) + \tilde{U}_2(M) = \iota(A(x, y, z)) + \tilde{U}_2(M) = 0.$$

This implies that $I^* \subseteq \operatorname{Ker} \tau$. Therefore, there is an epimorphism of Z-graded superalgebras $\tilde{\tau} : \tilde{T}(M)/I^* \to gr\tilde{U}(M)$ such that $\tilde{\tau}(m + I^*) = \tau(m) = \iota(m)$, for all $m \in M$. ■

From these two lemmas, we know that the composite map

$$\tilde{\tau}\tilde{\pi}^{-1} : V(M) \to gr\tilde{U}(M)$$

is an epimorphism of superalgebras satisfying $\tilde{\tau}\tilde{\pi}^{-1}(m) = \iota(m)$ for all $m \in M$. It is our aim, to prove that this epimorphism is in fact an isomorphism. For this we need to endow $V(M)$ with a convenient superalgebra structure.

Let $\{e_r : r \in \Delta\}$ be a basis of M indexed by the totally ordered set $\Delta = \Delta_0 \cup \Delta_1$ satisfying the following: $\{e_r : r \in \Delta_\alpha\}$ is a basis of M_α, $\alpha = 0, 1$, and $r < s$ if $r \in \Delta_0$, $s \in \Delta_1$. In these conditions, $\{e_{r_1}e_{r_2} : r_1 \leq r_2$, and $r_1 < r_2$ if $r_1, r_2 \in \Delta_1\}$ is a basis of $V^2(M)$ and $\{e_{r_1}e_{r_2}e_{r_3} : r_1 \leq r_2 \leq r_3$, and $r_p < r_{p+1}$ if $r_p, r_{p+1} \in \Delta_1\}$ is a basis of $V^3(M)$.

In the supervector space $V(M)$ we define a new multiplication denoted by $*$ in the following way: if $a \in V^i(M)$, $b \in V^j(M)$, then $a * b = a \otimes b$ if $i + j > 3$; if $i + j \leq 3$ the multiplication is defined on the basis elements by the following identities (for simplicity we use $\bar{r}, \bar{s}, \bar{k}$ to denote the degrees of e_r, e_s, e_k, respectively):

$$e_r * e_s = \begin{cases} e_r e_s, & \text{if } r \leq s \text{ and } r \neq s \text{ if } r \in \Delta_1; \\ 1/2[e_r, e_r], & \text{if } r = s \in \Delta_1; \\ (-1)^{\bar{r}\bar{s}} e_s e_r + [e_r, e_s], & \text{if } r > s. \end{cases}$$

$$(e_r e_s) * e_k =$$

$$= \begin{cases} e_r e_s e_k, & \text{if } r \leq s \leq k \text{ and } k \neq s \text{ if } s \in \Delta_1; \\ A(e_r, e_s, e_s) + 1/2 e_r * [e_s, e_s], & \text{if } r < s = k \text{ and } s \in \Delta_1; \\ (-1)^{\bar{k}\bar{s}} e_r e_k e_s + e_r * [e_s, e_k] + & \\ +A(e_r, e_s, e_k) - (-1)^{\bar{s}\bar{k}} A(e_r, e_k, e_s), & \text{if } r \leq k < s \text{ and } r \neq k \text{ if } r \in \Delta_1; \\ -1/2[e_r, e_r] * e_s + e_r * [e_s, e_r] + & \\ +A(e_r, e_s, e_r) + A(e_r, e_r, e_s), & \text{if } r = k < s \text{ and } r \in \Delta_1; \\ (-1)^{\bar{k}(\bar{r}+\bar{s})} e_k e_r e_s + & \\ +(-1)^{\bar{k}\bar{s}} [e_r, e_k] * e_s + e_r * [e_s, e_k] - & \\ -(-1)^{\bar{s}\bar{k}} A(e_r, e_k, e_s) + A(e_r, e_s, e_k), & \text{if } k < r \leq s. \end{cases}$$

$$e_r * (e_s e_k) =$$

$$= \begin{cases} e_r e_s e_k - A(e_r, e_s, e_k), & \text{if } r \leq s \leq k \text{ and } r \neq s \text{ if } s \in \Delta_1; \\ 1/2[e_r, e_r] * e_k - A(e_r, e_r, e_k), & \text{if } r = s < k \text{ and } r \in \Delta_1; \\ (-1)^{\bar{r}\bar{s}} e_s e_r e_k - A(e_r, e_s, e_k) + & \\ +[e_r, e_s] * e_k, & \text{if } s < r \leq k \text{ and } r \neq k \text{ if } r \in \Delta_1; \\ 1/2(-1)^{\bar{r}\bar{s}} e_s * [e_r, e_r] + [e_r, e_s] * e_r - & \\ -A(e_r, e_s, e_r) + (-1)^{\bar{r}\bar{s}} A(e_s, e_r, e_r), & \text{if } s < r = k \text{ and } r \in \Delta_1; \\ (-1)^{\bar{r}(\bar{k}+\bar{s})} e_s e_k e_r + & \\ +(-1)^{\bar{r}\bar{s}} e_s * [e_r, e_k] + [e_r, e_s] * e_k + & \\ +(-1)^{\bar{r}\bar{s}} A(e_s, e_r, e_k) - A(e_r, e_s, e_k) - & \\ -(-1)^{(\bar{s}+\bar{k})\bar{r}} A(e_s, e_k, e_r), & \text{if } s \leq k < r. \end{cases}$$

Note that if we consider a basis element $e_p e_q$ of $V^2(M)$ we always assume $p \leq q$ and $p \neq q$ if $p, q \in \Delta_1$.

With this multiplication $V(M)$ becomes a superalgebra that will be denoted by $\tilde{V}(M)$.

LEMMA 5.3. *There is a homomorphism of superalgebras $\hat{\epsilon} : \tilde{U}(M) \to \tilde{V}(M)$, of degree 0, satisfying $\hat{\epsilon}(\iota(m)) = m$, for all $m \in M$.*

Proof: Denote the operations in the Akivis superalgebra $\tilde{V}(M)^A$ by

$$< x, y >= x * y - (-1)^{\alpha\beta} y * x, \quad < x, y, z >= (x * y) * z - x * (y * z),$$

for $x \in M_\alpha$, $y \in M_\beta$, $z \in M_\gamma$.

We start by proving that the inclusion map $\epsilon : M \to \tilde{V}(M)^A$ is a homomorphism of Akivis superalgebras. For this, it is enough to show that

$$[e_r, e_s] =< e_r, e_s > \quad \text{and} \quad A(e_r, e_s, e_k) =< e_r, e_s, e_k >,$$

for the basis elements e_r, e_s, e_k considered above. It is quite simple to see that the first of these two inequalities holds. For the second one, we have to consider several cases. Here we present only four of them, being the other cases similar.

1. $r = s = k \in \Delta_1$:
$< e_r, e_r, e_r >= (e_r * e_r) * e_r - e_r * (e_r * e_r) = 1/2([e_r, e_r] * e_r - e_r * [e_r, e_r])$.
As $[e_r, e_r] \in M_0$, we have $[e_r, e_r] = \sum_{t\in\Delta_0} \alpha_t e_t$ (sum with finite support), for scalars $\alpha_t \in K$. Therefore, as $t < r$ for any $t \in \Delta_0$, there holds

$$[e_r, e_r] * e_r = \sum_{t\in\Delta_0} \alpha_t e_t * e_r = \sum_{t\in\Delta_0} \alpha_t e_t e_r.$$

In a similar way, we see that

$$e_r * [e_r, e_r] = \sum_{t\in\Delta_0} \alpha_t (e_t e_r + [e_r, e_t]) = \sum_{t\in\Delta_0} \alpha_t e_t e_r + [e_r, [e_r, e_r]].$$

Hence $< e_r, e_r, e_r >= -1/2[e_r, [e_r, e_r]] = 1/2[[e_r, e_r], e_r]$. On the other hand, from the definition of Akivis superalgebra, we have that $SJ(e_r, e_r, e_r) = 3[[e_r, e_r], e_r] = 6A(e_r, e_r, e_r)$. Thus $< e_r, e_r, e_r >= A(e_r, e_r, e_r)$.

2. $r \leq k < s$ and $r \neq k$ if $r \in \Delta_1$:

$$\begin{aligned}&< e_r, e_s, e_k >= (e_r e_s) * e_k - e_r * ((-1)^{\bar{s}\bar{k}} e_k e_s + [e_s, e_k]) =\\ &(-1)^{\bar{s}\bar{k}} e_r e_k e_s + A(e_r, e_s, e_k) - (-1)^{\bar{s}\bar{k}} A(e_r, e_k, e_s) +\\ &+ e_r * [e_s, e_k] - e_r * [e_s, e_k] - (-1)^{\bar{s}\bar{k}} e_r e_k e_s + (-1)^{\bar{s}\bar{k}} A(e_r, e_k, e_s) =\\ &= A(e_r, e_s, e_k).\end{aligned}$$

3. $r = k < s$ and $r \in \Delta_1$:

$$\begin{aligned}&< e_r, e_s, e_r >= -1/2[e_r, e_r] * e_s + e_r * [e_s, e_r] +\\ &+ A(e_r, e_s, e_r) + A(e_r, e_r, e_s) + e_r * (e_r e_s) - e_r * [e_s, e_r] =\\ &= A(e_r, e_s, e_r) + A(e_r, e_r, e_s) - 1/2[e_r, e_r] * e_s + 1/2[e_r, e_r] * e_s - A(e_r, e_r, e_s) =\\ &= A(e_r, e_s, e_r).\end{aligned}$$

4. $k < s < r$:

$$\begin{aligned}&< e_r, e_s, e_k >=\\ &(-1)^{\bar{s}\bar{r}} (e_s e_r) * e_k + [e_r, e_s] * e_k - (-1)^{\bar{s}\bar{k}} e_r * (e_k e_s) - e_r * [e_s, e_k] =\\ &= (-1)^{(\bar{s}+\bar{k})\bar{r}} [e_s, e_k] * e_r + (-1)^{\bar{s}\bar{r}} e_s * [e_r, e_k] + [e_r, e_s] * e_k - e_r * [e_s, e_k] -\\ &- (-1)^{(\bar{s}+\bar{r})\bar{k}} e_k * [e_r, e_s] - (-1)^{\bar{s}\bar{k}} [e_r, e_k] * e_s + (-1)^{\bar{s}\bar{r}} A(e_s, e_r, e_k) -\\ &- (-1)^{(\bar{s}+\bar{k})\bar{r}} A(e_s, e_k, e_r) + (-1)^{\bar{s}\bar{k}+(\bar{s}+\bar{k})\bar{r}} A(e_k, e_s, e_r) - (-1)^{(\bar{s}+\bar{r})\bar{k}} A(e_k, e_r, e_s) +\\ &+ (-1)^{\bar{s}\bar{k}} A(e_r, e_k, e_s) = \quad (\text{ since } [x, y] =< x, y >, \text{ all } x, y \in M)\\ &= SJ(e_r, e_s, e_k) - ((-1)^{\bar{s}\bar{r}} A(e_s, e_r, e_k) - (-1)^{(\bar{s}+\bar{k})\bar{r}} A(e_s, e_k, e_r) +\\ &+ (-1)^{\bar{s}\bar{k}+(\bar{s}+\bar{k})\bar{r}} A(e_k, e_s, e_r) - (-1)^{(\bar{s}+\bar{r})\bar{k}} A(e_k, e_r, e_s) + (-1)^{\bar{s}\bar{k}} A(e_r, e_k, e_s)) =\\ &= A(e_r, e_s, e_k) \ (\text{by the definition of Akivis superalgebra}).\end{aligned}$$

As ϵ is an Akivis homomorphism, from the definition of enveloping superalgebra, there is a homomorphism of superalgebras of degree 0, $\hat{\epsilon} : \tilde{U}(M) \to \tilde{V}(M)$ satisfying $\hat{\epsilon}(\iota(m)) = m$, all $m \in M$. ■

THEOREM 5.4. *The Z-graded superalgebras $V(M)$ and $gr\tilde{U}(M)$ are isomorphic.*

Proof: The superalgebra $\tilde{V}(M)$ has a natural filtration defined by the sequence of subspaces $\tilde{V}_n(M) = \oplus_{i=0}^n V^i(M)$. So we may consider the associated Z-graded superalgebra $gr\tilde{V}(M)$. (As usual we identify $(gr\tilde{V}(M))_1$ with M). Since the map $\hat{\epsilon}$, considered in the previous lemma, is a homomorphism of superalgebras, we have $\hat{\epsilon}(\tilde{U}_n(M)) \subseteq \tilde{V}_n(M)$. So, we may define $\tilde{\epsilon} : gr\tilde{U}(M) \to gr\tilde{V}(M)$ by $\tilde{\epsilon}(a_i + \tilde{U}_{i-1}(M)) = \hat{\epsilon}(a_i) + \tilde{V}_{i-1}(M)$. This map is a homomorphism of Z-graded superalgebras of degree 0 and satisfies $\tilde{\epsilon}(\iota(m)) = m$, all $m \in M$.

We now return to the superalgebra $V(M)$. For $n \geq 1$, the map $\mu_n : V^n(M) \to (gr\tilde{V}(M))_n$ defined by $\mu_n(v) = v + \tilde{V}_{n-1}(M)$ is an isomorphism of vector spaces. So taking $\mu_0 = Id_K$, $\mu = \oplus_{n\geq 0}\mu_n : V(M) \to gr\tilde{V}(M)$ is an isomorphism of Z-graded vector spaces. Looking at the formulas which define the multiplication in $\tilde{V}(M)$, it is easy to see that this is in fact an isomorphism of Z-graded superalgebras. The composite homomorphisms $\mu^{-1}\tilde{\epsilon} : gr\tilde{U}(M) \to gr\tilde{V}(M) \to V(M)$ and $\tilde{\tau}\tilde{\pi}^{-1} : V(M) \to \tilde{T}(M)/I^* \to gr\tilde{U}(M)$ (recall the preceeding lemmas) are inverse of each other. So the result follows.■

The following results are immediate consequence of this theorem.

COROLLARY 5.5. *The canonical map* $\iota : M \to \tilde{U}(M)$ *is injective.*

COROLLARY 5.6. *Any Akivis superalgebra, defined over a field of characteristic different from 2 and 3, is special.*

6. BIBLIOGRAPHY

[1] Akivis M. A., "Local algebras of a multidimensional three web" (Russian), *Sibirsk. Mat. Zh.* **17** (1), (1976) 5-11;

[2] Albuquerque H. and Shahn Majid, "Quasialgebra structure of the octonions", *Journal of Algebra* **220**, (1999) 188-224;

[3] Albuquerque H., A. Elduque and José Pérez-Izquierdo, "Z_2- quasialgebras", *Comm. in Algebra* **30** (5), (2002) 2161-2174;

[4] Pérez-Izquierdo J., "Algebras, hyperalgebras, nonassociative bialgebras and loops", *Advances in Mathematics* **208**, (2007) 834-876;

[5] Pérez-Izquierdo J. and Shestakov I., "An envelope for Malcev algebras", *J. of Algebra* **272** (1), (2005) 379-393;

[6] Shestakov I., "Every Akivis Algebra is Linear", *Geometriae Dedicata* **77** (2), (1999) 215-223;

[7] Shestakov I. and U. U. Umirbaev, "Free Akivis Algebras, primitive elements and hyperalgebras", *J. Algebra* **250**(2), (2002) 533-548.

[8] Sheunert M., " The theory of Lie superalgebras", *LNM* **716** Springer Verlag, Berlin 1979.

CMUC, DEPARTMENT OF MATHEMATICS, UNIVERSITY OF COIMBRA, 3001-454 COIMBRA, PORTUGAL

E-mail address: lena@mat.uc.pt

CMUC, DEPARTMENT OF MATHEMATICS, UNIVERSITY OF COIMBRA, 3001-454 COIMBRA, PORTUGAL

E-mail address: aps@mat.uc.pt

Contemporary Mathematics
Volume **483**, 2009

Properties of some semisimple Hopf algebras

V.A. Artamonov and I.A.Chubarov

To Prof. I.P. Shestakov on the occasion of his 60th anniversary

ABSTRACT. The paper considers properties of two classes of finite dimensional semisimple Hopf algebras from [**A**]. We show that for any positive integer $n > 1$ there exists a semisimple Hopf algebra of dimension $2n^2$ from [**A**].

Introduction

One of the most important problems in the theory of Hopf algebras is a classification of finite dimensional semisimple Hopf algebras.

The topic of this paper is motivated with a result by G.M.Seitz [**H**][Theorem 7.10] who characterized finite groups G having only one irreducible complex representation of degree $n > 1$. Such a group G is either an extraspecial 2-group of order 2^{2m+1}, $n = 2^m$, or $|G| = n(n+1)$, where $n + 1 = p^f$, p a prime. So we consider semisimple finite dimensional Hopf algebras H which have only one irreducible representation Φ of degree n greater than 1.

Generalizing this situation the paper [**A**] considers semisimple finite dimensional Hopf algebras H which has only one irreducible representation Φ of degree n greater than 1. Under assumption that the dimension of H is equal to $2n^2$ it is shown in [**A**] that there exist two series of semisimple Hopf algebras corresponding to symmetric and symplectic cases. In the paper [**ACh**] the structure of the dual algebras for each of these series was found. As a corollary it was shown that if $n > 2$ then none of Hopf algebras from these two series is self-dual.

The aim of the present paper is to simplify the construction of algebras from [**A**] and to show that for any $n > 1$ exists a Hopf algebra of dimension $2n^2$ in each of two classes of algebras.

Let k be the basic algebraically closed field of characteristic not dividing $2n$ and H a finite dimensional Hopf algebra over k such that H as an algebra has up to an isomorphism only one irreducible representation M of degree $n > 1$. One-dimensional direct summands in H correspond to group-like elements in the dual Hopf algebra H^* that is to algebra homomorphisms $H \to k$. So the number of these

1991 *Mathematics Subject Classification.* Primary 16W30; Secondary 20C20.
Key words and phrases. Hopf algebra, projective representations of groups.
Research is partially supported by grants RFFI 06-01-00037.
Research is partially supported by grants RFFI 06-01-00037.

summands is equal to the order of the group $G = G(H^*)$ of group-like elements in H^* and it is a divisor of $\dim H$ [**M**, § 3.1]. H as an algebra has a direct sum decomposition

$$H = \oplus_{g\in G} k e_g \oplus \mathrm{Mat}(n,k)E, \tag{0.1}$$

where $\{e_g,\ g \in G, E\}$ is a set of central indecomposable idempotents and E is the identity matrix. Since $\dim H = |G| + n^2$ the order of G is a divisor of n^2.

Recall that there are left and right actions $f \rightharpoonup x,\ x \leftharpoonup f$ of a dual Hopf algebra H^* on H, namely if $f \in H^*,\ x \in H$, and $\Delta(x) = \sum_x x_{(1)} \otimes x_{(2)}$ then

$$f \rightharpoonup x = \sum_x x_{(1)}\langle f, x_{(2)}\rangle,\ x \leftharpoonup f = \sum_x \langle f, x_{(1)}\rangle (x_{(2)}).$$

The convolution multiplication $f * g$ in H^* has the form

$$\begin{aligned}\langle f * g, x\rangle = \mu(f \otimes g)\Delta(x) &= \sum_x \langle f, x_{(1)}\rangle \langle g, x_{(2)}\rangle \\ &= \langle f, g \rightharpoonup x\rangle = \langle g, x \leftharpoonup f\rangle,\end{aligned}$$

where $x \in H,\ f, g \in H^*$ and $\mu : H \otimes H \to H$ – is the multiplication map in H. It is easy to see that $\langle gh, x\rangle = \langle g * h, x\rangle$ for all $x \in H,\ g, h \in G = G(H^*)$. Note that

$$f \rightharpoonup (x \leftharpoonup g) = \sum_x \left\langle g, x_{(1)}\right\rangle x_{(2)} \left\langle f, x_{(3)}\right\rangle = (f \rightharpoonup x) \leftharpoonup g$$

for all $x \in H$ and $f, g \in H^*$.

Everywhere in this paper the transpose of the matrix A is denoted as ${}^tA$.

THEOREM 0.1 ([**A**]). *Let $G = G(H^*)$ be the group of group-like elements in the dual Hopf algebra H^* and H has a direct sum decomposition* (0.1). *Then there exist elements $\Delta'(x)$, $\Delta_t \in \mathrm{Mat}(n,k)^{\otimes 2}$, and a (skew)symmetric matrix U of size n such that the comultiplication Δ, the counit ε and the antipode S in H have the form*

$$\Delta(x) = \sum_{g\in G} [(g \rightharpoonup x) \otimes e_g + e_g \otimes (x \leftharpoonup g)] + \Delta'(x),$$

$$\Delta(e_f) = \sum_{g,h\in G,\, gh=f} e_g \otimes e_h + \Delta_f,$$

$$\varepsilon(e_g) = \delta_{g,1}, \quad \varepsilon(x) = 0; \quad S(e_g) = e_{g^{-1}}, \quad S(x) = U\,{}^t x U^{-1}.$$

for all $g, h, f \in G$ and for $x \in \mathrm{Mat}(n,k)$. Moreover if $\mu : H \otimes H \to H$ is the map of multiplication in H then

$$\mu(1 \otimes S)\Delta_g = \mu(S \otimes 1)\Delta_g = \delta_{g,1}E,$$
$$\mu(1 \otimes S)\Delta'(x) = \mu(S \otimes 1)\Delta'(x) = 0$$

for all $g \in G$ and for all $x \in \mathrm{Mat}(n,k)$.

If $|G| = n^2$ then $G = G(H^*)$ is an Abelian group [**TY**, Corollary 3.3].

THEOREM 0.2 ([**A**]). *Let H be as above with direct sum decomposition* (0.1), *the order of the group $G = G(H^*)$ equal to n^2 and* $\mathrm{char}\, k$ *not a divisor of* $\dim H$. *Taking an isomorphic copy H we can assume that the comultiplication Δ, the counit ε and the antipode S in H are defined as follows:*

$$\Delta(e_g) = \sum_{h\in G} e_h \otimes e_{h^{-1}g} + \Delta_g;$$

$$\Delta(x) = \sum_{g\in G} [(g \rightharpoonup x) \otimes e_g + e_g \otimes (x \leftharpoonup g)];$$
$$\varepsilon(e_g) = \delta_{g,1}, \quad \varepsilon(x) = 0; \quad S(x) = U\,{}^t x U^{-1}$$

for all $x \in \mathrm{Mat}(n,k)$, *where* $U = (u_{ij})$, $\frac{1}{n}U^{-1} = (v_{ij}) \in \mathrm{Mat}(n,k)$ *and either* $U = E$, *or*

$$U = \mathcal{S} = \begin{pmatrix} T & 0 & \dots & 0 \\ 0 & T & \dots & 0 \\ \dots & \dots & \dots & \dots \\ 0 & 0 & \dots & T \end{pmatrix}, \quad T = \begin{pmatrix} 0 & -1 \\ 1 & 0 \end{pmatrix}. \tag{0.2}$$

Moreover there exists a projective representation $g \mapsto A_g \in \mathrm{GL}(n,k)$ *such that*

$$\begin{aligned} &g \rightharpoonup x = A_g x A_{g^{-1}}, \\ &x \leftharpoonup g = U\,{}^t A_g U^{-1} x\, U {}^t A_{g^{-1}} U^{-1}, \\ &A_g U\,{}^t A_h U^{-1} A_g^{-1} U\,{}^t A_h^{-1} U^{-1} = [A_g,\, U\,{}^t A_h U^{-1}] = \mu_{g,h} E, \\ &\mu_{g,h} \in k^*, \\ &\Delta_g = \sum_{i,j,p,q} \left(E_{ij} \leftharpoonup g^{-1}\right) \otimes u_{ip} v_{qj} E_{pq} \\ &\quad = \sum_{i,j,p,q} E_{ij} \otimes u_{ip} v_{qj} \left(g^{-1} \rightharpoonup E_{pq}\right) \end{aligned} \tag{0.3}$$

for all $g \in G$ *and* $\mathrm{tr}\, A_g = n\delta_{g,1}$. *Here* E_{ij}, $1 \leqslant i,j \leqslant n$, *stand for matrix units in* $\mathrm{Mat}(n,k)$. *Also*

$$\mathcal{R} = \sum_{i,j} E_{ij} \otimes E_{ji} = \frac{1}{n} \sum_{g\in G} U^t A_{g^{-1}} \otimes {}^t A_g U^{-1}. \tag{0.4}$$

We are assuming in (0.3) and (0.4) that $A_{g^{-1}} = A_g^{-1}$ for all $g \in G$ and $A_1 = E$.

REMARK 0.3. Instead of $U = \mathcal{S}$ in (0.2) in an even case $n = 2m$ we can also take as U the matrix

$$\mathcal{J} = \begin{pmatrix} 0 & -E \\ E & 0 \end{pmatrix} \tag{0.5}$$

where E is the unit matrix of the size m. Both matrices (0.2) and (0.5) are canonical forms of skew-symmetric matrices.

THEOREM 0.4 ([**A**], Theorem 7.1). *An element* $w = \sum_{g\in G} \chi_g e_g + x \in H$, $x \in \mathrm{Mat}(n,k)$, $\chi_g \in k$ *in Theorem 0.2 is a group-like element if and only if the following conditions are satisfied:*

1) $\chi_{gh} = \chi_g \chi_h$ *for all* $g, h \in G$;
2) $g \rightharpoonup x = \chi_g x = x \leftharpoonup g$ *for any* $g \in G$;
3) $xU\,{}^t x = U$.

PROPOSITION 0.5 ([**A**], Proposition 5.2). A Hopf algebra H from Theorem 0.2 is cocommutative if and only if $g \leftharpoonup x = x \rightharpoonup g$ for all $g \in G$ and $x \in \mathrm{Mat}(n,k)$.

COROLLARY 0.6. A Hopf algebra H from Theorem 0.2 is cocommutative if and only if $A_g = \xi_g U\,{}^t A_g U^{-1}$ for all $g \in G$ and $x \in \mathrm{Mat}(n,k)$ where $\xi_g = \pm 1$.

PROOF. By Proposition 0.5 and Theorem 0.2

$$A_g x A_g^{-1} = g \leftharpoonup x = x \rightharpoonup g = U\,{}^t A_g U^{-1} x U\,{}^t A_g^{-1} U^{-1}$$

for any matrix x. Hence the matrix $U\,{}^t A_g^{-1} U^{-1} A_g$ is a central matrix for any $g \in G$. Hence $A_g = \xi_g U\,{}^t A_g U^{-1}$. But ${}^t U = \pm U = U^{-1}$ and therefore $\xi_g = \pm 1$. □

As it was shown in [**ACh**, §2] the dual Hopf algebra H^* for H from Theorem 0.2 is a $\mathbb{Z}_2$-graded algebra

$$H^* = H_0^* \oplus H_1^*, \quad H_0^* = kG,\ H_1^* = \mathrm{Mat}(n,k).$$

The space of matrices $\mathrm{Mat}(n,k)$ is equipped with a symmetric bilinear form

$$\langle A, B\rangle = \mathrm{tr}\,(A \cdot S(B)) = \mathrm{tr}\left(A \cdot U\,{}^t B U\right).$$

It is shown in [**ACh**] that for all $g \in G$ and $X, Y \in \mathrm{Mat}(n,k)$ we have

$$\langle X, Y \leftharpoonup g\rangle = \langle g \rightharpoonup X, Y\rangle, \quad \langle X, Y\rangle = \langle Y, X\rangle.$$

Multiplication and comultiplication Δ^* in H^* are defined in the following way. If $g \in G$, $X \in \mathrm{Mat}(n,k)$ and $E_{ij} \in \mathrm{Mat}(n,k)$ are matrix units then

$$g * h = gh, \quad g * X = g \rightharpoonup X, \quad X * g = X \leftharpoonup g,$$
$$X * Y = \frac{1}{n}\sum_{g\in G}\langle Y \leftharpoonup g^{-1}, X\rangle g = \frac{1}{n}\sum_{g\in G}\langle Y * g^{-1}, X\rangle g;$$
$$\Delta^*(g) = g \otimes g, \quad \Delta^*(E_{ij}) = \sum_{t=1}^{n} E_{it} \otimes E_{tj}.$$

It is easy to see that

$$\Delta^*(g * X * h) = (g \otimes g) * \Delta^*(X) * (h \otimes h)$$

for all $g, h \in G$, and $X \in \mathrm{Mat}(n,k)$.

It follows from Theorem 0.2 that the Hopf algebra H depends only on a projective n-dimensional representation $g \mapsto A_g$ of the Abelian group G of order n^2 with the properties (0.3), (0.4) and $\mathrm{tr}\, A_g = n\delta_{g,1}$. We shall show that (0.4) is equivalent to the irreducibility of the representation. For an irreducible representation the equality $\mathrm{tr}\, A_g = n\delta_{g,1}$ holds. It is shown that for any positive integer $n > 1$ there exists an Abelian group G of order n^2 with an irreducible representation of dimension n satisfying (0.3), (0.4) and $\mathrm{tr}\, A_g = n\delta_{g,1}$. Hence for any $n > 1$ there exists a Hopf algebra from Theorem 0.2 with $U = E$.

It is necessary to mention some other paper considering the same class of Hopf algebras. In the paper [**T**] there is given an explicit form of H if the order of G is n^2 and either n is odd or the group G is an elementary Abelian 2-group.

In the paper [**TY**] it is shown that if $n = 2$ then there exist up to equivalence four classes of Hopf algebras H, namely group algebras of Abelian groups of order 8, the group algebras of the dihedral group D_4, of the quaternions Q_8, and G. Kac Hopf algebra H_8 [**KP**].

Recall that H_8 is generated as an algebra by elements x, y, z with defining relations

$$x^2 = y^2 = 1,\ xy = yx,\ zx = yz,\ zy = xz,\ z^2 = \frac{1}{2}(1 + x + y - xy)$$

where x, y are group-like elements and

$$\Delta(z) = \frac{1}{2}\left((1+y)z \otimes z + (1-y)z \otimes xz\right).$$

The group of group-like elements in H_8 is the group $\{1, x\, y, \, xy\}$. By [**Ma**] the only semisimple Hopf algebras of dimension 8 are H_8 and the group algebras of the dihedral group D_4 and of the quaternion group Q_8.

In the paper [**S**] it is assumed that the order of G is smaller than n^2 and the algebra H has only one irreducible module M of dimension n.

It is shown that $|G| \leqslant d+1$. If $|G| = d+1$, then G is a cyclic group of order $p^a - 1$ for some a. A complete classification is found in the cases when the order of G is 2,3,4.

The authors are grateful to Prof. E.S. Golod for valuable comments and suggestions and the referee for his very useful remarks.

1. Representations of G

In this section we shall consider more detailed properties of the projective representations of the Abelian group $G = G(H^*)$ used in the definition of Hopf algebras H from Theorem 0.2. First we need some properties of the element $\mathcal{R}$ from (0.4).

PROPOSITION 1.1. If $A, B \in \mathrm{Mat}(n, k)$ then $(A \otimes B)\mathcal{R} = \mathcal{R}(B \otimes A)$. If $P \in \mathrm{GL}(n, k)$ then $(P \otimes P)\mathcal{R}(P^{-1} \otimes P^{-1}) = \mathcal{R}$.

PROOF. In order to prove the first statement it suffices to consider the case $A = E_{pq}$, $B = E_{rs}$. It is a routine calculation to check the statement in this particular case.

The second statement follows from the the first one. □

COROLLARY 1.2. The equality (0.4) is equivalent to

$$\mathcal{R} = \frac{1}{n}\sum_{g\in G} A_g^{-1} \otimes A_g. \tag{1.1}$$

If $P \in \mathrm{GL}(n, k)$ and U, A_g are from Theorem 0.2 then

$$\frac{1}{n}\sum_{g\in G} PA_g^{-1}P^{-1} \otimes PA_gP^{-1} = \mathcal{R}.$$

PROOF. Using Proposition 1.1 we obtain

$$\frac{1}{n}\sum_{g\in G} {}^tA_g^{-1} \otimes {}^tA_g = (U^{-1} \otimes E)\left(\frac{1}{n}\sum_{g\in G} U\,{}^tA_g^{-1} \otimes {}^tA_gU^{-1}\right)(E \otimes U)$$
$$= (U^{-1} \otimes E)\mathcal{R}(E \otimes U) = \mathcal{R}(E \otimes U^{-1})(E \otimes U) = \mathcal{R}.$$

Taking transposes in both tensor factors in the last equality we obtain (1.1). Again by Proposition 1.1

$$\frac{1}{n}\sum_{g\in G} PA_g^{-1}P^{-1} \otimes PA_gP^{-1}$$
$$= (P \otimes P)\left(\frac{1}{n}\sum_{g\in G} A_g^{-1} \otimes A_g\right)(P^{-1} \otimes P^{-1})$$

$$= (P \otimes P)\mathcal{R}(P^{-1} \otimes P^{-1}) = \mathcal{R}.$$

$\square$

PROPOSITION 1.3. Let $\mathcal{H} \in \mathrm{Mat}(n,k) \otimes \mathrm{Mat}(n,k)$ be an element such that $(A \otimes B)\mathcal{H} = \mathcal{H}(B \otimes A)$ for all $A, B \in \mathrm{Mat}(n,k)$. Then $\mathcal{H} = \theta\mathcal{R}$ for some $\theta \in k$.

PROOF. Let $\mathcal{H} = \sum_{i,j,p,q=1}^{n} \gamma_{ijpq} E_{ij} \otimes E_{pq}$. Then

$$(E_{ab} \otimes E)\mathcal{H} = \sum_{j,p,q=1}^{n} \gamma_{bjpq} E_{aj} \otimes E_{pq} =$$

$$\mathcal{H}(E \otimes E_{ab}) = \sum_{i,j,p=1}^{n} \gamma_{ijpa} E_{ij} \otimes E_{pb}.$$

Hence γ_{ijpq} vanishes if $i \neq q$ and $\gamma_{bjpb} = \gamma_{ajpa}$.

Similarly the identity $(E \otimes E_{ab})\mathcal{H} = \mathcal{H}(E_{ab} \otimes E)$ implies that γ_{ijpi} vanishes if $j \neq p$ and $\gamma_{ijji} = \gamma_{ippi} = \theta$ for all i, p, j. $\square$

THEOREM 1.4. *Let G be a finite group whose order and an integer n are coprime with* $\operatorname{char} k$. *A projective representation $\Omega : G \to \mathrm{PGL}(n,k)$ is irreducible if and only if*

$$\mathcal{R} = \frac{n}{|G|} \sum_{g \in G} \Omega(g^{-1}) \otimes \Omega(g). \tag{1.2}$$

PROOF. Suppose that the projective representation Ω is reducible. Then there exists an invertible matrix $P \in \mathrm{GL}(n,k)$ such that

$$P\Omega(g)P^{-1} = \left(\begin{array}{c|c} B_g & 0 \\ \hline 0 & C_g \end{array}\right)$$

with square blocks B_g, C_g of a smaller size. Then $E_{1n} \otimes E_{n1}$ does not occur in

$$\frac{n}{|G|} \sum_{g \in G} P\Omega(g^{-1})P^{-1} \otimes P\Omega(g)P^{-1}$$

which contradicts (1.2).

Conversely let a representation Ω be irreducible. Then

$$\Omega(g)\Omega(h) = \mu_{g,h}\Omega(gh), \quad \Omega(g)^{-1} = \mu_{g,h}^{-1}\Omega(h)\Omega(gh)^{-1} \tag{1.3}$$

where $\mu_{g,h} \in k^*$. It follows from (1.3) that

$$\left(\sum_{g \in G} \Omega(g^{-1}) \otimes \Omega(g)\right)(E \otimes \Omega(h)) = \sum_{g \in G} \Omega(g^{-1}) \otimes \Omega(g)\Omega(h)$$

$$= \sum_{g \in G} \mu_{g,h}\Omega(g^{-1}) \otimes \Omega(gh) = \sum_{g \in G} \mu_{g,h}\mu_{g,h}^{-1}\Omega(h)\Omega(gh)^{-1} \otimes \Omega(gh)$$

$$= (\Omega(h) \otimes E)\left(\sum_{g \in G} \Omega(gh)^{-1} \otimes \Omega(gh)\right)$$

$$= (\Omega(h) \otimes E)\left(\sum_{f \in G} \Omega(f)^{-1} \otimes \Omega(f)\right)$$

Since the representation Ω is irreducible the linear span of all $\Omega(h),\ h \in G$, coincides with $\mathrm{Mat}(n,k)$. Thus

$$(A \otimes E)\left(\sum_{f\in G} \Omega(f)^{-1} \otimes \Omega(f)\right) = \left(\sum_{f\in G} \Omega(f)^{-1} \otimes \Omega(f)\right)(E \otimes A)$$

for any matrix A. Similarly

$$(E \otimes B)\left(\sum_{f\in G} \Omega(f)^{-1} \otimes \Omega(f)\right) = \left(\sum_{f\in G} \Omega(f)^{-1} \otimes \Omega(f)\right)(B \otimes E)$$

for any B. By Proposition 1.3 we obtain

$$\sum_{f\in G} \Omega(f)^{-1} \otimes \Omega(f) = \theta\mathcal{R}. \tag{1.4}$$

Applying the map of multiplication

$$m : \mathrm{Mat}(n,k) \otimes \mathrm{Mat}(n,k) \to \mathrm{Mat}(n,k)$$

to both sides of (1.4) we obtain $|G| = \theta n$. Thus $\theta = \frac{|G|}{n}$. □

According to Schur theory [**CR**, Chapter 7, §52] there exists a central group extension

$$1 \longrightarrow H^2(G,k^*) \longrightarrow G^* \longrightarrow G \longrightarrow 1 \tag{1.5}$$

such that each projective representation of G can be lifted to an ordinary linear representation of G^*. Here $H^2(G,k^*)$ is the second cohomology group of G with coefficients in the multiplicative group k^*. Corresponding representations of G and of G^* are irreducible simultaneously.

The field k is algebraically closed. By [**McL**, Theorem 7.1, Chapter IV and Theorem 7.4, Chapter X] the second cohomology group $H^2(G,k^*)$ is trivial if and only if the group G is cyclic. Suppose that the Abelian group G is a direct product

$$G = \langle a_1\rangle_{m_1} \times \cdots \times \langle a_s\rangle_{m_s}, \tag{1.6}$$

of cyclic groups of orders $m_1, m_2, \ldots, m_s$ and $|G| = m_1 \cdots m_s$. The next result is known [**H**, Satz 25.11]

THEOREM 1.5. *If $s > 1$ then the group G^* is generated by elements $b_1, \ldots, b_s$ with defining relations $b_i^{m_i} = [b_r, [b_i, b_j]] = 1$ for all $i, j, r = 1, \ldots, s$.*

PROOF. It is necessary to check that any group homomorphism $\Omega : G \to \mathrm{PGL}(n,k)$ can be lifted to a group homomorphism $\Psi : G^* \to \mathrm{GL}(n,k)$.

Let B_i be some lift of $\Omega(a_i)$, $1 \leqslant i \leqslant s$. Then $B_i^{m_i} = \lambda_i E$, $\lambda_i \in k^*$. Since k is algebraically closed there exists $\tau_i \in k^*$ such that $\tau_i^{m_i} = \lambda_i$. Define $\Psi(b_i) = \mu_i^{-1} B_i$. Then $\Psi(b_i)^{m_i} = E$ and $[\Psi(b_i), \Psi(b_j)]$ is a scalar matrix $\lambda_{ij}E$. Then $[\Psi(b_r), [\Psi(b_i), \Psi(b_j)]] = 1$. Hence $\Psi : G^* \to \mathrm{GL}(n,k)$ is a group homomorphism. Using universal property of G^* we complete the proof. □

The next statement follows from the defining relations of G^*.

PROPOSITION 1.6. The order of an element $[b_i, b_j]$, $i \neq j$ is equal to the greatest common divisor (m_i, m_j). The derived subgroup $[G^*, G^*]$ is central and has a direct decomposition

$$\prod_{1 \leqslant i<j \leqslant s} \langle c_{ij} \rangle_{(m_i,m_j)}.$$

In particular the group G^* is nilpotent of class 2 and the order of G^* is equal to

$$m_1 \cdots m_s \left(\prod_{i<j} (m_i, m_j) \right).$$

In particular if G is not cyclic then the derived subgroup $[G^*, G^*]$ coincides with the center of G^* and is isomorphic to $H^2(G, k)$.

We shall consider which projective representations of G (linear representations of G^*) of degree $n > 1$ can occur as projective representations in Theorem 0.2. The necessary and sufficient conditions for this representation are (0.3) and (1.1).

Applying Theorem 1.4 we obtain

PROPOSITION 1.7. If $n > 1$ then the group G is not cyclic.

PROOF. If G is cyclic then $G^* = G$ because $H^2(G, k^*)$ is trivial. Then all irreducible representations of G have dimension 1. But by Theorem 1.4 the representation Ψ is irreducible and has dimension n. □

Now we have the following situation. The group G is a non-cyclic Abelian group and the group G^* is an extension (1.5) having presentation from Theorem 1.5. The group G^* is nilpotent and its order is found in Proposition 1.6. Hence every irreducible representation of G^* is monomial [**CR**, Chapter 7, Theorem 52.1] which means that it is induced by a one-dimensional representation of some subgroup $D \subset G^*$. Clearly D contains the derived central subgroup $[G^*, G^*]$ and therefore D is normal in G^*.

Consider an irreducible linear representation Ψ of G^* in a space V of dimension n induced by a one-dimensional representation of a normal subgroup D, containing $[G^*, G^*]$. Then $n = \dim V = |G^*/D|$. Let $G/D = \{g_1 D, \ldots, g_n D\}$ where $g_1 = 1$. Then V has a base

$$e_1 = e,\ e_2 = \Psi(g_2)e,\ \ldots,\ e_n = \Psi(g_n)e \tag{1.7}$$

where $\Psi(d)e = \chi(d)e$, $d \in D$, with some one-dimensional character χ of D.

PROPOSITION 1.8. If $d \in D$ then $\Psi(d)e_m = \chi(d)\chi([d, g_m])e_m$

for any m.

PROOF. Since $(G^*)' \subseteq D$ for any $d \in D$ we obtain

$$\begin{aligned} \Psi(d)e_m &= \Psi(d)\Psi(g_m)e = \Psi(g_m)\Psi(d)\Psi([d^{-1}, g_m^{-1}])e \\ &= \Psi(g_m)\Psi(d)\Psi([d, g_m])e = \chi(d)\chi([d, g_m])\Psi(g_m)e \\ &= \chi(d)\chi([d, g_m])e_m \end{aligned}$$

for any $d \in D$. □

COROLLARY 1.9. If $d \in D$ then in the base (1.7)

$$\Psi(d) = \chi(d)\begin{pmatrix} 1 & 0 & \dots & 0 \\ 0 & \chi([d,g_2]) & \dots & 0 \\ \dots & \dots & \dots & \dots \\ 0 & 0 & \dots & \chi([d,g_n]) \end{pmatrix}.$$

In particular $\operatorname{tr}\Psi(d) = \chi(d)\,(\sum_{m=1}^{n}\chi([d,g_m]))$.

Take an element $g \in G^* \setminus D$ whose image in G^*/D has order $t > 1$ dividing n.

PROPOSITION 1.10. Choose a system of representatives g_j of cosets G^*/D in such a way that $g_{tj+r} = g^r g_{tj}$ for $j = 0, \dots, \frac{n}{t} - 2$ and for $r = 0, \dots, t-1$. Then

$$\Psi(g)e_{tj+r} = \begin{cases} e_{tj+r+1}, & r = 0, \dots, t-2, \\ \chi(g^t)\chi([g,g_{tj}])^t e_{tj}, & r = t-1. \end{cases}$$

PROOF. We have

$$\Psi(g)e_{tj+r} = \Psi(g)\Psi(g_{tj+r})e = \Psi(gg_{tj+r})e.$$

If $r = 0, \dots, t-2$, then $gg_{tj+r} = g_{tj+r+1}$ and therefore

$$\Psi(g)e_{tj+r} = \Psi(g_{tj+r+1})e = e_{tj+r+1}.$$

If $r = t-1$ then $gg_{tj+t-1} = g^t g_{tj}$ and by Proposition 1.8

$$\Psi(g)e_{tj+t-1} = \Psi(g^t)\Psi(g_{tj})e = \Psi(g^t)e_{tj} = \chi(g^t)\chi([g,g_{tj}])^t e_{tj}.$$

□

COROLLARY 1.11. If $g \in G^* \setminus D$ then $\Psi(g)$ has only zero diagonal entries.

PROPOSITION 1.12. Let Ψ be any irreducible monomial linear representation of G^*. If $g \in G$ and $\operatorname{tr}\Psi(g) \neq 0$ then $\Psi(g)$ is a scalar matrix.

PROOF. If $\operatorname{tr}\Psi(g) \neq 0$ then $\Psi(g)$ has a nonzero diagonal entry. By Corollary 1.11 we can conclude that $g \in D$. Applying Corollary 1.9 we see that

$$\operatorname{tr}\Psi(g) = \chi(g)\left(\sum_{r=1}^{m}\chi([g,g_r])\right).$$

Note that $\chi(d) = 1$ for all $d \in D' = [D,D]$. Each map $g_rD \mapsto \chi([g,g_r])$ is a group homomorphism from G^*/D to k^*. So if this homomorphism is nontrivial then $\sum_{r=1}^{m}\chi([g,g_r]) = 0 = \operatorname{tr}\Psi(g)$.

Suppose that $\chi([g,g_r]) = 1$ for all r. Then $\Psi(g) = \lambda E$. □

We have considered irreducible linear representation of G^* in a special base (1.7). If A_g, $g \in G$, is a projective representation from Theorem 0.2 then there exists a matrix $L \in \mathrm{GL}(n,k)$ such that

$$A_g = \lambda_g L\Psi(g)L^{-1}, \qquad \lambda_g \in k^*, \quad \lambda_1 = 1,$$

for all $g \in G$. Then $\operatorname{tr}A_g = \lambda_g \operatorname{tr}\Psi(g) = n\delta_{g,1}$ by Proposition 1.12.

Moreover by (0.3) in $\mathrm{PGL}(n,k)$ we have

$$\begin{aligned} 1 &= [A_g,\, U\,{}^tA_hU^{-1}] = [L\Psi(g)L^{-1},\, U\,{}^tL^{-1}\,{}^t\Psi(h)\,{}^tLU^{-1}] \\ &= L[\Psi(g),\, L^{-1}U\,{}^tL^{-1}\,{}^t\Psi(h)\,{}^tLU^{-1}L]L^{-1}. \end{aligned}$$

Put $\Lambda = L^{-1}U\,{}^tL^{-1} \in \mathrm{GL}(n,k)$.

We have proved

THEOREM 1.13. *The equalities* (0.3) *hold if and only if*

$$[\Psi(g), \Lambda^t\Psi(h)\Lambda^{-1}] = 1$$

in $\mathrm{PGL}(n,k)$ *for some matrix* $\Lambda \in GL(n,k)$ *which is symmetric if* $U = E$ *and is skew-symmetric if* U *is from* (0.2).

2. Construction

In this section we shall show that for any n there exists a Hopf algebra H as in Theorem 0.2 in which $U = E$ is the identity matrix and G is a direct product $G = \langle a\rangle \times \langle b\rangle$ of two cyclic groups $\langle a\rangle$, $\langle b\rangle$ of order n. By Theorem 1.5 the derived subgroup $[G^*, G^*] = \langle c\rangle$ is a central cyclic group of order n. Moreover as it follows from Theorem 1.5 the group G^* is a semidirect product of a normal subgroup $\langle b\rangle \times \langle c\rangle$ by a cyclic subgroup $\langle a\rangle$. More precisely

$$a^n = b^n = c^n = 1, \quad [a,b] = aba^{-1}b^{-1} = c, \quad [a,c] = [b,c] = 1.$$

In particular

$$ba^ib^{-1} = \left(bab^{-1}\right)^i = ([b,a]a)^i = (c^{-1}a)^i = a^ic^{-i} \tag{2.1}$$

Consider a linear representation Ψ of the group G^* of dimension n in a space V induced by a one-dimensional representation Ψ of $\langle b\rangle \times \langle c\rangle$ in one-dimensional space W with one basic element e such that $\Psi(b)e = \omega e$, $\Psi(c)e = \eta e$. Here ω, η are primitive roots of 1 of degree n. Then the base (1.7) in V has the form $e_1 = e$, $e_2 = \Psi(a)e$, $\ldots$, $e_n = \Psi(a)^{n-1}e$.

By Proposition 1.8

$$\Psi(b)e_i = \omega\eta^{-i+1}e_i, \quad \Psi(c)e_i = \eta e_i. \tag{2.2}$$

It follows from (2.2) that in the base (1.7) of the space V

$$\Psi(a^ib^j) = \sum_{p\in\mathbb{Z}_n} E_{p+i,p}\omega^j\eta^{-j(p-1)} \in \mathrm{GL}(n,k), \quad \Psi(c^l) = \eta^l E \tag{2.3}$$

for $0 \leqslant i,j \leqslant n-1$.

THEOREM 2.1. *The representation* Ψ *is irreducible and the set of matrices* $\Psi(a^ib^j)$ *satisfies the conditions of Theorem 1.13 with* $\Lambda = E$.

PROOF. In order to prove irreducibility of Ψ by [**CR**, Corollary 45.4] we need to show that for any a^i, $1 \leqslant i \leqslant n-1$ there exists an element $y \in H$ such that $\chi(y) \neq \chi(a^iya^{-i})$ or $\chi([a^i, y]) \neq 1$. Taking $y = b$ we obtain

$$\chi([a^i, b]) = \chi([a,b])^i = \chi(c)^i = \eta^i \neq 1$$

because η is a primitive root of 1 of degree n.

In view of Proposition 1.12 and Theorem 1.4 we need to show that each matrix $\Psi(a^ib^j)$ commutes in $\mathrm{PGL}(n,k)$ with the transpose of another matrix $\Psi(a^rb^s)$.

Each matrix $\Psi(a^j)$ is a permutation matrix. Hence its inverse coincides with its transpose and therefore they commute. Each matrix $\Psi(b^j)$ is diagonal by (2.2). Hence its transpose coincides with itself. It follows that the transpose of $\Psi(a^ib^j)$ is equal to $\Psi(b^ja^{-i})$ in $\mathrm{GL}(n,k)$ and to $\Psi(a^{-i}b^j)$ in $\mathrm{PGL}(n,k)$. Hence by (2.1) matrices $\Psi(a^{-i}b^j)$, $\Psi(a^rb^s)$ commute in $\mathrm{PGL}(n,k)$ and the first statement is proved. □

It is interesting to mention that the same projective representation of the same group $G = \langle a\rangle \times \langle b\rangle$ is used in [**BSZ**] for the classification of group gradings on full matrix algebras $\mathrm{Mat}(n, k)$ (by an Abelian group). Gradings by non-Abelian finite groups are considered in [**BZ**].

COROLLARY 2.2. For any n there exists an example of a Hopf algebra H in Theorem 0.2 with $U = E$.

PROOF. By Theorem 1.4, Proposition 1.12, Theorem 1.13 and Theorem 2.1 the conditions (0.3) and (0.4) are satisfied. □

THEOREM 2.3. *Let H be a Hopf algebra from Theorem 0.2 where n is an odd prime. Then $G = \langle a\rangle_2 \times \langle b\rangle_2$ is a direct product of two cyclic groups of order n and the projective representation $g \mapsto A_g$ coincides with the representation Ψ from (2.3) for some primitive roots ω, η of 1 of degree n in k. If n is odd then $U = E$ and therefore H is uniquely defined up to a choice of ω. If $n = 2$, then either $U = E$ or $U = \mathcal{S}$ from (0.2) and both cases occur.*

PROOF. Since G by Proposition 1.7 is a non-cyclic Abelian group of order n^2, then G has the required direct sum decomposition. Suppose that W is the n-dimensional irreducible linear representation of the group G^*. The Abelian subgroup $\langle a\rangle \times \langle b\rangle$ has common eigenvector $e_0 \in W \setminus 0$ such that $be_0 = \omega e_0$, $ce_0 = \eta e_0$ where ω, η are roots of 1 of degree n. If either $\eta = 1$ or $\omega = 1$ the $\ker \Psi$ contains either c or b and $[a, b] = c$. In both cases Ψ is reduced to an irreducible linear representation of Abelian group either G or $\langle a\rangle$. In these cases $\dim W = 1$, a contradiction.

Thus ω, η are primitive roots of 1 of degree n in k and the representation coincides with the representation Ψ from (2.3).

If n is odd then always $U = E$.

Suppose that $n = 2$. It the case $n = 2$ we have $\omega = \eta = -1$ and

$$\Psi(a) = \begin{pmatrix} 0 & 1 \\ 1 & 0 \end{pmatrix}, \quad \Psi(b) = \begin{pmatrix} -1 & 0 \\ 0 & 1 \end{pmatrix}. \tag{2.4}$$

So

$$\Psi(ba) = -\Psi(ab) = \begin{pmatrix} 0 & -1 \\ 1 & 0 \end{pmatrix} = T = \mathcal{S}.$$

Therefore all conditions from Theorem 0.2 are satisfied for $U = E$ and for $U = \mathcal{S}$. □

PROPOSITION 2.4. Let Ψ be the representation of the group $G = \langle a\rangle_2 \times \langle b\rangle_2$ from (2.4) and

$$e_1 + \chi_a e_a + \chi_b + \chi_{ab} ab + x, \quad x \in \mathrm{Mat}(n, k)),$$

be a group element from Theorem 0.4, where $\chi : G \to k^*$ is a character, $\chi_a \chi_b = \pm 1$. Then

$$x = \begin{pmatrix} \alpha & \beta \\ \chi_a \beta & \chi_a \alpha \end{pmatrix} = \chi_b \begin{pmatrix} \alpha & -\beta \\ -\chi_a \beta & \chi_a \alpha \end{pmatrix}. \tag{2.5}$$

PROOF. Let

$$x = \begin{pmatrix} \alpha & \beta \\ \gamma & \delta \end{pmatrix}$$

Direct calculations show by Theorem 0.4 that

$$\Psi(a)\begin{pmatrix}\alpha & \beta\\ \gamma & \delta\end{pmatrix}\Psi(a)^{-1}=\begin{pmatrix}\delta & \gamma\\ \beta & \alpha\end{pmatrix}=\chi_a\begin{pmatrix}\alpha & \beta\\ \gamma & \delta\end{pmatrix}.$$

Hence $\delta=\chi_a\alpha$, $\gamma=\chi_a\beta$ and therefore x has the required form.

Similarly

$$\Psi(b)\begin{pmatrix}\alpha & \beta\\ \gamma & \delta\end{pmatrix}\Psi(a)^{-1}=\begin{pmatrix}\alpha & -\beta\\ -\gamma & \delta\end{pmatrix}=\chi_b\begin{pmatrix}\alpha & \beta\\ \gamma & \delta\end{pmatrix}$$

and the proof is completed. □

THEOREM 2.5. *Let $n=2$ in Theorem 2.3. If $U=E$ then H is the group algebra of the dihedral group D_4. If $U=T$ then H is the group algebra of the quaternion group Q_8. If*

$$U=\begin{pmatrix}-1 & 1\\ 1 & 1\end{pmatrix}$$

then H is the Hopf algebra H_8 from [**KP**] [**SN**, Appendix A], [**Ma**] *If $n>2$ in Theorem 2.1 then H is not cocommutative.*

PROOF. Suppose first that $n=2$. Then $G=\langle a\rangle_2\times\langle b\rangle_2$ is a direct product of two cyclic groups of order 2. We shall find 8 group-like elements in H using Theorem 0.4. Let $\Psi(a)$, $\Psi(b)$ be from (2.4). Let

$$e_1+\chi_a e_a+\chi_b e_b+\chi_a\chi_b e_{ab}+x$$

be a group-like element from Theorem 0.4 where $\chi_a,\chi_b=\pm1$. Then (2.5) holds and $xU\,{}^tx=U$ by Theorem 0.4.

Suppose that $U=E$ and x is from (2.5). Then $x\,{}^tx=E$ and direct calculations show that the group $G(H)$ of group-like elements in H consists of 8 elements

$$\begin{aligned}
&e_1+e_a+e_b+e_{ab}\pm E;\\
&e_1+e_a-e_b-e_{ab}\pm\begin{pmatrix}0 & 1\\ 1 & 0\end{pmatrix};\\
&e_1-e_a+e_b-e_{ab}\pm\begin{pmatrix}-1 & 0\\ 0 & 1\end{pmatrix};\\
&e_1-e_a-e_b+e_{ab}\pm\begin{pmatrix}0 & -1\\ 1 & 0\end{pmatrix}.
\end{aligned}$$

Hence $G(H)$ is isomorphic to the group consisting of matrices

$$\pm E,\quad \pm\begin{pmatrix}0 & 1\\ 1 & 0\end{pmatrix},\quad \pm\begin{pmatrix}-1 & 0\\ 0 & 1\end{pmatrix},\quad \pm\begin{pmatrix}0 & -1\\ 1 & 0\end{pmatrix}$$

which is isomorphic to the group D_4. Hence H is the group algebra of D_4.

If $U=T$ then (2.5) and $xU\,{}^tx=U$. Moreover

$$\chi_g x=U\,{}^t\Psi(g)U^{-1}xU\,{}^t\Psi(g)^{-1}U^{-1},\quad g=a,b\in G.$$

Again direct calculations show that in this case the group $G(H)$ of group-like elements in H consists of 8 elements

$$\begin{aligned}
&e_1+e_a+e_b+e_{ab}\pm E;\\
&e_1+e_a-e_b-e_{ab}\pm\begin{pmatrix}0 & i\\ i & 0\end{pmatrix},\quad i^2=-1;
\end{aligned}$$

$$e_1 - e_a + e_b - e_{ab} \pm \begin{pmatrix} -i & 0 \\ 0 & i \end{pmatrix}, \quad i^2 = -1;$$

$$e_1 - e_a - e_b + e_{ab} \pm \begin{pmatrix} 0 & -1 \\ 1 & 0 \end{pmatrix}.$$

Hence $G(H)$ is isomorphic to the group consisting of matrices

$$\pm E, \quad \pm \begin{pmatrix} 0 & i \\ i & 0 \end{pmatrix}, \quad \pm \begin{pmatrix} -i & 0 \\ 0 & i \end{pmatrix}, \quad \pm \begin{pmatrix} 0 & -1 \\ 1 & 0 \end{pmatrix}$$

which is isomorphic to the quaternion group Q_8 and H is the group algebra of Q_8.

Suppose finally that

$$U = \begin{pmatrix} -1 & 1 \\ 1 & 1 \end{pmatrix}$$

Then

$$U\,{}^t\Psi(a)U^{-1} = \Psi(b), \quad U\,{}^t\Psi(b)U^{-1} = \Psi(a).$$

It follows from Theorem 0.4 that if x is from (2.5) then

$$a \rightharpoonup x = \Phi(a)x\Phi(a)^{-1} = \chi_a x = x \leftharpoonup a = \Psi(b)x\Psi(b)^{-1} = b \rightharpoonup a = \chi_b x.$$

Thus $\chi_a = \chi_b = \pm 1$. By direct calculation using the equation $xU\,{}^t x = U$ we can finally obtain that the group $G(H)$ consists of 4 elements

$$e_1 + e_a + e_b + a_{ab} \pm E,$$

$$e_1 - e_a - e_b + a_{ab} \pm \begin{pmatrix} 0 & 1 \\ 1 & 0 \end{pmatrix}.$$

So this case corresponds to unique Hopf algebra non-isomorphic to group algebras of D_4 and Q_8. Hence $H \simeq H_8$.

Finally if $n > 2$ and $\Psi(a)$ is from (2.1) then Corollary 0.6 is not satisfied and H is not cocommutative. □

Finally it is necessary to mention that S. Spiridonova has shown that for the group G and its representation from Theorem 2.1 with even n there exists a skew-symmetric invertible matrix U of size n which satisfies the condition of Theorem 1.13.

A Hopf algebra H is determined by Theorem 0.2 by a (skew-)symmetric matrix U and by an irreducible projective representation $g \mapsto A_g$ of the group G. If we reduce U to a canonical form E of $\mathcal{S}$ then by [**A**, Theorem 5.2] H is uniquely determined by the representation up to an orthogonal equivalence for $U = E$ and up to a symplectic equivalence if $U = \mathcal{S}$ of representations. If we fix an irreducible monomial representation of G then H is determined by a (skew-)symmetric matrix $U = \Lambda$ satisfying the equality from Theorem 1.13. The matrix Λ can be replaced by a matrix $R\Lambda R^{-1}$, $R \in \mathrm{GL}(n,k)$, where R is permutable in $\mathrm{PGL}(n,k)$ with any matrix A_g, $g \in G$, and $R\Lambda R^{-1}$ is satisfying the equality from Theorem 1.13.

Note also that according to [**ACh**, Corollary 4.6] the order of the group $G(H)$ of group-like elements in H does not exceed $3n$ and if G is Abelian then $|G(H)| \leqslant 2n$. If all cases if $n > 3$ then H from Theorem 0.2 is not a group algebra.

PROBLEM 2.6. Let G be a non-cyclic Abelian group of order n^2. Does there exist an irreducible linear monomial representation $\Psi : G^* \to \mathrm{GL}(n,k)$ such that the statement of Theorem 1.13 is satisfied for some symmetric invertible matrix Λ?

PROBLEM 2.7. Let n be an even positive number and G be a non-cyclic Abelian group of order n^2. Does there exist an irreducible linear monomial representation $\Psi : G^* \to \mathrm{GL}(n, k)$ such that the statement of Theorem 1.13 is satisfied for some skew-symmetric invertible matrix Λ?

References

[A] V.A.Artamonov, On semisimple finite dimensional Hopf algebras, Mat. Sbornik, – 198(2007), N 9, 3-28.

[ACh] V.A.Artamonov, I.A.Chubarov, Dual algebras of some semisimple finite dimensional Hopf algebras, Modules and Comodules Trends in Mathematics, 65-85, 2008 Birkhäuser Verlag Basel/Switzerland.

[CR] Curtis Ch.W., Reiner I., Representation theory of finite groups and associative algebras, Interscience Publ., John Wiley & Sons, New York, London, 1962.

[H] Huppert B. Endliche Gruppen, I, Berlin, Heidelberg, New York, Springer, 1972.

[M] Montgomery,S. *Hopf Algebras and Their Actions on Rings*, Regional Conf. Ser. Math. Amer. Math. Soc., Providence RI, 1993.

[Ma] Masuoka A., Semisimple Hopf algebras of dimension 6, 8, Israel J. Math., 92(1995), 361-373.

[McL] MacLane Saunders, Homology, Springer-Verlag, Berlin-Göttingen-Heidelberg, 1963.

[SN] Sonia Natale, Semisolvability of semisimple Hopf algebras of Low dimension, Memoirs of the Amer. Math. Soc., vol.186(2007), N 874.

[TY] Tambara D., Yamagami S., Tensor categories with fusion rules of self-duality for finite abelian groups, J.Algebra 209(1998), 692-707.

[KP] Kac, G., Paljutkin, V., Finite ring groups, Trudy Moscow Math. Obschestva, 15 (1966), 224-261.

[T] Tambara D., representations of tensor categories with fusion rules of self-duality for Abelian groups, Israel J. Math. 118(2000), 29-60.

[S] Siehler J., Near-group categoies, Algebraic and geometric topology, 3(2003), 719-775.

[BSZ] Bahturin Y.A,Sehgal S.K., Zaicev M.V., Group gradings on associative algebras, J.Algebra, 241(2001), 677-698.

[BZ] Bahturin Y.A, Zaicev M.V., Group gradings on matrix algebras,Canad.Math.Bull.,45 (2002),499-508.

DEPARTMENT OF ALGEBRA, FACULTY OF MECHANICS AND MATHEMATICS, MOSCOW, STATE UNIVERSITY, RUSSIA

E-mail address: artamon@mech.math.msu.su

DEPARTMENT OF ALGEBRA, FACULTY OF MECHANICS AND MATHEMATICS, MOSCOW STATE UNIVERSITY

E-mail address: igrek@dubki.ru

Contemporary Mathematics
Volume **483**, 2009

Classifying simple color Lie superalgebras

Yuri Bahturin and Dušan Pagon

Abstract. We classify simple finite-dimensional color Lie algebras over an algebraically closed field of characteristic zero. The color is given by a skew-symmetric bicharacter on the group $\mathbb{Z}_2 \times \mathbb{Z}_2$.

1. Introduction

In this paper we discuss the first steps toward the classification of simple generalized finite-dimensional Lie superalgebras over an algebraically closed field F of characteristic zero. To be precise, we consider Lie superalgebras over a field k graded by a finite abelian group G, written multiplicatively, of the form

$$L = \bigoplus_{g\in G} L_g \tag{1.1}$$

with bracket operation $[\ ,\ ]$ satisfying the generalized anticommutativity and Jacobi identities for any homogeneous $x \in L_g, y \in L_h, x \in L_k$:

$$[x, y] + \beta(x, y)[y, x] = 0, \tag{1.2}$$

$$[[x, y], z] = [x, [y, z]] - \beta(x, y)[y, [x, z]]. \tag{1.3}$$

Here $\beta : G \times G \to F^*$ is the *commutation factor*, i.e. a function satisfying

$$\beta(gh, k) = \beta(g, k)\beta(h, k),\ \beta(g, hk) = \beta(g, h)\beta(g, k),\ \beta(g, h)\beta(h, g) = 1. \tag{1.4}$$

We write $\beta(x, y) = \beta(g, h)$ if $x \in L_g, y \in L_h$. If $G = \{e\}$ we have usual Lie algebras, if $G = \mathbb{Z}_2 = \{-1, 1\}$ and $\beta(-1, -1) = -1$ we have *ordinary Lie superalgebras.* It is easy to see that $\beta(g, g) = \pm 1$ for any $g \in G$. We set $G_\pm = \{g \in G | \beta(g, g) = \pm 1\}$. We call the Lie superalgebras just defined (G, β)*-Lie superalgebras.* A particular case, of special importance to us, is when $G = G_+$. Such (G, β)*-Lie superalgebras* are called *color Lie algebras* For details see [**2**]. If $A = \oplus_{g\in G} A_g$ is a G-graded associative algebra, then setting

$$[x, y] = xy - \beta(x, y)yx \tag{1.5}$$

for $x \in A_g$, $y \in A_h$, we make A into (G, β)-Lie superalgebra $[A]_\beta$.

Two (G, β)-Lie superalgebras L and M are called *isomorphic* if there is a linear isomorphism $\varphi : L \to M$ such that $\varphi(L_g) = M_g$ for any $g \in G$ and also $\varphi([x, y]) = [\varphi(x), \varphi(y)]$, for any $x, y \in L$.

2000 *Mathematics Subject Classification.* 17B70, 17B75 .

2. M. Scheunert's Trick

One of the tools in the theory of (G,β)-superalgebras is a procedure of M. Scheunert's (often called Scheunert's trick, however see [**17**]) enabling one to pass from (G,β)-Lie superalgebras to ordinary Lie superalgebras with a number of properties preserved. If $\sigma \in Z^2(G,F^*)$ is a multiplicative 2-cocycle, and L a (G,β)-Lie superalgebra satisfying (1.2), (1.3) with the commutation factor β, then replacing the multiplication in L on homogeneous elements $x, y \in L$ by

$$[x,y]^\sigma = \sigma(x,y)[x,y] \tag{2.1}$$

we arrive at a (G,β')-Lie-superalgebra satisfying (1.2), (1.3) with the commutation factor $\beta' = \beta\delta$ where $\delta(x,y) = \sigma(x,y)/\sigma(y,x)$. Let $\varepsilon_0 : G \times G \to F^*$ be the ordinary superalgebra commutation factor, i.e. $\varepsilon_0(g,h) = 1$ except $\varepsilon_0(g,h) = -1$ for $g, h \in G_-$.

THEOREM 2.1. [**19**] *Let G be a finitely generated abelian group, $\beta : G\times G \to F^*$ a commutation factor with $G = G_+$. Then there exists a 2-cocycle $\sigma \in Z^2(G,F^*)$ such that $\beta\delta = 1$.* □

In other words L^σ is an ordinary Lie algebra (with the G-grading still preserved!). Since $\beta\varepsilon_0$ satisfies the conditions of the theorem we can find σ with $\beta\varepsilon_0\delta = 1$ whence $\beta\delta = \beta_0^{-1} = \varepsilon_0$. Thus from any (G,β)-Lie superalgebra by a change of the form (2.1) we can switch to a Lie superalgebra $L^\sigma = L_0^\sigma \oplus L_1^\sigma$ with $L_0^\sigma = L_+, L_1^\sigma = L_-$.

Of course this procedure is invertible and starting from a G-graded Lie algebra or an ordinary Lie superalgebra one can change multiplication by an inverse of σ to obtain a (G,β)-Lie superalgebra. Also, it is important to notice that the 2-cocycle σ here can be chosen a (non-symmetric) bicharacter, that is, a function with the first two conditions in (1.4).

PROPOSITION 2.2. *If A is a G-graded associative algebra, $[A]_\beta$ the respective (G,β)-Lie superalgebra given by (4′) then A^σ with multiplication*

$$(ab)^\sigma = \sigma(g,h)ab, \quad a \in A_g, \quad b \in A_h, \tag{2.2}$$

is a G-graded associative algebra and we have

$$[A^\sigma]_{\beta'} = [A]^\sigma,$$

where $\beta' = \beta\delta$ as after (2.1).

Proof. This is a simple computation. Take homogeneous $a, b \in A$ and compare the β'-bracket in A^σ and the σ-twisted β-bracket in [A]. That is,

$$\begin{aligned} [a,b]_{\beta'} &= (ab)^\sigma - \beta'(a,b)(ba)^\sigma \\ &= \sigma(a,b)ab - \beta'(a,b)\sigma(b,a)ba \\ &= \sigma(a,b)(ab - (\beta'\delta^{-1})(a,b)ba) \\ &= \sigma(a,b)(ab - \beta(a,b)ba) \\ &= \sigma(a,b)[a,b]_\beta = ([a,b]_\beta)^\sigma \end{aligned}$$

proving our claim. □

The second part of this theorem gives yet another way of replacing the study of color Lie superalgebras by the study of the ordinary ones.

The procedure described above is often called *discoloration.* It is immediate from the definitions and Theorem 2.1 that if we apply discoloration to a color (G,β)-Lie algebra then what we obtain is a G-graded algebra.

3. Introductory remarks about simple color Lie superalgebras

Let $L=\bigoplus_{g\in G}L_g$ be a G-graded algebra. Any element $x\in L$ can be uniquely written as the sum $x=\sum_{g\in G}x_g$ where $x_g\in L_g$. The element x_g is called a *homogeneous component of x of degree g.* We say that x is *homogeneous of degree* g if $x=x_g$. A subspace M of L is called *graded* if $M=\bigoplus_{g\in G}(M\cap L_g)$. In other words, M is graded if and only if the homogeneous components of any $x\in M$ are in M.

DEFINITION 3.1. A (G,β)-Lie algebra L is called *simple* if L has no proper nonzero graded ideals.

According to our previous remarks, the classification of simple (G,β)-Lie algebras reduces to the following steps

(1) Classification of simple G-graded Lie algebras and Lie superalgebras
(2) Classification of commutation factors on G

As we will see in Section 4, the classification of simple G-graded Lie (super)algebras in the case of elementary abelian groups reduces to the classification of K-gradings on simple Lie (super)algebras, where K is a subgroup of G. This, of course, includes the classification of gradings on the simple Lie algebras and superalgebras themselves. If we restrict ourselves to algebraically closed fields of characteristic zero, then the classification of simple Lie algebras is a classical matter. Now simple ordinary Lie superalgebras have been classified by V. Kac [**15**]. So to proceed with the classification of simple (G,β)-Lie algebras we need to know the gradings by groups on simple Lie algebras and simple Lie superalgebras. In the case of simple Lie algebras this work seems to be finished, except the cases of exceptional Lie algebras E_7 and E_8 [**6, 8, 11, 12, 3**]. The gradings on simple ordinary Lie superalgebras are less explored (see for example [**4**]). Therefore, the first sufficiently motivated step in the classification of simple (G,β)-Lie superalgebras is the *classification of (G,β)-Lie superalgebras in the case of color Lie algebras, which are the forms of classical simple graded Lie algebras.* By this we mean that such superalgebra after discoloration becomes a simple graded Lie algebra whose simple component (see Section 4 of one of the types $A_l, l\geq 1$, $B_l, l\geq 2$, $C_l, l\geq 3$, and $D_l, l\geq 4$. As for the description of the commutation factors, this is done, for example in a paper of A. Zolotykh [**22**].

In such approach to the classification of (G,β)-graded Lie algebras it seems reasonable to start with the case where G is cyclic, with generator g. However since $G=G_+$ we must have $\beta(g^i,g^j)=\beta(g,g)^{ij}=1$. In this case our superalgebras are just G-graded simple Lie algebras. The classification of gradings by cyclic groups on all simple Lie algebras, not necessarily classical was given by V. Kac (see [**16**] and the monograph by Vinberg - Onishchik [**21**]).

The first unexplored case, which is the contents of this research is the one where $G=\mathbb{Z}_2\times\mathbb{Z}_2$. The gradings by $\mathbb{Z}_2\times\mathbb{Z}_2$ on classical simple Lie algebras, except D_4

are dealt with in great detail in a recent paper [**1**]. This latter paper will be our main source of information in this work. Also the commutation factors β on $\mathbb{Z}_2\times\mathbb{Z}_2$ are very easy to classify and in the case $G=G_+$ they include just two cases: the trivial case where all the values are 1 and a non-trivial case. If $\mathbb{Z}_2\times\mathbb{Z}_2=\langle a\rangle_2\times\langle b\rangle_2$ then we have

$$\beta_0(a^ib^j,a^kb^l)=(-1)^{jk-il}. \tag{3.1}$$

One more preliminary rather general remark is important.

Proposition 3.2. *Let L and M are two simple (G,β) - Lie superalgebras and σ a 2-cocycle such that for any $g,h\in G$ we have $\beta(g,h)\sigma(g,h)\sigma(h,g)^{-1}=1$. Then L^σ and M^σ are two graded simple Lie algebras such that $L\cong M$ if and only if $L^\sigma\cong M^\sigma$. Thus for any fixed σ satisfying the above condition, the map $L\mapsto L^\sigma$ is the bijection between simple (G,β)-Lie superalgebras and G-graded simple Lie algebras.*

This result allows one in the classification of simple (G,β)-Lie superalgebras to use *just one* coloring method by any fixed 2-cocycle σ satisfying $\beta(g,h)\sigma(g,h)$ $\sigma(h,g)^{-1}=1$. For example, in the case of the bicharacter defined in (3.1) we can fix the bicharacter $\sigma(a^ib^j,a^kb^l)=(-1)^{jk}$.

4. Graded simple algebras

Let G be an abelian group. We call G *elementary* if for any nontrivial proper subgroup H of G there is a subgroup K such that $G=H\times K$. We also call a G-graded algebra L *graded simple* if L has no proper nonzero graded ideals.

Proposition 4.1. *Let L be a finite-dimensional G-graded simple Lie algebra over an algebraically closed field F of characteristic zero. Suppose that G is an elementary abelian group. Then there is a simple ideal M in L and subgroups H, K in G, $G=H\times K$ such that $L\cong F[H]\otimes M$ where the commutator is given by $[a\otimes x,b\otimes y]=(ab)\otimes[x,y]$ for any a,b in the group ring $F[H]$ and $x,y\in M$. The ideal M has a K-grading and an element $h\otimes x$ is homogeneous of degree hk provided that* $\deg x=k$.

Proof. Let $\widehat{G}$ be the dual group of G, that is, the group of multiplicative homomorphisms (characters) from G into the multiplicative group F^* of F. The natural action of $\widehat{G}$ by automorphisms on L is given by $\chi*x=\chi(g)x$, provided that x is homogeneous of degree g. Let M be a minimal nonzero ideal of L. If $L=M$ there is nothing to do. Otherwise let Λ be a subgroup of $\widehat{G}$ acting by those automorphisms which leave M invariant. There is a subgroup Π such that $\widehat{G}=\Lambda\times\Pi$. It is known that $\widehat{G}\cong G$, so $\widehat{G}$ is also elementary. Next we set $H=\Lambda^\perp=\{g\in G\,|\,\lambda(g)=1\,\forall\lambda\in\Lambda\}$. Similarly we define $K=\Pi^\perp$. Now we define an K-grading on M by saying $\deg x=k$ if $\lambda*x=\lambda(k)x$. Now we consider $\widetilde{L}=F[H]\otimes M$ as described in the theorem, with the operation and the grading defined therein. Let us define a linear map $\varphi:\widetilde{L}\to L$ by setting

$$\varphi(h\otimes x)=\sum_{\pi\in\Pi}(\pi(h))^{-1}(\pi*x). \tag{4.1}$$

We have to check that φ is an isomorphism of graded algebras. First of all, it is easy to check that $L = \sum_{\pi\in\Pi}(\pi * M)$. Then

$$\begin{aligned}\varphi([h\otimes x, h'\otimes y]) &= [\sum_{\pi\in\Pi}(\pi(h))^{-1}(\pi * x), \sum_{\rho\in\Pi}(\rho(h'))^{-1}(\rho * y)]\\ &= \sum_{\pi\in\Pi}[(\pi(h))^{-1}(\pi * x), (\pi(h'))^{-1}(\pi * y)]\\ &= \sum_{\pi\in\Pi}(\pi(hh'))^{-1}(\pi * [x,y] = \varphi((hh')\otimes[x,y]).\end{aligned}$$

To show φ preserves the grading, let us assume that $\deg x = k$. Then the degree on the left side of (4.1) is hk. To find the degree of the right side we have to apply $\lambda\rho$. Then we obtain

$$\begin{aligned}(\lambda\rho) * (\varphi(h\otimes x)) &= \sum_{\pi\in\Pi}(\pi(h))^{-1}(\rho\pi * (\lambda * x))\\ &= \lambda(k)\rho(h)\sum_{\pi\in\Pi}((\pi\rho)(h))^{-1}((\rho\pi) * x)\\ &= (\lambda\rho)(hk)\sum_{\sigma\in\Pi}(\sigma(h))^{-1}(\sigma * x),\end{aligned}$$

as needed. □

This proposition allows us to conclude that in the case of $\mathbb{Z}_2\times\mathbb{Z}_2$-simple graded algebras we have to consider three cases (in every one $\mathfrak{g}$ is a simple Lie algebra)

(1) $L = F[\mathbb{Z}_2\times\mathbb{Z}_2]\otimes\mathfrak{g}$, with no grading on M;
(2) $L = F[\mathbb{Z}_2]\otimes\mathfrak{g}$, with a $\mathbb{Z}_2$-grading on $\mathfrak{g}$;
(3) $L = \mathfrak{g}$, with a $\mathbb{Z}_2\times\mathbb{Z}_2$-grading on $\mathfrak{g}$.

Case 1 is quite obvious. In this case $L = F[\mathbb{Z}_2\times\mathbb{Z}_2]\otimes\mathfrak{g}$, $\deg(g\otimes x) = g$ and, up to equivalence of gradings, it subdivides into two cases:

(a) $[a^ib^j\otimes x, a^kb^l\otimes y]_L = a^{i+k}b^{j+l}\otimes[x,y]_{\mathfrak{g}}$;
(b) $[a^ib^j\otimes x, a^kb^l\otimes y]_L = (-1)^{jk}a^{i+k}b^{j+l}\otimes[x,y]_{\mathfrak{g}}$.

Case 2 involves $\mathbb{Z}_2$-gradings on simple Lie algebras, known since quite old times in the classification theory of the symmetric spaces (see for example [**16**]). If we write this grading as $\mathfrak{g} = \mathfrak{g}_e\oplus\mathfrak{g}_q$ then $L = F[\langle p\rangle]\otimes\mathfrak{g}$ and the $\mathbb{Z}_2\times\mathbb{Z}_2$ grading is given by $L_e = e\otimes\mathfrak{g}_e$, $L_p = p\otimes\mathfrak{g}_e$, $L_p = e\otimes\mathfrak{g}_q$, $L_{pq} = p\otimes\mathfrak{g}_q$. As in Case 1 there are two options for the bracket

(a) $[p^i\otimes x, p^k\otimes y]_L = p^{i+k}\otimes[x,y]_{\mathfrak{g}}$;
(b) $[p^i\otimes x, p^k\otimes y]_L = (-1)^{jk}p^{i+k}\otimes[x,y]_{\mathfrak{g}}$ where $x\in M_{q^j}, y\in M_{q^l}$.

Case 3 amounts to $\mathbb{Z}_2\times\mathbb{Z}_2$-gradings on simple Lie algebras and the application of color in one case. In other words, if $L = \mathfrak{g}$ is a simple Lie algebra endowed with a $\mathbb{Z}_2\times\mathbb{Z}_2$ grading $\mathfrak{g} = \mathfrak{g}_e\otimes\mathfrak{g}_a\otimes\mathfrak{g}_b\otimes\mathfrak{g}_{ab}$ then, up to equivalence (see below) we have one of two cases

(a) $L = \mathfrak{g}$ stays as it is, but its subspaces, subalgebras and ideals are the graded subspaces, etc. of $\mathfrak{g}$. The commutator of L is the same as in $\mathfrak{g} : [x,y]_L = [x,y]_{\mathfrak{g}}$;
(b) $L = \mathfrak{g}$ stays as it is, but its subspaces, subalgebras and ideals are the graded subspaces, etc. of $\mathfrak{g}$. The commutator of L is given by $[x,y]_L = (-1)^{jk}[x,y]_{\mathfrak{g}}$ for $x\in\mathfrak{g}_{a^ib^j}$, $y\in\mathfrak{g}_{a^kb^l}$.

Our main concern in this paper, will be Case 3. It follows from Proposition 3.2 that the superalgebras arising in different case are mutually non-isomorphic.

5. $(\mathbb{Z}_2 \times \mathbb{Z}_2, \beta)$-superalgebra structures on classical simple algebras

By definition, given two G-gradings $\mathfrak{g} = \oplus_{g\in G}\mathfrak{g}_g$ and $\mathfrak{g} = \oplus_{g\in G}\mathfrak{g}'_g$ of an algebra $\mathfrak{g}$ by a group G, we call them *equivalent* or *isomorphic* if there exists an automorphism α of $\mathfrak{g}$ such that $\mathfrak{g}'_g = \alpha(\mathfrak{g}_g)$. We also say that in this case two graded algebras in question are isomorphic.

In [**9**] the classification of gradings was presented with the use another equivalence relation on the gradings. One calls two G-gradings $\mathfrak{g} = \oplus_{g\in G}\mathfrak{g}_g$ and $\mathfrak{g} = \oplus_{g\in G}\mathfrak{g}'_a$ of an algebra $\mathfrak{g}$ by a group G *weakly equivalent* if there exists an automorphism π of $\mathfrak{g}$ and an automorphism ω of G such that $\mathfrak{g}'_g = \pi(\mathfrak{g}_{\omega(g)})$. The classification given in [**9**] is up *up to the weak equivalence.* To obtain all gradings up to isomorphism one has to apply to the representatives of the classes of weak equivalence the automorphisms of $\mathbb{Z}_2\times\mathbb{Z}_2$ or, which is the same, all possible permutations of nontrivial elements, six in total. We will see, however that in the whole number of cases there is no need to apply the group automorphisms since some of them produce the same result as the algebra automorphisms and the respective algebras are not just weakly equivalent but isomorphic as graded algebras. This, in particular, happens to many so called elementary gradings.

In what follows $G = \{e, a, b, c\}$ is the group $\mathbb{Z}_2 \times \mathbb{Z}_2$ with identity e and $a^2 = b^2 = c^2 = e,\ ab = c$. We will consider the $\mathbb{Z}_2 \times \mathbb{Z}_2$-grading on a complex simple Lie algebra $\mathfrak{g}$ one if the types $A_l,\ l \geq 1,\ B_l,\ l \geq 2,\ C_l,\ l \geq 3$ and $D_l,\ l \geq 4$. We are going to use some results of [**6**] and [**8**].

5.1. Connection with gradings on matrix algebras. According to [**1**], which quotes [**6**] and [**12**], any G-grading of a simple Lie algebra $\mathfrak{g} = \mathrm{so}(2l+1),\ l \geq 2,\ \mathfrak{g} = \mathrm{so}(2l),\ l \geq 4$ and $\mathrm{sp}(2l),\ l \geq 3$ is induced from an G-grading of the respective associative matrix algebra $R = M_{2l+1}$ in the first case, or M_{2l} in the second and the third case. This is not true for so(8) for general G but still true provided that $G = \mathbb{Z}_2 \times \mathbb{Z}_2$. In the case of sl$(n)$ there are two types of gradings: the first one is those induced from the gradings of M_n and the second where we still need to know the gradings of M_n but they should be modified with the help of a so called graded involution of M_n (see more details below in Section 6).

Two kinds of G-grading on the associative matrix algebra $M_n = R = \oplus_{g\in G} R_a$ are of special importance:

1) *Elementary gradings.* Each elementary grading is defined by an n-tuple $(g_1, ..., g_n)$ of elements of A in such a way that all matrix units E_{ij} are homogeneous with $E_{ij} \in R_g$ if and only if $g = g_i^{-1}g_j$. Two tuples $(g_1, ..., g_n)$ and $(g'_1, ..., g'_n)$ define isomorphic elementary gradings if and only if there is a permutation ν of indexes $1, 2, \ldots, n$ and $g_0 \in G$ such that $g_0 g_{\nu(i)} = g'_i$, for all $i = 1, 2, \ldots, n$ [**7**]. In the case $G = \mathbb{Z}_2 \times \mathbb{Z}_2$ the grading by any tuple is thus isomorphic to the grading by a tuple of the form

$$(e^{(k_1)}, a^{(k_2)}, b^{(k_3)}, c^{(k_4)}) \text{ where } g^{(k)} = \underbrace{g, \ldots, g}_{k} \text{ and } k_1 + k_2 + k_3 + k_4 = n.$$

The number of pairwise non isomorphic gradings can be evaluated as follows. The multiplication by $g_0 \in \mathbb{Z}_2 \times \mathbb{Z}_2$ does not change the homogeneous components, but causes 4 fixed point free permutations of k_1, k_2, k_3, k_4. Thus we have at most 6 pairwise nonisomorphic elementary gradings and if we assign on of the numbers to a fixed element of the group, we obtain 1-1 correspondence of the tuples and the gradings. For example, if we have one of k_i zero, we assign 0 to c and then we have the 1-1 correspondence between the gradings and different tuples of the form $(e^{(k_1)}, a^{(k_2)}, b^{(k_3)})$ while if none of k_i is zero we should consider all tuples $(e^{(k_1)}, a^{(k_2)}, b^{(k_3)}, c^{(k_4)})$ where the value of k_1 is the smallest possible. However if k_1 is equal to another k_i then we can permute the remaining two numbers and the total number of inequivalent gradings becomes 3 or even 1 if those remaining numbers are the same.

The same applies to the "elementary gradings" of a simple Lie algebra $\mathfrak{g}$, that is, the restriction of elementary gradings of M_n to $\mathfrak{g}$ embedded as a graded Lie subalgebra in M_n. It is obvious that if a conjugation by a matrix T maps one such grading onto another and if this conjugation leaves $\mathfrak{g}$ invariant then the gradings are equivalent on $\mathfrak{g}$. This condition on T obviously holds for $\mathfrak{g} = \mathrm{sl}(n)$. On the other hand, any grading of a simple algebra of one of the types B, C, D is induced from a grading of an appropriate M_n in such a way that $\mathfrak{g}$ is the stabilizer of an appropriate symmetric or skew-symmetric matrix Φ split into blocks in accordance with the tuple (k_1, k_2, k_3, k_r). But T is a permutational matrix of the blocks defined exactly by the same tuple. So T is orthogonal with respect to Φ and then it leaves $\mathfrak{g}$ invariant.

The converse is also true since any automorphism of $\mathfrak{g}$ of the types B, C, D, except D_4, can be induced from the matrix algebra. In the case of of the type A given two equivalent elementary gradings, one can find an outer automorphism (of the type $X \to -X^t$) which leaves the first elementary grading invariant and so the gradings are still equivalent in M_n.

2) *Fine gradings.* The characteristic property of such gradings is that for every $g \in \mathrm{Supp}\,(R)$, $\dim R_g = 1$ where $\mathrm{Supp}\,(R) = \{g \in G,\ \dim R_g \neq 0\}$. In the case of $G = \mathbb{Z}_2 \times \mathbb{Z}_2$, each fine grading is either trivial or weakly equivalent to the grading on $R \cong M_2$ given by the Pauli matrices

$$X_e = I = \begin{pmatrix} 1 & 0 \\ 0 & 1 \end{pmatrix},\ X_a = \begin{pmatrix} -1 & 0 \\ 0 & 1 \end{pmatrix},\ X_b = \begin{pmatrix} 0 & 1 \\ 1 & 0 \end{pmatrix},\ X_c = \begin{pmatrix} 0 & -1 \\ 1 & 0 \end{pmatrix}$$

in such a way that the graded component R_g of degree g is spanned by X_g, $g = e, a, b, c$. Actually, the components R_a, R_b and R_c (not the matrices X_a, X_b, X_c!) of the fine grading of M_2 can be arbitrarily permuted by conjugation by an appropriate nonsingular matrix. The same is true for the induced grading on $\mathfrak{g} = \mathrm{sl}(2)$.

According to [**5**] and [**10**] any G-grading of $R = M_n$ can be written as the tensor product of two graded matrix subalgebras $R = A \otimes B$, where $A \cong M_k$ and its grading is (equivalent to) elementary, and the grading of $B \cong M_l$ is fine with $\mathrm{Supp}\,A \cap \mathrm{Supp}\,B = \{e\}$, $kl = n$. Thus the only cases possible, when $G = \mathbb{Z}_2 \times \mathbb{Z}_2$, are

1) $B = F$ and $R = A \otimes F \cong A$

2) $B = M_2$ and the grading on A is trivial.

If R is graded by G as above, then an involution $*$ of $R = M_n$ is called *graded* if $(R_g)^* = R_g$ for any $g \in G$. In the case of such involution, the spaces $K(R, *) = \{X \in R, X^* = -X\}$ of skew - symmetric elements under $*$ and $H(R, *) = \{X \in$

$R, X^* = X\}$ of symmetric elements under $*$ are graded and the first is a simple Lie algebra of one of the types B, C, D under the bracket $[X, Y] = XY - YX$.

5.2. The case of the types B, C and D. It is proved in [**6**] that $\mathfrak{g}$ as a G-graded algebra is isomorphic to $K(R, *) = \{X \in R, X^* = -X\}$ for an appropriate graded involution $*$.

In general the involution does not need to respect A and B. But this is the case when $G = \mathbb{Z}_2 \times \mathbb{Z}_2$. Any involution has the form $* : X \longrightarrow X^* = \Phi^{-1}X^t\Phi$, for a nonsingular matrix Φ, which is either symmetric in the orthogonal case and skew-symmetric in the symplectic case. Since the elementary and fine components are invariant under the involution, we have that $\Phi = \Phi_1 \otimes \Phi_2$ where Φ_1 defines a graded involution on A and Φ_2 on B.

Up to a matrix conjugation, the possible forms for Φ_1 are as follows.

In the **symmetric** case we have one of three options

(i) The identity matrix $\Phi_1 = I_n$. This is compatible with an elementary grading defined by the tuple $\tau = (g_1^{(k_1)}, g_2^{(k_2)}, g_3^{(k_3)}, g_4^{(k_4)})$, $k_1 + k_2 + k_3 + k_4 = n$, $k_1 > 0$, $k_2, k_3, k_4 \geq 0$. Also here and in what follows we put $g_1 = e$ and denote by g_2, g_3, g_4 all pairwise different elements in the set $\{a, b, c\}$.

(ii) $\Phi_1 = \begin{pmatrix} 0 & I_{k_1} \\ I_{k_1} & 0 \end{pmatrix}$. This is compatible with an elementary grading defined by the tuple $\tau = (e^{(k_1)}, g_2^{(k_2)})$ where $k_1 = k_2$. Actually, this refers to the case of $\mathbb{Z}_2$-gradings.

(iii) $\Phi_1 = \operatorname{diag}\left\{\begin{pmatrix} 0 & I_{k_1} \\ I_{k_1} & 0 \end{pmatrix}, \begin{pmatrix} 0 & I_{k_3} \\ I_{k_3} & 0 \end{pmatrix}\right\}$. This is compatible with an elementary grading defined by the tuple $\tau = (e^{(k_1)}, g_2^{(k_2)}, g_3^{(k_3)}, g_4^{(k_3)})$ where $k_1 = k_2$ and $k_3 = k_4$.

In the **skew-symmetric** case we again have one of three options

(i) The matrix $\Phi_1 = \operatorname{diag}\{S_{k_1}, \ldots, S_{k_4}\}$. This is compatible with an elementary grading defined by the tuple $\tau = (g_1^{(2k_1)}, g_2^{(2k_2)}, g_3^{(2k_3)}, g_4^{(2k_4)})$, $2k_1 + 2k_2 + 2k_3 + 2k_4 = n$, $k_1 > 0$, $k_2, k_3, k_4 \geq 0$.

(ii) $\Phi_1 = \begin{pmatrix} 0 & I_{2k_1} \\ -I_{2k_1} & 0 \end{pmatrix}$. This is compatible with an elementary grading defined by the tuple $\tau = (e^{(2k_1)}, g_2^{(2k_2)})$ where $k_1 = k_2$. Again, this rather refers to the case of $\mathbb{Z}_2$-gradings;

(iii) $\Phi_1 = \operatorname{diag}\left\{\begin{pmatrix} 0 & I_{2k_1} \\ -I_{2k_1} & 0 \end{pmatrix}, \begin{pmatrix} 0 & I_{2k_3} \\ -I_{2k_3} & 0 \end{pmatrix}\right\}$. This is compatible with an elementary grading defined by the tuple

$$\tau = (e^{(2k_1)}, g_2^{(2k_2)}, g_3^{(2k_3)}, g_4^{(2k_4)}) \text{ where } k_1 = k_2 \text{ and } k_3 = k_4.$$

Here as usual, I_k stands for the identity matrix of order n and S_k for the matrix of order $2k$ of the form $S_k = \begin{pmatrix} 0 & I_k \\ -I_k & 0 \end{pmatrix}$.

So if we restrict ourselves to the case where the matrix algebra $R = M_n$ has no fine component B in the decomposition $R = A \otimes B$ then we obtain the following family of $(\mathbb{Z}_2 \times \mathbb{Z}_2, \beta)$-superalgebras. It consists of 10 classes of simple graded algebras, each producing 2 classes of $(\mathbb{Z}_2 \times \mathbb{Z}_2, \beta)$ - Lie superalgebras, as described in Case 3 in Section 4. If σ is the 2-cocycle performing the discoloration of the nontrivial commutation factor β_0 then in addition to each graded algebra L in the list we have to take L^σ. It will be shown in the isomorphism section (see Section 6) that in the most cases L is largely by its identity component.

The identity components of the algebras discussed above are in Table 1 below. In this table and in the following ones the minimal values for the arguments of the algebras on left column should be $n = 2$ for $\mathrm{sl}(n)$, $m = 2$ for $\mathrm{so}(2m+1)$, $m = 3$ for $\mathrm{sp}(2m)$ and $m = 4$ for $\mathrm{so}(2m)$. Also, the algebras $\mathrm{gl}(k)$, $\mathrm{sl}(k)$, $\mathrm{so}(k)$ and $\mathrm{sp}(2k)$ with small values for k on the right column should be replaced according to the well-known isomorphisms. Considering all these special cases here would dramatically increase the size of this paper.

$\mathfrak{g}$	$\mathfrak{g}_e$
$\mathrm{so}(k_1+k_2)$	$\mathrm{so}(k_1)\oplus\mathrm{so}(k_2)$
$\mathrm{so}(k_1+k_2+k_3)$	$\mathrm{so}(k_1)\oplus\mathrm{so}(k_2)\oplus\mathrm{so}(k_3)$
$\mathrm{so}(k_1+k_2+k_3+k_4)$	$\mathrm{so}(k_1)\oplus\mathrm{so}(k_2)\oplus\mathrm{so}(k_3)\oplus\mathrm{so}(k_4)$
$\mathrm{sp}(k_1+k_2)$	$\mathrm{sp}(k_1)\oplus\mathrm{sp}(k_2)$
$\mathrm{sp}(k_1+k_2+k_3)$	$\mathrm{sp}(k_1)\oplus\mathrm{sp}(k_2)\oplus\mathrm{sp}(k_3)$
$\mathrm{sp}(k_1+k_2+k_3+k_4)$	$\mathrm{sp}(k_1)\oplus\mathrm{sp}(k_2)\oplus\mathrm{sp}(k_3)\oplus\mathrm{sp}(k_4)$
$\mathrm{so}(2m)$	$\mathrm{gl}(m)$
$\mathrm{so}(2(k_1+k_2))$	$\mathrm{gl}(k_1)\oplus\mathrm{gl}(k_2)$
$\mathrm{sp}(2m)$	$\mathrm{gl}(m)$
$\mathrm{sp}(2(k_1+k_2))$	$\mathrm{gl}(k_1)\oplus\mathrm{gl}(k_2)$

Table 1

If $R = M_n = A \otimes B \cong M_m \otimes M_k$ and $B \neq F$, then $\text{Supp}\,(A) = \{e\}$. As before, we have $\Phi = \Phi_1 \otimes \Phi_2$ and the involution on R defines involutions on A and B. It follows that Φ is symmetric if and only if either Φ_1 and Φ_2 are both symmetric or they both are skew-symmetric. Similarly, Φ is skew-symmetric if one of Φ_1,Φ_2 is symmetric and the other is skew-symmetric. Since there is no grading on A we can take Φ_1 in its simplest canonical form. As for Φ_2, it was proved in [**6**] that M_2 with graded involution is isomorphic to M_2 with G-graded basis of Pauli matrices as above and the graded involution is given by one of $\Phi_2 = X_e, X_a, X_b, X_c$. Thus we always have the situation where $A \cong M_m$ and $B \cong M_2$.

We can quite easily calculate the identity component of the graded algebras arising. Indeed, we obviously have

$$\mathfrak{g} = K(R,\Phi) = K(A,\Phi_1) \otimes H(B,\Phi_2) \oplus H(A,\Phi_1) \otimes K(B,\Phi_2).$$

Hence $\mathfrak{g}_e = K(R_e, \Phi_1) \otimes I_2$. If Φ_1 is symmetric then $\mathfrak{g}_e = \text{so}(m)$, and if Φ_1 is skew symmetric, $g_e = \text{sp}(m)$.

In the case under consideration we have the following table of possible pairs $(\mathfrak{g}, \mathfrak{g}_e)$:

$\mathfrak{g}$	$\mathfrak{g}_e$
$\text{so}(2m)$	$\text{so}(m)$
$\text{so}(4m)$	$\text{sp}(2m)$
$\text{sp}(4m)$	$\text{sp}(2m)$
$\text{sp}(2m)$	$\text{so}(m)$

Table 2

As in the previous case of elementary grading, the explicit form of the grading can be determined if one lists all possible graded involutions.

If Φ is **symmetric** then it can be reduced to one of the forms

$$\Psi_1 = I_m \otimes I_2,\ \Psi_2 = I_m \otimes X_a,\ \Psi_3 = I_m \otimes X_b,\ \Psi_4 = S_m \otimes X_c.$$

If Φ is **skew-symmetric** then it can be reduced to one of the forms

$$\overline{\Psi}_1 = S_m \otimes I_2,\ \overline{\Psi}_2 = S_m \otimes X_a,\ \overline{\Psi}_3 = S_m \otimes X_b,\ \overline{\Psi}_4 = I_m \otimes X_c.$$

This shows that formally we have 16 families of (G,β)-Lie superalgebras. We will show in the isomorphism section that actually there are isomorphisms and actually the 1-1 correspondence with the identity components still exists.

5.3. The case of A-type simple Lie algebras. Now let $\mathfrak{g}$ be a simple Lie algebra of type A_l. We view $\mathfrak{g}$ as the set $\mathfrak{g} = \mathrm{sl}(n)$ of all matrices of trace zero in the matrix algebra $R = M_n(F)$ where $n = l+1$. In this case any G-grading of $\mathfrak{g}$ belongs to one of the following two classes (see [**8**]).

- For Class I gradings, any grading of $\mathfrak{g}$ is induced from a G-grading of $R = \bigoplus_{g\in G} R_g$ and one simply has to set $\mathfrak{g}_g = R_g$ for $g \neq e$ and $\mathfrak{g}_e = R_e \cap \mathfrak{g}$, otherwise. For $G = \mathbb{Z}_2 \times \mathbb{Z}_2$ we still have to distinguish between the cases $R = A \otimes \mathbb{C}$ with an elementary grading on $A = M_n$ or $R = A \otimes B$ with trivial grading on A and fine $\mathbb{Z}_2 \times \mathbb{Z}_2$-grading on $B = M_2$.

- For Class II gradings, we have to fix an element q of order 2 in G and an involution G-grading $R = \oplus_{g\in G} R_g$. Then for any $g \in G$ one has

$$\mathfrak{g}_g = K(R_g, *) \oplus H(R_{gq}, *) \cap \mathfrak{g}.$$

However in the case of $G = \mathbb{Z}_2 \times \mathbb{Z}_2$ a simplified approach is possible as shown in [**9**]; it is based on the results of [**8**].

5.3.1. *Classification of Class* I *gradings on A-type Lie algebras.* If we have the fine component of a $\mathbb{Z}_2 \times \mathbb{Z}_2$-grading on $R = M_n = M_m \otimes M_2$ is nontrivial then the grading on A is trivial and so the grading is given by

$$\begin{aligned}
\mathfrak{g} &= \mathfrak{g}_e \oplus \mathfrak{g}_a \oplus \mathfrak{g}_b \oplus g_c \\
&= \left\{ \begin{pmatrix} X & 0 \\ 0 & X \end{pmatrix} \middle| X \in sl(m) \right\} \oplus \left\{ \begin{pmatrix} X & 0 \\ 0 & -X \end{pmatrix} \middle| X \in M_m \right\} \\
&\oplus \left\{ \begin{pmatrix} 0 & X \\ X & 0 \end{pmatrix} \middle| X \in M_m \right\} \oplus \left\{ \begin{pmatrix} 0 & -X \\ X & 0 \end{pmatrix} \middle| X \in M_m \right\}.
\end{aligned}$$

Now if no fine component is present in $R = M_n \supset \mathfrak{g} = \mathrm{sl}(n)$, $n = l+1$, then all is defined by the n-tuple $(g_1^{(k_1)}, g_2^{(k_2)}, g_3^{(k_3)}, g_4^{(k_4)})$ where $n = k_1 + k_2 + k_3 + k_4$, $k_1 > 0$, $k_2, k_3, k_4 \geq 0$, g_2, g_3, g_4 are pairwise different elements in $\{b, c, d\}$. It is known that two elementary gradings defined by the tuple as above and also $(h_1^{(l_1)}, h_2^{(l_2)}, h_3^{(l_3)}, h_4^{(l_4)})$ give isomorphic gradings of M_n if and only if the sequences of upper indexes in both tuples are the same up to a permutation: $k_i = l_{\lambda(i)}$ and there is $g_0 \in \mathbb{Z}_2 \times \mathbb{Z}_2$ such that $g_0 g_i = h_{\lambda(i)}$.

Depending on how many of the numbers k_1, k_2, k_3, k_4 are nonzero, we have three principal cases differing by the number of simple components in $\mathfrak{g}_e$.

The arising families of gradings on $\mathrm{sl}(n)$ are in the Table 3 below.

$\mathfrak{g}$	$\mathfrak{g}_e$
$\mathrm{sl}(2n)$	$\mathrm{sl}(n)$
$\mathrm{sl}(k_1+k_2)$	$\mathrm{sl}(k_1)\oplus\mathrm{sl}(k_2)\oplus\mathbb{C}$
$\mathrm{sl}(k_1+k_2+k_3)$	$\mathrm{sl}(k_1)\oplus\mathrm{sl}(k_2)\oplus\mathrm{sl}(k_3)\oplus\mathbb{C}^2$
$\mathrm{sl}(k_1+k_2+k_3+k_4)$	$\mathrm{sl}(k_1)\oplus\mathrm{sl}(k_2)\oplus\mathrm{sl}(k_3)\oplus\mathrm{sl}(k_4)\oplus\mathbb{C}^3$

Table 3

5.3.2. *Classification of Class* II *gradings on A-type Lie algebras.* The general approach described in 5.3 enables one to classify the Class II gradings on $\mathfrak{g}=\mathrm{sl}(n)$, for any $n\geq 2$ and any grading group G. However, in the case $G=\mathbb{Z}_2\times\mathbb{Z}_2$ the amount of work can be significantly reduced if one uses the results of [**9**] and [**8**]. The gradings by $\mathbb{Z}_2\times\mathbb{Z}_2$ on $\mathrm{sl}(n)$ correspond to $\mathbb{Z}_2$-graded eigenspaces of a (negative to) a graded involution φ on a $\mathbb{Z}_2$-graded associative algebra $R=M_n$. Any $\mathbb{Z}_2$-grading on R is elementary, given by a tuple $\nu_1=(e^{(k_1)},q^{(k_2)})$ where q is the generator of $\mathbb{Z}_2$ and $k_1+k_2=n$. As described in [**8**, Theorem 3], any graded involution is graded equivalent to $X\to\Phi^{-1}X^t\Phi$ where Φ is of one the following types

(i) $\Phi=\mathrm{diag}\,\{I_{k_1},I_{k_2}\}$
(ii) $\Phi=\begin{pmatrix}0 & I_{k_1}\\ I_{k_1} & 0\end{pmatrix}$
(iii) $\Phi=\mathrm{diag}\,\{S_{k_1},S_{k_2}\}$
(iv) $\Phi=\begin{pmatrix}0 & I_{k_1}\\ -I_{k_1} & 0\end{pmatrix}$.

Now it remains to apply [**9**, Corollary 5.6], where $K=\langle p\rangle$ to obtain that all Class II gradings of $\mathfrak{g}$ have the following form

$$\begin{aligned}\mathfrak{g}_e &= K(R_e,\Phi)\\ \mathfrak{g}_q &= K(R_q,\Phi)\\ \mathfrak{g}_p &= H(R_e,\Phi)\cap\mathfrak{g}\\ \mathfrak{g}_{pq} &= H(R_q,\Phi)\end{aligned}$$

As a result, $(\mathfrak{g},\mathfrak{g}_e)$ given in the following table:

$\mathfrak{g}$	$\mathfrak{g}_e$
sl$(2k)$	gl(k)
sl(n)	so(n)
sl$(2m)$	sp$(2m)$
sl(k_1+k_2)	so$(k_1)\oplus$ so(k_2)
sl$(2(k_1+k_2))$	sp$(2k_1)\oplus$ sp$(2k_2)$

Table 4

We summarize the result obtained so far as follows.

THEOREM 5.1. *Every $(\mathbb{Z}_2\times\mathbb{Z}_2,\beta)$-Lie superalgebra can be obtained by coloring one of the graded Lie algebras listed in the previous sections. Their identity components are given in Tables 1 to 4, which splits the $(\mathbb{Z}_2\times\mathbb{Z}_2,\beta)$-Lie superalgebras into non-isomorphic classes.*

6. The Isomorphisms

Here we discuss the isomorphisms between the superalgebras introduced earlier. As noted before, we have only to consider the (graded) isomorphisms between the $\mathbb{Z}_2\times\mathbb{Z}_2$-graded algebras we have obtained.

6.1. Case 1. All gradings on $F[\mathbb{Z}_2\times\mathbb{Z}_2]$ by $\mathbb{Z}_2\times\mathbb{Z}_2$ are easily seen to be isomorphic to the canonical one, and these isomorphisms can be easily extended to the Lie algebra $F[\mathbb{Z}_2\times\mathbb{Z}_2]\otimes\mathfrak{g}$. So in this case we have only one graded Lie algebra $F[\mathbb{Z}_2\times\mathbb{Z}_2]\otimes\mathfrak{g}$ for each simple Lie algebra $\mathfrak{g}$ and one $(\mathbb{Z}_2\times\mathbb{Z}_2,\beta_0)$-Lie superalgebra L^σ. If $[a,b]|_M$ stands for the commutator between a,b in an algebra M then all the commutators in L^σ are the same as in L, except $[b\otimes x,a\otimes y]|_{L^\sigma}=-[b\otimes x,a\otimes y]|_L=-(ab)\otimes[x,y]_{\mathfrak{g}}$ and $[c\otimes x,a\otimes y]_{L^\sigma}=-[c\otimes x,a\otimes y]_L=-(ac)\otimes[x,y]_{\mathfrak{g}}$.

6.2. Case 2. In this case we have the subgroup decomposition $\mathbb{Z}_2\times\mathbb{Z}_2=H\times K$ where H and K are of order 2, and then $L=F[K]\otimes\mathfrak{g}$ where $\mathfrak{g}$ is ($\mathbb{Z}_2$-) graded by K. If $K=\langle a\rangle$, $H=\langle b\rangle$ then setting $\varepsilon_1=\frac{e+a}{2}$ and $\varepsilon_2=\frac{e-a}{2}$ and defining the isomorphic ideals $L_i=\varepsilon_i\otimes\mathfrak{g}$, $i=1,2$, we find that $L=L_1\oplus L_2$. We have $\mathfrak{g}=\mathfrak{g}_e\oplus\mathfrak{g}_b$. In this case the components of L are as follows $L_e=\{(x,x)\,|\,x\in\mathfrak{g}_e\}$, $L_a=\{(x,-x)\,|\,x\in\mathfrak{g}_e\}$, $L_b=\{(y,y)\,|\,y\in\mathfrak{g}_b\}$, $L_c=\{(y,-y)\,|\,y\in\mathfrak{g}_b\}$. The automorphisms of L are of one of the forms (α_1,α_2) or $\tau(\alpha_1,\alpha_2)$ there α_i is an automorphism of $L_i\cong\mathfrak{g}$ and τ is the twist between the isomorphic components. Since $\mathfrak{g}$ is $\mathbb{Z}_2$-graded there is an automorphism γ with $\gamma(x)=x$, $\gamma(y)=-y$. If we apply $(\mathrm{id}_{\mathfrak{g}}, \gamma)$ to L then L_b will be changed places with L_c, and the other components will be left fixed. It is quite clear then that there is no algebra homomorphism mapping L_a into L_b because it will induce an algebra automorphism of $\mathfrak{g}$ mapping $\mathfrak{g}_e$ onto $\mathfrak{g}_b$ which is not possible since the first component is a subalgebra while the second is not.

As a result we have three isomorphism classes of simple graded algebras:

(a) $L(a) = F[\langle a\rangle] \otimes \mathfrak{g}$ with simple $\mathfrak{g}$ endowed by an elementary $\langle b\rangle$-grading;
(a) $L(b) = F[\langle b\rangle] \otimes \mathfrak{g}$ with simple $\mathfrak{g}$ endowed by an elementary $\langle c\rangle$-grading;
(a) $L(c) = F[\langle c\rangle] \otimes \mathfrak{g}$ with simple $\mathfrak{g}$ endowed by an elementary $\langle a\rangle$-grading.

There is an obvious symmetry between the above algebras but when we pass to superalgebras we must apply the same (non-symmetric) 2-cocycle σ. As a result, for the "similarly-looking" elements in $L(a)^\sigma$ we have $[a\otimes x, e\otimes y]^\sigma = a\otimes[x,y]$, that is, $a\otimes x$ and $e\otimes y$ anticommute while in $L(b)^\sigma$ $[a\otimes x, e\otimes y]^\sigma = -a\otimes[x,y]$, that is, $a\otimes x$ and $e\otimes y$ commute.

6.3. Case 3. We start with the known case of $\mathbb{Z}_2$-gradings because they appear as a degenerate case in our classification of $\mathbb{Z}_2\times\mathbb{Z}_2$-graded algebras. In none of these cases there is a nontrivial superalgebra structure. We denote by d the generator of $\mathbb{Z}_2$. If we want to obtain $\mathbb{Z}_2\times\mathbb{Z}_2$-gradings we have to replace d by a, b, c one by one.

6.3.1. *$\mathbb{Z}_2$-gradings.*

(a) Here $\mathfrak{g} = \mathrm{sl}(n)$, $n \geq 2$, $n > k \geq n/2$, the grading is elementary given by the tuple $(e^{(k)}, d^{(n-k)})$;
(b) Here $\mathfrak{g} = \mathrm{sl}(n)$, $\mathfrak{g}_e = \{x \mid x^t = -x\}$, $\mathfrak{g}_d = \{x \mid x^t = -x\} \cap \mathfrak{g}$;
(c) Here $\mathfrak{g} = \mathrm{sl}(n)$, $\mathfrak{g}_e = \{x \mid x^* = -x\}$, $\mathfrak{g}_d = \{x \mid x^* = x\} \cap \mathfrak{g}$ where $*$ is the symplectic involution on M_n;
(d) Here $\mathfrak{g} = \mathrm{so}(n)$, the set of skew symmetric matrices under the ordinary transpose in M_n, $n \geq 5$, $n > k \geq n/2$, the grading is elementary given by the tuple $(e^{(k)}, d^{(n-k)})$;
(e) Here $m \geq 3$, $m > k \geq m/2$, $\mathfrak{g} = \mathrm{sp}(2m)$, the set of skew symmetric matrices under the involution with the matrix $\Phi = \mathrm{diag}\{S_k, S_{m-k}\}$ in M_{2m}, the grading is elementary given by the tuple $(e^{(2k)}, d^{(2(m-k))})$;
(f) Here $\mathfrak{g}$ is the set of matrices in M_{2n} skew-symmetric under the involution with the matrix $\Phi = \begin{pmatrix} 0 & I_n \\ I_n & 0 \end{pmatrix}$; the grading is elementary, given by the tuple $(e^{(n)}, d^{(n)})$;
(g) Here $\mathfrak{g}$ is the set of matrices in M_{2n} skew-symmetric under the involution with the matrix $\Phi = \begin{pmatrix} 0 & I_n \\ -I_n & 0 \end{pmatrix}$; the grading is elementary, given by the tuple $(e^{(n)}, d^{(n)})$;

6.3.2. *A_l-type algebras, $l \geq 1$.* In all cases below every time the grading is "elementary" some conditions apply, which are described in Section 5.1. We do not mention them every time.

(a) Here $\mathfrak{g} = \mathrm{sl}(n)$, $n = l+1$, the grading is induced from the elementary grading of M_n defined by the tuple $(e^{(k_1)}, a^{(k_2)}, b^{(k_3)})$, $k_1+k_2+k_3 = n$, all k_i are positive integers. The triple of the numbers (k_1, k_2, k_3) is defined uniquely. The identity component is $\mathrm{sl}(k_1)\oplus\mathrm{sl}(k_2)\oplus\mathrm{sl}(k_3)\oplus F^2$.

(b) Here $\mathfrak{g} = \mathrm{sl}(n)$, $n = l+1$, the grading is induced from the elementary grading of M_n defined by the tuple $(e^{(k_1)}, a^{(k_2)}, b^{(k_3)}, c^{(k_4)})$, $k_1+k_2+k_3+k_4 = n$, all k_i are positive integers. Additionally, k_1 is the smallest of the upper indexes.

(c) Here $\mathfrak{g}$ is the set of matrices with trace zero in $R = M_n \otimes M_2$. We have $\mathfrak{g}_e = \{A \otimes X_e \mid A \in M_n, \operatorname{tr} A = 0\}$, $\mathfrak{g}_e = \{A \otimes X_a \mid A \in M_n\}$, $\mathfrak{g}_e = \{A \otimes X_b \mid A \in M_n\}$, $\mathfrak{g}_e = \{A \otimes X_c \mid A \in M_n\}$. The identity component is $\mathrm{sl}(n)$.

(d) Here $\mathfrak{g} = \mathrm{sl}(n)$, $n = l+1$, $\mathfrak{g} \subset M_n$, M_n is given an elementary grading by the tuple $(e^{(k_1)}, a^{(k_2)})$. This grading is compatible with the ordinary transpose and we have

$$\mathfrak{g}_e = K(R_e),\ \mathfrak{g}_p = K(R_p),\ \mathfrak{g}_q = H(R_e),\ \mathfrak{g}_{pq} = H(R_p).$$

In the matrix form we have $\mathfrak{g}_e = \operatorname{diag}\{A_1, A_2\}$, $\mathfrak{g}_q = \operatorname{diag}\{B_1, B_2\} \cap \mathfrak{g}$, $\mathfrak{g}_p = \operatorname{diag}\left\{\begin{pmatrix} 0 & U \\ -U^t & 0 \end{pmatrix}\right\}$, $\mathfrak{g}_{pq} = \operatorname{diag}\left\{\begin{pmatrix} 0 & U \\ U^t & 0 \end{pmatrix}\right\}$. Here A_i are skew-symmetric, B_i symmetric, U any $k_1 \times k_2$-matrix. The conjugation by the matrix $T = \operatorname{diag}\{iI_{k_1}, I_{k_2}\}$ changes places $\mathfrak{g}_p$ and $\mathfrak{g}_{pq}$ while leaving other components invariant. Except for the case $k_1 = k_2 = 1$, $\dim \mathfrak{g}_p \neq \dim \mathfrak{g}_q$ and so we have 3 pairwise non isomorphic gradings for $p = a, q = b$; $p = a, q = c$ and $p = b, q = c$.

The identity component is $\mathrm{so}(k_1) \oplus \mathrm{so}(k_2)$.

(e) Here $\mathfrak{g} = \mathrm{sl}(2n)$, $2n = l+1$, $\mathfrak{g} \subset M_{2n}$, M_{2n} is given an elementary grading by the tuple $(e^{(2k_1)}, a^{(2k_2)})$, $k_1 + k_2 = n$, both k_1, k_2 are positive. This grading is compatible with the involutions defined by the matrix $\Phi = \operatorname{diag}\{S_{k_1}, S_{k_2}\}$ and we have

$$\mathfrak{g}_e = K(R_e, \Phi),\ \mathfrak{g}_p = K(R_p, \Phi),\ \mathfrak{g}_q = H(R_e, \Phi),\ \mathfrak{g}_{pq} = H(R_p, \Phi).$$

An argument similar to that in (d) produces three pairwise non isomorphic gradings for $p = a, q = b$; $p = a, q = c$ and $p = b, q = c$ The identity component is $\mathrm{sp}(2k_1) \oplus \mathrm{sp}(2k_2)$.

(f) Here $\mathfrak{g} = \mathrm{sl}(2k)$, $2k = l+1$, $\mathfrak{g} \subset M_{2k}$, M_{2k} is given an elementary grading by the tuple $(e^{(k)}, p^{(k)})$. This grading is compatible with the involutions defined by one of the two matrices $\Phi_\pm = \begin{pmatrix} 0 & I_k \\ \pm I_k & 0 \end{pmatrix}$ and we have

$$\mathfrak{g}_e = K(R_e, \Phi_\pm),\ \mathfrak{g}_p = K(R_p, \Phi_\pm),\ \mathfrak{g}_q = H(R_e, \Phi_\pm),\ \mathfrak{g}_{pq} = H(R_p, \Phi_\pm).$$

In the matrix form we have $\mathfrak{g}_e = \operatorname{diag}\{A, -A^t\}$, $\mathfrak{g}_p = \operatorname{diag}\{A, A^t\} \cap \mathfrak{g}$, $\mathfrak{g}_q = \operatorname{diag}\left\{\begin{pmatrix} 0 & U_1 \\ U_2 & 0 \end{pmatrix}\right\}$, $\mathfrak{g}_{pq} = \operatorname{diag}\left\{\begin{pmatrix} 0 & V_1 \\ V_2 & 0 \end{pmatrix}\right\}$. Here A is any matrix of order k, each U_i skew symmetric in the case of Φ_+ and symmetric in the case of Φ_-, while each V_i is symmetric in the case of Φ_+ and skew-symmetric in the case of Φ_-. This shows that we do not need to consider the case of Φ_- at all. On the other hand, if we compare the dimensions of homogeneous components $\mathfrak{g}_p$, $\mathfrak{g}_q$ and $\mathfrak{g}_{pq}$ then we will see that no automorphism can permute the components and so the replacement of p, q, pq by any permutation of a, b, c leads to pairwise non-isomorphic gradings. The identity component is $\mathrm{gl}(n)$.

The uniqueness in the cases (a) and (b) is discussed at the end of the next subsection. In the case (c) we refer to the uniqueness of fine gradings on M_2 (see Section 5.1).

6.3.3. *B_l-type algebras, $l \geq 2$.*

(a) Here $\mathfrak{g}$ is the set of skew symmetric matrices of order $2l+1$ with elementary grading given by the tuple $(e^{(k_1)}, a^{(k_2)}, b^{(k_3)})$, $k_1 + k_2 + k_3 = 2l + 1$, all k_i are positive integers. The identity component is $\mathrm{so}(k_1) \oplus \mathrm{so}(k_2) \oplus \mathrm{so}(k_3)$.

(b) Here $\mathfrak{g}$ is the set of skew symmetric matrices of order $2l+1$ with elementary grading given by the tuple $(e^{(k_1)}, a^{(k_2)}, b^{(k_3)}, c^{(k_4)})$, $k_1+k_2+k_3+k_4 = 2l+1$, all k_i are positive integers. The identity component is $\mathrm{so}(k_1) \oplus \mathrm{so}(k_2) \oplus \mathrm{so}(k_3) \oplus \mathrm{so}(k_4)$.

6.3.4. *C_l-type algebras, $l \geq 3$.*

(a) Here $\mathfrak{g}$ is the set of matrices of order $n = 2l$ which are skew symmetric with respect to the symplectic involution given by the matrix $\Phi = \mathrm{diag}\{S_{k_1}, S_{k_2}, S_{k_3}\}$ with $2k_1 + 2k_2 + 2k_3 = 2n$. The $\mathbb{Z}_2 \times \mathbb{Z}_2$-grading is induced from the elementary grading of M_n given by the tuple $(e^{(2k_1)}, a^{(2k_2)}, b^{(2k_3)})$. The identity component is $\mathrm{sp}(2k_1) \oplus \mathrm{sp}(2k_2) \oplus \mathrm{sp}(2k_3)$.

(b) Here $\mathfrak{g}$ is the set of matrices of order $n = 2l$ which are skew symmetric with respect to the symplectic involution given by the matrix $\Phi = \mathrm{diag}\{S_{k_1}, S_{k_2}, S_{k_3}, S_{k_4}\}$ with $2k_1 + 2k_2 + 2k_3 + 2k_4 = 2n$. The $\mathbb{Z}_2 \times \mathbb{Z}_2$-grading is induced from the elementary grading of M_n given by the tuple $(e^{(2k_1)}, a^{(2k_2)}, b^{(2k_3)}, c^{(2k_4)})$. The identity component is $\mathrm{sp}(2k_1) \oplus \mathrm{sp}(2k_2) \oplus \mathrm{sp}(2k_3) \oplus \mathrm{sp}(2k_4)$.

(c) Here $\mathfrak{g}$ is the set of matrices of order $n = 2l$ which are skew symmetric with respect to the symplectic involution given by the matrix

$$\Phi = \mathrm{diag}\left\{\begin{pmatrix} 0 & I_{k_1} \\ -I_{k_1} & 0 \end{pmatrix}, \begin{pmatrix} 0 & I_{k_2} \\ -I_{k_2} & 0 \end{pmatrix}\right\}$$

with $2k_1 + 2k_2 = 2n$ The $\mathbb{Z}_2 \times \mathbb{Z}_2$-grading is induced from the elementary grading of M_n given by the tuple $(e^{(2k_1)}, a^{(2k_1)}, b^{(2k_2)}, c^{(2k_2)})$. The identity component is $\mathrm{gl}(k_1) \oplus \mathrm{gl}(k_2)$.

(d) Here $\mathfrak{g}$ is the set of skew-symmetric matrices in $R = M_{2k} \otimes M_2$ with respect to the symplectic involution given by the Kronecker product matrix $\Phi = S_k \otimes I_2$. We have $\mathfrak{g}_e = \{A \otimes X_e \,|\, A \in K(M_{2k}, S_k)\}$, $\mathfrak{g}_a = \{A \otimes X_a \,|\, A \in K(M_{2k}, S_k)\}$, $\mathfrak{g}_b = \{A \otimes X_b | A \in K(M_{2k}, S_k)\}$, $\mathfrak{g}_c = \{B \otimes X_c | H(M_{2k}, S_k)\}$. There are altogether three matrices at the end of Section 5.2 which produce this kind of grading with the same identity component $\mathrm{so}(n)$. However, the conjugation by the matrices of the form $I_{2k} \otimes T$ where T is one of the matrices permuting the spaces spanned by X_a, X_b and X_c (see Section 5.1) make all these three types of gradings isomorphic. The identity component is $\mathrm{so}(n)$.

(e) Here $\mathfrak{g}$ is the set of skew-symmetric matrices in $R = M_n \otimes M_2$ with respect to the involution given by the Kronecker product matrix $\Phi = I_n \otimes X_c$. We have $\mathfrak{g}_e = \{A \otimes X_e \,|\, A \in K_n\}$, $\mathfrak{g}_a = \{B \otimes X_a \,|\, A \in H_n\}$, $\mathfrak{g}_b = \{A \otimes X_b \,|\, A \in H_n\}$, $\mathfrak{g}_c = \{B \otimes X_c \mid B \in H_n\}$. The identity component is $\mathrm{sp}(2k)$.

6.3.5. *D_l-type algebras, $l \geq 4$.*

(a) Here $\mathfrak{g}$ is the set of skew symmetric matrices of order $n = 2l$. The $\mathbb{Z}_2 \times \mathbb{Z}_2$-grading is induced from the elementary grading of M_n given by the tuple $(e^{(k_1)}, a^{(k_2)}, b^{(k_3)})$, where $k_1 + k_2 + k_3 = 2n$. The identity component is $\mathrm{so}(k_1) \oplus \mathrm{so}(k_2) \oplus \mathrm{so}(k_3)$.

(b) Here $\mathfrak{g}$ is the set of skew symmetric matrices of order $n = 2l$. The $\mathbb{Z}_2 \times \mathbb{Z}_2$-grading is induced from the elementary grading of M_n given by the tuple $(e^{(2k_1)}, a^{(2k_2)}, b^{(2k_3)}, c^{(2k_4)})$ where $k_1 + k_2 + k_3 + k_4 = 2n$. The identity component is $\mathrm{so}(k_1) \oplus \mathrm{so}(k_2) \oplus \mathrm{so}(k_3) \oplus \mathrm{so}(k_4)$.

(c) Here $\mathfrak{g}$ is the set of matrices of order $n = 2l$ which are skew symmetric with respect to the involution given by the matrix

$$\Phi = \mathrm{diag}\left\{ \begin{pmatrix} 0 & I_{k_1} \\ I_{k_1} & 0 \end{pmatrix}, \begin{pmatrix} 0 & I_{k_2} \\ I_{k_2} & 0 \end{pmatrix} \right\}$$

with $2k_1 + 2k_2 = 2n$ The $\mathbb{Z}_2 \times \mathbb{Z}_2$-grading is induced from the elementary grading of M_n given by the tuple $(e^{(2k_1)}, a^{(2k_1)}, b^{(2k_2)}, c^{(2k_2)})$. The identity component is $\mathrm{gl}(k_1) \oplus \mathrm{gl}(k_2)$.

(d) Here $\mathfrak{g}$ is the set of skew-symmetric matrices in $R = M_n \otimes M_2$ with respect to the involution given by the Kronecker product matrix $\Phi = I_n \otimes X_c$. We have $\mathfrak{g}_e = \{A \otimes X_e \,|\, A \in K_n\}$, $\mathfrak{g}_a = \{A \otimes X_a \,|\, A \in K_n\}$, $\mathfrak{g}_b = \{A \otimes X_b \,|\, A \in K_n\}$, $\mathfrak{g}_c = \{B \otimes X_c \mid B \in H_n\}$. The same argument as in Section 6.3.4 (d) applies also here. The identity component is $\mathrm{so}(n)$.

(e) Here $\mathfrak{g}$ is the set of skew-symmetric matrices in $R = M_{2k} \otimes M_2$ with respect to the involution given by the Kronecker product matrix $\Phi = S_k \otimes X_c$. We have $\mathfrak{g}_e = \{A \otimes X_e \mid A \in K(M_{2k}, S_k)\}$, $\mathfrak{g}_a = \{B \otimes X_a \mid B \in H(M_{2k}, S_k)\}$, $\mathfrak{g}_b = \{A \otimes X_b \mid B \in H(M_{2k}, S_k)\}$, $\mathfrak{g}_c = \{B \otimes X_c \mid B \in H(M_{2k}, S_k)\}$. The identity component is $\mathrm{sp}(2k)$.

REMARK 6.1. Some of these gradings (within respective families (a) to (e)) could be isomorphic in the case of D_4 algebras.

REMARK 6.2. Some further isomorphisms are possible thanks to the isomorphisms of simple Lie algebras in small dimensions.

Acknowledgment

This paper was written while the first author visited the University of Maribor, Slovenia in the years 2007 and 2008. It gives him pleasure to thank the Slovenian colleagues for their hospitality.

References

1. Yuri Bahturin, Michel Goze, *$\mathbb{Z}_2 \times \mathbb{Z}_2$-symmetric spaces*, Pacific J. Math., **236** (2008), 1 - 21.
2. Yu. A. Bahturin, A. Mikhalev, V. Petrogradskii, M. Zaicev, Infinte Dimensional Lie Superalgebras, Expos. Math. vol 7, Walter de Gruyter, Berlin, 1992.
3. Bahturin, Y.; M. Tvalavadze, Group gradings on G_2, Comm. Algebra, to appear.
4. Bahturin, Y.; M. Tvalavadze; T. Tvalavadze, Group gradings on superinvolution simple superalgebras, arxiv math/0611549.
5. Bahturin, Y., Sehgal, S, and M. Zaicev, *Group Gradings on Associative Algebras*, J. Algebra **241** (2001), 677–698.
6. Bahturin, Yuri; Shestakov, Ivan; Zaicev, Mikhail, *Gradings on simple Jordan and Lie algebras*, J. Algebra **283**(2005), 849 - 868.
7. Bahturin, Yuri; Zaicev, Mikhail, *Gradings on algebras of finitary matrices*, preprint.

8. Bahturin, Yuri; Zaicev, Mikhail, *Involutions on graded matrix algebras*, **315** (2007), 527540.
9. Bahturin, Yuri; Zaicev, Mikhail, *Group gradings on simple Lie algebras of type "A"*, J. Lie Theory **16**(2006), 719–742.
10. Bahturin, Yuri; Zaicev, Mikhail, *Graded algebras and graded identities*, Polynomial identities and combinatorial methods (Pantelleria, 2001), 101-139, Lecture Notes in Pure and Appl. Math., **235**, Dekker, New York, 2003
11. Draper, Cristina; Mártin, Candído., *Gradings on g_2*, Linear Algebra Appl. **418** (2006), 85 - 111.
12. Draper, C; Viruel, A., *Gradings on $o(8,\mathbb{C})$*, arXiv: 0709.0194
13. Havliček, Miroslav; Jiři Patera; Edita Pelantova, *On Lie gradings, II*, Linear Algebra Appl. **277** (1998), 97 - 125.
14. Havliček, Miroslav; Jiři Patera; Edita Pelantova, *On Lie gradings, III. Gradings of the real forms of classical Lie algebras*, Linear Algebra Appl. **314** (2000), 87 - 159.
15. V. Kac, Lie superalgebras, Advances in Math. **26** (1977), 8-96.
16. V. Kac, *Graded algebras and symmetric spaces*,Funct. Anal. Pril. 2 (1968), 93 - 94.
17. M. V. Mosolova, *Functions of noncommutative operators generating a graded Lie algebra*, Mat. Zametki 29 (1981), 35-44.
18. Patera, Jiři; Zassanhaus, Hans, *On Lie gradings, I*, Linear Algebra Appl. **112** (1989), 87 - 159.
19. M. Scheunert, Generalized Lie algebras, J. Math Physics 20 (1979), 712-720.
20. M. Scheunert, The Theory of Lie Superalgebras, Lecture Notes in Math vol 716, Springer-Verlag, Berlin, 1979.
21. Vinberg E.B., Onishchik A.L., Lie groups and lie algebras III (Encyclopedia of mathematical sciences vol.41, Springer-Verlag, 1994.
22. Zolotykh, A., Commutation factors and varieties of associative algebras (Russian), Fund. Prikl. Mat.,**3**(1997), 453 - 468.

DEPARTMENT OF ALGEBRA, MOSCOW STATE UNIVERSITY, 119899 MOSCOW, RUSSIAN FEDERATION

Current address: Department of Mathematics and Statistics, Memorial University of Newfoundland, St. John's, NL, A1C5S7, Canada

E-mail address: `yuri@math.mun.ca`

DEPARTMENT OF MATHEMATICS, UNIVERSITY OF MARIBOR, KOROŠKA CESTA 160, 2000 MARIBOR, SLOVENIA

E-mail address: `dusan.pagon@uni-mb.si`

Contemporary Mathematics
Volume **483**, 2009

Derived categories for algebras with radical square zero

Viktor Bekkert and Yuriy Drozd

Dedicated to Ivan Shestakov for his 60th birthday.

ABSTRACT. We determine the derived representation types of algebras with radical square zero and give a description of the indecomposable objects in their bounded derived categories.

Introduction

This paper is based on a talk given by the first author at the Conference "Algebras, Representations and Applications" (Sao Paulo, Brazil, August 2007).

Let $\mathcal{A}$ be a finite dimensional algebra over an algebraically closed field $\Bbbk$, $\mathcal{A}$-mod be the category of left finitely generated $\mathcal{A}$-modules and let $\mathcal{D}^b(\mathcal{A})$ be the bounded derived category of the category $\mathcal{A}$-mod.

The category $\mathcal{D}^b(\mathcal{A})$ is known for few algebras $\mathcal{A}$. For example, the structure of $\mathcal{D}^b(\mathcal{A})$ is well-known for hereditary algebras of finite and tame type [**H**] and for tubular algebras [**HR**].

In the present paper we investigate the derived category $\mathcal{D}^b(\mathcal{A})$ for the finite dimensional algebras with radical square zero.

The structure of the paper is as follows. In Section 1 preliminary results about derived categories are given. We replace finite dimensional algebras by *locally finite dimensional categories* (shortly *lofd*). If such a category only has finitely many indecomposable objects, this language is equivalent to that of finite dimensional algebras.

In Section 2 for a given lofd category $\mathcal{A}$ with radical square zero, we construct following [**BD**] a box such that its representations classify the objects of the derived category $\mathcal{D}^b(\mathcal{A})$, which is used in the next sections.

It follows from [**BD**] that every lofd category over an algebraically closed field is either derived tame or derived wild. In Section 3 we establish the derived representation type for lofd categories with radical square zero is given.

2000 *Mathematics Subject Classification.* Primary 16G60, 16G70; Secondary 15A21, 16E05, 18E30.

Key words and phrases. Derived categories, algebras with radical square zero, derived representation type.

The first author was supported by FAPESP (Grant N 98/14538-0) and CNPq (Grant 301183/00-7).

A description of indecomposables in $\mathcal{D}^b(\mathcal{A})$ is given in Section 4. Namely, we reduce this problem to the problem of description of indecomposables finite dimensional modules for some hereditary path algebra $\Bbbk\mathcal{Q}_{\mathcal{A}}$. In derived tame cases we describe indecomposables in $\mathcal{D}^b(\mathcal{A})$ explicitly.

After this paper was finished, we were told that similar results had been obtained by R. Bautista and S. Liu [**BL**]. Note that they use quite different methods.

1. Derived categories

We will follow in general the notations and terminology of [**BD**] (see also [**D3**], [**D4**]).

We consider categories and algebras over a fixed algebraically closed field $\Bbbk$. A $\Bbbk$-category $\mathcal{A}$ is called *locally finite dimensional* (shortly *lofd*) if the following conditions hold:

1. All spaces $\mathcal{A}(x,y)$ are finite dimensional for all objects x,y.
2. $\mathcal{A}$ is *fully additive*, i.e. it is additive and all idempotents in it split. Conditions 1,2 imply that the category $\mathcal{A}$ is *Krull–Schmidt*, i.e. each object uniquely decomposes into a direct sum of indecomposable objects; moreover, it is *local*, i.e. for each indecomposable object x the algebra $\mathcal{A}(x,x)$ is local. We denote by $\operatorname{ind}\mathcal{A}$ a set of representatives of isomorphism classes of indecomposable objects from $\mathcal{A}$.
3. For each object x the set $\{\, y \in \operatorname{ind}\mathcal{A} \mid \mathcal{A}(x,y) \neq 0 \text{ or } \mathcal{A}(y,x) \neq 0 \,\}$ is finite.

We denote by vec the category of finite dimensional vector spaces over $\Bbbk$ and by $\mathcal{A}$-mod the category of *finite dimensional $\mathcal{A}$-modules*, i.e. functors $M : \mathcal{A} \to \mathsf{vec}$ such that $\{\, x \in \operatorname{ind}\mathcal{A} \mid Mx \neq 0 \,\}$ is finite.

For an arbitrary category $\mathcal{C}$ we denote by $\operatorname{add}\mathcal{C}$ the minimal fully additive category containing $\mathcal{C}$. For instance, one can consider $\operatorname{add}\mathcal{C}$ as the category of finitely generated projective $\mathcal{C}$-modules; especially, $\operatorname{add}\Bbbk = \mathsf{vec}$. We denote by $\operatorname{Rep}(\mathcal{A},\mathcal{C})$ the category of functors $\operatorname{Fun}(\mathcal{A},\operatorname{add}\mathcal{C})$ and call them *representations* of the category $\mathcal{A}$ in the category $\mathcal{C}$. Obviously, $\operatorname{Rep}(\mathcal{A},\mathcal{C}) \simeq \operatorname{Rep}(\operatorname{add}\mathcal{A},\mathcal{C})$. If the category $\mathcal{A}$ is lofd, we denote by $\operatorname{rep}(\mathcal{A},\mathcal{C})$ the full subcategory of $\operatorname{Rep}(\mathcal{A},\mathcal{C})$ consisting of the representations M with *finite support* $\operatorname{supp} M = \{\, x \in \operatorname{ind}\mathcal{A} \mid Mx \neq 0 \,\}$. In particular, $\operatorname{rep}(\mathcal{A},\Bbbk) = \mathcal{A}$-mod.

We recall that a quiver is locally finite if at most finitely many arrows start or stop at each vertex. We recall also that every lofd category is equivalent to a quiver category, i.e. $\mathcal{A} = \operatorname{add}\Bbbk\mathcal{Q}/\mathcal{I}$, where $\mathcal{Q} = \mathcal{Q}_{\mathcal{A}}$ is the locally finite quiver of $\mathcal{A}$ and $\mathcal{I}$ is an admissible ideal in the path category $\Bbbk\mathcal{Q}$ of $\mathcal{Q}$.

We denote by $\mathcal{D}(\mathcal{A})$ (respectively, $\mathcal{D}^b(\mathcal{A})$) the *derived category* (respectively, (two-sided) bounded derived category) of the category $\mathcal{A}$-mod, where $\mathcal{A}$ is a lofd category. These categories are triangulated categories. We denote the shift functor by $[1]$, and its inverse by $[-1]$. Recall that $\mathcal{A}^{\mathrm{op}}$ embeds as a full subcategory into $\mathcal{A}$-mod. Namely, each object x corresponds to the functor $\mathcal{A}^x = \mathcal{A}(x,\_)$. These functors are projective in the category $\mathcal{A}$-mod; if $\mathcal{A}$ is fully additive, these are all projectives (up to isomorphism). On the other hand, $\mathcal{A}$-mod embeds as a full subcategory into $\mathcal{D}^b(\mathcal{A})$: a module M is treated as the complex only having a unique nonzero component equal M at the 0-th position. It is also known that $\mathcal{D}^b(\mathcal{A})$ can be identified with the category $\mathcal{K}^{-,b}(\mathcal{A})$ whose objects are right bounded complexes of projective modules with bounded homology (that is, complexes of finitely generated projective modules with the property that the homology groups are non zero only at a finite number of places) and morphisms are homomorphisms of complexes modulo homotopy [**GM**]. If gl.dim $\mathcal{A} < \infty$, every bounded complex has a bounded projective resolution, hence $\mathcal{D}^b(\mathcal{A})$ can identified with $\mathcal{K}^b(\mathcal{A})$, the category of bounded projective complexes modulo homotopy, but that is not the case if gl.dim $\mathcal{A} = \infty$. Moreover, if $\mathcal{A}$ is lofd, we can confine the considered complexes by *minimal* ones, i.e. always suppose that $\operatorname{Im} d_n \subseteq \operatorname{rad} P_{n-1}$ for all n. We denote by $\mathcal{P}^b_{\min}(\mathcal{A})$ the category of minimal bounded complexes of projective $\mathcal{A}$-modules.

Given $M \in \mathcal{D}^b(\mathcal{A})$, we denote by P_M the minimal projective resolution of M.

For $P \neq 0 \in \mathcal{K}^b(\mathcal{A})$, let t be the minimal number such that $P_i = 0$ for $i > t$. Then, $\beta(P)$ denotes the *(good) truncation* of P below t, i.e. the complex given by

$$\beta(P)_i = \begin{cases} P_i & \text{, if } i \leq t; \\ \operatorname{Ker} d(P)_t & \text{, if } i = t+1; \\ 0 & \text{, otherwise,} \end{cases}$$

$$d(\beta(P))_i = \begin{cases} d(P)_i & \text{, if } i \leq t; \\ i_{\operatorname{Ker} d(P)_t} & \text{, if } i = t+1; \\ 0 & \text{, otherwise,} \end{cases}$$

where $i_{\operatorname{Ker} d_P^t}$ is the obvious embedding.

Let $\overline{\mathcal{X}(\mathcal{A})} = \{\, M \in \operatorname{ind} \mathcal{P}^b_{\min}(\mathcal{A}) \mid P_{\beta(M)} \notin \mathcal{K}^b(\mathcal{A}) \,\}$. Let $\cong_{\mathcal{X}}$ be the equivalence relation on the set $\overline{\mathcal{X}(\mathcal{A})}$ defined by $M \cong_{\mathcal{X}} N$ if and only if $P_{\beta(M)} \cong P_{\beta(N)}$ in $\mathcal{K}^{-,b}(\mathcal{A})$. We use the notation $\mathcal{X}(\mathcal{A})$ for a fixed set of representatives of the quotient set $\overline{\mathcal{X}(\mathcal{A})}$ over the equivalence relation $\cong_{\mathcal{X}}$.

PROPOSITION 1.1. [**BM**] $\operatorname{ind} \mathcal{D}^b(\mathcal{A}) = \operatorname{ind} \mathcal{P}^b_{\min}(\mathcal{A}) \cup \{\beta(M) \mid M \in \mathcal{X}(\mathcal{A})\}$.

REMARK 1.2. If $\mathcal{A}$ has finite global dimension, we have $\mathcal{X}(\mathcal{A}) = \emptyset$ and hence $\operatorname{ind} \mathcal{D}^b(\mathcal{A}) = \operatorname{ind} \mathcal{P}^b_{\min}(\mathcal{A})$.

We recall the definitions of derived tame and derived wild lofd categories from [**BD**].

DEFINITION 1.3.
1. The *rank* of an object $x \in \mathcal{A}$ (or of the corresponding projective module $\mathcal{A}^x$) is the function $\mathbf{r}(x) : \operatorname{ind} \mathcal{A} \to \mathbb{Z}$ such that $x \simeq \bigoplus_{y \in \operatorname{ind} \mathcal{A}} \mathbf{r}(x)(y) y$. The *vector rank* $\mathbf{r}_\bullet(P_\bullet)$ of a bounded complex of projective $\mathcal{A}$-modules is the sequence $(\dots, \mathbf{r}(P_n), \mathbf{r}(P_{n-1}), \dots)$ (actually it has only finitely many nonzero entries).
2. We call a *rational family* of bounded minimal complexes over $\mathcal{A}$ a bounded complex $(P_\bullet, d_\bullet)$ of finitely generated projective $\mathcal{A} \otimes \mathsf{R}$-modules, where R is a *rational algebra*, i.e. $\mathsf{R} = \Bbbk[t, f(t)^{-1}]$ for a nonzero polynomial $f(t)$, and $\operatorname{Im} d_n \subseteq \mathsf{J} P_{n-1}$ For such a complex we define $P_\bullet(m, \lambda)$, where $m \in \mathbb{N}$, $\lambda \in \Bbbk$, $f(\lambda) \neq 0$, the complex $(P_\bullet \otimes_{\mathsf{R}} \mathsf{R}/(t-\lambda)^m, d_\bullet \otimes 1)$. It is indeed a complex of projective $\mathcal{A}$-modules. We put $\mathbf{r}_\bullet(P_\bullet) = \mathbf{r}_\bullet(P_\bullet(1, \lambda))$ (this vector rank does not depend on λ).
3. We call a lofd category $\mathcal{A}$ *derived tame* if there is a set $\mathfrak{P}$ of rational families of bounded complexes over $\mathcal{A}$ such that:
   (a) For each vector rank $\mathbf{r}_\bullet$ the set $\mathfrak{P}(\mathbf{r}_\bullet) = \{\, P_\bullet \in \mathfrak{P} \mid \mathbf{r}_\bullet(P_\bullet) = \mathbf{r} \,\}$ is finite.

(b) For each vector rank $\mathbf{r}_\bullet$ all indecomposable complexes $(P_\bullet, d_\bullet)$ of projective $\mathcal{A}$-modules of this vector rank, except finitely many isomorphism classes, are isomorphic to $P_\bullet(m,\lambda)$ for some $P_\bullet \in \mathfrak{P}$ and some m, λ.

The set $\mathfrak{P}$ is called a *parameterising set* of $\mathcal{A}$-complexes.

4. We call a lofd category $\mathcal{A}$ *derived wild* if there is a bounded complex $P_\bullet$ of projective modules over $\mathcal{A} \otimes \Sigma$, where Σ is the free $\Bbbk$-algebra in 2 variables, such that, for every finite dimensional Σ-modules L, L',
   (a) $P_\bullet \otimes_\Sigma L \simeq P_\bullet \otimes_\Sigma L'$ if and only if $L \simeq L'$.
   (b) $P_\bullet \otimes_\Sigma L$ is indecomposable if and only if so is L.
   (It is well-known that then an analogous complex of $\mathcal{A} \otimes \Gamma$-modules exists for every finitely generated $\Bbbk$-algebra Γ.)

Note that, according to these definitions, every *derived discrete* (in particular, *derived finite*) lofd category [V] is derived tame (with the empty set $\mathfrak{P}$).

It was proved in [BD] that every lofd category over an algebraically closed field is either derived tame or derived wild.

2. Related boxes

Recall (see [D1], [D2]) that a *box* is a pair $\mathfrak{A} = (\mathcal{A}, \mathcal{V})$ consisting of a category $\mathcal{A}$ and an $\mathcal{A}$-coalgebra $\mathcal{V}$. We denote by μ the comultiplication in $\mathcal{V}$, by ε its counit and by $\overline{\mathcal{V}} = \ker \varepsilon$ its *kernel.* We always suppose that $\mathfrak{A}$ is *normal,* i.e. there is a *section* $\omega : x \to \omega_x$ ($x \in \operatorname{ob} \mathcal{A}$) such that $\varepsilon(\omega_x) = 1_x$ and $\mu(\omega_x) = \omega_x \otimes \omega_x$ for all x. A category $\mathcal{A}$ is called *free* if it is isomorphic to a path category $\Bbbk\mathcal{Q}$ of a quiver $\mathcal{Q}$. A normal box $\mathfrak{A} = (\mathcal{A}, \mathcal{V})$ is called *free* if so is the category $\mathcal{A}$, while the kernel $\overline{\mathcal{V}}$ is a free $\mathcal{A}$-bimodule.

Recall that the *differential* of a normal box $\mathfrak{A} = (\mathcal{A}, \mathcal{V})$ is the pair $\partial = (\partial_0, \partial_1)$ of mappings, $\partial_0 : \mathcal{A} \to \overline{\mathcal{V}}$, $\partial_1 : \overline{\mathcal{V}} \to \overline{\mathcal{V}} \otimes_{\mathcal{A}} \overline{\mathcal{V}}$, namely

$$\partial_0 a = a\omega_x - \omega_y a \text{ for } a \in \mathcal{A}(x,y),$$

$$\partial_1 v = \mu(v) - v \otimes \omega_x - \omega_y \otimes v \text{ for } v \in \overline{\mathcal{V}}(x,y).$$

A *representation* of a box $\mathfrak{A} = (\mathcal{A}, \mathcal{V})$ over a category $\mathcal{C}$ is defined as a functor $M : \mathcal{A} \to \operatorname{add} \mathcal{C}$. A *morphism* of such representations $f : M \to N$ is defined as a homomorphisms of $\mathcal{A}$-modules $\mathcal{V} \otimes_{\mathcal{A}} M \to N$. If $g : N \to L$ is another morphism, there product is defined as the composition

$$\mathcal{V} \otimes_{\mathcal{A}} M \stackrel{\mu\otimes 1}{\to} \mathcal{V} \otimes_{\mathcal{A}} \mathcal{V} \otimes_{\mathcal{A}} M \stackrel{1\otimes f}{\to} \mathcal{V} \otimes_{\mathcal{A}} N \stackrel{g}{\to} L\,.$$

Thus we obtain the *category of representations* $\operatorname{Rep}(\mathfrak{A}, \mathcal{C})$. If $\mathfrak{A}$ is a free box, we denote by $\operatorname{rep}(\mathfrak{A}, \mathcal{C})$ the full subcategory of $\operatorname{Rep}(\mathfrak{A}, \mathcal{C})$ consisting of representations with finite support $\operatorname{supp} M = \{x \in \operatorname{ob} \mathcal{A} | Mx \neq 0\}$. If $\mathcal{C} = \mathsf{vec}$, we write $\operatorname{Rep}(\mathcal{A})$ and $\operatorname{rep}(\mathcal{A})$.

Given a lofd $\mathcal{A}$ with radical square zero, we are going to construct a box such that its representations classify the objects of the derived category $\mathcal{D}^b(\mathcal{A})$ (see [BD], [D3] for the case of an arbitrary lofd category).

Let $\mathcal{Q} = \mathcal{Q}_{\mathcal{A}}$ be a quiver of $\mathcal{A}$. Given two vertices a and b we define $\mathcal{Q}_1[a,b]$ as the set of all arrows from a to b. Given an arrow a of $\mathcal{Q}$, let us denote by a^{-1} a formal inverse of a, and let us set $s(a^{-1}) = t(a)$ and $t(a^{-1}) = s(a)$. By a *walk* w of length n we mean a sequence $w_1 w_2 \cdots w_n$ where each w_i is either of the form a

or a^{-1}, a being an arrow in $\mathcal{Q}$ and where $s(w_{i+1}) = t(w_i)$ for $1 \leq i < n$. For each walk $w = w_1 w_2 \cdots w_n$ we define $s(w) = s(w_1)$ and $t(w) = t(w_n)$. By definition, a *closed walk* is a walk w such that $s(w) = t(w)$.

Consider the path category $\mathcal{A}^{\square} = \Bbbk\mathcal{Q}^{\square}$, where $\mathcal{Q}^{\square}$ is the quiver with the set of points $\mathcal{Q}_0^{\square} = \mathcal{Q}_0 \times \mathbb{Z}$ and with the set of arrows $\mathcal{Q}_1^{\square} = \mathcal{Q}_1 \times \mathbb{Z}$, where for given $\alpha : a \to b$ in $\mathcal{Q}$ we set $s((\alpha, i)) = (b, i)$ and $t((\alpha, i)) := (a, i-1)$.

Consider the normal free box $\mathfrak{A} = \mathfrak{A}(\mathcal{A}) = (\mathcal{A}^{\square}, \mathcal{W})$, with the kernel $\overline{\mathcal{W}}$ freely generated by the set $\{\varphi_{\alpha,i} \,|\, \alpha \in \mathcal{Q}_1, i \in \mathbb{Z}\}$, where $s(\varphi_{\alpha,i}) = (t(\alpha), i)$, $t(\varphi_{\alpha,i}) = (s(\alpha), i)$ and with zero differential ∂.

Given a box $\mathfrak{A}$, we denote by $\operatorname{rep}(\mathfrak{A})$ the category of finite dimensional representations of $\mathfrak{A}$.

Let us consider the following functor $\mathbf{F} : \operatorname{rep}(\mathfrak{A}(\mathcal{A})) \to \mathcal{P}^b_{\min}(\mathcal{A})$.

A representation $M \in \operatorname{rep}(\mathfrak{A})$ is given by vector spaces $M(x,n)$ and linear mappings $M(\alpha, n) : M(y,n) \to M(x, n-1)$, where $\alpha \in \mathcal{Q}_1[x,y]$ and $x, y \in \mathcal{Q}_0$, $n \in \mathbb{Z}$. For such a representation, set $P_n = \bigoplus_{x \in \mathcal{Q}_0} \mathcal{A}^x \otimes M(x,n)$ and $d_n = \bigoplus_{x,y \in \mathcal{Q}_0} \sum_{\alpha \in \mathcal{Q}[x,y]} \mathcal{A}^{\alpha} \otimes M(\alpha, n)$. A morphism $\Psi : M \to M'$ is given by linear mappings $\Psi(z,n) : M(z,n) \to M'(z,n)$ and $\Psi(\varphi_{\alpha,m}) : M(y,n) \to M'(x,n)$, where $x, y, z \in \mathcal{Q}_0$ and $\alpha \in \mathcal{Q}_1[x,y]$. We define a homomorphism $F(\Psi) : F(M) \to F(M')$ by the following rule. Given $x, y \in \mathcal{Q}_0$ we set

$$\widetilde{\mathcal{Q}}_1[x,y] = \begin{cases} \mathcal{Q}_1[x,y] & \text{, if } x \neq y \\ \mathcal{Q}_1[x,y] \cup \{1_x\} & \text{, otherwise.} \end{cases}$$

For given $\alpha \in \widetilde{\mathcal{Q}}_1[x,y]$ we set

$$\Psi_{\alpha,n} = \begin{cases} \Psi(\varphi_{\alpha,n}) & \text{, if } \alpha \in \mathcal{Q}_1 \\ \Psi(x,n) & \text{, otherwise.} \end{cases}$$

Then $F(\Psi)$ is defined by $F(\Psi)_n = \bigoplus_{x,y \in \mathcal{Q}_0} \sum_{\alpha \in \widetilde{\mathcal{Q}}[x,y]} \mathcal{A}^{\alpha} \otimes \Psi_{\alpha,n}$.

The following Theorem follows from the Theorem 2.2 in [**BD**].

THEOREM 2.1. $\mathbf{F}$ *is an equivalence of categories.*

We define a shift functor in $\operatorname{rep}(\mathfrak{A}(\mathcal{A}))$ by the following rule. Given $M \in \operatorname{rep}(\mathfrak{A}(\mathcal{A}))$ and $j \in \mathbb{Z}$ we define $M[j] \in \operatorname{rep}(\mathfrak{A}(\mathcal{A}))$ by $M[j](a,i) = M(a, i-j)$ and $M[j](\alpha, i) = M(\alpha, i-j)$. In the same way we can define $[j]$ for morphisms in $\operatorname{rep}(\mathfrak{A}(\mathcal{A}))$.

LEMMA 2.2. $\mathbf{F}(M[i]) = \mathbf{F}(M)[i]$ *and* $\mathbf{F}(\varphi[i]) = \mathbf{F}(\varphi)[i]$ *for any object* M *and any morphism* φ *in* $\operatorname{rep}(\mathfrak{A}(\mathcal{A}))$.

PROOF. Straightforward. □

Given a quiver $\mathcal{Q}$ we fix some vertex $a \in \mathcal{Q}_0$ and denote by $\mathcal{Q}[i]$ the connected component of $\mathcal{Q}^{\square}$ which contain the vertex (a,i). Given a walk $w = w_1 \cdots w_n$ in $\mathcal{Q}$ we denote by $\varepsilon^+(w)$ (resp., $\varepsilon^-(w)$) the number of w_i of the form p (resp., p^{-1}), p being an arrow. We set $\varepsilon(w) = |\varepsilon^+(w) - \varepsilon^-(w)|$. We denote by $\mathcal{Q}^c$ the set of all closed walks in $\mathcal{Q}$ and set $\varepsilon(\mathcal{Q}) = \min_{w \in \mathcal{Q}^c} \varepsilon(w)$ in case of $\mathcal{Q}^c \neq \emptyset$ and $\varepsilon(\mathcal{Q}) = 0$ otherwise. We say that a quiver $\mathcal{Q}$ satisfies the *walk condition* provided $\varepsilon(\mathcal{Q}) = 0$ (= the number of clockwise oriented arrows is the same as the number of counterclockwise oriented arrows for any closed walk w of $\mathcal{Q}$).

LEMMA 2.3. *Let* $b \in \mathcal{Q}_0$, $i, j \in \mathbb{Z}$ *and* a *as above. Then* $(b,j) \in \mathcal{Q}[i]$ *if and only if there exists a walk* w *from* a *to* b *in* $\mathcal{Q}$ *such that* $j = i + \varepsilon^+(w) - \varepsilon^-(w)$.

PROOF. Straightforward. □

COROLLARY 2.4. *Let $\mathcal{Q}$ be a connected quiver. Then $\mathcal{Q}[i] = \mathcal{Q}[j]$ if and only if there exists a closed walk w in $\mathcal{Q}$ such that $i \equiv j \pmod{\varepsilon(w)}$.*

PROOF. Let $a \in \mathcal{Q}$ be as above.

" $\Longrightarrow$." Suppose that $\mathcal{Q}[i] = \mathcal{Q}[j]$ for some $i \neq j \in \mathbb{Z}$. Then $(a, j) \in \mathcal{Q}[i]$ and by Lemma 2.3 there exists a walk w from a to a in $\mathcal{Q}$ such that $j = i + \varepsilon^+(w) - \varepsilon^-(w)$, hence $i \equiv j \pmod{\varepsilon(w)}$.

" $\Longleftarrow$." Let w be a closed walk in Q such that $i \equiv j \pmod{\varepsilon(w)}$. Then $j = i + m\varepsilon(w)$ for some $m \in \mathbb{Z}$. Since $\mathcal{Q}$ is connected, there exists a walk u from a to $s(w)$. Then for the closed walk $v = u^{-1}w^m u$ in $\mathcal{Q}$ we have $j = i + \varepsilon^+(v) - \varepsilon^-(v)$. Therefore $(a, j) \in \mathcal{Q}[i]$ by Lemma 2.3 and hence $\mathcal{Q}[j] = \mathcal{Q}[i]$. □

COROLLARY 2.5. *Let $\mathcal{Q}$ be a connected quiver. Then $\mathcal{Q}[i] = \mathcal{Q}[j]$ if and only if $i \equiv j \pmod{\varepsilon(\mathcal{Q})}$.*

PROOF. It is easy to see that if $\mathcal{Q}$ is connected and $\mathcal{Q}^c \neq \emptyset$, then for any closed walk w we have $\varepsilon(w) = m\varepsilon(\mathcal{Q})$ for some $m \in \mathbb{N}$. Hence the statement follows from Corollary 2.4. □

LEMMA 2.6.
1. *Let $\mathcal{Q}$ be a connected quiver which satisfies the walk condition. Then $\mathcal{Q}^{\square}$ is a disjoint union $\bigsqcup_{i\in\mathbb{Z}} \mathcal{Q}[i]$, where $\mathcal{Q}[i] \simeq \mathcal{Q}^{\mathrm{op}}$ for all i.*
2. *Let $\mathcal{Q}$ be a quiver which not satisfy the walk condition. Then $\mathcal{Q}^{\square}$ is the disjoint union $\bigsqcup_{0\le i<\varepsilon(\mathcal{Q})} \mathcal{Q}[i]$, where $\mathcal{Q}[i] \simeq \mathcal{Q}^{\diamond}$ for all i for some quiver $\mathcal{Q}^{\diamond}$.*

PROOF.
1. It follows from Corollary 2.4 that if $i \neq j$ we have $\mathcal{Q}[i] \neq \mathcal{Q}[j]$. It is easy to see that in this case $\mathcal{Q}[i] \simeq \mathcal{Q}^{\mathrm{op}}$.
2. It is easy to see that $\mathcal{Q}[i] \simeq \mathcal{Q}[j]$ for all $i, j \in \mathbb{Z}$. Therefore the statement follows from Corollary 2.4.

□

3. Derived representation type

THEOREM 3.1. *Let $\mathcal{A}$ be a lofd connected category with radical square zero.*
1. *$\mathcal{A}$ is derived tame if and only if $\mathcal{Q}_{\mathcal{A}}$ is a Dynkin quiver (of types $\mathbb{A}_n$ $(n \ge 1)$, $\mathbb{D}_n$ $(n \ge 4)$, $\mathbb{E}_n$ $(8 \ge n \ge 6)$) or an Euclidian quiver (of types $\tilde{\mathbb{A}}_n$ $(n \ge 1)$, $\tilde{\mathbb{D}}_n$ $(n \ge 4)$, $\tilde{\mathbb{E}}_n$ $(8 \ge n \ge 6)$) or a quiver of types $\mathbb{A}_\infty$, $\mathbb{A}_\infty^\infty$ or $\mathbb{D}_\infty$.*
2. *$\mathcal{A}$ is derived discrete if and only if $\mathcal{Q}_{\mathcal{A}}$ is a Dynkin quiver or an Euclidian quiver $\tilde{\mathbb{A}}_n$ $(n \ge 1)$ which does not satisfy the walk condition or a quiver of types $\mathbb{A}_\infty$, $\mathbb{A}_\infty^\infty$ or $\mathbb{D}_\infty$.*
3. *$\mathcal{A}$ is derived finite if and only if $\mathcal{Q}_{\mathcal{A}}$ is a Dynkin quiver.*

PROOF. Let $\mathcal{Q} = \mathcal{Q}_{\mathcal{A}}$. We distinguish three cases.

(a) $\mathcal{Q}$ has no cycles.

Then by Lemma 2.6 we have in this case $\mathcal{Q}^{\square} = \bigsqcup_{i\in\mathbb{Z}} \mathcal{Q}[i]$, where $\mathcal{Q}[i] = \mathcal{Q}^{\mathrm{op}}$. Hence the statements of the Theorem in this case follow from [**G**], [**N**] and Proposition 1.1.

(b) $\mathcal{Q}$ is an Euclidian quiver $\tilde{\mathbb{A}}_n$.

It follows from Lemma 2.6 that if $\mathcal{Q}$ satisfies the walk condition, then $\mathcal{Q}^{\square} = \bigsqcup_{i\in\mathbb{Z}} \mathcal{Q}[i]$, where $\mathcal{Q}[i] = \mathcal{Q}^{\mathrm{op}}$, hence $\mathcal{A}$ is derived tame, but is not derived discrete by [**N**] and Proposition 1.1; and if $\mathcal{Q}$ does not satisfy the walk condition, then by Lemma 2.6 we have $\mathcal{Q}^{\square} = \bigsqcup_{0\leq i<\varepsilon(\mathcal{Q})} \mathcal{Q}[i]$, where $\mathcal{Q}[i] \simeq \mathbb{A}_{\infty}^{\infty}$ for all i, hence $\mathcal{A}$ is derived discrete by [**G**] and Proposition 1.1.

(c) $\mathcal{Q}$ has an Euclidian sub-quiver $\mathcal{Q}' \neq \mathcal{Q}$ of type $\tilde{\mathbb{A}}_n$.

It follows from (b) that $\mathcal{Q}'^{\square}$ has connected sub-quiver X of type $\tilde{\mathbb{A}}_n$ or $\mathbb{A}_{\infty}^{\infty}$. Let X' be the connected sub-quiver of $\mathcal{Q}^{\square}$ which contains X. Since $\mathcal{Q}' \neq \mathcal{Q}$, we have $X' \neq X$. Therefore $\Bbbk\mathcal{Q}^{\square}$ is wild by [**N**] and hence $\mathcal{A}$ is derived wild. □

4. Indecomposable objects

Let $\mathbf{F}$ be as in Section 2 and $\mathcal{X}$ as in Section 1.

THEOREM 4.1. *Let $\mathcal{A}$ be a lofd category with radical square zero. Then the complexes $\mathbf{F}(M)$ and $\beta(N)$, where $M \in \operatorname{ind}\operatorname{rep}(\Bbbk\mathcal{Q}^{\square})$ and $N \in \mathcal{X}(\Bbbk\mathcal{Q}^{\square})$, constitute an exhaustive list of pairwise non-isomorphic indecomposable objects of $\mathcal{D}^b(\mathcal{A})$.*

PROOF. Since $\partial = 0$ for the box $\mathfrak{A} = \mathfrak{A}(\mathcal{A})$, we have that $\operatorname{ind}\operatorname{rep}(\mathfrak{A}) = \operatorname{ind}\operatorname{rep}(\Bbbk\mathcal{Q}^{\square})$. Hence the statement follows from Theorem 2.1 and Proposition 1.1. □

For a quiver $\mathcal{Q}$ which satisfies the walk condition, we denote by $\imath : \operatorname{rep}(\mathcal{Q}^{\mathrm{op}}) \to \operatorname{rep}(\mathcal{Q}^{\square})$ the inclusion functor which sent $\mathcal{Q}^{\mathrm{op}}$ to $\mathcal{Q}[0]$. It follows from Lemma 2.6 that in this case gl.dim $\mathcal{A} < \infty$ (because the quiver $\mathcal{Q}$ has no oriented cycle) and it is the disjoint union $\bigsqcup_{i\in\mathbb{Z}} \mathcal{Q}[i]$, where $\mathcal{Q}[i] = \mathcal{Q}^{\mathrm{op}}$ for all i. Hence we obtain the following Corollary.

COROLLARY 4.2. *Let $\mathcal{A}$ be a lofd category with radical square zero whose quiver $\mathcal{Q} = \mathcal{Q}_{\mathcal{A}}$ satisfies the walk condition. Then the complexes $\mathbf{F}(\imath(M))[i]$, where $M \in \operatorname{ind}\operatorname{rep}(\Bbbk\mathcal{Q}^{\mathrm{op}})$ and $i \in \mathbb{Z}$, constitute an exhaustive list of pairwise non-isomorphic indecomposable objects of $\mathcal{D}^b(\mathcal{A})$.*

References

[BD] V. Bekkert and Yu. Drozd, *Tame-wild dichotomy for derived categories*, arXiv:math. RT/0310352.

[BL] R. Bautista and S. Liu, *The bounded derived category of an algebra with radical squared zero*, Preprint.

[BM] V. Bekkert and H. Merklen, *Indecomposables in derived categories of gentle algebras*, Algebras and Representation Theory **6** (2003), 285–302.

[D1] Yu. A. Drozd, *Tame and wild matrix problems*, Representations and quadratic form, Institute of Mathematics, Kiev, 1979, pp. 39-74; English translation: Amer. Math. Soc. Translations **128** (1986), 31-55.

[D2] Yu. A. Drozd, *Reduction algorithm and representations of boxes and algebras*, Comptes Rendues Math. Acad. Sci. Canada **23** (2001), 97–125.

[D3] Yu. A. Drozd, *Derived tame and derived wild algebras*, Algebra and Discrete Math. **3** (2004), 57–74.

[D4] Yu. A. Drozd, *Semi-continuity for derived categories*, Algebras and Representation Theory **8** (2005), 239–248.

[G] P. Gabriel, *Unzerlegbare Darstellungen* I, Manuscr. Math. **6** (1972), 71–103.

[GM] S. I. Gelfand and Yu. I. Manin, *Methods of Homological Algebra*, Springer–Verlag, 1996.

[H] D. Happel, *Triangulated Categories in the Representation Theory of Finite Dimensional Algebras*, London Mathematical Society Lecture Notes Series **119**, Cambridge University Press, Cambridge, 1988.

[HR] D. Happel and C. M. Ringel, *The derived category of a tubular algebra*, Springer LNM **1177** (1984), 156–180.

[N] L. A. Nazarova, *Representations of quivers of infinite type*, Izv. Akad. Nauk SSSR, Ser. Mat. **37** (1973), 752–791; English transl.: Math. USSR. Izv. **7** (1973), 749–792.

[V] D. Vossieck, *The algebras with discrete derived category*, J. Algebra **243** (2001), 168–176.

Departamento de Matemática, ICEx, Universidade Federal de Minas Gerais, Av. Antônio Carlos, 6627, CP 702, CEP 30123-970, Belo Horizonte-MG, Brasil

E-mail address: bekkert@mat.ufmg.br

Institute of Mathematics, National Academy of Sciences of Ukraine, Tereschenkivska 3, 01601 Kiev, Ukraine

E-mail address: drozd@imath.kiev.ua

Contemporary Mathematics
Volume **483**, 2009

Tits construction, triple systems and pairs

Pilar Benito and Fabián Martín-Herce

ABSTRACT. The paper is concerned with the description of certain well-known models of triple systems and simple Jordan pairs by reviewing a slightly generalised version of Tits' classical construction of Lie algebras from composition and simple Jordan algebras.

1. Introduction

In 1962, J. Tits [**18**, Theorem 1] gave a construction of Lie algebras in characteristic different from 2 and 3, taking as basic ingredients a simple 3-dimensional Lie algebra $(\mathcal{S}, [ab])$ and a unital Jordan algebra $(\mathcal{J}, xy)$ (the unity element is denoted by $1 \in \mathcal{J}$). The algebras $\mathcal{S}$ and $\mathcal{J}$ are glued toghether to build the Lie algebra:

$$\mathcal{L}_{\mathcal{S}}(\mathcal{J}, \mathcal{D}) = \mathcal{D} \oplus \mathcal{S} \otimes \mathcal{J} \tag{1.1}$$

where $\mathcal{D}$ is any subalgebra of derivations of $\mathcal{J}$ which contains the subspace $D_{\mathcal{J},\mathcal{J}}$ spanned by the maps

$$d_{x,y}(z) = (yz)x - y(zx) = (y, z, x) = [R_x, R_y](z) \tag{1.2}$$

where R_x denotes the right multiplication by x. The operators $d_{x,y}$ are the so called *inner derivations of* $\mathcal{J}$ and from $[[R_x, R_y], R_z] = R_{(y,z,x)}$ we have that the vector space $D_{\mathcal{J},\mathcal{J}} = \mathrm{span}\langle d_{x,y} : x, y \in \mathcal{J}\rangle$ is a Lie algebra named *inner derivation Lie algebra* of $\mathcal{J}$. The product which defines a Lie structure on $\mathcal{L}_{\mathcal{S}}(\mathcal{J}, \mathcal{D})$ is given by the formulae,

$$\begin{aligned} &[d, a \otimes x] = a \otimes d(x) \\ &[a \otimes x, b \otimes y] = \tfrac{1}{2}\,\mathrm{tr}(\mathrm{ad}\, a\, \mathrm{ad}\, b) d_{x,y} + [a, b] \otimes xy \end{aligned} \tag{1.3}$$

for $d \in \mathcal{D}, a, b \in \mathcal{S}, x, y \in \mathcal{J}$, $\mathrm{tr}(\mathrm{ad}\, a\, \mathrm{ad}\, b)$ is the usual linear trace and $\mathcal{D}$ is viewed as a Lie subalgebra of $\mathcal{L}_{\mathcal{S}}(\mathcal{J}, \mathcal{D})$. In fact, $\mathcal{S} \otimes 1 \simeq \mathcal{S}$ also is a subalgebra and centralizes $\mathcal{D}$, that is $[\mathcal{D}, \mathcal{S} \otimes 1] = 0$.

Any 3-dimensional simple Lie algebra $\mathcal{S}$ can be realized through a suitable quaternion algebra $(\mathcal{Q}, ab)$, as either the zero trace elements $\mathcal{Q}_0$ of $\mathcal{Q}$ under the product $[a, b] = ab - ba = \mathrm{ad}\, a(b)$ or as the vector space $\mathrm{ad}\, \mathcal{Q} = \mathrm{span}\langle \mathrm{ad}\, a : a \in \mathcal{Q}\rangle$ with product $[\mathrm{ad}\, a, \mathrm{ad}\, b] = \mathrm{ad}\, a\, \mathrm{ad}\, b - \mathrm{ad}\, b\, \mathrm{ad}\, a$. So, if we set $S = \mathcal{Q}_0$ and pay

2000 *Mathematics Subject Classification.* 17B60, 17A30.
Key words and phrases. Lie-Yamaguti algebra, Lie triple system, Jordan pair, Tits and Tits-Kantor-Koecher construction.
Supported by the Spanish Ministerio de Educación y Ciencia and FEDER (MTM 2007-67884-C04-03) and from the Comunidad Autónoma de La Rioja (ANGI2005/05,06).

attention to (1.1) in case $\mathcal{D} = \mathcal{D}_{\mathcal{J},\mathcal{J}}$, identifying $a \otimes 1 \in \mathcal{Q}_0$ with $\operatorname{ad} a$, we can display $\mathcal{L}_{\mathcal{Q}_0}(\mathcal{J}, \mathcal{D}_{\mathcal{J},\mathcal{J}})$ into the form

$$\mathcal{T}(\mathcal{Q}, \mathcal{J}) = \mathcal{D}_{\mathcal{J},\mathcal{J}} \oplus \mathcal{Q}_0 \otimes \mathcal{J}_0 \oplus \operatorname{ad} \mathcal{Q} \tag{1.4}$$

where $\mathcal{J}_0$ denotes the set of trace zero elements in the Jordan algebra $\mathcal{J}$. In characteristic zero, we have the relationship $\operatorname{tr}(\operatorname{ad} a \operatorname{ad} b) = 4t(ab)$ for $a, b \in \mathcal{Q}_0$ and $t(a)$ the trace in $\mathcal{Q}$ defined by $t(a) = n(a, 1)$, $n(a)$ the norm on the quaternion algebra $\mathcal{Q}$. Moreover, for $x, y \in \mathcal{J}$ and $T(x)$ the generic trace on $\mathcal{J}$, we can consider the projection product on $\mathcal{J}_0$

$$x \star y = xy - \frac{1}{\operatorname{degree} \mathcal{J}} T(xy)1 \tag{1.5}$$

which allows us to rewrite (1.3) as

$$\begin{aligned} &[D + d, a \otimes x] = D(a) \otimes x + a \otimes d(x) \\ &[a \otimes x, b \otimes y] = 2t(ab) d_{x,y} + [a, b] \otimes x \star y + \tfrac{1}{\operatorname{degree} \mathcal{J}} T(xy) \operatorname{ad}[a, b] \end{aligned} \tag{1.6}$$

for $D \in \operatorname{ad} \mathcal{Q}, d \in \mathcal{D}_{\mathcal{J},\mathcal{J}}$ and $a \in \mathcal{Q}_0$, $x \in \mathcal{J}_0$. Here $\mathcal{D}_{\mathcal{J},\mathcal{J}}$ and $\operatorname{ad} \mathcal{Q}$ are viewed as subalgebras centralizing each other, therefore $[\mathcal{D}_{\mathcal{J},\mathcal{J}}, \operatorname{ad} \mathcal{Q}] = 0$. We also note that any α-product-deformation in (1.6)

$$[a \otimes x, b \otimes y]_\alpha = 2\alpha^2\, t(ab) d_{x,y} + \alpha\, [a, b] \otimes x \star y + \frac{\alpha^2}{n} T(xy) \operatorname{ad}[a, b] \tag{1.7}$$

for a nonzero $\alpha \in k$ and $n = T(1)$ the degree of $\mathcal{J}$, provides isomorphic Lie algebras.

When the quaternion algebra $\mathcal{Q}$ is chosen to be split, so $\mathcal{Q} = M_2(k)$ is the associative algebra of 2×2 matrices, the algebra $\mathcal{L}_{\mathcal{S}}(\mathcal{J}, \mathcal{D}_{\mathcal{J},\mathcal{J}})$ matches the well-known Tits-Kantor-Koecher construction ($\mathcal{TKK}$-construction for short) of a Lie algebra from the Jordan algebra $\mathcal{J}$ (see Chapter VIII, Section 5 in [**8**] for a complete description of the $\mathcal{TKK}$-construction). In case $\mathcal{J}$ is a central simple Jordan algebra of degree $n = 3$, from (1.4) and (1.6) we find the Classical Tits construction given in [**19**] (see Chapter IV, Section 4 in [**17**] for a straightforward description) which works using not only quaternions but any composition algebra. In the Classical Tits construction, if an arbitrary composition algebra $\mathcal{C}$ is used, the quaternionic derivation $\operatorname{ad}[a, b]$ which appears in (1.6) must be changed by the general derivation of composition algebras:

$$D_{a,b}(c) = \operatorname{ad}[a, b] + 3((ac)b - a(cb)) \tag{1.8}$$

and the set $\operatorname{ad} \mathcal{Q}$ in (1.4) replaced by the derivation algebra of $\mathcal{C}$ which is given by $D_{\mathcal{C},\mathcal{C}} = \operatorname{span}\langle D_{a,b} : a, b \in \mathcal{C}\rangle$. In the quaternionic case we have $D_{a,b} = \operatorname{ad}[a, b]$, so $D_{\mathcal{Q},\mathcal{Q}} = \operatorname{ad} \mathcal{Q}$ and therefore $\mathcal{L}_{\mathcal{Q}_0}(\mathcal{J}, \mathcal{D}_{\mathcal{J},\mathcal{J}})$ is viewed as the particular case $\mathcal{C} = \mathcal{Q}$ in the the more general setting

$$\mathcal{T}(\mathcal{C}, \mathcal{J}) = \mathcal{D}_{\mathcal{J},\mathcal{J}} \oplus \mathcal{Q}_0 \otimes \mathcal{J}_0 \oplus \mathcal{D}_{\mathcal{C},\mathcal{C}} \tag{1.9}$$

where (1.6) is replaced by

$$[a \otimes x, b \otimes y] = 2t(ab) d_{x,y} + [a, b] \otimes x \star y + \frac{1}{\operatorname{degree} \mathcal{J}} T(xy) D_{a,b} \tag{1.10}$$

The ingredients in this Lie construction recipe are composition and Jordan algebras as well as their (inner) derivation Lie algebras. In [**4**, Theorem 3.1] it is pointed out that, over the real numbers, the vector space $\mathcal{T}(\mathcal{C}, \mathcal{J})$ described in

(1.9) but taking any composition algebra $\mathcal{C}$ together with any $\mathcal{J}$ central simple Jordan algebra of degree ≥ 3 under the product (1.7) for $\alpha = \frac{1}{2}$ and $D_{a,b}$ instead $\mathrm{ad}[a,b]$, provides a Lie algebra if and only if either $\mathcal{C}$ is an associative algebra or the Jordan algebra $\mathcal{J}$ satisfies a certain cubic identity. In other words, octonions can only be mixed with Jordan algebras of degree three to get a Lie algebra structure on the vector space (1.9) by means of the (α-product-deformation) product given in (1.10). In Section 2 we shall see that the result is true in a more general context in case we are looking for $[\mathcal{C}_0 \otimes \mathcal{J}_0, \mathcal{C}_0 \otimes \mathcal{J}_0] \neq 0$ (in the section,we shall outline the result without proof, a more complete description will be given in a forthcoming paper). We will also explain why the construction in (1.9) under the product (1.10) is esentially the only one we can expect in order to get a Lie algebra from a composition algebra and a Jordan algebra. The last section is devoted to relate the Lie algebras $\mathcal{T}(\mathcal{C},\mathcal{J})$ to generalized Lie triple systems and Jordan pairs.

2. Lie algebras *à la* Tits

From now on, all the vector spaces and algebras are defined over a ground field k of characteristic zero. Let us denote by $\mathcal{C}$ a composition algebra and by $\mathcal{J}$ a unital central simple Jordan algebra. For the previous algebras we establish the following terminology:

$\mathcal{C}$ composition algebra:

- $n(a)$ denotes the *norm*
- $t(a) = n(a+1) - n(a) - n(1)$ is the *trace* and $\mathcal{C}_0$ is the set of trace zero elements, so $\mathcal{C}_0 = \{a \in \mathcal{C} : t(a) = 0\}$
- The different posibilities for $\mathcal{C}$ are denoted by $k, \mathcal{K}$ (composition algebras of dimension 1 and 2), $\mathcal{Q}$ *quaternions* and $\mathcal{O}$ for *generalized octonions*, which are algebras of dimension 4 and 8 respectively
- The *derivation Lie algebra* of $\mathcal{C}$, which is spanned by the standard derivations $D_{a,b}$ in (1.8), will be denoted by $D_{\mathcal{C},\mathcal{C}}$

$\mathcal{J}$ central simple Jordan algebra:

- $T(x)$ denotes the *generic trace*, $n = T(1)$ is the degree and $\mathcal{J}_0$ the trace zero elements, so $\mathcal{J}_0 = \{a \in \mathcal{J} : T(a) = 0\}$
- For the different type descriptions of $\mathcal{J}$ we follow [**17**, Chapter IV, Section 2]. In case $\mathcal{J}$ is of Type **A**, we will use the more general description $\mathcal{J} = \mathcal{H}(\mathcal{A}, \jmath)$ for a central simple involutorial associative algebra $(\mathcal{A}, \jmath)$ of the second kind with center $P = k[q]$ either a quadratic extension of k (case $\mathbf{A_{II}}$ in [**17**]) or isomorphic to $k \times k$ (related to the case $\mathbf{A_I}$)
- The *derivation Lie algebra* $\mathrm{Der}\,\mathcal{J}$ of $\mathcal{J}$ is spanned by the standard derivations $d_{x,y} = [R_x, R_y]$ in (1.2), so $\mathrm{Der}\,\mathcal{J}$ agrees with the inner derivation algebra $D_{\mathcal{J},\mathcal{J}}$

In both classes of algebras, the associator of any three elements is given by

$$(x, y, z) = (xy)z - x(yz) \tag{2.1}$$

For Lie algebras we shall use basic terminology in [**6**]. Over an algebraically closed ground field and up to isomorphisms, simple Lie algebras fall into either the *classical* families *special* $\mathfrak{sl}_m(k)$ (type A), *orthogonal* $\mathfrak{so}_m(k)$ (types B or D) and *symplectic* $\mathfrak{sp}_{2m}(k)$ (type C) or the *exceptionals types* G_2, F_4, E_6, E_7 and E_8. A Lie algebra $\mathcal{L}$

is said to be of type X (that is A, B, C, D or exceptional) in case after extending scalars up to the algebraic clousure $\bar{k}$, the algebra $\mathcal{L}_{\bar{k}} = \mathcal{L} \otimes_k \bar{k}$ is of this type.

In the process of building a Lie algebra over the vector space $\mathcal{T}(\mathcal{C}, \mathcal{J})$ given in (1.9) apart from anticommutativity, the original construction given by Tits imposes the following conditions:

$$\begin{array}{l} D_{\mathcal{C},\mathcal{C}} \text{ and } D_{\mathcal{J},\mathcal{J}} \text{ are Lie subalgebras of } \mathcal{T}(\mathcal{C}, \mathcal{J}) \\ [D_{\mathcal{C},\mathcal{C}}, D_{\mathcal{J},\mathcal{J}}] = 0 \\ [D, a \otimes x] = D(a) \otimes x \\ [d, a \otimes x] = a \otimes d(x) \end{array} \tag{2.2}$$

for any $a \in \mathcal{C}$, $x \in \mathcal{J}$, $D \in D_{\mathcal{C},\mathcal{C}}$ and $d \in D_{\mathcal{J},\mathcal{J}}$. Tits construction ended by expressing the product $[a \otimes x, b \otimes y]$ as an α-product-deformation of (1.10) (see [**19**] or [**17**, Chapter IV, Section 4]). A closer look at the formula in (1.10) show us that $[a \otimes x, b \otimes y]$ is determined through symmetric and skew-symmetric maps $\mathcal{X}_0 \otimes \mathcal{X}_0 \to \mathcal{Y}$, suitably chosen for $\mathcal{X} = \mathcal{C}, \mathcal{J}$ and $\mathcal{Y} = \mathcal{X}_0, D_{\mathcal{X},\mathcal{X}}$ or k where $\mathcal{X}_0$ and $D_{\mathcal{X},\mathcal{X}}$ are as previously defined for composition and Jordan algebras. This requires a deep knowledge of the vector space $\mathrm{Hom}_{D_{\mathcal{X},\mathcal{X}}}(\mathcal{X}_0 \otimes \mathcal{X}_0, \mathcal{Y})$. For the selected ingredients, this vector space can be easily described (see [**14**] for the complete proof or [**1**, Theorem 4.3] in case $\mathcal{X}$ Jordan and $\mathcal{Y} = \mathcal{X}_0$):

LEMMA 2.1. *Let $\mathcal{X}$ be a unital composition or central simple Jordan algebra such that $\mathcal{X}_0 \neq 0$ and let $\mathcal{Y}$ denote $\mathcal{X}_0, D_{\mathcal{X},\mathcal{X}}$ or k. Then:*

(i) *If $\mathcal{X}$ is a composition algebra, the vector space $\mathrm{Hom}_{D_{\mathcal{X},\mathcal{X}}}(\mathcal{X}_0 \otimes \mathcal{X}_0, \mathcal{Y})$ is spanned by*
- *the skew-symmetric product $a \otimes b \mapsto [a, b] = ab - ba$ in case $\mathcal{Y} = \mathcal{X}_0$,*
- *the skew-symmetric product $a \otimes b \mapsto D_{a,b} = \mathrm{ad}[a, b] + 3(a, \cdot, b)$ in case $\mathcal{Y} = D_{\mathcal{X},\mathcal{X}}$, where (a, c, b) is as in (2.1),*
- *the symmetric product $a \otimes b \mapsto t(ab)$ in case $\mathcal{Y} = k$.*

(ii) *If $\mathcal{X}$ is a Jordan algebra of type different from* **A**, *the vector space $\mathrm{Hom}_{D_{\mathcal{X},\mathcal{X}}}(\mathcal{X}_0 \otimes \mathcal{X}_0, \mathcal{Y})$ is spanned by*
- *the symmetric product $x \otimes y \mapsto x \star y = xy - \frac{1}{n}T(xy)1$ in case $\mathcal{Y} = \mathcal{X}_0$, where n is the degree of $\mathcal{J}$,*
- *the skew-symmetric product $x \otimes y \mapsto d_{x,y} = (y, \cdot, x)$ in case $\mathcal{Y} = D_{\mathcal{X},\mathcal{X}}$, where (y, z, x) is as in (2.1),*
- *the symmetric product $x \otimes y \mapsto T(xy)$ in case $\mathcal{Y} = k$*

(iii) *If $\mathcal{X} = H(\mathcal{A}, \jmath)$ with $\mathcal{A}$ a central simple involutorial associative of second kind and center $P = k[q]$, we have that as $D_{\mathcal{X},\mathcal{X}}$-modules $\mathcal{X}_0 \cong D_{\mathcal{X},\mathcal{X}}$ and:*
- *$\mathrm{Hom}_{D_{\mathcal{X},\mathcal{X}}}(\mathcal{X}_0 \otimes \mathcal{X}_0, \mathcal{X}_0)$ is spanned by the skew-symmetric product $a \otimes b \mapsto d_{x,y} = (y, \cdot, x)$ and the symmetric $x \otimes y \mapsto q[x, y]$,*
- *$\mathrm{Hom}_{D_{\mathcal{X},\mathcal{X}}}(\mathcal{X}_0 \otimes \mathcal{X}_0, k)$ is spanned by the product $a \otimes b \mapsto T(xy)$.*

Moreover, the displayed products are nonzero except the skew-symmetric products $[a, b]$ and $D_{a,b}$ in case $\mathcal{X}$ is a 2-dimensional composition algebra, and the symmetric product $x \star y$ in case $\mathcal{X}$ is a Jordan algebra of degree 2. ◊

The previous Lemma 2.1 provides an explanation to the formula in (1.10) and points out that, for a composition algebra or a simple Jordan algebra, the vector space of symmetric or skew-symmetric products in $\mathrm{Hom}_{D_{\mathcal{X},\mathcal{X}}}(\mathcal{X}_0 \otimes \mathcal{X}_0, \mathcal{Y})$ have dimension ≤ 1 in case $\mathcal{Y} = \mathcal{X}_0, D_{\mathcal{X},\mathcal{X}}, k$ (at most 2 if we consider the whole vector space of products). Now, we will use this information to introduce a Lie algebra structure into the space $\mathcal{T}(\mathcal{C},\mathcal{J})$ described in (1.9), following Tits original idea but as general as possible. In this way, rules in (2.2) must be regarded and, for $a,b\in\mathcal{C}$, $x,y\in\mathcal{J}$, taking into account the projections of the product $[a\otimes x, b\otimes y]$ over $D_{\mathcal{C},\mathcal{C}}\oplus D_{\mathcal{J},\mathcal{J}}$ and $\mathcal{C}_0\otimes\mathcal{J}_0$, together with the anticommutativity and Lemma 2.1, we arrive at the following conclusions:

- the projection of $[a\otimes x, b\otimes y]$ over $D_{\mathcal{C},\mathcal{C}}\oplus D_{\mathcal{J},\mathcal{J}}$ must be described by a skew-symmetric product in the set

$$\mathrm{Hom}(\mathcal{C}_0\otimes\mathcal{C}_0, D_{\mathcal{C},\mathcal{C}})\otimes\mathrm{Hom}(\mathcal{J}_0\otimes\mathcal{J}_0,k)\oplus\mathrm{Hom}(\mathcal{C}_0\otimes\mathcal{C}_0,k)\otimes\mathrm{Hom}(\mathcal{J}_0\otimes\mathcal{J}_0,D_{\mathcal{J},\mathcal{J}})$$

- the projection of $[a\otimes x, b\otimes y]$ over $\mathcal{C}_0\otimes\mathcal{J}_0$ must be described by a skew-symmetric product in the set

$$\mathrm{Hom}(\mathcal{C}_0\otimes\mathcal{C}_0,\mathcal{C}_0)\otimes\mathrm{Hom}(\mathcal{J}_0\otimes\mathcal{J}_0,\mathcal{C}_0)$$

So, the most general form that this product can take is:

$$[a\otimes x, b\otimes y]=\gamma t(ab)d_{x,y}+\alpha[a,b]\otimes(xy-\frac{1}{n}T(xy))+\beta T(xy)D_{a,b} \tag{2.3}$$

for some scalars α, β and γ and n the degree of the Jordan algebra $\mathcal{J}$. Now, by imposing Jacobi identity in a similar vein to Tits' proof (see [**17**, Theorem 4.13] for a sketch), we obtain the following result which appears in [**14**]:

THEOREM 2.2. *Given $\alpha,\beta,\gamma\in k$, a composition algebra $\mathcal{C}$ and a central simple Jordan algebra $\mathcal{J}$, the vector space $\mathcal{T}(\mathcal{C},\mathcal{J};\alpha,\beta,\gamma))=D_{\mathcal{C},\mathcal{C}}\oplus\mathcal{C}_0\otimes\mathcal{J}_0\oplus D_{\mathcal{J},\mathcal{J}}$ defined in (1.9) with products as in (2.2) and (2.3) is a Lie algebra if and only if one of the following holds:*

(i) *either (2.3) is a zero product. In this case $\alpha=\beta=\gamma=0$ and $\mathcal{T}(\mathcal{C},\mathcal{J})$ is the Lie algebra obtained as the split null extension of $D_{\mathcal{C},\mathcal{C}}\oplus D_{\mathcal{J},\mathcal{J}}$ by the natural tensor module $\mathcal{C}_0\otimes\mathcal{J}_0$*

or

(ii) *the product in (2.3) is not zero and:*

(a) *$\mathcal{C}=k$ and $\gamma\neq 0$ or $\mathcal{J}=k$ and $\beta\neq 0$.*
In this case, $\mathcal{T}(\mathcal{C},k)$ provides the Lie derivation algebra of the composition algebra $\mathcal{C}$ and $\mathcal{T}(k,\mathcal{J})$ the Lie derivation algebra of the Jordan algebra $\mathcal{J}$. Hence the construction gives Lie algebras of type A,B,C or D and the exceptionals of types G_2 and F_4.

(b) *$\mathcal{C}=\mathcal{K}$, $\gamma\neq 0$, $\mathcal{K}_0=ks$ with $t(s^2)=\alpha\neq 0$ and $\mathcal{J}$ is a Jordan algebra of degree ≥ 2.*
In this case, $\mathcal{T}(\mathcal{K},\mathcal{J})$ is the λ-product-deformation Lie algebra $\mathfrak{Strl}_0(\mathcal{J})^\lambda=R_{\mathcal{J}_0}\oplus D_{\mathcal{J},\mathcal{J}}$ given by $[R_x,R_y]_\lambda=\lambda(R_xR_y-R_yR_x)$ where $\lambda=\gamma\alpha$. Hence the construction gives Lie algebras of types A,B,D and the exceptional of type E_6 or a direct sum of two copies of a Lie algebra of type A.

(c) *$\mathcal{C}=\mathcal{Q}$, $\gamma\neq 0$ and $\mathcal{J}$ is a Jordan algebra of degree $n\geq 2$ and*

$$\frac{\beta}{\gamma}=\frac{1}{2n}\quad\text{and}\qquad \gamma=2\alpha^2\quad\text{in case } n\geq 3 \tag{2.4}$$

In this case, from $\mathcal{T}(\mathcal{Q},\mathcal{J})$ we arrive at an isomorphic copy of the Lie algebra $\mathcal{L}_{\mathcal{Q}_0}(\mathcal{J},\mathcal{D}_{\mathcal{J},\mathcal{J}})=\mathcal{D}_{\mathcal{J},\mathcal{J}}\oplus\mathcal{Q}_0\otimes\mathcal{J}$ described in (1.1) and *we get Lie algebras of type A, B, C, D and the exceptional of type E_7. Moreover, in the particular case $\mathcal{Q}$ is the split quaternions algebra, the construction matches the well-known $\mathcal{TKK}$-construction of the corresponding Jordan algebra.*

(d) *$\mathcal{C} = \mathcal{O}$ and $\mathcal{J}$ is a Jordan algebra of degree 3 and the scalars α, β and γ satisfying the conditions given in (2.4).*
In this case, $\mathcal{T}(\mathcal{O}, \mathcal{J})$ provides the classical Tits cosntruction of exceptional Lie algebras and therefore models of exceptional Lie algebras of type F_4, E_6, E_7 and E_8. ♢

From the previous Theorem 2.2, the best and unique way in order to get Lie algebras using composition algebras and Jordan algebras together with their derivations algebras and preserving natural actions, goes parallel to the $\mathcal{L}_{\mathcal{S}}(\mathcal{J}, \mathcal{D})$ construction process due to Tits in [**18**].

3. Related non-associative systems

Apart from the known models of Lie algebras displayed in last section, the reviewed Tits construction $\mathcal{T}(\mathcal{C}, \mathcal{J})$ allows us to find another well-known systems related to triple products such as Lie triple systems and Jordan pairs.

A. Generalized Lie triple systems or Lie-Yamaguti algebras. All the Lie algebras obtained from the construction $\mathcal{T}(\mathcal{C}, \mathcal{J}; \alpha, \frac{1}{2n}\gamma, \gamma)$ given in Theorem 2.2 decomposes in a *reductive way*, that is, the Lie algebra $\mathcal{T}(\mathcal{C}, \mathcal{J})$ splits into subspaces $\mathfrak{h} = D_{\mathcal{C},\mathcal{C}} \oplus D_{\mathcal{J},\mathcal{J}}$ and $\mathfrak{m} = \mathcal{C}_0 \otimes \mathcal{J}_0$ satisfying

$$\begin{aligned} &[\mathfrak{h}, \mathfrak{h}] \subset \mathfrak{h} \quad (\mathfrak{h} \text{ is a subalgebra}) \\ &[\mathfrak{h}, \mathfrak{m}] \subset \mathfrak{m} \quad (\mathfrak{m} \text{ is an ad}\, \mathfrak{h}\text{-module}) \end{aligned} \tag{3.1}$$

In this situation, examples of binary-ternary algebras called generalized Lie triple systems (g.L.t.s. for short) can be obtained. These non-associative systems were introduced by Yamaguti in [**20**] and renamed as Lie-Yamaguti algebras in [**10**] (see [**2**] for a general definition) and appear by considering on the vector space $\mathfrak{m} = \mathcal{C}_0 \otimes \mathcal{J}_0$, in case $\mathfrak{m} \neq 0$, the binary and ternary products:

$$\begin{aligned} &(a \otimes x) \cdot (b \otimes y) = \alpha\, [a, b] \otimes (xy - \tfrac{1}{n} T(xy) 1) \\ &[(a \otimes x)(b \otimes y)(c \otimes z)] = \tfrac{1}{2n}\, \gamma\, T(xy)\, D_{a,b}(c) \otimes z + \gamma\, t(ab)\, c \otimes d_{x,y}(z) \end{aligned} \tag{3.2}$$

with $\gamma = 2\alpha^2$ for $n \geq 3$. As the Lie algebras $\mathcal{T}(\mathcal{C}, \mathcal{J})$ are simple (excluding split null extensions described in Theorem 2.2, item (i)), following [**2**, Proposition 1.2], from the vector space $\mathfrak{m} = \mathcal{C}_0 \otimes \mathcal{J}_0$ and the products (3.2), we get models of simple g.L.t.s.

In [**3**] irreducible g.L.t.s. are studied. Over an algebraically closed field, three different situations appear for these systems: adjoint, non-simple and generic. In the paper the so called g.L.t.s. of adjoint type (which basically are simple Lie algebras) and those of non-simple type are completely classified. For the non-simple type, the classification includes the triple systems $\mathfrak{m} = \mathcal{C}_0 \otimes \mathcal{J}_0$ described by the product in (3.2), and two further classes that can be realized by considering certain generalized extensions of the Tits construction $\mathcal{T}(\mathcal{C}, \mathcal{J})$.

B. Lie triple systems. The g.L.t.s. with trivial binary product are the so called Lie triple systems (L.t.s. for short) introduced by Jacobson in [**7**]. Without any doubt, they are the best known and studied class of g.L.t.s. due to its connection with symmetric spaces (see [**5**] for a clear explanation of this assertion). Very important examples of L.t.s. appear from non-associative systems such as Lie and Jordan algebras, Jordan pairs and Freudenthal triple systems among others, but in fact, taking associative algebras as starting point for L.t.s., Lie and Jordan algebras are the systems which allow to define Lie triple systems in the easiest way.

Any associative algebra $(\mathcal{A}, xy)$ becomes a Jordan or Lie algebra by considering the products $x \bullet y = xy + yx$ or $\{x, y\} = xy - yx$. The previous products are related by

$$\{\{x, y\}, z\} = ((y \bullet z) \bullet x) - (y \bullet (z \bullet x)) \tag{3.3}$$

and allow us to introduce a L.t.s. structure on the vector space $\mathcal{A}$ by setting $xyz = \{\{x, y\}, z\}$ as triple product (or the associator (y, z, x) with respect to the binary product $\bullet$). In this vein, two natural classes of L.t.s. appear:

L.t.s. of *Lie type*: $\mathcal{L}$ Lie algebra with triple product

$$abc = \mathrm{ad}[a, b](c) = [[a, b], c] \tag{3.4}$$

L.t.s. of *Jordan type*: $\mathcal{J}_0$ trace zero elements of $\mathcal{J}$ Jordan algebra with triple product

$$xyz = d_{x,y}(z) = (y, z, x) \tag{3.5}$$

The complete Jordan algebra $\mathcal{J}$ also works, but trace zero elements $\mathcal{J}_0$ is more convenient if we are looking for simple L.t.s. We also note that, for a given associative algebra $\mathcal{A}$, in case $\mathcal{L}$ is the Lie algebra $\mathcal{A}^-$ with respect to $\{x, y\}$, and $\mathcal{J}$ is the Jordan algebra $\mathcal{A}^+$ with respect to $x \bullet y$, Jordan and Lie types of L.t.s. agree according to (3.3). We shall refer to this particular case as L.t.s. of *Lie-Jordan type*.

On the other hand, for a given L.t.s. $(\mathcal{T}, xyz)$ and $\lambda \in k$, we can consider the triple product $[xyz]_\lambda = \lambda\, xyz$. Then, the vector space $\mathcal{T}_\lambda = \mathcal{T}$ with the previous λ-product-deformation is a L.t.s. Over an algebraically closed ground field, from [**7**, Chapter VIII, Section 1, Exercice 4], the triple systems $\mathcal{T}$ and $\mathcal{T}_\lambda$ are isomorphic for any nonzero scalar λ.

Following [**11**, Theorem 1.1], a L.t.s. can be viewed as the set of skew-symmetric elements relative to a certain automorphisms of order two (involutive authomorphism in the sequel) in a Lie algebra. In other words, a L.t.s. is nothing else but the odd part of a $\mathbb{Z}_2$-graded Lie algebra. For such an algebra $\mathcal{L} = \mathcal{L}_{\bar{0}} \oplus \mathcal{L}_{\bar{1}}$ with product $[ab]$, the triple product is defined on the vector space $\mathfrak{m} = \mathcal{L}_{\bar{1}}$ by means of :

$$abc = \mathrm{ad}_{\mathcal{L}}[ab](c) = [[ab]c] \tag{3.6}$$

Moreover, the graded-simplicity of $\mathcal{L}$ (that is, $\mathcal{L}$ simple as graded algebra), is equivalent to the simplicity of $\mathcal{L}_{\bar{1}}$ as L.t.s.

The Lie algebras $\mathcal{T}(\mathcal{K}, \mathcal{J}; \gamma)$ in item (ii)-(b) in Theorem 2.2 (note that $D_{\mathcal{K},\mathcal{K}} = 0$, so α and β can be dropped) are graded-simple with $\mathfrak{m} = \mathcal{K}_0 \otimes \mathcal{J}_0$ as odd part if they are not split null extensions. Since the characteristic is 0 (so different from 2), any 2-dimensional composition algebra is spanned by 1 and s with $t(s) = 0$ (thus

$\mathcal{K}_0 = ks$) and $s^2 = \eta 1 \neq 0$ and therefore $t(s^2) = 2\eta \neq 0$. Then, from (3.2) the triple product in $\mathfrak{m} = \mathcal{K}_0 \otimes \mathcal{J}_0$ is given by:

$$(s \otimes x)(s \otimes y)(s \otimes z) = 2\gamma\eta\ s \otimes d_{x,y}(z) = (2\gamma\eta)\ s \otimes (y, z, x) \tag{3.7}$$

Hence, identifying the subspaces $\mathcal{K}_0 \otimes \mathcal{J}_0$ and $\mathcal{J}_0$ by means of $s \otimes x \mapsto x$, the resulting triple systems are $(2\gamma\eta)$-deformations of the *Jordan type* L.t.s. $\mathcal{J}_0$. In the particular case $\mathcal{J}$ is a Jordan algebra of type **A**, we get *Lie-Jordan type* triple systems.

In Theorem 2.2, item (ii)-(c), the Lie algebras $\mathcal{T}(\mathcal{Q}, \mathcal{J};, \frac{1}{4}\gamma, \gamma)$ for Jordan algebras $\mathcal{J}$ of degree 2 (note that $xy = \frac{1}{2}T(xy)1$, so α can be dropped), provide another family of graded-simple Lie algebras and therefore, $\mathfrak{m} = \mathcal{Q}_0 \otimes \mathcal{J}_0$ becomes a simple L.t.s. Over algebraically closed fields, triple systems of this type match those which appear from $\mathbb{Z}_2$-graduations of simple orthogonal Lie algebras $\mathfrak{so}(V, b) \cong \mathfrak{so}_{\dim V}(k)$, so b is a nondegenerate symmetric form, which are inherit by orthogonal gradings $V = V_{\bar{0}} \oplus V_{\bar{1}}$, in the particular case $V_{\bar{0}}$ is a 3-dimensional vector space.

C. Jordan linear pairs. The well known $\mathcal{TKK}$-construction obtained by Tits, Koecher and Kantor in the mid sixties provides 3-graded Lie algebras

$$\mathcal{L} := \mathcal{TKK}(\mathcal{J}) = \mathcal{L}_{-1} \oplus \mathcal{L}_0 \oplus \mathcal{L}_1 \tag{3.8}$$

by using two copies of a Jordan algebra $\mathcal{J}$ as $\mathcal{L}_{\pm 1}$(see [**8**] for a complete description of this construction). From the construction, the notions of *Jordan triple systems* and *Jordan linear pairs* (J.l.p. for short) have raised (see [**16**] and [**12**] for a more general definition of Jordan pair which agrees with that of of J.l.p. over fields of characteristic different from 2, 3 following [**16**, Proposition 2.2, Chapter I]).

From [**5**, Proposition III.3.10], there is a canonical bijection between 3-graded Lie algebras and J.l.p.: in a quite natural way, any 3-graded Lie algebra produces a J.l.p. and conversely (for a short outline of this relation, see the introduction in [**13**, Section 1]). There also is a trivial relationship between 3-graded and $\mathbb{Z}_2$-graded Lie algebras: setting $\mathcal{L}_{\bar{0}} = \mathcal{L}_0$ and $\mathcal{L}_{\bar{1}} = \mathcal{L}_{-1} \oplus \mathcal{L}_1$ in (3.8), we get a $\mathbb{Z}_2$-graded Lie algebra and therefore $\mathfrak{m} = \mathcal{L}_{\bar{1}}$ becomes a *polarized L.t.s.* (for the 3-graded algebra, the pair of vector spaces $(\mathcal{L}_{-1}, \mathcal{L}_1)$ is the J.l.p. associated to the 3-graded Lie algebra $\mathcal{L}$). In fact, following [**5**, Theorem III.3.12] we have that *polarized* L.t.s., *polarized Jordan triple systems* and *Jordan pairs* are equivalent categories.

In the special case of Lie algebras in item (ii)-(c) in Theorem 2.2 with $\mathcal{Q}$ a split quaternion algebra, it is possible to find a basis $\{e, f, h\}$ in $\mathcal{Q}_0$ for which $[h, e] = he - ef = 2e, [h, f] = -2f, [e, f] = h$ (that is, the Lie algebra of zero trace split quaternions agree with the *special* Lie algebra $\mathfrak{sl}_2(k)$). Then, following [**15**], the Lie algebra

$$\mathcal{L} = \mathcal{T}(\mathcal{Q}, \mathcal{J}; \alpha, \frac{1}{2n}\gamma, \gamma) = D_{\mathcal{Q},\mathcal{Q}} \oplus \mathcal{Q}_0 \otimes \mathcal{J}_0 \oplus D_{\mathcal{J},\mathcal{J}} \tag{3.9}$$

arranged in the form:

$$\begin{aligned} \mathcal{L}_{-1} &= \operatorname{ad} f \otimes k1 \oplus f \otimes \mathcal{J}_0 && \simeq \mathcal{J} \\ \mathcal{L}_0 &= \operatorname{ad} h \otimes k1 \oplus h \otimes \mathcal{J}_0 \oplus D_{\mathcal{J},\mathcal{J}} && \\ \mathcal{L}_1 &= \operatorname{ad} e \otimes k1 \oplus e \otimes \mathcal{J}_0 && \simeq \mathcal{J} \end{aligned} \tag{3.10}$$

is nothing but a clone of the $\mathcal{TKK}$-construction of the Jordan algebra $\mathcal{J}$ (for Jordan algebras of degree $n \geq 3$, we can assume $\alpha = 1$ and $\gamma = 2$). Note that the Lie algebra $\mathcal{L}_0$ splits into a one-dimensional center $Z(\mathcal{L}_0) = \operatorname{ad} h \otimes k1$ and the derived algebra

$[\mathcal{L}_0, \mathcal{L}_0] = h \otimes \mathcal{J}_0 \oplus D_{\mathcal{J},\mathcal{J}} \simeq R_{\mathcal{J}_0} \oplus D_{\mathcal{J},\mathcal{J}} = \mathfrak{Strl}_0(\mathcal{J})$. Moreover, $\operatorname{ad} h \otimes 1$ acts as 2ε-scalar operator on $\mathcal{L}_\varepsilon$ for $\varepsilon = \pm 1$ and, for the whole action of $\mathcal{L}_0$, the subspaces $\mathcal{L}_{-1}$ and $\mathcal{L}_1$ become irreducible and contragradient $\mathcal{L}_0$-modules. So, the associated J.l.p. $(\mathcal{L}_{-1}, \mathcal{L}_1)$ is a simple Jordan pair. In fact, grading $\mathcal{L}$ as in (3.10), we get that $\mathfrak{m} = \mathcal{L}_{-1} \oplus \mathcal{L}_1$ is a simple (polarized) L.t.s.

REMARKS 3.1. Over algebraically closed fields, a complete classification of simple L.t.s. can be found in [**11**] using automorphisms of simple Lie algebras. The classification can be split into those triple systems of *Lie type* (up to isomorphism one for each family of simple Lie algebras: classical $\mathfrak{sl}_m(k)$, $\mathfrak{so}_m(k)$, $\mathfrak{sp}_{2m}(k)$ and exceptionals G_2, F_4, E_6, E_7, E_8) and those obtained as skew-symmetric elements of involutive authomorphisms in simple Lie algebras. In the former case, triple systems are given from the graduation induced through the vector spaces given by skew-symmetric and symmetric elements of the corresponding automorphism and using the triple product rule (3.6). In this way, simple L.t.s. in [**11**] are modeled as skew-symmetric elements of involutive automorphisms with triple product inherited from the Lie product of the corresponding simple Lie algebra. For classical Lie algebras, the related triple systems appear through sets of matrices.

A closer look at the classification in [**11**] shows that, for involutive automorphisms of special Lie algebras $\mathfrak{sl}_m(k)$, triple systems of *Jordan type* (one for each simple Jordan algebra of degree $n \geq 3$) together with polarized L.t.s. obtained from the serie of simple J.l.p. $(M_{p,q}(k), M_{p,q}(k))$ (the so called Jordan pairs of type $\mathrm{I}_{p,q}$ following the classification in [**12**, Chapter III, Section 17]) exahust all possibilities. These triple systems agree with the ones that appear as $\mathcal{K}_0 \otimes \mathcal{J}_0$ in (3.7) ($2\gamma\eta = 1$ could be taken), for $\mathcal{J}$ a degree $n \geq 3$ Jordan algebra of type different from **A**. The case $n = 2$ is related to some of the automorphisms of orthogonal Lie algebras $\mathfrak{so}_m(k)$.

On the other hand, pairs $(\mathcal{L}_{-1}, \mathcal{L}_1)$ obtained by decomposing the Lie algebra $\mathcal{T}(\mathcal{Q}, \mathcal{J})$ as in (3.10) produce models for all simple Jordan pairs except those of type $\mathrm{I}_{p,q}$ with $p \neq q$ and of type V described by the couple $(M_{1,2}(\mathcal{O}), M_{1,2}(\mathcal{O}^{\mathrm{op}}))$. Hence from $\mathfrak{m} = \mathcal{L}_{-1} \oplus \mathcal{L}_1$ we get all polarized simple L.t.s. up to those related with J.l.p. of types mentioned above.

References

[1] P. Benito, C. Draper, and A. Elduque, *On some algebras related to simple Lie triple systems*, J. Algebra **219** (1999), no. 1, 234–254.

[2] P. Benito, C. Draper, and A. Elduque, *Lie-Yamaguti algebras related to G_2*, J. Pure Appl. Algebra **202** (2005), 22–54.

[3] P. Benito, A. Elduque, and F. Martín-Herce, *Irreducible Lie-Yamaguti algebras*. Preprint.

[4] C.H.. Barton, A. Sudbery, *Magic squares and matrix models of Lie algebras*, Advances in Math. **180** (2003), 596–647.

[5] W. Bertram, *The geometry of Jordan and Lie structures*, Lectures Notes in Math. **1754**. Springer-Verlag, Berlin, 2003.

[6] J. E. Humphreys, *Introduction to Lie algebras and representation theory*, Springer-Verlag, New York, 1972.

[7] N. Jacobson, *General representation theory of Jordan algebras*, Trans. Amer. Math. Soc. **70** (1951), 509–530.

[8] ———, *Structure and representation of Jordan algebras*, Amer. Math. Soc. Colloquium Publications, Vol. XXXiX. Amer. Math. Soc., Providence, R.I., 1968.

[9] ———, *Lie algebras*, Dover Publications Inc., New York, 1979, Republication of the 1962 original.

[10] M. K. Kinyon and A. Weinstein, *Leibniz algebras, Courant algebroids, and multiplications on reductive homogeneous spaces*, Amer. J. Math. **123** (2001), no. 3, 525–550.

[11] W.G. Lister, *A structure theory of Lie triple systems*, Trans. Amer. Math. Soc. **72** (1952), 217–242.

[12] O. Loos, *Jordan pairs*, Springer-Verlag, Berlin 1975, Lectures Notes in Mathematics, Vol. 460.

[13] E. Neher, *3-graded Lie algebras and Jordan pairs*, Lecture Notes in Pure and Appl. Math., **211**, 296-299, Dekker, New York, 2000. Nonassociative algebra and its applications (So Paulo, 1998), 21–33.

[14] F. Martín-Herce, *Algebras de Lie-Yamaguti y sistemas algebraicos no asociativos*, Doctoral Thesis, Universidad de La Rioja, 2006.

[15] K. McCrimmon, *A taste of Jordan algebras*, Universitytext, Springer-Verlag, New York, 2004.

[16] K. Meyberg, *Lectures on algebras and triple systems*, Notes of a course given during the academic year 1971-1972. University of Virginia, Charlottesville, 1972.

[17] R.D. Schafer, *An introduction to nonassociative algebras*, Dover Publications Inc., New York, 1995, Corrected reprint of the 1966 original.

[18] J. Tits, *Une class d'algèbres de Lie en relation avec les algèbres de Jordan*, Nederl. Akad. Wetensch. proc. Ser. A **65**=Indag. Math. A **24** (1962),530–535.

[19] ———, *Algèbres alternatives, algèbres de de Jordan et algèbres de Lie exceptionelles*, Nederl. Akad. Wetensch. proc. Ser. A **69**=Indag. Math. A **28** (1966),223–237.

[20] K. Yamaguti, *On the Lie triple system and its generalization*, J. Sci. Hiroshima Univ. Ser. A **21** (1957/1958), 155–160.

Departamento de Matemáticas y Computación, Universidad de La Rioja, 26004 Logroño, Spain

E-mail address: pilar.benito@unirioja.es

Departamento de Matemáticas y Computación, Universidad de La Rioja, 26004 Logroño, Spain

E-mail address: fabian.martin@dmc.unirioja.es

Contemporary Mathematics
Volume **483**, 2009

Universal Enveloping Algebras of the Four-Dimensional Malcev Algebra

Murray R. Bremner, Irvin R. Hentzel, Luiz A. Peresi, and Hamid Usefi

Dedicated to Professor Ivan P. Shestakov in honor of his sixtieth birthday

ABSTRACT. We determine structure constants for the universal nonassociative enveloping algebra $U(\mathbb{M})$ of the four-dimensional non-Lie Malcev algebra $\mathbb{M}$ by constructing a representation of $U(\mathbb{M})$ by differential operators on the polynomial algebra $P(\mathbb{M})$. These structure constants involve Stirling numbers of the second kind. This work is based on the recent theorem of Pérez-Izquierdo and Shestakov which generalizes the Poincaré-Birkhoff-Witt theorem from Lie algebras to Malcev algebras. We use our results for $U(\mathbb{M})$ to determine structure constants for the universal alternative enveloping algebra $A(\mathbb{M}) = U(\mathbb{M})/I(\mathbb{M})$ where $I(\mathbb{M})$ is the alternator ideal of $U(\mathbb{M})$. The structure constants for $A(\mathbb{M})$ were obtained earlier by Shestakov using different methods.

1. Introduction

A Malcev algebra M over a field $\mathbb{F}$ is a vector space with a bilinear product $M \times M \to M$ denoted $(x, y) \mapsto [x, y]$, satisfying the anticommutative identity $[x, x] = 0$ and the Malcev identity $[J(x, y, z), x] = J(x, y, [x, z])$, where $J(x, y, z) = [[x, y], z] + [[y, z], x] + [[z, x], y]$. These two identities hold for the commutator $[x, y] = xy - yx$ in any alternative algebra. Basic references on Malcev algebras are [**1, 2, 3, 4, 6**].

The Poincaré-Birkhoff-Witt (PBW) theorem constructs, for any Lie algebra L, a universal associative enveloping algebra $U(L)$ together with an injective Lie algebra morphism $\iota\colon L \to U(L)^-$; thus L is isomorphic to a subalgebra of the commutator algebra of an associative algebra. It is an open problem whether every Malcev algebra is special (isomorphic to a subalgebra of the commutator algebra of an alternative algebra); see Shestakov [**7, 8, 9**], Shestakov and Zhukavets [**10, 11, 12, 13**]. A solution to a closely related problem was given a few years ago

2000 *Mathematics Subject Classification.* Primary 17D10; Secondary 17D05, 17B35, 16S32, 16W30.

We thank J. M. Pérez-Izquierdo and I. P. Shestakov for helpful comments; in particular, Shestakov sent us his structure constants for the universal alternative enveloping algebra $A(\mathbb{M})$. We thank A. Behn for telling us about the alternative algebra $\mathbb{A}$ of Table 3. Bremner, Hentzel and Peresi thank BIRS for its hospitality during our Research in Teams program in May 2005. Bremner and Usefi were partially supported by NSERC. Peresi was partially supported by CNPq.

TABLE 1. The four-dimensional Malcev algebra $\mathbb{M}$

$[\,,]$	a	b	c	d
a	0	$-b$	$-c$	d
b	b	0	$2d$	0
c	c	$-2d$	0	0
d	$-d$	0	0	0

by Pérez-Izquierdo and Shestakov [**5**]: they constructed universal nonassociative enveloping algebras for Malcev algebras.

In dimension 4, there is (up to isomorphism) a unique non-Lie Malcev algebra over any field of characteristic $\neq 2, 3$. This algebra is solvable; its structure constants appear in Table 1. We write $\mathbb{M}$ for this algebra, and M for an arbitrary Malcev algebra. In this paper we determine: (1) explicit structure constants for the universal nonassociative enveloping algebra $U(\mathbb{M})$; (2) a finite set of generators for the alternator ideal $I(\mathbb{M}) \subset U(\mathbb{M})$; (3) explicit structure constants for the universal alternative enveloping algebra $A(\mathbb{M}) = U(\mathbb{M})/I(\mathbb{M})$. Shestakov [**8**, Example 1] found the structure constants for $A(\mathbb{M})$ as an application of Malcev Poisson algebras. Zhelyabin and Shestakov [**14**] proved that if M is finite dimensional and semisimple then $U(M)$ is a free module over its center and that the center is isomorphic to a polynomial algebra on n variables where n is the dimension of the Cartan subalgebra; they also calculate the center of $U(M)$ for several Malcev algebras of small dimension. In the case $M = \mathbb{M}$, the center can be obtained as a corollary to our structure constants for $U(\mathbb{M})$.

2. Preliminary results

All multilinear structures are over a field $\mathbb{F}$ with $\operatorname{char}\mathbb{F} \neq 2, 3$.

DEFINITION 2.1. The *generalized alternative nucleus* of a nonassociative algebra A is

$$N_{\text{alt}}(A) = \{\, a \in A \mid (a, x, y) = -(x, a, y) = (x, y, a),\ \forall\, x, y \in A \,\},$$

where the *associator* is $(x, y, z) = (xy)z - x(yz)$.

LEMMA 2.2. *In general $N_{\text{alt}}(A)$ is not a subalgebra of A, but it is a subalgebra of A^- and is a Malcev algebra.*

THEOREM 2.3 (Pérez-Izquierdo and Shestakov). *For every Malcev algebra M there is a universal nonassociative enveloping algebra $U(M)$ and an injective morphism $\iota\colon M \to U(M)^-$ with $\iota(M) \subseteq N_{alt}(U(M))$.*

Let $F(M)$ be the unital free nonassociative algebra on a basis of M. Let $R(M)$ be the ideal generated by the elements $ab - ba - [a, b]$, $(a, x, y) + (x, a, y)$, $(x, a, y) + (x, y, a)$ for all $a, b \in M$, $x, y \in F(M)$. Define $U(M) = F(M)/R(M)$, and the mapping $\iota\colon M \to U(M)$ by $a \mapsto \iota(a) = \overline{a} = a + R(M)$. Since ι is injective, we identify M with $\iota(M) \subseteq N_{\text{alt}}(U(M))$. Let $B = \{a_i \mid i \in \mathcal{I}\}$ be a basis of M with $<$ a total order on the index set $\mathcal{I}$. Define $\Omega = \{\, (i_1, \dots, i_n) \mid i_1 \leq \cdots \leq i_n \,\}$. The empty tuple $\emptyset$ ($n = 0$) gives $\overline{a}_\emptyset = 1 \in U(M)$. The n-tuple $I = (i_1, \dots, i_n) \in \Omega$ ($n \geq 1$) defines a left-tapped monomial $\overline{a}_I = \overline{a}_{i_1}(\overline{a}_{i_2}(\cdots(\overline{a}_{i_{n-1}}\overline{a}_{i_n})\cdots))$ of degree

$|\overline{a}_I| = n$. The set $\{\, \overline{a}_I \mid I \in \Omega \,\}$ is a basis of $U(M)$. For details, see Pérez-Izquierdo and Shestakov [**5**].

For any $f, g \in M$ and $y \in U(M)$, since $f, g \in N_{\mathrm{alt}}(U(M))$ we obtain

$$(f, g, y) = \tfrac{1}{6}[[y, f], g] - \tfrac{1}{6}[[y, g], f] - \tfrac{1}{6}[[y, [f, g]].$$

This equation implies the next three lemmas, which are implicit in [**5**]. We first compute $[x, f]$ in $U(M)$; for $|x| = 1$ we use the bracket in M.

LEMMA 2.4. *Let x be a basis monomial of $U(M)$ with $|x| \geq 2$, and let f be an element of M. Write $x = gy$ with $g \in M$. Then*

$$[x, f] = [gy, f] = [g, f]y + g[y, f] + \tfrac{1}{2}[[y, f], g] - \tfrac{1}{2}[[y, g], f] - \tfrac{1}{2}[y, [f, g]].$$

We next compute fx in $U(M)$; for $|x| = 1$ we have two cases: if $f \leq x$ in the ordered basis, then fx is a basis monomial; otherwise, $fx = xf + [f, x]$ where $[f, x] \in M$.

LEMMA 2.5. *Let x be a basis monomial of $U(M)$ with $|x| \geq 2$, and let f be an element of M. Write $x = gy$ with $g \in M$. Then*

$$fx = f(gy) = g(fy) + [f, g]y - \tfrac{1}{3}[[y, f], g] + \tfrac{1}{3}[[y, g], f] + \tfrac{1}{3}[y, [f, g]].$$

We finally compute yz in $U(M)$; for $|y| = 1$ we use Lemma 2.5.

LEMMA 2.6. *Let y and z be basis monomials of $U(M)$ with $|y| \geq 2$. Write $y = fx$ with $f \in M$. Then*

$$yz = (fx)z = 2f(xz) - x(fz) - x[z, f] + [xz, f].$$

Expansion in the free nonassociative algebra establishes the identity

$$(pq, r, s) - (p, qr, s) + (p, q, rs) = p(q, r, s) + (p, q, r)s.$$

From this equation the next lemma easily follows.

LEMMA 2.7. *For all $g \in M$ and $x \in U(M)$ we have*

$$(g^i, g, x) = (g^i, x, g) = (g, g^i, x) = (g, x, g^i) = (x, g^i, g) = (x, g, g^i) = 0.$$

From this, induction gives $(g^j, g^i, x) = 0$ and hence $[g^k x, g] = g^k[x, g]$.

The algebra $\mathbb{M}$ has solvable Lie subalgebras with bases $\{a, b\}$, $\{a, c\}$, $\{a, d\}$, and a nilpotent Lie subalgebra with basis $\{b, c, d\}$. The next two lemmas are standard computations in enveloping algebras.

LEMMA 2.8. *For $e \in \{b, c\}$ these equations hold in $U(\mathbb{M})$:*

$$(a^i e^j)(a^k e^\ell) = a^i (a{+}j)^k e^{j+\ell}, \quad (a^i d^j)(a^k d^\ell) = a^i (a{-}j)^k d^{j+\ell}.$$

LEMMA 2.9. *These equations hold in $U(\mathbb{M})$:*

$$(b^i c^j d^k)(b^\ell c^m d^n) = \sum_{h=0}^{\ell} (-1)^h 2^h \binom{\ell}{h} \frac{j!}{(j{-}h)!} b^{i+\ell-h} c^{j+m-h} d^{k+n+h},$$

$$[b^i c^j d^k, b] = -2j b^i c^{j-1} d^{k+1}, \quad [b^i c^j d^k, c] = 2i b^{i-1} c^j d^{k+1}, \quad [b^i c^j d^k, d] = 0.$$

The following representation will play an important role in our computation of the structure constants for $U(\mathbb{M})$.

TABLE 2. Differential operators $\rho(x)$ and $L(x)$ on $P(\mathbb{M})$

x	$\rho(x)$	$L(x)$
a	$M_bD_b + M_cD_c - M_dD_d - 3M_dD_bD_c$	M_a
b	$(I{-}S)M_b + \left(S{-}I{-}2S^{-1}\right) M_dD_c$	$SM_b + \left(S^{-1}{-}S\right) M_dD_c$
c	$(I{-}S)M_c + \left(S{-}I{+}2S^{-1}\right) M_dD_b$	$SM_c - \left(S^{-1}{+}S\right) M_dD_b$
d	$\left(I{-}S^{-1}\right) M_d$	$S^{-1}M_d$

DEFINITION 2.10. Let M be a Malcev algebra, and let $P(M)$ be the polynomial algebra on a basis of M. By Theorem 2.3 we have a linear isomorphism $\phi\colon U(M) \to P(M)$ defined by

$$\overline{a}_{i_1}(\cdots(\overline{a}_{i_{n-2}}(\overline{a}_{i_{n-1}}\overline{a}_{i_n}))\cdots) \longmapsto a_{i_1}\cdots a_{i_{n-2}}a_{i_{n-1}}a_{i_n}.$$

For $x \in U(M)$, $f \in P(M)$ we define the *right bracket operator* ρ and the *left multiplication operator* L as follows:

$$\rho(x)(f) = \phi\big([\phi^{-1}(f), x]\big), \qquad L(x)(f) = \phi\big(x\phi^{-1}(f)\big).$$

Thus $\rho(x)$ (respectively $L(x)$) is the operator on $P(M)$ induced by the mapping $y \mapsto [y,x]$ (respectively $y \mapsto xy$) in $U(M)$. We also have the *right multiplication operator* $R(x) = \rho(x) + L(x)$.

3. Representation of $\mathbb{M}$ by differential operators

DEFINITION 3.1. We have these operators on $P(\mathbb{M})$: I is the *identity*; M_x is *multiplication* by $x \in \{a,b,c,d\}$; D_x is *differentiation* with respect to $x \in \{a,b,c,d\}$; S is the *shift* $a \mapsto a{+}1$: $S(a^ib^jc^kd^\ell) = (a{+}1)^ib^jc^kd^\ell$. Since S is invertible, S^t is defined for all $t \in \mathbb{Z}$.

In this Section we determine $\rho(x)$ and $L(x)$ for $x \in \{a,b,c,d\}$ as differential operators on $P(\mathbb{M})$. We summarize our results in Table 2.

LEMMA 3.2. *For $x,y \in \{a,b,c,d\}$ we have*

$$\begin{aligned}
&[D_x,M_x] = I, \qquad [D_x,M_y] = 0\ (x \neq y),\ [D_x,D_y] = 0,\ [M_x,M_y] = 0,\\
&[M_a,S] = -S, \qquad [M_x,S] = 0\ (x \neq a), \quad [D_x,S] = 0,\ [D_x,S^{-1}] = 0,\\
&[M_a,S^{-1}] = S^{-1},\ [M_x,S^{-1}] = 0\ (x \neq a).
\end{aligned}$$

PROOF. These follow easily from Definition 3.1. □

LEMMA 3.3. *We have $[b^nc^pd^q,a] = (n{+}p{-}q)b^nc^pd^q - 3npb^{n-1}c^{p-1}d^{q+1}$.*

PROOF. Induction on n; the basis $n=0$ is $[c^pd^q,a] = (p{-}q)c^pd^q$, which follows since a,c,d span a Lie subalgebra of $\mathbb{M}$. We now let $n \geq 0$ and use Lemma 2.4 with $f=a$, $g=b$; we see that $[b^{n+1}c^pd^q,a]$ equals

$$[b,a]b^nc^pd^q + b[b^nc^pd^q,a] + \tfrac{1}{2}\big([[b^nc^pd^q,a],b] - [[b^nc^pd^q,b],a] - [b^nc^pd^q,[a,b]]\big).$$

We apply Lemma 2.9 to the right side:

$$b^{n+1}c^pd^q + b[b^nc^pd^q,a] + \tfrac{1}{2}[[b^nc^pd^q,a],b] + p[b^nc^{p-1}d^{q+1},a] - pb^nc^{p-1}d^{q+1}.$$

The inductive hypothesis gives

$$\begin{aligned}&b^{n+1}c^pd^q + (n+p-q)b^{n+1}c^pd^q - 3npb^nc^{p-1}d^{q+1} + \tfrac{1}{2}(n+p-q)[b^nc^pd^q, b]\\&- \tfrac{3}{2}np[b^{n-1}c^{p-1}d^{q+1}, b] + (n+p-q-2)pb^nc^{p-1}d^{q+1}\\&- 3np(p-1)b^{n-1}c^{p-2}d^{q+2} - pb^nc^{p-1}d^{q+1}.\end{aligned}$$

We use Lemma 2.9 again to get

$$\begin{aligned}&b^{n+1}c^pd^q + (n+p-q)b^{n+1}c^pd^q - 3np\,b^nc^{p-1}d^{q+1} - (n+p-q)pb^nc^{p-1}d^{q+1}\\&+ 3np(p-1)b^{n-1}c^{p-2}d^{q+2} + (n+p-q-2)pb^nc^{p-1}d^{q+1}\\&- 3np(p-1)b^{n-1}c^{p-2}d^{q+2} - pb^nc^{p-1}d^{q+1}.\end{aligned}$$

Combining terms gives $(n+1+p-q)b^{n+1}c^pd^q - 3(n+1)pb^nc^{p-1}d^{q+1}$. □

LEMMA 3.4. *We have*

$$\rho(a) = M_bD_b + M_cD_c - M_dD_d - 3M_dD_bD_c, \qquad L(a) = M_a.$$

PROOF. Lemma 2.7 gives $[a^mb^nc^pd^q, a] = a^m[b^nc^pd^q, a]$, and now Lemma 3.3 gives the formula for $\rho(a)$. The formula for $L(a)$ is clear. □

LEMMA 3.5. *We have*

$$\rho(b) = \big(I-S\big)M_b + \big(S-I-2S^{-1}\big)M_dD_c, \;\; L(b) = SM_b + \big(S^{-1}-S\big)M_dD_c.$$

PROOF. Induction on m where $y = a^mb^nc^pd^q$. We prove the formulas together, since each requires the inductive hypothesis of the other. The basis $m = 0$ for $\rho(b)$ is Lemma 2.9, and for $L(b)$ it is clear. We assume both formulas for $m \geq 0$. Lemma 2.4 ($f = b$, $g = a$) gives

$$\begin{aligned}\rho(b)(ay) &= -by + a[y,b] + \tfrac{1}{2}\big([[y,b],a] - [[y,a],b] - [y,b]\big)\\&= \big(-L(b) + M_a\rho(b) + \tfrac{1}{2}[\rho(a),\rho(b)] - \tfrac{1}{2}\rho(b)\big)(y).\end{aligned}$$

The inductive hypothesis for $\rho(b)$, Lemma 3.4 and Lemma 3.2 give

$$[\rho(a),\rho(b)](y) = \big((I-S)M_b + (S-I+4S^{-1})M_dD_c\big)(y).$$

Combining this with both inductive hypotheses we get

$$\begin{aligned}\rho(b)(ay) &= -\big(SM_b + (S^{-1}-S)M_dD_c\big)(y)\\&\quad + M_a\big((I-S)M_b + (S-I-2S^{-1})M_dD_c\big)(y)\\&\qquad + \tfrac{1}{2}\big((I-S)M_b + (S-I+4S^{-1})M_dD_c\big)(y)\\&\qquad\quad - \tfrac{1}{2}\big((I-S)M_b + (S-I-2S^{-1})M_dD_c\big)(y)\\&= \big(M_a - (M_a+I)S\big)M_b(y)\\&\quad + \big((M_a+I)S - M_a - 2(M_a-I)S^{-1}\big)M_dD_c(y)\\&= (I-S)M_b(ay) + (S-I-2S^{-1})M_dD_c(ay),\end{aligned}$$

which completes the proof for $\rho(b)$. Lemma 2.5 with $f = b$, $g = a$ gives

$$\begin{aligned}L(b)(ay) &= (a+1)(by) + \tfrac{1}{3}\big([[y,a],b] - [[y,b],a] + [y,b]\big)\\&= \big((a+1)L(b) - \tfrac{1}{3}[\rho(a),\rho(b)] + \tfrac{1}{3}\rho(b)\big)\,(y).\end{aligned}$$

Using the inductive hypotheses for $L(b)$ and $\rho(b)$ we get

$$L(b)(ay) = (a+1)SM_b(y) + (a+1)(S^{-1}-S)M_dD_c(y)$$

$$
\begin{aligned}
&- \tfrac{1}{3}\big((I-S)M_b + (S-I+4S^{-1})M_dD_c\big)(y)\\
&\qquad + \tfrac{1}{3}\big((I-S)M_b + (S-I-2S^{-1})M_dD_c\big)(y)\\
&= (a+1)SM_b(y) + \big((a-1)S^{-1} - (a+1)S\big)M_dD_c(y)\\
&= SM_b(ay) + (S^{-1}-S)M_dD_c(ay),
\end{aligned}
$$

which completes the proof for $L(b)$. □

LEMMA 3.6. *We have*

$$\rho(c) = \big(I-S\big)M_c + \big(S-I+2S^{-1}\big)M_dD_b, \quad L(c) = SM_c - \big(S+S^{-1}\big)M_dD_b.$$

PROOF. Similar to the proof of Lemma 3.5. □

LEMMA 3.7. *We have* $\rho(d) = \big(I-S^{-1}\big)M_d$ *and* $L(d) = S^{-1}M_d$.

PROOF. Induction on m where $y = a^m b^n c^p d^q$. We prove both formulas together. The basis $m = 0$ is Lemma 2.9. We assume both formulas for $m \geq 0$. Lemma 2.4 with $f = d$, $g = a$ gives

$$
\begin{aligned}
\rho(d)(ay) &= dy + a[y,d] + \tfrac{1}{2}\big([[y,d],a] - [[y,a],d] + [y,d]\big)\\
&= \big(L(d) + M_a\rho(d) + \tfrac{1}{2}[\rho(a),\rho(d)] + \tfrac{1}{2}\rho(d)\big)(y).
\end{aligned}
$$

The inductive hypothesis gives $[\rho(a),\rho(d)](y) = -\rho(d)(y)$ and so

$$\rho(d)(ay) = \big(L(d) + M_a\rho(d)\big)(y) = (I-S^{-1})M_d(ay),$$

which completes the proof for $\rho(d)$. Lemma 2.5 with $f = d$, $g = a$ gives

$$
\begin{aligned}
L(d)(ay) &= a(dy) - dy - \tfrac{1}{3}[[y,d],a] + \tfrac{1}{3}[[y,a],d] - \tfrac{1}{3}[y,d]\\
&= \big(M_aL(d) - L(d) + \tfrac{1}{3}[\rho(d),\rho(a)] - \tfrac{1}{3}\rho(d)\big)(y)\\
&= \big(M_aL(d) - L(d)\big)(y) = (M_a-I)S^{-1}M_d(y) = S^{-1}M_d(ay),
\end{aligned}
$$

which completes the proof for $L(d)$. □

4. Representation of $U(\mathbb{M})$ by differential operators

In this Section we determine $L(x)$ for $x = a^i b^j c^k d^\ell$ as a differential operator on $P(\mathbb{M})$. We often use the facts that linear operators E, F, G satisfy $[E, FG] = [E,F]G + F[E,G]$, and that if $[[E,F],F] = 0$ then $[E,F^k] = k[E,F]F^{k-1}$ for every $k \geq 1$.

Since c, d span an Abelian Lie subalgebra $\mathbb{A} \subset \mathbb{M}$, associativity gives $L(c^k d^\ell) = L(c)^k L(d)^\ell$ on $U(\mathbb{A})$; this is also true on $U(\mathbb{M})$.

LEMMA 4.1. *In* $U(\mathbb{M})$ *we have* $L(c^k d^\ell) = L(c)^k L(d)^\ell$.

PROOF. We first prove $L(d^\ell) = L(d)^\ell$ by induction. For $\ell \geq 1$ we get

$$(dd^\ell)(a^m b^n c^p d^q) = (d, d^\ell, a^m b^n c^p d^q) + d\big((d^\ell)(a^m b^n c^p d^q)\big).$$

The associator is zero by Lemma 2.7. We now use induction on k. Lemma 2.6 with $f = c$, $x = c^k d^\ell$ gives

$$(c^{k+1}d^\ell)z = 2c((c^k d^\ell)z) - (c^k d^\ell)(cz) - (c^k d^\ell)[z,c] + [(c^k d^\ell)z, c],$$

which can be written as

$$L(c^{k+1}d^\ell) = L(c)L(c^k d^\ell) + [L(c), L(c^k d^\ell)] + [\rho(c), L(c^k d^\ell)].$$

The inductive hypothesis gives

$$[\rho(c), L(c^k d^\ell)] = L(c)^k [\rho(c), L(d)^\ell] + [\rho(c), L(c)^k] L(d)^\ell = 0,$$

and similarly $[L(c), L(c^k d^\ell)] = 0$. □

Since b, c, d span a nilpotent Lie subalgebra $\mathbb{N} \subset \mathbb{M}$, associativity gives $L(b^j c^k d^\ell) = L(b)^j L(c)^k L(d)^\ell$ on $U(\mathbb{N})$; this is not true on $U(\mathbb{M})$.

LEMMA 4.2. *In $U(\mathbb{M})$ the operator $L(b^j c^k d^\ell)$ equals*

$$\sum_{\alpha=0}^{\min(j,k)} \sum_{\beta=0}^{\alpha} (-1)^{\alpha-\beta} \alpha! \binom{\alpha}{\beta} \binom{j}{\alpha} \binom{k}{\alpha} S^{-\beta} L(b)^{j-\alpha} L(c)^{k-\alpha} M_d^\alpha L(d)^\ell.$$

PROOF. Induction on j; the basis is Lemma 4.1. Lemma 2.6 with $f = b$, $x = b^j c^k d^\ell$, $z = a^m b^n c^p d^q$ gives

$$\begin{aligned}
(b^{j+1} c^k d^\ell)(a^m b^n c^p d^q) &= 2b(xz) - x(bz) - x[z, b] + [xz, b] \\
&= 2L(b)L(x)z - L(x)L(b)z - L(x)\rho(b)z + \rho(b)L(x)z \\
&= L(b)L(x)z + [L(b), L(x)]z + [\rho(b), L(x)]z \\
&= L(b)L(x)z + [R(b), L(x)]z.
\end{aligned}$$

Induction and $[R(b), L(b)] = [R(b), M_d] = [R(b), L(d)] = 0$ show that $[R(b), L(x)]$ equals

$$\sum_{\alpha=0}^{\min(j,k)} \alpha! \binom{j}{\alpha} \binom{k}{\alpha} \left(S^{-1} - I\right)^\alpha L(b)^{j-\alpha} \left[R(b), L(c)^{k-\alpha}\right] M_d^\alpha L(d)^\ell =$$

$$\sum_{\alpha=0}^{\min(j,k)} \alpha! \binom{j}{\alpha} (k-\alpha) \binom{k}{\alpha} \left(S^{-1} - I\right)^{\alpha+1} L(b)^{j-\alpha} L(c)^{k-\alpha-1} M_d^{\alpha+1} L(d)^\ell.$$

Replacing α by $\alpha-1$ gives

$$\sum_{\alpha=1}^{\min(j+1,k)} \alpha! \binom{j}{\alpha-1} \binom{k}{\alpha} \left(S^{-1} - I\right)^\alpha L(b)^{j+1-\alpha} L(c)^{k-\alpha} M_d^\alpha L(d)^\ell.$$

We use Pascal's identity $\binom{j}{\alpha} + \binom{j}{\alpha-1} = \binom{j+1}{\alpha}$ to combine $L(b)L(x)$ and $[R(b), L(x)]$, and obtain this formula for $L(b^{j+1} c^k d^\ell)$:

$$\sum_{\alpha=0}^{\min(j+1,k)} \alpha! \binom{j+1}{\alpha} \binom{k}{\alpha} \left(S^{-1} - I\right)^\alpha L(b)^{j+1-\alpha} L(c)^{k-\alpha} M_d^\alpha L(d)^\ell.$$

We now expand $(S^{-1} - I)^\alpha$ with the binomial theorem. □

LEMMA 4.3. *We have*

$$\begin{aligned}
&[R(a), L(a)^s S^t L(b)^u D_b^v D_c^w L(c)^x M_d^y L(d)^z] = \\
&\quad - (t+v+w+y) L(a)^s S^t L(b)^u D_b^v D_c^w L(c)^x M_d^y L(d)^z \\
&\qquad - u L(a)^s S^{t-1} L(b)^{u-1} D_b^v D_c^{w+1} L(c)^x M_d^{y+1} L(d)^z \\
&\qquad\quad + x L(a)^s S^{t-1} L(b)^u D_b^{v+1} D_c^w L(c)^{x-1} M_d^{y+1} L(d)^z.
\end{aligned}$$

PROOF. Table 2 and Lemma 3.2 give

$$\begin{aligned} &[R(a), L(a)] = 0, && [R(a), L(b)] = -S^{-1}M_dD_c, \\ &[R(a), L(c)] = S^{-1}M_dD_b, && [R(a), L(d)] = 0, \\ &[R(a), D_b] = -D_b, && [R(a), D_c] = -D_c, \\ &[R(a), M_d] = -M_d, && [R(a), S] = -S. \end{aligned}$$

From these equations we get

$$\begin{aligned} &[R(a), L(a)^sS^tL(b)^uD_b^vD_c^wL(c)^xM_d^yL(d)^z] = \\ &\quad L(a)^s\big[R(a), S^t\big]L(b)^uD_b^vD_c^wL(c)^xM_d^yL(d)^z \\ &\quad + L(a)^sS^t\big[R(a), L(b)^u\big]D_b^vD_c^wL(c)^xM_d^yL(d)^z \\ &\quad + L(a)^sS^tL(b)^u\big[R(a), D_b^v\big]D_c^wL(c)^xM_d^yL(d)^z \\ &\quad + L(a)^sS^tL(b)^uD_b^v\big[R(a), D_c^w\big]L(c)^xM_d^yL(d)^z \\ &\quad + L(a)^sS^tL(b)^uD_b^vD_c^w\big[R(a), L(c)^x\big]M_d^yL(d)^z \\ &\quad + L(a)^sS^tL(b)^uD_b^vD_c^wL(c)^x\big[R(a), M_d^y\big]L(d)^z. \end{aligned}$$

The right side simplifies to

$$\begin{aligned} &-tL(a)^sS^tL(b)^uD_b^vD_c^wL(c)^xM_d^yL(d)^z \\ &-uL(a)^sS^tL(b)^{u-1}S^{-1}D_cM_dD_b^vD_c^wL(c)^xM_d^yL(d)^z \\ &-vL(a)^sS^tL(b)^uD_b^vD_c^wL(c)^xM_d^yL(d)^z \\ &-wL(a)^sS^tL(b)^uD_b^vD_c^wL(c)^xM_d^yL(d)^z \\ &+xL(a)^sS^tL(b)^uD_b^vD_c^wL(c)^{x-1}S^{-1}D_bM_dM_d^yL(d)^z \\ &-yL(a)^sS^tL(b)^uD_b^vD_c^wL(c)^xM_d^yL(d)^z, \end{aligned}$$

which gives the result. □

LEMMA 4.4. *In $U(\mathbb{M})$ the operator $L(a^ib^jc^kd^\ell)$ equals*

$$\sum_{\alpha=0}^{\min(j,k)} \sum_{\beta=0}^{\alpha} \sum_{\gamma=0}^{i} \sum_{\delta=0}^{i-\gamma} \sum_{\epsilon=0}^{i-\gamma-\delta} (-1)^{i+\alpha-\beta-\gamma-\delta} \alpha!\delta!\epsilon! \binom{\alpha}{\beta}\binom{j}{\alpha,\epsilon}\binom{k}{\alpha,\delta} \times$$
$$X_i(\gamma,\delta,\epsilon)L(a)^\gamma S^{-\beta-\delta-\epsilon}L(b)^{j-\alpha-\epsilon}D_b^\delta D_c^\epsilon L(c)^{k-\alpha-\delta}M_d^{\alpha+\delta+\epsilon}L(d)^\ell,$$

where $X_i(\gamma,\delta,\epsilon)$ is a polynomial in $\alpha-\beta$ satisfying the recurrence

$$\begin{aligned} &X_{i+1}(\gamma,\delta,\epsilon) = \\ &(\alpha-\beta+\delta+\epsilon)X_i(\gamma,\delta,\epsilon) + X_i(\gamma-1,\delta,\epsilon) + X_i(\gamma,\delta-1,\epsilon) + X_i(\gamma,\delta,\epsilon-1), \end{aligned}$$

with $X_0(0,0,0) = 1$ and $X_i(\gamma,\delta,\epsilon) = 0$ unless $0 \le \gamma \le i$, $0 \le \delta \le i-\gamma$, $0 \le \epsilon \le i-\gamma-\delta$.

PROOF. Induction on i; the basis $i = 0$ is Lemma 4.2. Lemma 2.6 with $f = a$, $x = a^ib^jc^kd^\ell$, $z = a^mb^nc^pd^q$ gives

$$(a^{i+1}b^jc^kd^l)(a^mb^nc^pd^q) = L(a)L(x)z + [R(a), L(x)]z.$$

Induction and Lemma 4.3 give $[R(a), L(x)] = A + B + C$ where

$$A = -\sum_{\alpha=0}^{\min(j,k)} \sum_{\beta=0}^{\alpha} \sum_{\gamma=0}^{i} \sum_{\delta=0}^{i-\gamma} \sum_{\epsilon=0}^{i-\gamma-\delta} (-1)^{i+\alpha-\beta-\gamma-\delta} \times$$
$$(\alpha-\beta+\delta+\epsilon)\alpha!\delta!\epsilon! \binom{\alpha}{\beta} \binom{j}{\alpha, \epsilon} \binom{k}{\alpha, \delta} X_i(\gamma, \delta, \epsilon) \times$$
$$L(a)^\gamma S^{-\beta-\delta-\epsilon} L(b)^{j-\alpha-\epsilon} D_b^\delta D_c^\epsilon L(c)^{k-\alpha-\delta} M_d^{\alpha+\delta+\epsilon} L(d)^\ell,$$

$$B = -\sum_{\alpha=0}^{\min(j,k)} \sum_{\beta=0}^{\alpha} \sum_{\gamma=0}^{i} \sum_{\delta=0}^{i-\gamma} \sum_{\epsilon=0}^{i-\gamma-\delta} (-1)^{i+\alpha-\beta-\gamma-\delta} \times$$
$$(j-\alpha-\epsilon)\alpha!\delta!\epsilon! \binom{\alpha}{\beta} \binom{j}{\alpha, \epsilon} \binom{k}{\alpha, \delta} X_i(\gamma, \delta, \epsilon) \times$$
$$L(a)^\gamma S^{-\beta-\delta-\epsilon-1} L(b)^{j-\alpha-\epsilon-1} D_b^\delta D_c^{\epsilon+1} L(c)^{k-\alpha-\delta} M_d^{\alpha+\delta+\epsilon+1} L(d)^\ell,$$

$$C = \sum_{\alpha=0}^{\min(j,k)} \sum_{\beta=0}^{\alpha} \sum_{\gamma=0}^{i} \sum_{\delta=0}^{i-\gamma} \sum_{\epsilon=0}^{i-\gamma-\delta} (-1)^{i+\alpha-\beta-\gamma-\delta} \times$$
$$(k-\alpha-\delta)\alpha!\delta!\epsilon! \binom{\alpha}{\beta} \binom{j}{\alpha, \epsilon} \binom{k}{\alpha, \delta} X_i(\gamma, \delta, \epsilon) \times$$
$$L(a)^\gamma S^{-\beta-\delta-\epsilon-1} L(b)^{j-\alpha-\epsilon} D_b^{\delta+1} D_c^\epsilon L(c)^{k-\alpha-\delta-1} M_d^{\alpha+\delta+\epsilon+1} L(d)^\ell.$$

We write $D = L(a)L(x)$ and obtain

$$D = \sum_{\alpha=0}^{\min(j,k)} \sum_{\beta=0}^{\alpha} \sum_{\gamma=0}^{i} \sum_{\delta=0}^{i-\gamma} \sum_{\epsilon=0}^{i-\gamma-\delta} (-1)^{i+\alpha-\beta-\gamma-\delta} \times$$
$$\alpha!\delta!\epsilon! \binom{\alpha}{\beta} \binom{j}{\alpha, \epsilon} \binom{k}{\alpha, \delta} X_i(\gamma, \delta, \epsilon) \times$$
$$L(a)^{\gamma+1} S^{-\beta-\delta-\epsilon} L(b)^{j-\alpha-\epsilon} D_b^\delta D_c^\epsilon L(c)^{k-\alpha-\delta} M_d^{\alpha+\delta+\epsilon} L(d)^\ell.$$

In A, we include the term (which is zero) for $\epsilon = i+1-\gamma-\delta$, and absorb the minus sign. In B we replace ϵ by $\epsilon-1$, include the term for $\epsilon = 0$, simplify the coefficient using $(j-\alpha-\epsilon+1)(\epsilon-1)!\binom{j}{\alpha,\epsilon-1} = \epsilon!\binom{j}{\alpha,\epsilon}$, and absorb the minus sign. In C we replace δ by $\delta-1$, include the term for $\delta = 0$, and simplify the coefficient using $(k-\alpha-\delta+1)(\delta-1)!\binom{k}{\alpha,\delta-1} = \delta!\binom{k}{\alpha,\delta}$. In D we replace γ by $\gamma-1$, and include the term for $\gamma = 0$. We find that $A + B + C + D$ equals

$$\sum_{\alpha=0}^{\min(j,k)} \sum_{\beta=0}^{\alpha} \sum_{\gamma=0}^{i+1} \sum_{\delta=0}^{i+1-\gamma} \sum_{\epsilon=0}^{i+1-\gamma-\delta} (-1)^{i+1+\alpha-\beta-\gamma-\delta} \times$$
$$\alpha!\delta!\epsilon! \binom{\alpha}{\beta} \binom{j}{\alpha, \epsilon} \binom{k}{\alpha, \delta} X_{i+1}(\gamma, \delta, \epsilon) \times$$
$$L(a)^\gamma S^{-\beta-\delta-\epsilon} L(b)^{j-\alpha-\epsilon} D_b^\delta D_c^\epsilon L(c)^{k-\alpha-\delta} M_d^{\alpha+\delta+\epsilon} L(d)^\ell,$$

where $X_i(\gamma, \delta, \epsilon)$ satisfies the stated recurrence relation. □

DEFINITION 4.5. The *Stirling numbers of the second kind* are

$$\left\{ {r \atop s} \right\} = \frac{1}{s!} \sum_{t=0}^{s} (-1)^{s-t} \binom{s}{t} t^r.$$

LEMMA 4.6. *The unique solution to the recurrence of Lemma 4.4 is*

$$X_i(\gamma, \delta, \epsilon) = \binom{\delta+\epsilon}{\epsilon} \sum_{\zeta=0}^{i-\gamma-\delta-\epsilon} \binom{i}{\gamma, \zeta} \left\{ {i-\gamma-\zeta \atop \delta+\epsilon} \right\} (\alpha-\beta)^\zeta.$$

PROOF. The right side of the recurrence is the sum of these five terms:

$$(\alpha-\beta) X_i(\gamma, \delta, \epsilon) = \binom{\delta+\epsilon}{\epsilon} \sum_{\zeta=1}^{i+1-\gamma-\delta-\epsilon} \binom{i}{\gamma, \zeta-1} \left\{ {i+1-\gamma-\zeta \atop \delta+\epsilon} \right\} (\alpha-\beta)^\zeta,$$

$$(\delta+\epsilon) X_i(\gamma, \delta, \epsilon) = \binom{\delta+\epsilon}{\epsilon} \sum_{\zeta=0}^{i-\gamma-\delta-\epsilon} \binom{i}{\gamma, \zeta} (\delta+\epsilon) \left\{ {i-\gamma-\zeta \atop \delta+\epsilon} \right\} (\alpha-\beta)^\zeta,$$

$$X_i(\gamma-1, \delta, \epsilon) = \binom{\delta+\epsilon}{\epsilon} \sum_{\zeta=0}^{i+1-\gamma-\delta-\epsilon} \binom{i}{\gamma-1, \zeta} \left\{ {i+1-\gamma-\zeta \atop \delta+\epsilon} \right\} (\alpha-\beta)^\zeta,$$

$$X_i(\gamma, \delta-1, \epsilon) = \binom{\delta-1+\epsilon}{\epsilon} \sum_{\zeta=0}^{i+1-\gamma-\delta-\epsilon} \binom{i}{\gamma, \zeta} \left\{ {i-\gamma-\zeta \atop \delta-1+\epsilon} \right\} (\alpha-\beta)^\zeta,$$

$$X_i(\gamma, \delta, \epsilon-1) = \binom{\delta+\epsilon-1}{\epsilon-1} \sum_{\zeta=0}^{i+1-\gamma-\delta-\epsilon} \binom{i}{\gamma, \zeta} \left\{ {i-\gamma-\zeta \atop \delta+\epsilon-1} \right\} (\alpha-\beta)^\zeta.$$

Pascal's formula shows that $X_i(\gamma, \delta-1, \epsilon) + X_i(\gamma, \delta, \epsilon-1)$ equals

$$\binom{\delta+\epsilon}{\epsilon} \sum_{\zeta=0}^{i+1-\gamma-\delta-\epsilon} \binom{i}{\gamma, \zeta} \left\{ {i-\gamma-\zeta \atop \delta+\epsilon-1} \right\} (\alpha-\beta)^\zeta.$$

The Stirling numbers satisfy the recurrence

$$\left\{ {r \atop s} \right\} = s \left\{ {r-1 \atop s} \right\} + \left\{ {r-1 \atop s-1} \right\},$$

and therefore $(\delta+\epsilon) X_i(\gamma, \delta, \epsilon) + X_i(\gamma, \delta-1, \epsilon) + X_i(\gamma, \delta, \epsilon-1)$ equals

$$\binom{\delta+\epsilon}{\epsilon} \sum_{\zeta=0}^{i+1-\gamma-\delta-\epsilon} \binom{i}{\gamma, \zeta} \left\{ {i+1-\gamma-\zeta \atop \delta+\epsilon} \right\} (\alpha-\beta)^\zeta.$$

The complete sum of five terms now reduces to

$$\binom{\delta+\epsilon}{\epsilon} \sum_{\zeta=0}^{i+1-\gamma-\delta-\epsilon} \binom{i+1}{\gamma, \zeta} \left\{ {i+1-\gamma-\zeta \atop \delta+\epsilon} \right\} (\alpha-\beta)^\zeta = X_{i+1}(\gamma, \delta, \epsilon),$$

and this completes the proof. $\square$

5. The universal nonassociative enveloping algebra

LEMMA 5.1. *The powers of* $L(b)$ *and* $L(c)$ *are*

$$L(b)^u = \sum_{\eta=0}^{u} \sum_{\theta=0}^{u-\eta} (-1)^{u-\eta-\theta} \binom{u}{\eta, \theta} S^{u-2\theta} M_b^\eta M_d^{u-\eta} D_c^{u-\eta},$$

$$L(c)^x = \sum_{\lambda=0}^{x} \sum_{\mu=0}^{x-\lambda} (-1)^{x-\lambda} \binom{x}{\lambda, \mu} S^{x-2\mu} M_c^\lambda M_d^{x-\lambda} D_b^{x-\lambda}.$$

PROOF. We apply the trinomial theorem to the formulas for $L(b)$ and $L(c)$ in Table 2, since the terms in each operator commute:

$$L(b)^u = \sum_{\eta=0}^{u} \sum_{\theta=0}^{u-\eta} \binom{u}{\eta, \theta} (SM_b)^\eta (S^{-1}M_dD_c)^\theta (-SM_dD_c)^{u-\eta-\theta},$$

$$L(c)^x = \sum_{\lambda=0}^{x} \sum_{\mu=0}^{x-\lambda} \binom{x}{\lambda, \mu} (SM_c)^\lambda (-S^{-1}M_dD_b)^\mu (-SM_dD_b)^{x-\lambda-\mu}.$$

These formulas simplify as required using Lemma 3.2. □

LEMMA 5.2. *The operator monomial of Lemma 4.3 equals*

$$L(a)^s S^t L(b)^u D_b^v D_c^w L(c)^x M_d^y L(d)^z = \sum_{\eta=0}^{u} \sum_{\theta=0}^{u-\eta} \sum_{\lambda=0}^{x} \sum_{\mu=0}^{x-\lambda} (-1)^{u-\eta-\theta+x-\lambda} \times$$

$$\binom{u}{\eta, \theta} \binom{x}{\lambda, \mu} M_a^s S^{t+u-2\theta+x-2\mu-z} M_b^\eta D_b^{v+x-\lambda} D_c^{u-\eta+w} M_c^\lambda M_d^{u-\eta+x-\lambda+y+z}.$$

PROOF. Table 2 and Lemma 5.1 show that the operator monomial equals

$$\sum_{\eta=0}^{u} \sum_{\theta=0}^{u-\eta} \sum_{\lambda=0}^{x} \sum_{\mu=0}^{x-\lambda} M_a^s S^t (-1)^{u-\eta-\theta} \binom{u}{\eta, \theta} S^{u-2\theta} M_b^\eta M_d^{u-\eta} D_c^{u-\eta} D_b^v \times$$

$$D_c^w (-1)^{x-\lambda} \binom{x}{\lambda, \mu} S^{x-2\mu} M_c^\lambda M_d^{x-\lambda} D_b^{x-\lambda} M_d^y (S^{-1}M_d)^z,$$

which simplifies as required using Lemma 3.2. □

LEMMA 5.3. $L(a^ib^jc^kd^\ell)$ *expands in terms of* M_x, D_x *and* S *to*

$$\sum_{\alpha=0}^{\min(j,k)} \sum_{\beta=0}^{\alpha} \sum_{\gamma=0}^{i} \sum_{\delta=0}^{i-\gamma} \sum_{\epsilon=0}^{i-\gamma-\delta} \sum_{\zeta=0}^{i-\gamma-\delta-\epsilon} \sum_{\eta=0}^{j-\alpha-\epsilon} \sum_{\theta=0}^{j-\alpha-\epsilon-\eta} \sum_{\lambda=0}^{k-\alpha-\delta} \sum_{\mu=0}^{k-\alpha-\delta-\lambda}$$

$$(-1)^{i+j+k+\alpha-\beta-\gamma-\epsilon-\eta-\theta-\lambda} \times$$

$$(\alpha-\beta)^\zeta \alpha! \binom{\alpha}{\beta} (\delta+\epsilon)! \binom{i}{\gamma, \zeta} \left\{ {i-\gamma-\zeta \atop \delta+\epsilon} \right\} \binom{j}{\alpha, \epsilon, \eta, \theta} \binom{k}{\alpha, \delta, \lambda, \mu} \times$$

$$M_a^\gamma S^{j+k-\ell-2\alpha-\beta-2\delta-2\epsilon-2\theta-2\mu} M_b^\eta D_b^{k-\alpha-\lambda} D_c^{j-\alpha-\eta} M_c^\lambda M_d^{j+k+\ell-\alpha-\eta-\lambda}.$$

PROOF. In Lemma 5.2 we set $s = \gamma$, $t = -\beta-\delta-\epsilon$, $u = j-\alpha-\epsilon$, $v = \delta$, $w = \epsilon$, $x = k-\alpha-\delta$, $y = \alpha+\delta+\epsilon$, $z = \ell$ and obtain

$$L(a)^\gamma S^{-\beta-\delta-\epsilon} L(b)^{j-\alpha-\epsilon} D_b^\delta D_c^\epsilon L(c)^{k-\alpha-\delta} M_d^{\alpha+\delta+\epsilon} L(d)^\ell =$$

$$\sum_{\eta=0}^{j-\alpha-\epsilon}\sum_{\theta=0}^{j-\alpha-\epsilon-\eta}\sum_{\lambda=0}^{k-\alpha-\delta}\sum_{\mu=0}^{k-\alpha-\delta-\lambda}(-1)^{j-\epsilon-\eta-\theta+k-\delta-\lambda}\binom{j-\alpha-\epsilon}{\eta,\theta}\binom{k-\alpha-\delta}{\lambda,\mu}\times$$
$$M_a^\gamma S^{j+k-\ell-2\alpha-\beta-2\delta-2\epsilon-2\theta-2\mu}M_b^\eta D_b^{k-\alpha-\lambda}D_c^{j-\alpha-\eta}M_c^\lambda M_d^{j+k+\ell-\alpha-\eta-\lambda}.$$

We now combine this with Lemma 4.4 and Lemma 4.6. □

DEFINITION 5.4. The *differential coefficients* are

$$\begin{bmatrix} r\\ 0\end{bmatrix}=1,\ \begin{bmatrix} r\\ s\end{bmatrix}=r(r-1)\cdots(r-s+1),\text{ so that } D_x^s(x^r)=\begin{bmatrix} r\\ s\end{bmatrix}x^{r-s}.$$

In the next theorem we set $(\alpha-\beta)^\zeta=1$ when $\alpha=\beta$ and $\zeta=0$.

THEOREM 5.5. *The product* $(a^ib^jc^kd^\ell)(a^mb^nc^pd^q)$ *in* $U(\mathbb{M})$ *equals*

$$\sum_{\alpha=0}^{\min(j,k)}\sum_{\beta=0}^{\alpha}\sum_{\gamma=0}^{i}\sum_{\delta=0}^{i-\gamma}\sum_{\epsilon=0}^{i-\gamma-\delta}\sum_{\zeta=0}^{i-\gamma-\delta-\epsilon}\sum_{\eta=0}^{j-\alpha-\epsilon}\sum_{\theta=0}^{j-\alpha-\epsilon-\eta}\sum_{\lambda=0}^{k-\alpha-\delta}\sum_{\mu=0}^{k-\alpha-\delta-\lambda}\sum_{\nu=0}^{m}$$
$$(-1)^{i+j+k+\alpha-\beta-\gamma-\epsilon-\eta-\theta-\lambda}(\alpha-\beta)^\zeta\alpha!\binom{\alpha}{\beta}(\delta+\epsilon)!\omega^\nu\times$$
$$\binom{i}{\gamma,\zeta}\left\{\begin{matrix} i-\gamma-\zeta\\ \delta+\epsilon\end{matrix}\right\}\binom{j}{\alpha,\epsilon,\eta,\theta}\binom{k}{\alpha,\delta,\lambda,\mu}\binom{m}{\nu}\begin{bmatrix} n\\ k-\alpha-\lambda\end{bmatrix}\begin{bmatrix} p+\lambda\\ j-\alpha-\eta\end{bmatrix}\times$$
$$a^{m+\gamma-\nu}b^{-k+n+\alpha+\eta+\lambda}c^{-j+p+\alpha+\eta+\lambda}d^{j+k+\ell+q-\alpha-\eta-\lambda},$$

where $\omega=j+k-\ell-2\alpha-\beta-2\delta-2\epsilon-2\theta-2\mu$.

PROOF. Apply the M_x, D_x, S operators in Lemma 5.3 to $a^mb^nc^pd^q$:

$$\begin{bmatrix} p+\lambda\\ j-\alpha-\eta\end{bmatrix}\begin{bmatrix} n\\ k-\alpha-\lambda\end{bmatrix}a^\gamma(a+\omega)^m b^{-k+n+\alpha+\eta+\lambda}c^{-j+p+\alpha+\eta+\lambda}d^{j+k+\ell+q-\alpha-\eta-\lambda}.$$

Use this in Lemma 5.3 and expand $(a+\omega)^m$. □

6. The universal alternative enveloping algebra

DEFINITION 6.1. The *alternator ideal* in a nonassociative algebra A is generated by the elements (x,x,y) and (y,x,x) for all $x,y\in A$.

DEFINITION 6.2. Let M be a Malcev algebra, $U(M)$ its universal enveloping algebra, and $I(M)\subseteq U(M)$ the alternator ideal. The *universal alternative enveloping algebra* of M is $A(M)=U(M)/I(M)$.

LEMMA 6.3. *We have the following nonzero alternators in* $U(\mathbb{M})$*:*

$$(a,bc,bc)=2d^2,\qquad (b,ac,ac)=cd,\qquad (c,ab,ab)=-bd.$$

PROOF. Theorem 5.5 gives

$$\big(a(bc)\big)(bc)=ab^2c^2-2abcd+2d^2,\qquad a\big((bc)(bc)\big)=ab^2c^2-2abcd,$$

which imply the first result. The other two are similar. □

DEFINITION 6.4. Let $J\subseteq U(\mathbb{M})$ be the ideal generated by d^2, cd, bd. In $U(\mathbb{M})/J$ it suffices to consider two types of monomials, a^id and $a^ib^jc^k$, which we call *type 1* and *type 2* respectively. If m is one of these monomials, we write m when we mean $m+J$ in the next lemma.

LEMMA 6.5. *In* $U(\mathbb{M})/J$ *we have*

$$(a^i d)(a^m d) = 0, \tag{1}$$

$$(a^i b^j c^k)(a^m d) = \delta_{j0}\delta_{k0}\, a^{i+m} d, \tag{2}$$

$$(a^i d)(a^m b^n c^p) = \delta_{n0}\delta_{p0}\, a^i (a-1)^m d, \tag{3}$$

$$(a^i b^j c^k)(a^m b^n c^p) = a^i (a+j+k)^m b^{j+n} c^{k+p} + \delta_{j+n,1}\delta_{k+p,1} T^{im}_{jk}, \tag{4}$$

where

$$T^{im}_{jk} = \begin{cases} 0 & \text{if } (j,k) = (0,0), \\ (a-1)^{i+m} d - a^i (a+1)^m d & \text{if } (j,k) = (1,0), \\ -(a-1)^{i+m} d - a^i (a+1)^m d & \text{if } (j,k) = (0,1), \\ a^i (a-1)^m d - a^i (a+2)^m d & \text{if } (j,k) = (1,1). \end{cases}$$

PROOF. We only need the terms in Theorem 5.5 in which the d-exponent is 0, or the d-exponent is 1 and the b- and c-exponents are 0.

For equation (1), we have $j = k = n = p = 0$, $\ell = q = 1$; hence $\min(j,k) = 0$, so $\alpha = 0$. The sums on η and λ are empty unless $\delta = 0$ and $\epsilon = 0$; hence $\eta = \lambda = 0$. Now each term in Theorem 5.5 has d-exponent $j+k+\ell+q-\alpha-\eta-\lambda = 2$; but $d^2 = 0$.

For equation (2), we have $\ell = n = p = 0$, $q = 1$. The d-exponent is $j+k+1-\alpha-\eta-\lambda$. This is 0 if and only if $\alpha+\eta+\lambda = j+k+1$; since $\alpha+\eta \leq j$ and $\lambda \leq k$ there are no solutions. The d-exponent is 1 if and only if $\alpha+\eta+\lambda = j+k$. Since $\eta \leq j$, $\alpha+\eta \leq j$, $\lambda \leq k$, $\alpha+\lambda \leq k$, the solution has $\eta = j$, $\lambda = k$. Therefore $\alpha = 0$, $\beta = 0$, and the sums on η, λ are empty unless $\delta = 0$, $\epsilon = 0$ so we get $\theta = \mu = 0$. We need $\zeta = 0$ to make the power of $\alpha-\beta$ nonzero. But $\zeta = i-\gamma$ since $\left\{ {r \atop 0} \right\} = \delta_{r0}$, and so $\gamma = i$. The sum collapses to

$$\sum_{\nu=0}^{m} (j+k)^\nu \binom{m}{\nu} a^{i+m-\nu} b^j c^k d = a^i (a+j+k)^m b^j c^k d.$$

Since $bd = cd = 0$, this is 0 unless $j = k = 0$.

For equation (3), we have $j = k = q = 0$, $\ell = 1$; hence $\min(j,k) = 0$, so $\alpha = \beta = 0$. The power of $\alpha-\beta$ is zero unless $\zeta = 0$. Since $j = \alpha = 0$, the sum on η is empty unless $\epsilon = 0$, so $\eta = \theta = 0$. Since $k = \alpha = 0$, the sum on λ is empty unless $\delta = 0$, so $\lambda = \mu = 0$. We are left with

$$\sum_{\gamma=0}^{i} \sum_{\nu=0}^{m} (-1)^{i-\gamma} (-1)^\nu \binom{i}{\gamma} \left\{ {i-\gamma \atop 0} \right\} \binom{m}{\nu} a^{m+\gamma-\nu} b^n c^p d.$$

The Stirling number is 0 unless $\gamma = i$, so we get

$$\delta_{n0}\delta_{p0} \sum_{\nu=0}^{m} (-1)^\nu \binom{m}{\nu} a^{i+m-\nu} d = \delta_{n0}\delta_{p0}\, a^i (a-1)^m d,$$

since the monomial vanishes unless $n = p = 0$.

For equation (4), we have $\ell = q = 0$; the d-exponent is $j+k-\alpha-\eta-\lambda$. This is 0 if and only if $\alpha+\eta+\lambda = j+k$. As before $\eta = j$, $\lambda = k$; hence $\alpha = 0$, $\beta = 0$, and so $\delta = 0$, $\epsilon = 0$, $\theta = 0$, $\mu = 0$ and $\zeta = 0$. But $\zeta = i-\gamma$ since $\left\{ {r \atop 0} \right\} = \delta_{r0}$, and so $\gamma = i$. The sum collapses to

$$\sum_{\nu=0}^{m} (j+k)^\nu \binom{m}{\nu} a^{i+m-\nu} b^{j+n} c^{k+p} = a^i (a+j+k)^m b^{j+n} c^{k+p}.$$

If the d-exponent is 1, the b- and c-exponents are 0:

$$-k+n+\alpha+\eta+\lambda = 0, \quad -j+p+\alpha+\eta+\lambda = 0, \quad j+k-\alpha-\eta-\lambda = 1.$$

Adding the first and third (resp. second and third) gives $j+n = 1$ (resp. $k+p = 1$), so we have four cases: $(a^i)(a^m bc)$, $(a^i b)(a^m c)$, $(a^i c)(a^m b)$, $(a^i bc)(a^m)$.

Case 1: $jknp = 0011$. We have $(a^i)(a^m bc) = a^{i+m}bc$, so there is no term with d-exponent 1.

Case 2: $jknp = 1001$. We have $\alpha = \beta = 0$ and hence $\zeta = 0$. The λ-sum is empty unless $\delta = 0$, and then $\lambda = \mu = 0$. The η-sum is empty unless $\epsilon \in \{0,1\}$, so we have four subcases: $(\epsilon,\eta,\theta) = (0,0,0)$, $(0,0,1)$, $(0,1,0)$, $(1,0,0)$; the last case occurs only when $\gamma < i$. For $(0,0,0)$ the exponent of -1 is $i+1-\gamma$; otherwise it is $i-\gamma$. For $(0,0,0)$, $(0,1,0)$ the factor ω^ν is 1; otherwise it is $(-1)^\nu$. If $\gamma < i$ then the Stirling number is $\delta_{\epsilon 1}$ (so $\eta = \theta = 0$); otherwise it is $\delta_{\epsilon 0}$. The monomial for $(0,1,0)$ when $\gamma = i$ has d-exponent 0, contradicting our assumption, so this term does not appear. The sum collapses to

$$\sum_{\gamma=0}^{i}\sum_{\nu=0}^{m}(-1)^{i-\gamma}(-1)^\nu\binom{i}{\gamma}\binom{m}{\nu}a^{\gamma+m-\nu}d - \sum_{\nu=0}^{m}\binom{m}{\nu}a^{i+m-\nu}d,$$

which gives the result.

Case 3: $jknp = 0110$. Similar to Case 2.

Case 4: $jknp = 1100$. We have $\alpha \in \{0,1\}$. There are three cases: $(\alpha,\beta) = (0,0)$, $(1,0)$, $(1,1)$. The d-exponent is $2-\alpha-\eta-\lambda$; by assumption this is 1, so $\alpha+\eta+\lambda = 1$. For $(\alpha,\beta) = (1,1)$ we must have $\delta = 0$ and then $\lambda = \mu = 0$; likewise $\epsilon = 0$ and then $\eta = \theta = 0$. Furthermore $\zeta = 0$ and $\gamma = i$. The sum collapses to

$$\sum_{\nu=0}^{m}(-1)^\nu\binom{m}{\nu}a^{i+m-\nu}d = a^i(a-1)^m d.$$

For $(\alpha,\beta) = (1,0)$ the sum collapses to

$$-\sum_{\gamma=0}^{i}(-1)^{i-\gamma}\binom{i}{\gamma}a^{\gamma+m}d = -(a-1)^i a^m d.$$

For $(\alpha,\beta) = (0,0)$ the sum collapses to

$$\sum_{\gamma=0}^{i}\sum_{\epsilon=0}^{i-\gamma}\sum_{\theta=0}^{1-\epsilon}\sum_{\nu=0}^{m}(-1)^{i-\gamma-\epsilon-\theta-1}(2-2\epsilon-2\theta)^\nu\binom{i}{\gamma}\left\{{i-\gamma \atop \epsilon}\right\}\binom{m}{\nu}a^{\gamma-\nu+m}d.$$

The sum on θ gives $\epsilon \in \{0,1\}$. If $\gamma < i$ then $\epsilon = 1$; hence $\theta = 0$ and $\nu = 0$. If $\gamma = i$ then $\epsilon = 0$. We separate the last term of the γ-sum:

$$\sum_{\gamma=0}^{i-1}(-1)^{i-\gamma}\binom{i}{\gamma}a^{\gamma+m}d + \left[\sum_{\theta=0}^{1}\sum_{\nu=0}^{m}(-1)^{-\theta-1}(2-2\theta)^\nu\binom{m}{\nu}a^{i-\nu+m}d\right].$$

The first term cancels with the result for $(\alpha,\beta) = (1,0)$. □

The following theorem was first established by Shestakov using different methods; a similar result appears in [**8**, Example 1].

THEOREM 6.6. *The universal alternative enveloping algebra* $A(\mathbb{M})$ *is isomorphic to the algebra with basis* $\{\, a^i d,\ a^i b^j c^k \mid i,j,k \geq 0 \,\}$ *and structure constants of Lemma 6.5.*

PROOF. Once we show that $U(\mathbb{M})/J$ is alternative, it follows that J equals the alternator ideal $I(M)$ and hence that $U(\mathbb{M})/J$ is isomorphic to $A(\mathbb{M})$. We prove alternativity by showing that the associator alternates. Since the associator is multilinear, it suffices to consider monomials. We use Lemma 6.5 repeatedly. Since the product of a monomial of type 1 with any monomial is a linear combination of monomials of type 1, every associator with two monomials of type 1 vanishes. We next consider one monomial of type 1 and two of type 2. Since the T-term in Equation (4) contains only monomials of type 1, $(a^i d, a^m b^n c^p, a^r b^s c^t)$ equals

$$\begin{aligned}&\big[\delta_{n0}\delta_{p0}a^i(a-1)^m d\big](a^r b^s c^t) - (a^i d)\big[a^m(a+n+p)^r b^{n+s}c^{p+t} + T^{**}_{**}\big] = \\ &\delta_{n0}\delta_{p0}\delta_{s0}\delta_{t0}a^i(a-1)^{m+r}d - \delta_{n+s,0}\delta_{p+t,0}a^i(a-1)^m(a-1+n+p)^r d = 0.\end{aligned}$$

Similarly $(a^i b^j c^k, a^m d, a^r b^s c^t) = (a^i b^j c^k, a^m b^n c^p, a^r d) = 0$. We finally consider three monomials of type 2: $(a^i b^j c^k, a^m b^n c^p, a^r b^s c^t)$ equals

$$\begin{aligned}&\big[a^i(a+j+k)^m b^{j+n}c^{k+p} + \delta_{j+n,1}\delta_{k+p,1}T^{im}_{jk}\big](a^r b^s c^t) \\ &\quad - (a^i b^j c^k)\big[a^m(a+n+p)^r b^{n+s}c^{p+t} + \delta_{n+s,1}\delta_{p+t,1}T^{mr}_{np}\big].\end{aligned}$$

We write this as $A - B + C - D$ where

$$\begin{aligned}A &= \big[a^i(a+j+k)^m b^{j+n}c^{k+p}\big](a^r b^s c^t), \\ B &= (a^i b^j c^k)\big[a^m(a+n+p)^r b^{n+s}c^{p+t}\big], \\ C &= \delta_{j+n,1}\delta_{k+p,1}T^{im}_{jk}(a^r b^s c^t), \qquad D = \delta_{n+s,1}\delta_{p+t,1}(a^i b^j c^k)T^{mr}_{np}.\end{aligned}$$

Expanding $(a+j+k)^m$ and $(a+n+p)^r$ we see that $A - B$ equals

$$\delta_{j+n+s,1}\delta_{k+p+t,1}\left[\sum_{\nu=0}^{m}\binom{m}{\nu}(j+k)^\nu T^{i+m-\nu,r}_{j+n,k+p} - \sum_{\xi=0}^{r}\binom{r}{\xi}(n+p)^\xi T^{i,m+r-\xi}_{jk}\right].$$

For $jknpst = 110000$ we get

$$\begin{aligned}A - B &= \sum_{\nu=0}^{m}\binom{m}{\nu}2^\nu T^{i+m-\nu,r}_{11} - T^{i,m+r}_{11} \\ &= a^i(a+2)^m(a-1)^r d - a^i(a+2)^{m+r}d - a^i(a-1)^{m+r}d + a^i(a+2)^{m+r}d \\ &= a^i(a-1)^r(a+2)^m d - a^i(a-1)^{m+r}d.\end{aligned}$$

Similar calculations give

$$\begin{aligned}jknpst = 100100&: \quad A - B = a^i(a-1)^r(a+1)^m d - a^r(a-1)^{i+m}d, \\ jknpst = 100001&: \quad A - B = a^m(a-1)^{i+r}d - (a-1)^{i+m+r}d, \\ jknpst = 011000&: \quad A - B = a^i(a-1)^r(a+1)^m d + (a-1)^{i+m}a^r d, \\ jknpst = 001100&: \quad A - B = a^{i+m}(a-1)^r d - a^{i+m}(a+2)^r d, \\ jknpst = 001001&: \quad A - B = (a-1)^{i+m+r}d - a^{i+m}(a+1)^r d, \\ jknpst = 010010&: \quad A - B = -a^m(a-1)^{i+r}d + (a-1)^{i+m+r}d, \\ jknpst = 000110&: \quad A - B = -(a-1)^{i+m+r}d - a^{i+m}(a+1)^r d, \\ jknpst = 000011&: \quad A - B = 0.\end{aligned}$$

For C and D we obtain

$$jknp = 1100: \quad C = \delta_{s0}\delta_{t0}a^i(a-1)^{m+r}d - \delta_{s0}\delta_{t0}a^i(a-1)^r(a+2)^m d,$$

$$
\begin{aligned}
jknp = 1001&: \quad C = \delta_{s0}\delta_{t0}(a-1)^{i+m+r}d - \delta_{s0}\delta_{t0}a^i(a-1)^r(a+1)^m d,\\
jknp = 0110&: \quad C = -\delta_{s0}\delta_{t0}(a-1)^{i+m+r}d - \delta_{s0}\delta_{t0}a^i(a-1)^r(a+1)^m d,\\
jknp = 0011&: \quad C = 0,\\
npst = 1100&: \quad D = \delta_{j0}\delta_{k0}a^{i+m}(a-1)^r d - \delta_{j0}\delta_{k0}a^{i+m}(a+2)^r d,\\
npst = 1001&: \quad D = \delta_{j0}\delta_{k0}a^i(a-1)^{m+r}d - \delta_{j0}\delta_{k0}a^{i+m}(a+1)^r d,\\
npst = 0110&: \quad D = -\delta_{j0}\delta_{k0}a^i(a-1)^{m+r}d - \delta_{j0}\delta_{k0}a^{i+m}(a+1)^r d,\\
npst = 0011&: \quad D = 0.
\end{aligned}
$$

We combine these results to get $A - B + C - D$:

$$
\begin{aligned}
jknpst = 110000&: \quad (a^ibc, a^m, a^r) = 0,\\
jknpst = 100100&: \quad (a^ib, a^mc, a^r) = (a-1)^{i+m+r}d - a^r(a-1)^{i+m}d,\\
jknpst = 100001&: \quad (a^ib, a^m, a^rc) = a^m(a-1)^{i+r}d - (a-1)^{i+m+r}d,\\
jknpst = 011000&: \quad (a^ic, a^mb, a^r) = -(a-1)^{i+m+r}d + (a-1)^{i+m}a^r d,\\
jknpst = 001100&: \quad (a^i, a^mbc, a^r) = 0,\\
jknpst = 001001&: \quad (a^i, a^mb, a^rc) = (a-1)^{i+m+r}d - a^i(a-1)^{m+r}d,\\
jknpst = 010010&: \quad (a^ic, a^m, a^rb) = -a^m(a-1)^{i+r}d + (a-1)^{i+m+r}d,\\
jknpst = 000110&: \quad (a^i, a^mc, a^rb) = -(a-1)^{i+m+r}d + a^i(a-1)^{m+r}d,\\
jknpst = 000011&: \quad (a^i, a^m, a^rbc) = 0.
\end{aligned}
$$

The alternativity property is now clear. □

7. Conclusion

Since the alternator ideal $I(\mathbb{M})$ contains no elements of degree 1, the natural mapping from $\mathbb{M}$ to $A(\mathbb{M})$ is injective, and hence $\mathbb{M}$ is special. This also follows directly from the isomorphism $\mathbb{M} \cong \mathbb{A}^-$ where $\mathbb{A}$ is the algebra in Table 3. For any $x, y, z \in \mathbb{A}$ we write $x = (x_1, \ldots, x_4)$ etc. and calculate the associator to prove that $\mathbb{A}$ is alternative:

$$
(xy)z - x(yz) = \left[0, 0, 0, -\det \begin{pmatrix} x_1 & x_2 & x_3 \\ y_1 & y_2 & y_3 \\ z_1 & z_2 & z_3 \end{pmatrix} \right].
$$

This is analogous to the construction of the split simple Lie algebra $sl_2(\mathbb{F})$ as a subalgebra (the trace-zero matrices) of the commutator algebra of the associative algebra $M_2(\mathbb{F})$ of 2×2 matrices over $\mathbb{F}$.

TABLE 3. The 4-dimensional alternative algebra $\mathbb{A}$

$\cdot$	a	b	c	d
a	a	0	0	d
b	b	0	d	0
c	c	$-d$	0	0
d	0	0	0	0

References

[1] M. R. Bremner, L. I. Murakami and I. P. Shestakov: *Nonassociative Algebras*, in *Handbook of Linear Algebra*, edited by L. Hogben, pages 69-1 to 69-26, Chapman & Hall / CRC, Boca Raton, 2007.

[2] E. N. Kuzmin: *Malcev algebras and their representations*, Algebra Logika **7**, 4 (1968) 48–69.

[3] E. N. Kuzmin and I. P. Shestakov: *Nonassociative Structures*, in *Algebra VI*, edited by R. V. Gamkrelidze, pages 197–280, Encyclopaedia of Mathematical Sciences **57**, Springer, Berlin, 1995.

[4] A. I. Malcev: *Analytic loops*, Mat. Sbornik N.S. **36/78** (1955) 569–576. See also an English translation at: `http://math.usask.ca/~bremner/research/translations/malcev.pdf`

[5] J. M. Pérez-Izquierdo and I. P. Shestakov: *An envelope for Malcev algebras*, J. Algebra **272**, 1 (2004) 379–393.

[6] A. A. Sagle: *Malcev algebras*, Trans. Amer. Math. Soc. **101** (1961) 426–458.

[7] I. P. Shestakov: *Speciality and deformations of algebras*, in *Algebra (Moscow, 1998)*, pages 345–356, de Gruyter, Berlin, 2000.

[8] I. P. Shestakov: *Speciality problem for Malcev algebras and Poisson Malcev algebras*, in *Nonassociative Algebra and its Applications (São Paulo, 1998)*, pages 365–371, Lecture Notes in Pure and Applied Mathematics **211**, Dekker, New York, 2000.

[9] I. P. Shestakov: *Free Malcev superalgebra on one odd generator*, J. Algebra Appl. **2**, 4 (2003) 451–461.

[10] I. P. Shestakov and N. Zhukavets: *The universal multiplicative envelope of the free Malcev superalgebra on one odd generator*, Comm. Algebra **34**, 4 (2006) 1319–1344.

[11] I. P. Shestakov and N. Zhukavets: *Speciality of Malcev superalgebras on one odd generator*, J. Algebra **301**, 2 (2006) 587–600.

[12] I. P. Shestakov and N. Zhukavets: *The Malcev Poisson superalgebra of the free Malcev superalgebra on one odd generator*, J. Algebra Appl. **5**, 4 (2006) 521–535.

[13] I. P. Shestakov and N. Zhukavets: *The free alternative superalgebra on one odd generator*, Internat. J. Algebra Comput. **17**, 5-6 (2007) 1215–1247.

[14] V. N. Zhelyabin and I. P. Shestakov: *The Chevalley and Kostant theorems for Malcev algebras*, Algebra Logika **46**, 5 (2007) 560–584.

Department of Mathematics and Statistics, University of Saskatchewan, Canada
E-mail address: `bremner@math.usask.ca`

Department of Mathematics, Iowa State University, U.S.A.
E-mail address: `hentzel@iastate.edu`

Departamento de Matemática, Universidade de São Paulo, Brasil
E-mail address: `peresi@ime.usp.br`

Department of Mathematics and Statistics, University of Saskatchewan, Canada.
Current address: Department of Mathematics, University of British Columbia, Canada
E-mail address: `usefi@math.ubc.ca`

Contemporary Mathematics
Volume **483**, 2009

The Supermagic Square in characteristic 3 and Jordan superalgebras

Isabel Cunha[◇] and Alberto Elduque[⋆]

Dedicated to Ivan Shestakov, on the occasion of his 60th birthday

Abstract. Recently, the classical Freudenthal Magic Square has been extended over fields of characteristic 3 with two more rows and columns filled with (mostly simple) Lie superalgebras specific to this characteristic. This *Supermagic Square* will be reviewed and some of the simple Lie superalgebras that appear will be shown to be isomorphic to the Tits-Kantor-Koecher Lie superalgebras of some Jordan superalgebras.

Introduction

The classical Freudenthal Magic Square, which contains in characteristic 0 all the exceptional simple finite dimensional Lie algebras other than G_2, is usually constructed based on two ingredients: a unital composition algebra and a central simple degree 3 Jordan algebra (see [**Sch95**, Chapter IV]). This construction, due to Tits, does not work in characteristic 3.

A more symmetric construction, based on two unital composition algebras which play symmetric roles, and their triality Lie algebras, has been given recently by several authors ([**AF93**], [**BS**], [**LM02**] [**LM04**]). Among other things, this construction has the advantage of being valid in characteristic 3. Simpler formulas for triality appear if symmetric composition algebras are used, instead of the more classical unital composition algebras ([**Eld04, Eld07a**]).

But characteristic 3 presents an exceptional feature, as only over fields of this characteristic there are nontrivial composition superalgebras of dimension 3 and 6 (the unital ones were discovered by Shestakov [**She97**]). This fact allows one to extend Freudenthal Magic Square [**CE07a**] by the addition in characteristic 3 of two further rows and columns, filled with (mostly simple) Lie superalgebras, specific

---

2000 *Mathematics Subject Classification.* Primary 17B50; Secondary 17B60, 17B25.
Key words and phrases. Superalgebra, Lie, Jordan, Freudenthal, magic, Tits-Kantor-Koecher construction.
◇ Supported by CMUC, Department of Mathematics, University of Coimbra.
⋆ Supported by the Spanish Ministerio de Educación y Ciencia and FEDER (MTM 2007-67884-C04-02) and by the Diputación General de Aragón (Grupo de Investigación de Álgebra).

of characteristic 3, which had appeared first (with one exception) in [**Eld06**] and [**Eld07b**].

Most of the Lie superalgebras in characteristic 3 that appear in the Supermagic Square have been shown to be related to degree three simple Jordan algebras in [**CE07b**].

The aim of this paper is to show that some of the Lie superalgebras in the Supermagic Square are isomorphic to the Tits-Kantor-Koecher Lie superalgebras of some distinguished Jordan superalgebras.

More specifically, let S_i denote the split para-Hurwitz algebra of dimension $i = 1, 2$ or 4, and let S be the para-Hurwitz superalgebra associated to the unital composition superalgebra C (see Section 1 for definitions and notations). Let $\mathfrak{g}(S_i, S)$ be the corresponding entry in the Supermagic Square. Then $\mathfrak{g}(S_1, S)$ was shown in [**CE07b**] to be isomorphic to the Lie superalgebra of derivations of the Jordan superalgebra $J = H_3(C)$ of hermitian 3×3 matrices over C. We will prove the following results:

(i) The Lie superalgebras $\mathfrak{g}(S_2, S)$ in the second row of the Supermagic Square will be shown to be isomorphic to the projective structure superalgebras of the Jordan superalgebras $J = H_3(C)$. Here the structure superalgebra is $\mathfrak{str}(J) = L_J \oplus \mathfrak{der}(J)$ and the projective structure superalgebra $\mathfrak{pstr}(J)$ is the quotient of $\mathfrak{str}(J)$ by its center. (See Theorem 3.3 and Corollary 3.5.)

(ii) The Lie superalgebras $\mathfrak{g}(S_4, S)$ in the third row of the Supermagic Square will be shown to be isomorphic to the Tits-Kantor-Koecher Lie superalgebras of the Jordan superalgebras $J = H_3(C)$. (See Theorem 3.8 and Corollary 3.10.)

(iii) The Lie superalgebra $\mathfrak{g}(S_{1,2}, S_{1,2})$ will be shown to be isomorphic to the Tits-Kantor-Koecher Lie superalgebra of the 9-dimensional Kac Jordan superalgebra K_9. Note that the 10-dimensional Kac Jordan superalgebra K_{10} is no longer simple in characteristic 3, but contains a 9-dimensional simple ideal, which is K_9. (See Theorem 4.2.)

(iv) The Lie superalgebra $\mathfrak{g}(S_1, S_{1,2})$ will be shown to be isomorphic to the Tits-Kantor-Koecher Lie superalgebra of the 3-dimensional Kaplansky superalgebra K_3. (See Corollary 4.3.)

The paper is structured as follows. In Section 1 the construction of the Supermagic Square in terms of two symmetric composition superalgebras will be reviewed. Then the relationship of the Lie superalgebras in the first row of the Supermagic Square with the Lie superalgebras of derivations of the Jordan superalgebras $J = H_3(C)$ above, proven in [**CE07b**], will be reviewed in Section 2. Section 3 will be devoted to the Lie superalgebras in the second and third rows of the Supermagic Square, while Section 4 will deal with the Lie superalgebra $\mathfrak{g}(S_{1,2}, S_{1,2})$ and the 9-dimensional Kac Jordan superalgebra K_9. It was Shestakov [**She96**] who first noticed that K_9 is isomorphic to the tensor product (in the graded sense) of two copies of the 3-dimensional Kaplansky Jordan superalgebra K_3 (this was further developed in [**BE02**]), and this is the key for the results in Section 4.

Unless otherwise stated, all the vector spaces and superspaces considered will be assumed to be finite dimensional over a ground field k of characteristic $\neq 2$.

1. The Supermagic Square

A quadratic superform on a $\mathbb{Z}_2$-graded vector space $U = U_{\bar{0}} \oplus U_{\bar{1}}$ over a field k is a pair $q = (q_{\bar{0}}, \mathrm{b})$ where $q_{\bar{0}} : U_{\bar{0}} \to k$ is a quadratic form, and $\mathrm{b} : U \times U \to k$ is a supersymmetric even bilinear form such that $\mathrm{b}|_{U_{\bar{0}} \times U_{\bar{0}}}$ is the polar form of $q_{\bar{0}}$:

$$\mathrm{b}(x_{\bar{0}}, y_{\bar{0}}) = q_{\bar{0}}(x_{\bar{0}} + y_{\bar{0}}) - q_{\bar{0}}(x_{\bar{0}}) - q_{\bar{0}}(y_{\bar{0}})$$

for any $x_{\bar{0}}, y_{\bar{0}} \in U_{\bar{0}}$.

The quadratic superform $q = (q_{\bar{0}}, \mathrm{b})$ is said to be *regular* if the bilinear form b is nondegenerate.

Then a superalgebra $C = C_{\bar{0}} \oplus C_{\bar{1}}$ over k, endowed with a regular quadratic superform $q = (q_{\bar{0}}, \mathrm{b})$, called the *norm*, is said to be a *composition superalgebra* (see [**EO02**]) in case

$$q_{\bar{0}}(x_{\bar{0}} y_{\bar{0}}) = q_{\bar{0}}(x_{\bar{0}}) q_{\bar{0}}(y_{\bar{0}}), \tag{1.1a}$$

$$\mathrm{b}(x_{\bar{0}} y, x_{\bar{0}} z) = q_{\bar{0}}(x_{\bar{0}}) \mathrm{b}(y, z) = \mathrm{b}(y x_{\bar{0}}, z x_{\bar{0}}), \tag{1.1b}$$

$$\mathrm{b}(xy, zt) + (-1)^{xy+xz+yz} \mathrm{b}(zy, xt) = (-1)^{yz} \mathrm{b}(x, z) \mathrm{b}(y, t), \tag{1.1c}$$

for any $x_{\bar{0}}, y_{\bar{0}} \in C_{\bar{0}}$ and homogeneous elements $x, y, z, t \in C$. (As we are working in characteristic $\neq 2$, it is enough to consider equation (1.1c).)

As usual, the expression $(-1)^{yz}$ equals -1 if the homogeneous elements y and z are both odd, otherwise, it equals 1.

The unital composition superalgebras are termed *Hurwitz superalgebras*, while a composition superalgebra is said to be *symmetric* in case its bilinear form is associative, that is,

$$\mathrm{b}(xy, z) = \mathrm{b}(x, yz),$$

for any x, y, z.

Hurwitz algebras are the well-known algebras that generalize the classical real division algebras of the real and complex numbers, quaternions and octonions. Over any algebraically closed field k, there are exactly four of them: k, $k \times k$, $\mathrm{Mat}_2(k)$ and $C(k)$ (the split Cayley algebra), with dimensions 1, 2, 4 and 8.

Nontrivial Hurwitz superalgebras appear only over fields of characteristic 3 (see [**EO02**]):

- Let V be a 2-dimensional vector space over a field k, endowed with a nonzero alternating bilinear form $\langle .|. \rangle$ (that is, $\langle v|v \rangle = 0$ for any $v \in V$). Consider the superspace $B(1,2)$ (see [**She97**]) with

$$B(1,2)_{\bar{0}} = k1, \qquad \text{and} \qquad B(1,2)_{\bar{1}} = V, \tag{1.2}$$

  endowed with the supercommutative multiplication given by

$$1x = x1 = x \qquad \text{and} \qquad uv = \langle u|v \rangle 1$$

  for any $x \in B(1,2)$ and $u, v \in V$, and with the quadratic superform $q = (q_{\bar{0}}, \mathrm{b})$ given by:

$$q_{\bar{0}}(1) = 1, \quad \mathrm{b}(u, v) = \langle u|v \rangle, \tag{1.3}$$

  for any $u, v \in V$. If the characteristic of k is equal to 3, then $B(1,2)$ is a Hurwitz superalgebra ([**EO02**, Proposition 2.7]).

- Moreover, with V as before, let $f \mapsto \bar{f}$ be the associated symplectic involution on $\mathrm{End}_k(V)$ (so $\langle f(u)|v\rangle = \langle u|\bar{f}(v)\rangle$ for any $u, v \in V$ and $f \in \mathrm{End}_k(V)$). Consider the superspace $B(4,2)$ (see [**She97**]) with

$$B(4,2)_{\bar{0}} = \mathrm{End}_k(V), \qquad \text{and} \qquad B(4,2)_{\bar{1}} = V, \tag{1.4}$$

with multiplication given by the usual one (composition of maps) in $\mathrm{End}_k(V)$, and by

$$v \cdot f = f(v) = \bar{f} \cdot v \in V,$$
$$u \cdot v = \langle .|u\rangle v \in \mathrm{End}_k(V)$$

for any $f \in \mathrm{End}_k(V)$ and $u, v \in V$, where $\langle .|u\rangle v$ denotes the endomorphism $w \mapsto \langle w|u\rangle v$; and with quadratic superform $q = (q_{\bar{0}}, \mathrm{b})$ such that

$$q_{\bar{0}}(f) = \det(f), \qquad \mathrm{b}(u, v) = \langle u|v\rangle,$$

for any $f \in \mathrm{End}_k(V)$ and $u, v \in V$. If the characteristic is equal to 3, $B(4,2)$ is a Hurwitz superalgebra ([**EO02**, Proposition 2.7]).

Given any Hurwitz superalgebra C with norm $q = (q_{\bar{0}}, \mathrm{b})$, its standard involution is given by

$$x \mapsto \bar{x} = \mathrm{b}(x, 1)1 - x.$$

A new product can be defined on C by means of

$$x \bullet y = \bar{x}\bar{y}. \tag{1.5}$$

The resulting superalgebra, denoted by $\bar{C}$, is called the *para-Hurwitz superalgebra* attached to C, and it turns out to be a symmetric composition superalgebra.

Given a symmetric composition superalgebra $(S, \bullet)$, its *triality Lie superalgebra* $\mathfrak{tri}(S) = \mathfrak{tri}(S)_{\bar{0}} \oplus \mathfrak{tri}(S)_{\bar{1}}$ is defined by:

$$\mathfrak{tri}(S)_{\bar{i}} = \{(d_0, d_1, d_2) \in \mathfrak{osp}(S, q)^3_{\bar{i}} :$$
$$d_0(x \bullet y) = d_1(x) \bullet y + (-1)^{ix} x \bullet d_2(y)\ \forall x, y \in S_{\bar{0}} \cup S_{\bar{1}}\},$$

where $\bar{i} = \bar{0}, \bar{1}$, and $\mathfrak{osp}(S, q)$ denotes the associated orthosymplectic Lie superalgebra. The bracket in $\mathfrak{tri}(S)$ is given componentwise.

Now, given two symmetric composition superalgebras S and S', one can form (see [**CE07a**, §3], or [**Eld04**] for the non-super situation) the Lie superalgebra:

$$\mathfrak{g} = \mathfrak{g}(S, S') = \big(\mathfrak{tri}(S) \oplus \mathfrak{tri}(S')\big) \oplus \big(\oplus_{i=0}^2 \iota_i(S \otimes S')\big), \tag{1.6}$$

where $\iota_i(S \otimes S')$ is just a copy of $S \otimes S'$ ($i = 0, 1, 2$), with bracket given by:

- the Lie bracket in $\mathfrak{tri}(S) \oplus \mathfrak{tri}(S')$, which thus becomes a Lie subsuperalgebra of $\mathfrak{g}$,
- $[(d_0, d_1, d_2), \iota_i(x \otimes x')] = \iota_i\big(d_i(x) \otimes x'\big)$,
- $[(d'_0, d'_1, d'_2), \iota_i(x \otimes x')] = (-1)^{d'_i x} \iota_i\big(x \otimes d'_i(x')\big)$,
- $[\iota_i(x \otimes x'), \iota_{i+1}(y \otimes y')] = (-1)^{x' y} \iota_{i+2}\big((x \bullet y) \otimes (x' \bullet y')\big)$ (indices modulo 3),
- $[\iota_i(x \otimes x'), \iota_i(y \otimes y')] = (-1)^{xx' + xy' + yy'} \mathrm{b}'(x', y') \theta^i(t_{x,y})$
$$+ (-1)^{yx'} \mathrm{b}(x, y) \theta'^i(t'_{x',y'}),$$

for any $i = 0, 1, 2$ and homogeneous $x, y \in S$, $x', y' \in S'$, $(d_0, d_1, d_2) \in \mathfrak{tri}(S)$, and $(d'_0, d'_1, d'_2) \in \mathfrak{tri}(S')$. Here θ denotes the natural automorphism $\theta : (d_0, d_1, d_2) \mapsto (d_2, d_0, d_1)$ in $\mathfrak{tri}(S)$, while $t_{x,y}$ is defined by

$$t_{x,y} = \left(\sigma_{x,y}, \tfrac{1}{2}\mathrm{b}(x,y)1 - r_x l_y, \tfrac{1}{2}\mathrm{b}(x,y)1 - l_x r_y\right) \tag{1.7}$$

with $l_x(y) = x \bullet y$, $r_x(y) = (-1)^{xy} y \bullet x$, and

$$\sigma_{x,y}(z) = (-1)^{yz}\mathrm{b}(x,z)y - (-1)^{x(y+z)}\mathrm{b}(y,z)x \tag{1.8}$$

for homogeneous $x, y, z \in S$. Also θ' and $t'_{x',y'}$ denote the analogous elements for $\mathfrak{tri}(S')$.

Over a field k of characteristic 3, let S_r ($r = 1$, 2, 4 or 8) denote the para-Hurwitz algebra attached to the split Hurwitz algebra of dimension r (this latter algebra being either k, $k \times k$, $\mathrm{Mat}_2(k)$ or $C(k)$). Also, denote by $S_{1,2}$ the para-Hurwitz superalgebra $\overline{B(1,2)}$, and by $S_{4,2}$ the para-Hurwitz superalgebra $\overline{B(4,2)}$. Then the Lie superalgebras $\mathfrak{g}(S, S')$, where S, S' run over $\{S_1, S_2, S_4, S_8, S_{1,2}, S_{4,2}\}$, appear in Table 1, which has been obtained in [**CE07a**].

	S_1	S_2	S_4	S_8	$S_{1,2}$	$S_{4,2}$
S_1	$\mathfrak{sl}_2$	$\mathfrak{pgl}_3$	$\mathfrak{sp}_6$	$\mathfrak{f}_4$	$\mathfrak{psl}_{2,2}$	$\mathfrak{sp}_6 \oplus (14)$
S_2		$\mathfrak{pgl}_3 \oplus \mathfrak{pgl}_3$	$\mathfrak{pgl}_6$	$\tilde{\mathfrak{e}}_6$	$(\mathfrak{pgl}_3 \oplus \mathfrak{sl}_2) \oplus (\mathfrak{psl}_3 \otimes (2))$	$\mathfrak{pgl}_6 \oplus (20)$
S_4			$\mathfrak{so}_{12}$	$\mathfrak{e}_7$	$(\mathfrak{sp}_6 \oplus \mathfrak{sl}_2) \oplus ((13) \otimes (2))$	$\mathfrak{so}_{12} \oplus spin_{12}$
S_8				$\mathfrak{e}_8$	$(\mathfrak{f}_4 \oplus \mathfrak{sl}_2) \oplus ((25) \otimes (2))$	$\mathfrak{e}_7 \oplus (56)$
$S_{1,2}$					$\mathfrak{so}_7 \oplus 2spin_7$	$\mathfrak{sp}_8 \oplus (40)$
$S_{4,2}$						$\mathfrak{so}_{13} \oplus spin_{13}$

TABLE 1. Supermagic Square (characteristic 3)

Since the construction of $\mathfrak{g}(S, S')$ is symmetric, only the entries above the diagonal are needed. In Table 1, $\mathfrak{f}_4, \mathfrak{e}_6, \mathfrak{e}_7, \mathfrak{e}_8$ denote the simple exceptional classical Lie algebras, $\tilde{\mathfrak{e}}_6$ denotes a 78-dimensional Lie algebras whose derived Lie algebra is the 77-dimensional simple Lie algebra $\mathfrak{e}_6$ in characteristic 3. The even and odd parts of the nontrivial superalgebras in the table which have no counterpart in the classification in characteristic 0 ([**Kac77a**]) are displayed, *spin* denotes the spin module for the corresponding orthogonal Lie algebra, while (n) denotes a module of dimension n, whose precise description is given in [**CE07a**]. Thus, for example, $\mathfrak{g}(S_4, S_{1,2})$ is a Lie superalgebra whose even part is (isomorphic to) the direct sum of the symplectic Lie algebra $\mathfrak{sp}_6$ and of $\mathfrak{sl}_2$, while its odd part is the tensor product of a 13-dimensional module for $\mathfrak{sp}_6$ and the natural 2-dimensional module for $\mathfrak{sl}_2$.

A precise description of these modules and of the Lie superalgebras as Lie superalgebras with a Cartan matrix is given in [**CE07a**]. All the inequivalent Cartan matrices for these simple Lie superalgebras are listed in [**BGL**].

With the exception of $\mathfrak{g}(S_{1,2}, S_{4,2})$, all these superalgebras have appeared previously in [**Eld06**] and [**Eld07b**].

2. Jordan superalgebras

Given any Hurwitz superalgebra C over our ground field k, with norm $q = (q_{\bar 0}, \mathrm{b})$ and standard involution $x \mapsto \bar{x}$, the superalgebra $H_3(C)$ of 3×3 hermitian matrices over C, under the superinvolution given by $(a_{ij})^* = (\bar{a}_{ji})$, is a Jordan superalgebra under the symmetrized product

$$x \circ y = \frac{1}{2}\big(xy + (-1)^{xy} yx\big). \tag{2.1}$$

Let us consider the associated para-Hurwitz superalgebra $S = \bar{C}$, with multiplication $a \bullet b = \bar{a}\bar{b}$ for any $a, b \in C$. The multiplication rules for 3×3-matrices are more uniformly described using the labels $0, 1, 2$ (read modulo 3) in place of the usual labels $1, 2, 3$ for the idempotents, rows, and columns. With these we have:

$$\begin{aligned} J = H_3(C) &= \left\{ \begin{pmatrix} \alpha_0 & \bar{a}_2 & a_1 \\ a_2 & \alpha_1 & \bar{a}_0 \\ \bar{a}_1 & a_0 & \alpha_2 \end{pmatrix} : \alpha_0, \alpha_1, \alpha_2 \in k,\ a_0, a_1, a_2 \in S \right\} \\ &= \big(\oplus_{i=0}^2 k e_i\big) \oplus \big(\oplus_{i=0}^2 \iota_i(S)\big), \end{aligned} \tag{2.2}$$

where

$$\begin{gathered} e_0 = \begin{pmatrix} 1 & 0 & 0 \\ 0 & 0 & 0 \\ 0 & 0 & 0 \end{pmatrix}, \qquad e_1 = \begin{pmatrix} 0 & 0 & 0 \\ 0 & 1 & 0 \\ 0 & 0 & 0 \end{pmatrix}, \qquad e_2 = \begin{pmatrix} 0 & 0 & 0 \\ 0 & 0 & 0 \\ 0 & 0 & 1 \end{pmatrix}, \\ \iota_0(a) = 2\begin{pmatrix} 0 & 0 & 0 \\ 0 & 0 & \bar{a} \\ 0 & a & 0 \end{pmatrix}, \quad \iota_1(a) = 2\begin{pmatrix} 0 & 0 & a \\ 0 & 0 & 0 \\ \bar{a} & 0 & 0 \end{pmatrix}, \quad \iota_2(a) = 2\begin{pmatrix} 0 & \bar{a} & 0 \\ a & 0 & 0 \\ 0 & 0 & 0 \end{pmatrix}, \end{gathered} \tag{2.3}$$

for any $a \in S$. Identify $ke_0 \oplus ke_1 \oplus ke_2$ to k^3 by means of $\alpha_0 e_0 + \alpha_1 e_1 + \alpha_2 e_2 \simeq (\alpha_0, \alpha_1, \alpha_2)$. Then the supercommutative multiplication (2.1) becomes:

$$\begin{cases} (\alpha_0, \alpha_1, \alpha_2) \circ (\beta_0, \beta_1, \beta_2) = (\alpha_0\beta_0, \alpha_1\beta_1, \alpha_2\beta_2), \\ (\alpha_0, \alpha_1, \alpha_2) \circ \iota_i(a) = \frac{1}{2}(\alpha_{i+1} + \alpha_{i+2})\iota_i(a), \\ \iota_i(a) \circ \iota_{i+1}(b) = \iota_{i+2}(a \bullet b), \\ \iota_i(a) \circ \iota_i(b) = 2\mathrm{b}(a, b)\big(e_{i+1} + e_{i+2}\big), \end{cases} \tag{2.4}$$

for any $\alpha_i, \beta_i \in k$, $a, b \in S$, $i = 0, 1, 2$, and where indices are taken modulo 3.

In [**CE07b**] it is shown that the Lie superalgebra of derivations of J is naturally isomorphic to the Lie superalgebra $\mathfrak{g}(S_1, S)$ in the first row of the Supermagic Square.

This is well-known for algebras, as $\mathfrak{g}(S_1, S)$ is isomorphic to the Lie algebra $\mathcal{T}(k, H_3(C))$ obtained by means of Tits construction (see [**Eld04**] and [**BS03**]), and this latter algebra is, by its own construction, the derivation algebra of $H_3(C)$. What was done in [**CE07b**, Section 3] is to make explicit this isomorphism $\mathfrak{g}(S_1, S) \cong \mathfrak{Der}\, J$ and extend it to superalgebras.

To begin with, (2.4) shows that J is graded over $\mathbb{Z}_2 \times \mathbb{Z}_2$ with:

$$J_{(0,0)} = k^3, \quad J_{(1,0)} = \iota_0(S), \quad J_{(0,1)} = \iota_1(S), \quad J_{(1,1)} = \iota_2(S)$$

and, therefore, $\mathfrak{Der}\, J$ is accordingly graded over $\mathbb{Z}_2 \times \mathbb{Z}_2$:

$$(\mathfrak{Der}\, J)_{(i,j)} = \{d \in \mathfrak{Der}\, J : d\big(J_{(r,s)}\big) \subseteq J_{(i+r, j+s)}\ \forall r, s = 0, 1\}.$$

Moreover, the zero component is ([**CE07b**, Lemmas 3.4 and 3.5]):

$$(\mathfrak{Der}\, J)_{(0,0)} = \{d \in \mathfrak{Der}\, J : d(e_i) = 0 \ \forall i = 0,1,2\},$$

and the linear map given by

$$\begin{aligned} \mathfrak{tri}(S) &\longrightarrow (\mathfrak{Der}\, J)_{(0,0)} \\ (d_0,d_1,d_2) &\mapsto D_{(d_0,d_1,d_2)}, \end{aligned}$$

where

$$\begin{cases} D_{(d_0,d_1,d_2)}(e_i) = 0, \\ D_{(d_0,d_1,d_2)}\big(\iota_i(a)\big) = \iota_i\big(d_i(a)\big) \end{cases} \tag{2.5}$$

for any $i = 0,1,2$ and $a \in S$, is an isomorphism.

Given any two elements x, y in a Jordan superalgebra, the commutator (in the graded sense) of the left multiplications by x and y:

$$d_{x,y} = [L_x, L_y] \tag{2.6}$$

is a derivation. These derivations are called inner derivations. For any $i = 0,1,2$ and $a \in S$, consider the following inner derivation of the Jordan superalgebra J:

$$D_i(a) = 2d_{\iota_i(a),e_{i+1}} = 2\big[L_{\iota_i(a)}, L_{e_{i+1}}\big] \tag{2.7}$$

(indices modulo 3), where L_x denotes the multiplication by x in J. Note that the restriction of L_{e_i} to $\iota_{i+1}(S) \oplus \iota_{i+2}(S)$ is half the identity, so the inner derivation $\big[L_{\iota_i(a)}, L_{e_i}\big]$ is trivial on $\iota_{i+1}(S) \oplus \iota_{i+2}(S)$, which generates J. Hence

$$\big[L_{\iota_i(a)}, L_{e_i}\big] = 0 \tag{2.8}$$

for any $i = 0,1,2$ and $a \in S$. Also, $L_{e_0+e_1+e_2}$ is the identity map, so the bracket $\big[L_{\iota_i(a)}, L_{e_0+e_1+e_2}\big]$ is 0 and hence

$$D_i(a) = 2\big[L_{\iota_i(a)}, L_{e_{i+1}}\big] = -2\big[L_{\iota_i(a)}, L_{e_{i+2}}\big]. \tag{2.9}$$

A straightforward computation with (2.4) gives

$$\begin{aligned} &D_i(a)(e_i) = 0,\ D_i(a)(e_{i+1}) = \frac{1}{2}\iota_i(a),\ D_i(a)(e_{i+2}) = -\frac{1}{2}\iota_i(a), \\ &D_i(a)\big(\iota_{i+1}(b)\big) = -\iota_{i+2}(a \bullet b), \\ &D_i(a)\big(\iota_{i+2}(b)\big) = (-1)^{|a||b|}\iota_{i+1}(b \bullet a), \\ &D_i(a)\big(\iota_i(b)\big) = 2\mathrm{b}(a,b)(-e_{i+1} + e_{i+2}), \end{aligned} \tag{2.10}$$

for any $i = 0,1,2$ and any homogeneous elements $a, b \in S$.

Denote by $D_i(S)$ the linear span of the $D_i(a)$'s, $a \in S$. Then the remaining components of the $\mathbb{Z}_2 \times \mathbb{Z}_2$-grading of $\mathfrak{Der}\, J$ are given by ([**CE07b**, Lemma 3.11]):

$$(\mathfrak{Der}\, J)_{(1,0)} = D_0(S), \quad (\mathfrak{Der}\, J)_{(0,1)} = D_1(S), \quad (\mathfrak{Der}\, J)_{(1,1)} = D_2(S).$$

Therefore, the $\mathbb{Z}_2 \times \mathbb{Z}_2$-grading of $\mathfrak{Der}\, J$ becomes

$$\mathfrak{Der}\, J = D_{\mathfrak{tri}(S)} \oplus \big(\oplus_{i=0}^2 D_i(S)\big)$$

On the other hand, $S_1 = k1$, with $1 \bullet 1 = 1$ and $\mathrm{b}(1,1) = 2$, so $\mathfrak{tri}(S_1) = 0$ and for the para-Hurwitz superalgebra S:

$$\begin{aligned} \mathfrak{g}(S_1, S) &= \mathfrak{tri}(S) \oplus \big(\oplus_{i=0}^2 \iota_i(S_1 \otimes S)\big) \\ &= \mathfrak{tri}(S) \oplus \big(\oplus_{i=0}^2 \iota_i(1 \otimes S)\big). \end{aligned} \tag{2.11}$$

THEOREM 2.12. *(See* [**CE07b**, Theorem 3.13]*) Let S be a para-Hurwitz superalgebra over k and let J be the Jordan superalgebra of 3×3 hermitian matrices over the associated Hurwitz superalgebra. Then the linear map:*

$$\Phi : \mathfrak{g}(S_1, S) \longrightarrow \mathfrak{der}\, J, \tag{2.13}$$

such that

$$\Phi\big((d_0, d_1, d_2)\big) = D_{(d_0,d_1,d_2)},$$
$$\Phi\big(\iota_i(1\otimes a)\big) = D_i(a),$$

for any $i = 0,1,2$, $a \in S$ and $(d_0, d_1, d_2) \in \mathfrak{tri}(S)$, is an isomorphism of Lie superalgebras.

The Lie superalgebra $d_{J,J} = [L_J, L_J]$ is the Lie superalgebra $\mathfrak{inder}\, J$ of inner derivations of J. It turns out that $(\mathfrak{der}\, J)_{(r,s)} = (\mathfrak{inder}\, J)_{(r,s)}$ for $(r,s) \neq (0,0)$, while

$$(\mathfrak{inder}\, J)_{(0,0)} = \sum_{i=0}^{2} \big[L_{\iota_i(S)}, L_{\iota_i(S)}\big] = D_{\sum_{i=0}^{2}\theta^i(t_{S,S})} = \Phi\big(\sum_{i=0}^{2}\theta^i(t_{S,S})\big)$$

(recall that $\theta\big((d_0, d_1, d_2)\big) = (d_2, d_0, d_1)$ for any $(d_0, d_1, d_2) \in \mathfrak{tri}(S)$).

In characteristic 3, $\mathfrak{tri}(S) = \sum_{i=0}^{2}\theta^i(t_{S,S})$ if $\dim S = 1, 4$ or 8 ([**Eld04**]), and the same happens with $\mathfrak{tri}(S_{1,2})$ and $\mathfrak{tri}(S_{4,2})$, because of [**CE07a**, Corollaries 2.12 and 2.23], while for $\dim S = 2$, $\mathfrak{tri}(S)$ has dimension 2 and $\sum_{i=0}^{2}\theta^i(t_{S,S}) = t_{S,S}$ has dimension 1 (see [**Eld04**]). In characteristic $\neq 3$, $\mathfrak{tri}(S) = \sum_{i=0}^{2}\theta^i(t_{S,S})$ always holds.

COROLLARY 2.14. *(See* [**CE07b**, Corollary 3.15]*) Let S be a para-Hurwitz (super)algebra over k, and let J be the Jordan (super)algebra of 3×3 hermitian matrices over the associated Hurwitz (super)algebra. Then $\mathfrak{der}\, J$ is a simple Lie (super)algebra that coincides with $\mathfrak{inder}\, J$ unless the characteristic is 3 and $\dim S = 2$. In this latter case $\mathfrak{inder}\, J$ coincides with $[\mathfrak{der}\, J, \mathfrak{der}\, J]$, which is a codimension 1 simple ideal of $\mathfrak{der}\, J$.*

3. The second and third rows in the Supermagic Square

The second row of the Supermagic Square is formed by the Lie superalgebras $\mathfrak{g}(S, S')$, where S and S' are symmetric composition superalgebras with $\dim S = 2$. Let S_2 be the split 2-dimensional para-Hurwitz algebra. Thus, S_2 is the para-Hurwitz algebra attached to the unital composition algebra $K = k \times k$, whose standard involution is given by $\overline{(\alpha,\beta)} = (\beta,\alpha)$ for any $\alpha, \beta \in k$. Then with $1 = (1,1)$ and $u = (1,-1)$, the multiplication and norm in S_2 are given by:

$$1\bullet 1 = 1, \qquad 1\bullet u = u\bullet 1 = -u, \quad u\bullet u = 1,$$
$$q(1) = 1, \quad \mathrm{b}(1,u) = \mathrm{b}(u,1) = 0, \quad q(u) = -1.$$

Moreover, the triality Lie algebra of S_2 is (see [**Eld04**, Corollary 3.4]) the Lie algebra:

$$\mathfrak{tri}(S_2) = \{(\alpha_0\sigma, \alpha_1\sigma, \alpha_2\sigma) : \alpha_0, \alpha_1, \alpha_2 \in k \text{ and } \alpha_0 + \alpha_1 + \alpha_2 = 0\},$$

where the linear map

$$\sigma = \frac{1}{2}\sigma_{1,u} \tag{3.1}$$

(recall (1.8)) satisfies $\sigma(1) = u$ and $\sigma(u) = 1$. Note that the element $t_{1,u} \in \mathfrak{tri}(S_2)$ defined in (1.7) is given by:

$$t_{1,u} = (2\sigma, -\sigma, -\sigma). \tag{3.2}$$

Let now J be the Jordan superalgebra considered in equation (2.2). Its *structure Lie superalgebra* (or Lie multiplication superalgebra) $\mathfrak{str}J$ is the subalgebra of the general Lie superalgebra $\mathfrak{gl}(J)$ spanned by the Lie superalgebra of derivations $\mathfrak{der}\, J$ of J and by the space L_J of left multiplications by elements in J. Since $[L_J, L_J]$ is a subalgebra of $\mathfrak{der}\, J$, it follows that

$$\mathfrak{str}J = \mathfrak{der}\, J \oplus L_J.$$

Then the center of this Lie superalgebra is spanned by L_1 (the identity map). We will consider too the projective structure Lie superalgebra $\mathfrak{pstr}J$, which is defined as the quotient of $\mathfrak{str}J$ modulo its center:

$$\mathfrak{pstr}J = \mathfrak{str}J/kL_1 = \mathfrak{str}J/kI,$$

where I denotes the identity map on J.

Then with the notations introduced in Section 2, we have:

THEOREM 3.3. *Let S be a para-Hurwitz superalgebra over k and let J be the Jordan superalgebra of 3×3 hermitian matrices over the associated Hurwitz superalgebra. Then the isomorphism Φ in equation (2.13) extends to the following isomorphism of Lie superalgebras*

$$\Phi_2 : \mathfrak{g}(S_2, S) \longrightarrow \mathfrak{pstr}J \tag{3.4}$$

where:

- *for any $(d_0, d_1, d_2) \in \mathfrak{tri}(S)$, $\Phi_2\big((d_0, d_1, d_2)\big) = D_{(d_0,d_1,d_2)} + kI$,*
- *for any $\alpha_0, \alpha_1, \alpha_2 \in k$ with $\alpha_0 + \alpha_1 + \alpha_2 = 0$, the image under Φ of the element $(\alpha_0\sigma, \alpha_1\sigma, \alpha_2\sigma) \in \mathfrak{tri}(S_2)$ is $L_{\alpha_2 e_1 - \alpha_1 e_2} + kI$ $\big(= L_{\alpha_1 e_0 - \alpha_0 e_1} + kI = L_{\alpha_0 e_2 - \alpha_2 e_1} + kI\big)$,*
- *and for any $i = 0, 1, 2$, $a \in S$, $\Phi_2\big(\iota_i\big((\alpha 1 + \beta u) \otimes a\big)\big) = \big(\alpha D_i(a) + \beta L_{\iota_i(a)}\big) + kI$.*

PROOF. The proof is obtained by straightforward computations using the results in Section 2. Thus, for instance, for any $\alpha_i \in k$ $(i = 0, 1, 2)$ with $\alpha_0 + \alpha_1 + \alpha_2 = 0$, and any $\alpha, \beta \in k$ and $a \in S$, we get:

$$\big[(\alpha_0\sigma, \alpha_1\sigma, \alpha_2\sigma), \iota_0\big((\alpha 1 + \beta u) \otimes a\big)\big] = \alpha_0 \iota_0\big((\beta 1 + \alpha u) \otimes a\big),$$

which maps under Φ_2 to

$$\alpha_0\big(\beta D_0(a) + \alpha L_{\iota_0(a)}\big) + kI.$$

On the other hand, we have:

$$\begin{aligned}
&\big[\Phi_2\big((\alpha_0\sigma,\alpha_1\sigma,\alpha_2\sigma)\big),\Phi_2\big(\iota_0\big((\alpha 1+\beta u)\otimes a\big)\big)\big]\\
&\qquad\qquad = \big[L_{\alpha_2 e_1-\alpha_1 e_2},\alpha D_0(a)+\beta L_{\iota_0(a)}\big]+kI\\
&\qquad\qquad = \big(-\alpha[D_0(a),L_{\alpha_1 e_0-\alpha_0 e_1}]-\beta[L_{\iota_0(a)},L_{\alpha_1 e_0-\alpha_0 e_1}]\big)+kI\\
&\qquad\qquad = \alpha_0\left(\alpha L_{D_0(a)(e_1)}+\beta D_0(a)\right)+kI\\
&\qquad\qquad = \alpha_0\left(\beta D_0(a)+\alpha L_{\iota_0(a)}\right)+kI,
\end{aligned}$$

where equations (2.7), (2.8), (2.9) and (2.10) have been used.

In a similar vein, for any homogeneous $a,b\in S$, the element

$$[\iota_0(1\otimes a),\iota_0(u\otimes b)]=\mathrm{b}(a,b)t_{1,u}=\mathrm{b}(a,b)(2\sigma,-\sigma,\sigma)$$

(recall equations (3.1) and (3.2)) maps under Φ_2 to $\mathrm{b}(a,b)L_{e_2-e_1}+kI$, while we have

$$\begin{aligned}
&\big[\Phi_2\big(\iota_0(1\otimes a)\big),\Phi_2\big(\iota_0(u\otimes b)\big)\big]\\
&\qquad\qquad = [D_{\iota_0(a)},L_{\iota_0(b)}]+kI\\
&\qquad\qquad = L_{D_0(a)(\iota_0(b))}+kI=\mathrm{b}(a,b)L_{-e_1+e_2}+kI,
\end{aligned}$$

as required.

The remaining computations needed to prove that Φ_2 is an isomorphism are similar to the ones above, and will be omitted. □

COROLLARY 3.5. *The Lie superalgebras $\mathfrak{g}(S_2,S_{1,2})$ and $\mathfrak{g}(S_2,S_{4,2})$ in the Supermagic Square in characteristic* 3 *are isomorphic, respectively, to the projective structure Lie superalgebras of the Jordan superalgebras of hermitian* 3×3 *matrices over the unital composition superalgebras $B(1,2)$ and $B(4,2)$.*

Let us turn now our attention to the third row of the Supermagic Square.

Thus, let Q be a quaternion algebra (that is, a 4-dimensional unital composition algebra) over k, with multiplication denoted by juxtaposition, and denote by $\bar Q$ the para-Hurwitz algebra with multiplication given by $x\bullet y=\overline{xy}=\bar y\bar x$. Note that $\bar Q$ is the para-Hurwitz algebra attached to the opposite algebra of Q (which is isomorphic to Q). In case Q is split, then it is isomorphic to the algebra of 2×2 matrices $\mathrm{Mat}_2(k)$.

According to [**Eld04**, Corollary 3.4], the triality Lie algebra of $\bar Q$ splits as:

$$\mathfrak{tri}(\bar Q)=\ker\pi_0\oplus\ker\pi_1\oplus\ker\pi_2, \tag{3.6}$$

where $\pi_i:\mathfrak{tri}(\bar Q)\to\mathfrak{so}(Q)$ is the projection onto the ith component. Moreover, let Q^0 denote the subspace of zero trace elements in Q, that is, the subspace orthogonal to the unity element. Then ([**Eld04**, Corollary 3.4]) we have:

$$\ker\pi_0=\{(0,l_a\tau,-r_a\tau):a\in Q^0\},$$

where l_a and r_a denote the left and right multiplications in $\bar Q$ and $\tau:x\mapsto\bar x$ is the standard involution of Q. Therefore, for any $a\in Q^0$ and $x\in Q$, since $\bar a=\tau(a)=-a$, we get:

$$\begin{aligned}
l_a\tau(x)&=a\bullet\bar x=\overline{a\bar x}=x\bar a=-xa=-R_a(x),\\
r_a\tau(x)&=\bar x\bullet a=\overline{\bar x a}=\bar a x=-ax=-L_a(x),
\end{aligned}$$

where L_a and R_a denote the left and right multiplications by a in Q. Hence the ideal $\ker\pi_0$ above becomes:

$$\ker\pi_0 = \{(0, -R_a, L_a) : a \in Q^0\}.$$

and, similarly, $\ker\pi_1 = \{(L_a, 0, -R_a) : a \in Q^0\}$ and $\ker\pi_2 = \{(-R_a, L_a, 0) : a \in Q^0\}$.

Now, for any $a, b, x \in Q$, $q(x) = x\bar{x} = \bar{x}x$, so $\mathrm{b}(a,b) = a\bar{b} + b\bar{a} = \bar{a}b + \bar{b}a$ and hence we have:

$$\begin{aligned}
&r_a l_b(x) = \overline{\overline{b\bar{x}}a} = \overline{\bar{x}\bar{b}}a = \bar{a}bx = L_{\bar{a}b}(x),\\
&l_a r_b(x) = \overline{a\overline{\bar{x}b}} = \overline{a\bar{b}\bar{x}} = xb\bar{a} = R_{b\bar{a}}(x),\\
&\sigma_{a,b}(x) = \mathrm{b}(a,x)b - \mathrm{b}(b,x)a = (a\bar{x} + x\bar{a})b - a(\bar{b}x + \bar{x}b) = (-L_{a\bar{b}} + R_{\bar{a}b})(x),\\
&\sigma_{a,b}(x) = \mathrm{b}(a,x)b - \mathrm{b}(b,x)a = b(\bar{a}x + \bar{x}a) - (b\bar{x} + x\bar{b})a = (L_{b\bar{a}} - R_{\bar{b}a})(x),\\
&\frac{1}{2}\mathrm{b}(a,b)x - r_a l_b(x) = \frac{1}{2}(\bar{a}b + \bar{b}a)x - L_{\bar{a}b}(x) = \frac{1}{2}L_{\bar{b}a - \bar{a}b}(x),\\
&\frac{1}{2}\mathrm{b}(a,b)x - l_a r_b(x) = \frac{1}{2}x(a\bar{b} + b\bar{a}) - R_{b\bar{a}}(x) = \frac{1}{2}R_{a\bar{b} - b\bar{a}}(x).
\end{aligned}$$

Therefore, the element $t_{a,b} \in \mathfrak{tri}(\bar{Q})$ in (1.7) becomes:

$$\begin{aligned}
t_{a,b} &= \left(\sigma_{a,b}, \frac{1}{2}\mathrm{b}(a,b)1 - r_a l_b, \frac{1}{2}\mathrm{b}(a,b) - l_a r_b\right)\\
&= \frac{1}{2}\left(-L_{a\bar{b} - b\bar{a}} + R_{\bar{a}b - \bar{b}a}, -L_{\bar{a}b - \bar{b}a}, R_{a\bar{b} - b\bar{a}}\right).
\end{aligned}$$

It must be noticed that the subspace Q^0 of trace zero elements is a 3-dimensional simple Lie algebra under the commutator, and that any 3-dimensional simple Lie algebra appears in this way. Moreover, given any Jordan algebra H, Tits considered in [**Tit62**] the Lie algebra defined on the vector space

$$\mathcal{T}(Q, H) = \left(Q^0 \otimes H\right) \oplus \mathfrak{Der}\, H \tag{3.7}$$

endowed with the bracket given by:

- the restriction to $\mathfrak{Der}\, H$ is the commutator in $\mathfrak{Der}\, H$,
- $[d, a \otimes x] = a \otimes d(x)$,
- $[a \otimes x, b \otimes y] = \left([a,b] \otimes xy\right) - 2\mathrm{b}(a,b)d_{x,y}$,

for any $a, b \in Q^0$, $x, y \in H$, and $d \in \mathfrak{Der}\, H$. (Recall the definition of $d_{x,y}$ in (2.6).)

In the split case $Q = \mathrm{Mat}_2(k)$, the resulting Lie algebra is the well-known Tits-Kantor-Koecher Lie algebra $\mathcal{TKK}(H)$ of the Jordan algebra H. An analogous proof applies to any Jordan superalgebra.

Let us return our attention to the Jordan superalgebra J of hermitian 3×3 matrices over a unital composition superalgebra C, with associated para-Hurwitz superalgebra denoted by S, as in equation (2.2). Consider the Lie superalgebra $\mathfrak{g}(\bar{Q}, S)$ in equation (1.6). As a vector space, $\bar{Q}$ splits into the direct sum $\bar{Q} = k1 \oplus Q^0$, and together with (1.6), (2.11) and (3.6) gives the following decomposition of $\mathfrak{g}(\bar{Q}, S)$:

$$\mathfrak{g}(\bar{Q}, S) = \mathfrak{g}(S_1, S) \oplus \left(\oplus_{i=0}^2 \ker\pi_i\right) \oplus \left(\oplus_{i=0}^2 \iota_i(Q^0 \otimes S)\right)$$

Then, as in Theorem 3.3, the isomorphism Φ in equation (2.13) can be extended to $\mathfrak{g}(\bar{Q}, S)$:

THEOREM 3.8. *Let S be a para-Hurwitz superalgebra over k and let J be the Jordan superalgebra of 3×3 hermitian matrices over the associated Hurwitz superalgebra. Then the isomorphism Φ in equation (2.13) extends to the following isomorphism of Lie superalgebras*

$$\Phi_3 : \mathfrak{g}(\bar{Q}, S) \longrightarrow \mathcal{T}(Q, J) = \big(Q^0 \otimes J\big) \oplus \mathfrak{der}\, J, \tag{3.9}$$

where:

- *the restriction of Φ_3 to $\mathfrak{g}(S_1, S)$ coincides with Φ in Theorem 2.12,*
- *for any $a \in Q^0$, the elements $(0, -R_a, L_a) \in \ker \pi_0$, $(L_a, 0, -R_a) \in \ker \pi_1$ and $(-R_a, L_a, 0) \in \ker \pi_2$ map, respectively, to $\frac{1}{2}(a \otimes e_0)$, $\frac{1}{2}(a \otimes e_1)$, and $\frac{1}{2}(a \otimes e_2)$,*
- *for any $i = 0, 1, 2$, $a \in Q^0$ and $x \in S$, $\Phi_3\big(\iota_i(a \otimes x)\big) = -\frac{1}{2} a \otimes \iota_i(x)$.*

The proof is obtained by straightforward computations and thus will be omitted.

Since $\mathcal{T}(Q, J)$ is the Tits-Kantor-Koecher Lie superalgebra of the Jordan superalgebra J in case Q is the split quaternion algebra, the next corollary follows at once:

COROLLARY 3.10. *The Lie superalgebras $\mathfrak{g}(S_4, S_{1,2})$ and $\mathfrak{g}(S_4, S_{4,2})$ in the Supermagic Square in characteristic 3 are isomorphic, respectively, to the Tits-Kantor-Koecher Lie superalgebras of the Jordan superalgebras of hermitian 3×3 matrices over the unital composition superalgebras $B(1, 2)$ and $B(4, 2)$.*

4. The Lie superalgebra $\mathfrak{g}(S_{1,2}, S_{1,2})$

The tiny *Kaplansky superalgebra* ([**Kap75, McC94**]) is the 3-dimensional Jordan superalgebra $K_3 = K_{\bar{0}} \oplus K_{\bar{1}}$, with $K_{\bar{0}} = ke$ and $K_{\bar{1}} = kx + ky$, and with multiplication given by:

$$e^2 = e, \quad ex = \frac{1}{2}x = xe, \quad ey = \frac{1}{2}y = ye,$$
$$xy = e = -yx, \quad x^2 = 0 = y^2.$$

On the other hand, the simple 10-dimensional Kac superalgebra K_{10} was originally constructed in [**Kac77b**] over algebraically closed fields of characteristic 0 by Lie-theoretical methods from a 3-grading of the exceptional Lie superalgebra $F(4)$. In characteristic 3, K_{10} is no longer simple but possesses a simple ideal K_9 of dimension 9. Shestakov ([**She96**] unpublished) noticed that K_9 is isomorphic to the tensor product (as superalgebras) $K_3 \otimes K_3$. Later on, it was proven in [**BE02**] that K_{10} appears as a direct sum $k1 \oplus (K_3 \otimes K_3)$, with a natural multiplication, in any characteristic, and in particular, if the characteristic is 3, then $K_9 = K_3 \otimes K_3$.

Assume in the remainder of this section that the characteristic of our ground field k is 3.

Take the unital composition superalgebra $B(1, 2)$ in (1.2) and its para-Hurwitz counterpart $S_{1,2}$. Then take a symplectic basis $\{u, v\}$ of $(S_{1,2})_{\bar{1}} = V$, so that

$\{1, u, v\}$ is a basis of $S_{1,2}$. Since the characteristic is 3, the multiplication of $S_{1,2}$ is given by (see (1.5)):

$$1 \bullet 1 = \bar{1}\bar{1} = 1, \quad 1 \bullet x = \bar{1}\bar{x} = 1(-x) = -x = \frac{1}{2}x = x \bullet 1 \quad \text{for any } x \in (S_{1,2})_{\bar{1}},$$
$$u \bullet v = \bar{u}\bar{v} = (-u)(-v) = 1, \quad u \bullet u = 0 = v \bullet v,$$

and therefore $S_{1,2}$ is just the tiny Kaplansky superalgebra K_3.

Hence, Kac superalgebra K_9 can be identified with the superalgebra $S_{1,2} \otimes S_{1,2}$.

Now, the triality Lie superalgebra of $S_{1,2}$ is computed in [**EO02**, Theorem 5.6] (see also [**CE07a**, Corollary 2.12]):

$$\mathfrak{tri}(S_{1,2}) = \{(d, d, d) : d \in \mathfrak{osp}(S_{1,2}, \mathrm{b})\}. \tag{4.1}$$

That is, $\mathfrak{tri}(S_{1,2})$ is isomorphic to the orthosymplectic Lie superalgebra on the 3-dimensional vector superspace $S_{1,2}$ relative to the polar form of its norm. Also, in [**EO02**, Theorem 5.8] it is proven that the Lie superalgebra of derivations of $S_{1,2}$ is the whole orthosymplectic Lie superalgebra $\mathfrak{osp}(S_{1,2}, \mathrm{b})$.

In [**BE02**, Theorem 2.8] it is proven that the Lie superalgebra of derivations of the Kac superalgebra $K_9 = K_3 \otimes K_3$ is the direct sum of the Lie superalgebras of derivations of the two copies of the tiny Kaplansky superalgebra involved. That is, we have:

$$\mathfrak{der}\, K_9 = (\mathfrak{der}\, K_3 \otimes I) \oplus (I \otimes \mathfrak{der}\, K_3),$$

where I denotes the identity map and given any homogeneous linear maps $\varphi, \psi \in \mathfrak{gl}(K_3)$, $\varphi \otimes \psi$ is the linear endomorphism in $\mathfrak{gl}(K_9) = \mathfrak{gl}(K_3 \otimes K_3)$ given by:

$$(\varphi \otimes \psi)(x \otimes y) = (-1)^{\psi x}(\varphi(x) \otimes \psi(y))$$

for any homogeneous elements $x, y \in K_3$.

Consider now the quaternion algebra Q over k with a basis $\{1, e_0, e_1, e_2\}$, where 1 is the unity element, and with

$$e_i^2 = -1, \qquad e_i e_{i+1} = -e_{i+2} = -e_{i+1} e_i,$$

(indices modulo 3). Its norm is the regular quadratic form q with $q(1) = 1 = q(e_i)$, for $i = 0, 1, 2$ and where the basis above is orthogonal. Since the characteristic is 3, $q(e_0 + e_1 + e_2) = 0$, so the norm of Q represents 0 and hence Q is the split quaternion algebra, that is, it is isomorphic to $\mathrm{Mat}_2(k)$. (An explicit isomorphism can be easily constructed.)

To avoid confusion, let us denote by b_q the polar form of q. The subspace of zero trace elements is $Q^0 = ke_0 + ke_1 + ke_2$.

The Lie superalgebra $\mathcal{T}(Q, K_9)$ (see equation (3.7)), which is isomorphic to the Tits-Kantor-Koecher Lie superalgebra of K_9 as Q is split, is given by

$$\mathcal{T}(Q, J) = \big(Q^0 \otimes K_9\big) \oplus \mathfrak{der}\, K_9,$$

and hence it decomposes as:

$$\mathcal{T}(Q, K_9) = \big(\oplus_{i=0}^2 e_i \otimes (K_3 \otimes K_3)\big) \oplus \big((\mathfrak{der}\, K_3 \otimes I) \oplus (I \otimes \mathfrak{der}\, K_3)\big),$$

that is, the direct sum of three copies of the tensor product $K_3 \otimes K_3$ (or $S_{1,2} \otimes S_{1,2}$) and two copies of $\mathfrak{der}\, K_3$, which is isomorphic to the orthoysimplectic Lie superalgebra $\mathfrak{osp}(S_{1,2}, \mathrm{b})$, exactly the situation that occurs for the Lie superalgebra $\mathfrak{g}(S_{1,2}, S_{1,2}) = \mathfrak{tri}(S_{1,2}) \oplus \mathfrak{tri}(S_{1,2}) \oplus \big(\oplus_{i=0}^2 \iota_i(S_{1,2} \otimes S_{1,2}\big)$.

THEOREM 4.2. *Let k be a field of characteristic 3. Then the Lie superalgebra $\mathfrak{g}(S_{1,2}, S_{1,2})$ is isomorphic to the Tits-Kantor-Koecher Lie superalgebra of the Kac superalgebra K_9.*

PROOF. It has been checked above that both Lie superalgebras, $\mathfrak{g}(S_{1,2}, S_{1,2})$ and the Tits-Kantor-Koecher Lie superalgebra $\mathcal{T}(Q, K_9)$, split as vector superspaces into direct sums of isomorphic summands. Let us consider the explicit linear isomorphism:

$$\Psi : \mathfrak{g}(S_{1,2}, S_{1,2}) \longrightarrow \mathcal{T}(Q, K_9),$$

given by:

- $\Psi\big((d, d, d)\big) = d \otimes I \in \mathfrak{Der}\, K_9$, for any (d, d, d) in the first copy of $\mathfrak{tri}(S_{1,2})$ in $\mathfrak{g}(S_{1,2}, S_{1,2})$,
- $\Psi\big(((d', d', d')\big) = I \otimes d'$, for any (d', d', d') in the second copy of $\mathfrak{tri}(S_{1,2})$ in $\mathfrak{g}(S_{1,2}, S_{1,2})$,
- $\Psi\big(\iota_i(x \otimes x') = e_i \otimes (x \otimes x')$, for any $i = 0, 1, 2$ and $x, x' \in S_{1,2}$,

and let us check that it is an isomorphism of Lie superalgebras.

In order to prove that Ψ is indeed a Lie superalgebra isomorphism, the only nontrivial point is to prove that

$$\Psi\big([\iota_i(x \otimes x'), \iota_i(y \otimes y')]\big) = [\Psi(\iota_i(x \otimes x')), \Psi(\iota_i(y \otimes y'))]$$

for any homogeneous, $x, x', y, y' \in S_{1,2} = K_3$ and $i = 0, 1, 2$. The symmetry of the constructions shows that it is enough to deal with $i = 0$. But the description of $\mathfrak{tri}(S_{1,2})$ in (4.1), together with equations (1.6), (1.7) and (1.8) give:

$$\begin{aligned}
[\iota_0(x \otimes x'), \iota_0(y \otimes y')] &= (-1)^{xx'+xy'+yy'} \mathrm{b}(x', y') t_{x,y} + (-1)^{x'y} \mathrm{b}(x, y) t_{x',y'} \\
&= (-1)^{xx'+xy'+yy'} \mathrm{b}(x', y')(\sigma_{x,y}, \sigma_{x,y}, \sigma_{x,y}) \\
&\qquad + (-1)^{x'y} \mathrm{b}(x, y)(\sigma_{x',y'}, \sigma_{x',y'}, \sigma_{x',y'}),
\end{aligned}$$

which maps under Ψ to:

$$\begin{aligned}
&\Psi\Big([\iota_0(x \otimes x'), \iota_0(y \otimes y')]\Big) \\
&\qquad = (-1)^{xx'+xy'+yy'} \mathrm{b}(x', y')(\sigma_{x,y} \otimes I) + (-1)^{x'y} \mathrm{b}(x, y)(I \otimes \sigma_{x',y'}), \\
&\qquad = (-1)^{x'y} \left((\sigma_{x,y} \otimes \mathrm{b}(x', y')I) + (\mathrm{b}(x, y)I \otimes \sigma_{x',y'})\right) \in \mathfrak{Der}\, K_9.
\end{aligned}$$

On the other hand, we have:

$$\begin{aligned}
&[\Psi\big(\iota_0(x \otimes x')\big), \Psi\big(\iota_0(y \otimes y')\big)] \\
&\qquad = [e_0 \otimes (x \otimes x'), e_0 \otimes (y \otimes y')] \\
&\qquad = -2b_q(e_0, e_0)[L_{x \otimes x'}, L_{y \otimes y'}] \quad \text{(recall the product in (3.7))} \\
&\qquad = -[L_{x \otimes x'}, L_{y \otimes y'}] \quad (\text{as } q(e_0) = 1, \text{ so } \mathrm{b}_q(e_0, e_0) = 2 = -1) \\
&\qquad = -(-1)^{x'y} \frac{1}{2} \left(([L_x, L_y] \otimes \mathrm{b}(x', y')I) + (\mathrm{b}(x, y)I \otimes [L_{x'}, L_{y'}])\right) \\
&\qquad\qquad\qquad \text{(by [\textbf{BE02}, (2.3)])} \\
&\qquad = (-1)^{x'y} \left((\sigma_{x,y} \otimes \mathrm{b}(x', y')I) + (\mathrm{b}(x, y)I \otimes \sigma_{x',y'})\right) \\
&\qquad\qquad\qquad \text{(because of [\textbf{BE02}, (1.6)])} \\
&\qquad = \Psi\Big([\iota_0(x \otimes x'), \iota_0(y \otimes y')]\Big),
\end{aligned}$$

as required. □

The Lie superalgebra $\mathfrak{g}(S_1, S_{1,2})$ is a subalgebra of $\mathfrak{g}(S_{1,2}, S_{1,2})$. The restriction of the isomorphism Ψ in the proof of Theorem 4.2 gives our last result:

COROLLARY 4.3. *Let k be a field of characteristic* 3. *Then the Lie superalgebra $\mathfrak{g}(S_1, S_{1,2})$ in the Supermagic Square is isomorphic to the Tits-Kantor-Koecher Lie superalgebra of the tiny Kaplansky superalgebra K_3.*

Note that the Lie superalgebra $\mathfrak{g}(S_1, S_{1,2})$ is known to be isomorphic to the projective special Lie superalgebra $\mathfrak{psl}_{2,2}$ (see [**CE07a**]).

References

[AF93] Bruce N. Allison and John R. Faulkner, *Nonassociative coefficient algebras for Steinberg unitary Lie algebras*, J. Algebra **161** (1993), no. 1, 1–19.

[BS] Chris H. Barton and Anthony Sudbery, *Magic Squares of Lie Algebras*, arXiv:math.RA/0001083.

[BS03] ———, *Magic squares and matrix models of Lie algebras*, Adv. Math. **180** (2003), no. 2, 596–647.

[BE02] Georgia Benkart and Alberto Elduque, *A new construction of the Kac Jordan superalgebra*, Proc. Amer. Math. Soc. **130**, (2002), no. 11, 3209–3217.

[BGL] Sofiane Bouarroudj, Pavel Grozman and Dimitry Leites, *Cartan matrices and presentations of Cunha and Elduque Superalgebras*, arXiv.math.RT/0611391.

[CE07a] Isabel Cunha and Alberto Elduque, *An extended Freudenthal Magic Square in characteristic 3*, J. Algebra **317** (2007), 471–509.

[CE07b] ———, *The extended Freudenthal magic square and Jordan algebras*, manuscripta math. **123** (2007), no. 3, 325–351.

[Eld04] Alberto Elduque, *The magic square and symmetric compositions*, Rev. Mat. Iberoamericana **20** (2004), no. 2, 475–491.

[Eld06] ———, *New simple Lie superalgebras in characteristic 3*, J. Algebra **296** (2006), no. 1, 196–233.

[Eld07a] ———, *The Magic Square and Symmetric Compositions II*, Rev. Mat. Iberoamericana **23** (2007), 57–84.

[Eld07b] ———, *Some new simple modular Lie superalgebras*, Pacific J. Math. **231** (2007), no. 2, 337–359.

[EO02] Alberto Elduque and Susumu Okubo, *Composition superalgebras*, Comm. Algebra **30** (2002), no. 11, 5447–5471.

[Kac77a] Victor G. Kac, *Lie superalgebras*, Adv. Math. **26** (1977), no. 1, 8–96.

[Kac77b] ———, *Classification of $\mathbb{Z}$-graded Lie superalgebras and simple Jordan superalgebras*, Comm. Algebra **5** (1977), 1375–1400.

[Kap75] Irvin Kaplansky, *Graded Jordan Algebras I*, preprint 1975. (Available at `http://www.justpasha.org/math/links/subj/lie/kaplansky/`)

[LM02] Joseph M. Landsberg and Laurent Manivel, *Triality, exceptional Lie algebras and Deligne dimension formulas*, Adv. Math. **171** (2002), no. 1, 59–85.

[LM04] ———, *Representation theory and projective geometry*, Algebraic transformation groups and algebraic varieties, Encyclopaedia Math. Sci., vol. 132, Springer, Berlin, 2004, pp. 71–122.

[McC94] Kevin McCrimmon, *Kaplansky superalgebras*, J. Algebra **164** (1994), 656–694.

[Sch95] Richard D. Schafer, *An introduction to nonassociative algebras*, Dover Publications Inc., New York, 1995.

[She96] Ivan P. Shestakov, Lecture at the Jordan Algebren Conference, Oberwolfach, Germany, 1996.

[She97] ———, *Prime alternative superalgebras of arbitrary characteristic*, Algebra i Logika **36** (1997), no. 6, 675–716, 722.

[Tit62] Jacques Tits, *Une classe d'algèbres de Lie en relation avec les algèbres de Jordan*, Nederl. Akad. Wetensch. Proc. Ser. A 65 = Indag. Math. **24** (1962), 530–535.

Departamento de Matemática, Universidade da Beira Interior, 6200 Covilhã, Portugal
E-mail address: icunha@mat.ubi.pt

Departamento de Matemáticas e Instituto Universitario de Matemáticas y Aplicaciones, Universidad de Zaragoza, 50009 Zaragoza, Spain
E-mail address: elduque@unizar.es

Contemporary Mathematics
Volume **483**, 2009

Monoidal categories of comodules for coquasi Hopf algebras and Radford's formula

Walter Ferrer Santos and Ignacio López Franco

Dedicated to I. Shestakov on the occasion of his 60th birthday

ABSTRACT. We study the basic monoidal properties of the category of Hopf modules for a coquasi Hopf algebra H. In particular we discuss the so called fundamental theorem that establishes a monoidal equivalence between the category of comodules and the category of Hopf modules. We present a categorical proof of Radford's S^4 formula for the case of a finite dimensional coquasi Hopf algebra, by establishing a monoidal isomorphism between certain double dual functors.

1. Introduction

The main purpose of this paper is to study the monoidal category of Hopf modules for a coquasi Hopf algebra. As a consequence we obtain a proof of Radford's S^4 formula valid for finite dimensional coquasi Hopf algebras. Inspired in [**9**] we show that this formula is intimately related to the existence of certain natural transformation relating the left and the right double dual functors for the category of right H–comodules. This natural transformation comes from the application of the structure theorem for Hopf modules, to ${}^*H$ viewed as a right H–Hopf module.

Coquasi Hopf algebras are the dual notion of the quasi Hopf algebras defined in [**8**]. The main difference with Hopf algebras is that for coquasi Hopf algebras the role of the multiplicative and comultiplicative structures is not longer interchangeable. In a coquasi Hopf algebra the multiplicative structure is no longer one dimensional, but two dimensional (in the sense of higher category theory); this is expressed in the fact that the multiplication is no longer associative but only up to isomorphism, provided by a functional ϕ. The antipode is also defined as a two dimensional structure, the extra dimension provided by two functionals α, β.

The category of Hopf modules in the context of (co)quasi Hopf algebras has been considered by different authors and it was initially studied in [**11, 25**].

2000 *Mathematics Subject Classification.* Primary 16W30; Secondary 18D10.

The first author would like to thank, Csic-UDELAR, Conicyt-MEC, Uruguay.

The second author acknowledges the support of a Internal Graduate Studentship at Trinity College, Cambridge.

In the case of Hopf algebras, Radford's formula for S^4 was first proved in full generality in [**22**], with predecessors in [**18**] and [**27**]. A more recent proof, appears in [**26**]. There are many generalizations of the formula from the case of Hopf algebras to other situations, *e.g.*: braided Hopf algebras, bF algebras –braided and classical–, quasi Hopf algebras, weak Hopf algebras, Hopf algebras over rings and co–Frobenius Hopf algebras. The following is a partial list of references for some proofs of these generalizations: [**3**], [**7**], [**10**], [**11**], [**14**], [**15**], [**21**] and [**1**].

In particular, in [**11**, Corollary 6.3] a formulation of Radford's formula for quasi Hopf algebras is presented. It involves a conjugation by a linear functional given by the modular function ("modulus" in the nomenclature of the authors) and by an element of H involving the modular element ("comodulus") and the element $f \in H \otimes H$ constructed by Drinfel'd in [**8**] to codify the quasi-anticomultiplicativity of the antipode. Their proof, is in spirit very close to the original proof due to Radford that appeared in [**22**], but the categorical perspective that we exploit is not developed. Being coquasi and quasi Hopf algebras dual to each other in the case of finite dimension, a dualization of the formula in [**11**] produces a formula with some elements in common with ours. But, as we mentioned before, the authors' methods are very different form ours, making it difficult to say if after dualization the formulas are exactly the same. Moreover, as the categories of modules over H and comodules over H^* are naturally equivalent (in the finite dimensional case), it is clear that a categorical proof of Radford's formula in the context of quasi Hopf algebras could be easily obtained applying our methods.

An analogue of Radford's formula for finite tensor categories, closer to the spirit of the present paper, appears in [**9**]. The construction of the isomorphism in Definition 11 is, for the most part, the same as the construction given in [**9**]. However, the authors use Deligne's tensor product of abelian categories and a version of the structure theorem on Hopf modules for finite tensor categories; this makes not obvious the explicit identification of the isomorphism in [**9**, Theorem 3.3], even in the simpler case of a Hopf algebra. In the present paper, by using a version of the structure theorem that is adequate for categories of comodules, we are able to be more precise about this point. Moreover, we do not stop at the categorical level but give a version of the formula for a coquasi Hopf algebra (see Theorem 8) and show that in the case of a Hopf algebra the formula specializes to the classical one due to Radford. We also provide a full proof of the monoidality of the isomorphism in Definition 11, and thus the monoidality of the invertible functional in the formula.

Now we describe the organization of the paper. In this Introduction we briefly recall the definition and the first basic properties of a coquasi bialgebra and of a coquasi Hopf algebra. We define the notion of monoidal morphism that is the adequate notion of morphism between coquasi bialgebras.

In Section 2, that is the technical core of the paper, we present the basic properties of the monoidal categories used later. We work with the categories of comodules and Hopf modules for coquasi bialgebras and recall the properties of the monoidal structures induced by the cotensor product over H and by the tensor product over $\Bbbk$. We look at some basic monoidal functors associated to monoidal morphisms given by corestriction of scalars and its adjoints given by coinduction. We also consider other monoidal functors –*e.g.*, the left adjoint comodule and right adjoint comodule functors– that will be used later.

In Section 3 we recall that in the case of coquasi Hopf algebras with invertible antipode the monoidal categories of finite dimensional comodules –or bicomodules– are rigid, *i.e.*, each object has a left and a right dual. We use this rigidity in order to describe explicitly –for finite dimensional Hopf algebras– the monoidal structure of the antipode.

In Section 4 we present a proof of the version of the fundamental theorem on Hopf modules for coquasi Hopf algebras that we need later in the paper. By applying some general results on Hopf modules over autonomous pseudomonoids to our context we prove that the free right Hopf module functor is a monoidal equivalence from the category of comodules into the category of Hopf modules.

In Section 5, we apply the fundamental theorem on Hopf modules to $^*H$ and obtain the Frobenius isomorphism that is a morphism in the category of Hopf modules between H and $^*H$. Along the way we identify the one dimensional object of cointegrals in this context.

In Section 6, using the categorical machinery constructed above and in the same vein than in [**9**], we prove the existence of a natural monoidal isomorphism between the double duals on the left and on the right of a finite dimensional left H–module. This isomorphism will yield Radford's formula.

It is not obvious *a priori* that the formula obtained for finite dimensional coquasi Hopf algebras is related to the classical Radford's formula for Hopf algebras. Thus, in Section 7, we apply the previously developed techniques to the case that H is a classical Hopf algebra in order to deduce the original Radford's formula for S^4 –see [**22**]–.

In Section 8 we collect in an Appendix some categorical background we use in the rest of the work.

1.1. Basic definitions. Recall that the category of coalgebras and morphisms of coalgebras has a monoidal structure such that the forgetful functor into the category of vector spaces is monoidal. In other words, the tensor product over $\Bbbk$ of two coalgebras is a coalgebra and $\Bbbk$ is a coalgebra, in a canonical way.

Assume (C, Δ, ε) is a coalgebra. The maps $\Delta : C \to C \otimes C$ and $\varepsilon : C \to \Bbbk$ are the comultiplication and the counit respectively. We will use Sweedler's notation as introduced in [**28**], and write $\Delta(c) = \sum c_1 \otimes c_2$. We use the notation Δ^2 for the morphism $\Delta^2(c) = \sum c_1 \otimes c_2 \otimes c_3$. Moreover, the convolution product will be denoted by the symbol $\star$.

DEFINITION 1. A *coquasi bialgebra* structure on the coalgebra (C, Δ, ε) is a triple (p, u, ϕ) where $p : C \otimes C \to C$ *–the product–* and $u : \Bbbk \to C$ *–the unit–* are coalgebra morphisms, and $\phi : C \otimes C \otimes C \to \Bbbk$ *–the associator–* is a convolution–invertible functional, satisfying the following axioms.

$$p(u \otimes \mathrm{id}) = \mathrm{id} = p(\mathrm{id} \otimes u) \tag{1}$$

$$\sum (c_1 d_1) e_1 \phi(c_2 \otimes d_2 \otimes e_2) = \sum \phi(c_1 \otimes d_1 \otimes e_1) c_2 (d_2 e_2) \tag{2}$$

$$\sum \phi(c_1 d_1 \otimes e_1 \otimes f_1)\phi(c_2 \otimes d_2 \otimes e_2 f_2) =$$
$$= \sum \phi(c_1 \otimes d_1 \otimes e_1)\phi(c_2 \otimes d_2 e_2 \otimes f_1)\phi(d_3 \otimes e_3 \otimes f_2) \tag{3}$$

$$\phi(c \otimes 1 \otimes d) = \varepsilon(c)\varepsilon(d) \tag{4}$$

The quadruple (C,p,u,ϕ) is called a *coquasi bialgebra.* In the above equations we have written $1 \in C$ for the image under u of the unit of $\Bbbk$, and $p(c,d) = cd$. Moreover, when multiplying three elements of C we used parenthesis in order to establish the way we performed the operations.

Along this paper coalgebras and coquasi bialgebras will be denoted with the letters, C, D, etc.

The concept of quasi bialgebra was originally defined in [**8**] and the above dual concept of coquasi bialgebra has since then been extensively studied by different authors, see for example [**23**], [**24**] and the predecessor [**19**] as well as the references in the bibliography.

For later use we record below an easy consequence of the axioms of a coquasi Hopf algebra.

$$\phi(1\otimes c\otimes d) = \varepsilon(c)\varepsilon(d) \qquad \phi(c\otimes d\otimes 1) = \varepsilon(c)\varepsilon(d) \tag{5}$$

OBSERVATION 1. If (C,p,u,ϕ) is a coquasi bialgebra, then C^{cop} has a structure of a coquasi bialgebra with unit u, multiplication $p\,\mathrm{sw}$ and associator $\phi(\mathrm{id}\otimes\mathrm{sw})(\mathrm{sw}\otimes\mathrm{id})(\mathrm{id}\otimes\mathrm{sw})$. We shall denote this coquasi bialgebra by C°. In the literature C° is denoted by C^{copop}.

Next we define the concept of *monoidal morphism* between coquasi bialgebras.

Monoidal morphisms are the right kind of morphisms for our category as they preserve multiplication and unit up to coherent isomorphisms. Although we will only need this monoidal morphisms, for the sake of completeness we also give the definition of *lax monoidal* morphism.

DEFINITION 2. Let C and D be coquasi bialgebras and $f : C \to D$ be a morphism of coalgebras. A *lax monoidal structure* on f is a functional $\chi : C\otimes C \to \Bbbk$ and a scalar $\rho \in \Bbbk$ satisfying

$$(\chi\otimes p(f\otimes f))\Delta_{C\otimes C} = (fp\otimes\chi)\Delta_{C\otimes C} \qquad u = fu \tag{6}$$

$$(\phi_D(f\otimes f\otimes f))\star(\chi\otimes\varepsilon)\star(\chi(p\otimes\mathrm{id})) = (\varepsilon\otimes\chi)\star(\chi(\mathrm{id}\otimes p))\star\phi_C \tag{7}$$

$$\rho\chi(u\otimes\mathrm{id}) = \varepsilon = \rho\chi(\mathrm{id}\otimes u). \tag{8}$$

The lax monoidal structure (χ,ρ) is called a *monoidal structure* when χ is invertible, – notice that ρ is always invertible–. A morphism of coalgebras between coquasi bialgebras equipped with a (lax) monoidal structure is called a *(lax) monoidal morphism.*

The above terminology on monoidal structures comes from category theory. In fact, the concept of monoidal morphism as defined above is a special instance of the concept of monoidal 1-cell between pseudomonoids.

In particular, in the case that the monoidal structure is $\chi = \varepsilon\otimes\varepsilon$ and $\rho = 1$ equations (6), (7) and (8) simply say that f preserves the product, the unit and that $\phi_D(f\otimes f\otimes f) = \phi_C$. Hence, it is clear that a map of coquasi bialgebras that preserves the product, the coproduct and the associator, is a monoidal morphism.

OBSERVATION 2. If $f : C \to D$ and $g : D \to E$ are monoidal morphisms with monoidal structures (χ^f,ρ^f) and (χ^g,ρ^g) respectively, then gf has canonical monoidal structure, namely, $(\chi^g(f\otimes f)\star\chi^f,\rho^f\rho^g)$. Also, the identity morphism $\mathrm{id} : C \to C$ is equipped with a monoidal structure given by $(\varepsilon\otimes\varepsilon,1)$.

In Proposition 3 we show that the antipode S –see the definition below– of a finite dimensional coquasi Hopf algebra H is a monoidal morphism from $H^{\mathbf{copop}}$ to H.

DEFINITION 3. An *antipode* for the coquasi bialgebra H is a triple (S, α, β) where $S : H^{\mathrm{cop}} \to H$ is a coalgebra morphism and the functionals $\alpha, \beta : H \to \Bbbk$ satisfy the following equations.

$$\sum S(h_1)\alpha(h_2)h_3 = \alpha(h)1 \qquad \sum h_1\beta(h_2)S(h_3) = \beta(h)1 \tag{9}$$

$$\sum \phi^{-1}(h_1 \otimes Sh_3 \otimes h_5)\beta(h_2)\alpha(h_4) = \varepsilon(h) \tag{10}$$

$$\sum \phi(Sh_1 \otimes h_3 \otimes Sh_5)\alpha(h_2)\beta(h_4) = \varepsilon(h) \tag{11}$$

A *coquasi Hopf algebra* is a coquasi bialgebra equipped with an antipode.

Along this paper coquasi Hopf algebras will be denoted as H.

OBSERVATION 3. If (S, α, β) is an antipode for the coquasi bialgebra H then (S, β, α) is an antipode for the coquasi bialgebra H° considered in Observation 1.

OBSERVATION 4. For future use we record the following fact. If $a \in H$ is a group like element then $S(a) = a^{-1}$. Indeed, from the equality $\phi(Sa \otimes a \otimes Sa)\alpha(a)\beta(a) = 1$, we deduce that $\alpha(a) \neq 0$. Then, from the equality $S(a)\alpha(a)a = \alpha(a)1$ we deduce that $S(a)a = 1$. The equality $aS(a) = 1$ can be proved in a similar manner.

Moreover, if $b \in H$ satisfies that $ab = ba = 1$, then $Sa = (ba)Sa$. Reassociating the product, if we call $\gamma : H \to \Bbbk$ the functional $\gamma(x) = \phi(x \otimes a \otimes Sa)$, we have that: $Sa = (ba)Sa = \gamma^{-1} \rightharpoonup b \leftharpoonup \gamma$. Then $b = \gamma \rightharpoonup Sa \leftharpoonup \gamma^{-1}$ and since Sa is a group like element, $b = \gamma \rightharpoonup Sa \leftharpoonup \gamma^{-1} = Sa$.

In the particular case of the group like element 1, we have that $S(1) = 1$.

Along this paper we study the basic properties of the categories of modules and comodules for a coquasi Hopf algebra. These categories have been considered by many authors –see for example [**16**] and more specifically [**11**] and [**25**].

Our main interest lays in the case that the coquasi Hopf algebra is finite dimensional as a vector space. In this case, it is known (see [**5**] and [**25**]) that the antipode S is a bijective linear transformation. The composition inverse of S will be denoted as $\overline{S}$.

We finish this Introduction by describing some of the notations we use.

We denote the usual duality functor in the category of vector spaces as $V \mapsto V^\vee$ and the usual evaluation and coevaluation maps as e and c.

Let $\mathcal{C} = (\mathcal{C}, \otimes, \Bbbk, \Phi, l, r)$ be a monoidal category with monoidal structure $\otimes : \mathcal{C} \times \mathcal{C} \to \mathcal{C}$, unit object object $\Bbbk$, associativity constraint with components $\Phi_{M,N,L} : (M \otimes N) \otimes L \to M \otimes (N \otimes L)$ and left and right unit constraints l and r. We denote as $\mathcal{C}^{\mathrm{rev}}$ the monoidal category $(\mathcal{C}, \otimes^{\mathrm{rev}} = \otimes \mathrm{sw}, \Bbbk, \widehat{\Phi}, \widehat{r}, \widehat{l})$ where $\otimes^{\mathrm{rev}}(M \otimes N) = N \otimes M$, $\widehat{\Phi}_{M,N,L} = \Phi^{-1}_{L,N,M}$, $\widehat{r} = l$ and $\widehat{l} = r$. Let $\mathcal{C}$ and $\mathcal{D}$ be monoidal categories and $T : \mathcal{C} \to \mathcal{D}$ a functor. A monoidal structure on T is a natural isomorphism $\otimes(T \times T) \Rightarrow T\otimes : \mathcal{C} \times \mathcal{C} \to \mathcal{D}$ and an isomorphism $\Bbbk \to T(\Bbbk)$ satisfying a certain natural list of coherence axioms (see [**13**] for details). A monoidal functor is a functor equipped with a monoidal structure.

We assume that the reader is familiar with the basic concepts concerning rigidity for monoidal categories as presented for example in [**13**] or [**16**]. Recall that monoidal functors preserve duals. In other words, if $T : \mathcal{C} \to \mathcal{D}$ is a monoidal functor and $M \in \mathcal{C}$ is a left rigid object in $\mathcal{C}$, then $T(M)$ is also left rigid and there is a canonical natural isomorphism $\eta : T({}^*M) \to {}^*T(M)$. This isomorphism is the unique arrow such that the following two compositions are equal

$$T({}^*M) \otimes T(M) \xrightarrow{\cong} T({}^*M \otimes M) \xrightarrow{T\mathrm{ev}_M} T(\Bbbk)$$

$$T({}^*M) \otimes T(M) \xrightarrow{\eta \otimes \mathrm{id}} {}^*T(M) \otimes T(M) \xrightarrow{\mathrm{ev}_{T(M)}} \Bbbk \xrightarrow{\cong} T(\Bbbk)$$

where the unlabelled isomorphisms are the natural isomorphisms of the monoidal structure of T.

Also, following the standard usage we write ${}^C\mathcal{M}$ (${}^C\mathcal{M}_f$), $\mathcal{M}^D(\mathcal{M}^D_f)$ and ${}^C\mathcal{M}^D$ (${}^C\mathcal{M}^D_f$) for the categories of left (finite dimensional) C–comodules, right (finite dimensional) D–comodules and (finite dimensional) (C, D)–bicomodules, where C and D are coalgebras.

2. The categories of bicomodules and of Hopf modules

2.1. The cotensor product. We start by briefly reviewing some of the basic properties of the cotensor product. Given coalgebras C, D and E, one can define the *cotensor product* functor ${}^C\mathcal{M}^D \times {}^D\mathcal{M}^E \to {}^C\mathcal{M}^D$ as follows. If M and N are objects of ${}^C\mathcal{M}^D$ and ${}^D\mathcal{M}^E$ respectively, its cotensor product over D, denoted by $M\square_D N$ is the equalizer of the following diagram

$$M \otimes N \underset{\mathrm{id}\otimes((\mathrm{id}\otimes\mathrm{id}\otimes\varepsilon)\chi_N)}{\overset{((\varepsilon\otimes\mathrm{id}\otimes\mathrm{id})\chi_M)\otimes\mathrm{id}}{\rightrightarrows}} M \otimes D \otimes N$$

endowed with the bicomodule structure induced by the left coaction of M and the right coaction of N.

If F is another coalgebra, there is a natural isomorphism between the two obvious functors ${}^C\mathcal{M}^D \times {}^D\mathcal{M}^E \times {}^E\mathcal{M}^F \to {}^C\mathcal{M}^F$, with components $L\square_D(M\square_E N) \cong (L\square_D M)\square_E N$ induced by the universal property of the equalizers. Also, the functors $C\square_C -$ and $-\square_D D : {}^C\mathcal{M}^D \to {}^C\mathcal{M}^D$ are canonically isomorphic to the identity functor. All these data satisfy coherence conditions; in categorical terminology we say that the categories of bicomodules form a *bicategory* [**2**].

From the above, it is clear that the cotensor product provides a monoidal structure to ${}^C\mathcal{M}^C$, with unit object (C, Δ^2).

OBSERVATION 5. The cotensor product functor ${}^C\mathcal{M}^D \times {}^D\mathcal{M}^E \to {}^C\mathcal{M}^D$ preserves filtered colimits in each variable. This is because finite limits commute with finite colimits.

Next we consider corestriction functors.

DEFINITION 4. If $f : C \to D$ is a morphism of coalgebras, we shall denote by $f_+ = C_f \in {}^C\mathcal{M}^D$ the object obtained from the regular bicomodule C by correstriction with f on the right, *i.e.*, the coaction in $f_+ = C_f$ is given by $x \mapsto \sum x_1 \otimes x_2 \otimes f(x_3)$. Similarly, we shall denote by $f^+ = {}_fC \in {}^D\mathcal{M}^C$ the object obtained form C by correstriction on the left.

Taking cotensor products with bicomodules that are induced by morphisms of coalgebras has convenient properties.

OBSERVATION 6. Suppose that f, C and D are as above and that A is an arbitrary coalgebra.
(1) For any bicomodule $M \in {}^A\mathcal{M}^C$, with coaction χ_M, the cotensor product $M\Box_C f_+ = M\Box_C C_f \in {}^A\mathcal{M}^D$ is canonically isomorphic with the bicomodule –sometimes called also M_f– with underlying space M and coaction $(\mathrm{id}_A \otimes \mathrm{id}_M \otimes f)\chi_M : M \to A \otimes M \otimes D$. In a completely analogous way, if $N \in {}^C\mathcal{M}^A$, then the cotensor product $f^+\Box_C N = {}_fC\Box_C N \in {}^D\mathcal{M}^A$ is canonically isomorphic to the bicomodule –sometimes called ${}_fN$– with underlying space N and coaction $(f \otimes \mathrm{id}_N \otimes \mathrm{id}_C)\chi_N : N \to D \otimes N \otimes A$. Hence, $-\Box_C f_+ = -\Box_C C_f$ and $f^+\Box_C - = {}_fC\Box_C -$ are the functors $M \mapsto M_f : {}^A\mathcal{M}^C \to {}^A\mathcal{M}^D$ and $N \mapsto {}_fN : {}^C\mathcal{M}^A \to {}^D\mathcal{M}^A$ given by correstriction with f.
(2) Given a morphism of coalgebras $f : C \to D$, it is clear that the functor considered above $-\Box_C f_+ = -\Box_C C_f : {}^A\mathcal{M}^C \to {}^A\mathcal{M}^D$ is left adjoint to the so called coinduction functor $-\Box_D f^+ = -\Box_D {}_fC : {}^A\mathcal{M}^D \to {}^A\mathcal{M}^C$. Similarly, the other correstriction functor $f^+\Box_C - = {}_fC\Box_C - : {}^C\mathcal{M}^A \to {}^D\mathcal{M}^A$ is left adjoint to the so called coinduction functor $f_+\Box_D - = C_f\Box_C - : {}^D\mathcal{M}^A \to {}^C\mathcal{M}^A$
(3) Assume that we have two morphisms of coalgebras $f : C \to D$ and $g : D \to E$. In that situation we have canonical isomorphisms between $(gf)_+ \cong f_+\Box_D g_+$ and $(gf)^+ \cong g^+\Box_D f^+$.

DEFINITION 5. Assume that the coalgebra D has a group like element that we call $1 \in D$. We apply the above construction to the morphism of coalgebras $u : \Bbbk \to D$. In this case we abbreviate $(-)_0 = -\Box_\Bbbk u_+ = (-)_u : {}^A\mathcal{M} \to {}^A\mathcal{M}^D$. Similarly we call ${}_0(-) = u^+\Box_\Bbbk - = {}_u(-) : \mathcal{M}^A \to {}^D\mathcal{M}^A$.

OBSERVATION 7. The underlying space for M_0 is M and the explicit formula for the associated coaction is $\chi_0(m) = \sum m_{-1} \otimes m_0 \otimes 1$ if $(M, \chi) \in {}^A\mathcal{M}$. As we observed before this functor is left adjoint to $-\Box_\Bbbk u^+ : {}^A\mathcal{M}^D \to {}^A\mathcal{M}$. It is clear that if $M \in {}^A\mathcal{M}^D$, then $N\Box_\Bbbk u^+ = N^{\mathrm{co}D}$. In other words, the functor $(-)_0$ that produces from a left A–comodule the (A, D)–bicomodule with trivial right structure is left adjoint to the fixed point functor.

The underlying space to ${}_0M$ is the same than M and the coaction is is ${}_0\chi(m) = \sum 1 \otimes m_0 \otimes m_1$. This functor is left adjoint to $u_+\Box_\Bbbk - : {}^D\mathcal{M}^A \to \mathcal{M}^A$, that is the functor that takes the left coinvariants, i.e., sends $M \mapsto {}^{\mathrm{co}D}M$. In other words, the functor $(-)_0$ that produces from a right A–comodule the (D, A)–bicomodule with the trivial left structure is left adjoint to the left fixed point functor.

THEOREM 1. *Let $g, h : C \to D$ be two morphisms of coalgebras, and consider the following structures.*

(1) *Functionals $\gamma : C \to \Bbbk$ satisfying $(\gamma \otimes g)\Delta = (h \otimes \gamma)\Delta$.*
(2) *Morphisms of bicomodules $\theta : g_+ \to h_+$.*
(3) *Natural transformations $\Theta : (-\Box_C g_+) \Rightarrow (-\Box_C h_+) : \mathcal{M}^C \to \mathcal{M}^D$.*

Each structure of type (1) induces a structure of type (2) by $\theta = (\mathrm{id}_C \otimes \gamma)\Delta$ and each structure of type (2) induces a structure of type (3) by $\Theta = -\Box_C \theta$. Moreover, if C is finite dimensional these correspondences are bijections, with inverses given by $\gamma = \varepsilon\theta$ and $\theta = \Theta_C$. The identity and the composition of the natural transformations in (3) correspond to the identity and the composition of the morphisms in (2) and to the counit ε and the convolution product of the functionals in (1). In particular, the natural transformation Θ is invertible iff the associated morphism θ is invertible iff the corresponding functional γ is convolution–invertible.

PROOF. That each structure (1) induces a structure (2) and each structure in (2) induces a structure (3) is easily verified.

Next we prove that the above described maps are indeed a bijection between (1) and (2). The map θ satisfies

$$(\mathrm{id} \otimes \theta \otimes g)\Delta^2 = (\mathrm{id} \otimes \mathrm{id} \otimes h)\Delta^2\theta. \tag{12}$$

Composing the above equality with $\mathrm{id}\otimes\mathrm{id}\otimes\varepsilon$ we deduce that $\Delta\theta = (\mathrm{id}\otimes\theta)\Delta$. Now, if we compose equation (12) with $\varepsilon \otimes \varepsilon \otimes \mathrm{id}$ we obtain $(\gamma \otimes g)\Delta = h\theta$, with $\gamma = \varepsilon\theta$. A direct calculation shows that $(h \otimes \gamma)\Delta = (h \otimes \varepsilon)(\mathrm{id} \otimes \theta)\Delta = (h \otimes \varepsilon)\Delta\theta = h\theta$. Hence, $\gamma = \varepsilon\theta$ satisfies condition (1). Clearly, the correspondences given above between elements γ and θ are inverses of each other.

If we assume that C is finite dimensional, the bijection between the structures in (2) and (3) is a consequence of Observation 21 in the Appendix. □

For a functional γ as in Theorem 1.3, we will denote as $\gamma_+ : g_+ \to h_+$ the associated morphism of comodules and as Γ the corresponding natural transformation.

OBSERVATION 8. In the case that γ is convolution invertible, the condition (3) that relates g and h in Theorem 1, can be written as $g = \gamma^{-1} \star h \star \gamma$ or as any of the equalities below valid for all $c \in C$:

$$g(c \leftharpoonup \gamma) = h(\gamma \rightharpoonup c) \qquad g(c) = h(\gamma \rightharpoonup c \leftharpoonup \gamma^{-1}) \tag{13}$$

2.2. The tensor product over $\Bbbk$. When we consider coalgebras that have the additional structure of a coquasi bialgebra, the corresponding categories of comodules and of bicomodules have –besides the monoidal structure given by the cotensor product– another monoidal structure. This monoidal structure is based upon the tensor product over the base field $\Bbbk$ with associativity constraint defined in terms of the corresponding functional ϕ. For example if C and D are coquasi bialgebras with associators ϕ_C and ϕ_D respectively, if $L, M, N \in {}^C\mathcal{M}^D$ the associativity constraint is the map

$\Phi_{L,M,N} : (L \otimes M) \otimes N \to L \otimes (M \otimes N)$

given by the formula

$$\Phi((l \otimes m) \otimes n) = \sum \phi_C(l_{-1} \otimes m_{-1} \otimes n_{-1}) l_0 \otimes (m_0 \otimes n_0)\phi_D^{-1}(l_1 \otimes m_1 \otimes n_1) \tag{14}$$

Here we view $M\otimes N$ as an object in ${}^C\mathcal{M}^D$ with the usual structure: $\chi_{M\otimes N}(m\otimes n) = \sum m_{-1}n_{-1} \otimes m_0 \otimes n_0 \otimes m_1 n_1 \in C \otimes M \otimes N \otimes D$. Notice that the above formula for the associativity constraint can be written using the standard actions associated to coactions as follows:

$\Phi_{L,M,N}((l \otimes m) \otimes n) = \phi_D^{-1} \rightharpoonup l \otimes m \otimes n \leftharpoonup \phi_C \in L \otimes (M \otimes N)$.

The unit constraints $M \otimes \Bbbk \cong M$ and $\Bbbk \otimes M \cong M$ are the same than in the category of $\Bbbk$–vector spaces.

In case that the categories are ${}^C\mathcal{M}$ or $\mathcal{M}^C$, the constraints are defined similarly but using only the action by ϕ^{-1} on the left for $\mathcal{M}^C$ and of ϕ on the right for ${}^C\mathcal{M}$.

OBSERVATION 9. If C is a coquasi bialgebra with unit u and multiplication p, the triple (C, p, u) is an associative algebra in ${}^C\mathcal{M}^C$. This can be proved directly using equation (2), which can be rewritten as $p(\mathrm{id}\otimes p)\Phi_{C,C,C} = p(p\otimes \mathrm{id}) : (C \otimes C) \otimes C \to C$.

DEFINITION 6. The category of right C–modules within ${}^C\mathcal{M}^C$ will be denoted as ${}^C\mathcal{M}^C_C$ and called the category of Hopf modules. Similarly we define the category ${}^C_C\mathcal{M}^C$.

Notice that the unit object $\Bbbk$ of the monoidal structure $\otimes$ is canonically a Hopf module with action given by the counit ε.

The category of Hopf modules in this context was first considered in [**11**] and [**25**].

OBSERVATION 10. a) The $\Box_C$ monoidal structure of ${}^C\mathcal{M}^C$ lifts to a monoidal structure on ${}^C\mathcal{M}^C_C$ in such a way that the forgetful functor ${}^C\mathcal{M}^C_C \to {}^C\mathcal{M}^C$ is monoidal.

Indeed, if M, N, L, R are in ${}^C\mathcal{M}^C$, one easily can define –using the universal property of equalizers– a natural morphism of bicomodules $(M\,\Box_C N)\otimes(L\,\Box_C R)\to (M\otimes L)\,\Box_C(N\otimes R)$ relating both monoidal structures on ${}^C\mathcal{M}^C$. If $L=R=C$, we obtain a map $(M\,\Box_C N)\otimes C\to(M\otimes C)\,\Box_C(N\otimes C)$ that composed with the right C–actions on M and N —$a_M : M\otimes C\to M$ and $a_N : N\otimes C\to N$—endows $M\,\Box_C N$ with the structure of a Hopf module

$$(M\,\Box_C N)\otimes C\to(M\otimes C)\,\Box_C(N\otimes C)\xrightarrow{a_M\otimes a_N} M\,\Box_C N. \tag{15}$$

Clearly the unit object C of $\Box_C$ in ${}^C\mathcal{M}^C$ is also a unit object in ${}^C\mathcal{M}^C_C$.
b) If $f:\Bbbk\to C$ and $g:C\to C$ are morphisms of coalgebras –in particular this means that $f(1)\in C$ is a group like element– then $f_+\otimes(M\Box_C\, g_+)\cong M\Box_C\, p(f\otimes g)_+$ and $(M\Box_C\, g_+)\otimes f_+\cong M\Box_C\, p(g\otimes f)_+$.

2.3. Monoidal functors induced by monoidal morphisms.

THEOREM 2. *Let $f : C\to D$ be a coalgebra morphism, and consider the following structures.*

1. *Monoidal structures on f,*
2. *Monoidal structures on the functor $(-\Box_C f_+):\mathcal{M}^C\to\mathcal{M}^D$,*
3. *Monoidal structures on the functor $(f^+\Box_C -):{}^C\mathcal{M}\to{}^D\mathcal{M}$.*

Each structure (1) induces structures (2) and (3). Moreover, if C is finite dimensional there are bijections between the three types of structures.

PROOF. First, we consider the relationship of structures of type (1) with structures of type (2). If $\chi : C\otimes C\to\Bbbk$, $\rho\in\Bbbk$ is a monoidal structure on $f:C\to D$, then the transformation with components $\Theta_{M,N} : M\otimes N\to M\otimes N$ given by $\Theta_{M,N}(m\otimes n)=\chi\rightharpoonup(m\otimes n)=\sum\chi(m_1\otimes n_1)m_0\otimes n_0$ together with the isomorphism $\Bbbk\to\Bbbk$ given by multiplication by ρ is a monoidal structure as in (2). Indeed, for example the condition that the map $\Theta_{M,N} : M_f\otimes N_f\to(M\otimes N)_f$ is a morphism of H–comodules –recall the notations of Observation 6– is equivalent with condition (6) in Definition 2. A structure as in (3) is obtained in a similar way, using χ^{-1} and ρ^{-1}.

If we now assume that C is finite dimensional we can proceed backwards in order to go from (2) to (1). Let Θ be a natural transformation as depicted below.

$$\begin{array}{ccc} \mathcal{M}^C\times\mathcal{M}^C & \xrightarrow{(-\Box_C f_+)\times(-\Box_C f_+)} & \mathcal{M}^D\times\mathcal{M}^D \\ {\scriptstyle\otimes}\downarrow & \Downarrow\Theta & \downarrow{\scriptstyle\otimes} \\ \mathcal{M}^C & \xrightarrow[(-\Box_C f_+)]{} & \mathcal{M}^D \end{array} \tag{16}$$

Since all the functors in this diagram preserve filtered colimits, Θ is determined by its restriction to the categories of finite dimensional comodules. So we can substitute the categories of comodules in diagram (16) by the corresponding categories of finite dimensional comodules. In the appendix –Section 8– we prove that for a finite dimensional coalgebra C, composition with the tensor product functor $\otimes_{\Bbbk} : \mathcal{M}_f^C \times \mathcal{M}_f^C \to \mathcal{M}_f^{C\otimes C}$ induces an equivalence $\mathrm{Lex}[\mathcal{M}_f^{C\otimes C}, \mathcal{M}_f^D] \simeq \mathrm{Lex}[\mathcal{M}_f^C, \mathcal{M}_f^C; \mathcal{M}_f^D]$. The category $\mathrm{Lex}[\mathcal{M}_f^C, \mathcal{M}_f^C; \mathcal{M}_f^D]$ appearing on the right hand side of the equivalence is the category of functors from $\mathcal{M}_f^C \otimes \mathcal{M}_f^C$ to $\mathcal{M}_f^D$ which are left exact in each variable–see also the Appendix for the definition of $\mathcal{M}_f^C \otimes \mathcal{M}_f^C$–. Using this fact, we deduce that natural transformations as in (16) are in bijection with natural transformations as in the diagram below:

$$\begin{array}{ccc}
\mathcal{M}_f^{C\otimes C} & \xrightarrow{-\square_{C\otimes C}(f\otimes f)_+} & \mathcal{M}_f^{D\otimes D} \\
{\scriptstyle -\square_{C\otimes C}p_+}\downarrow & \Downarrow\Theta' & \downarrow{\scriptstyle -\square_{D\otimes D}p_+} \\
\mathcal{M}_f^C & \xrightarrow[-\square_C f_+]{} & \mathcal{M}_f^D .
\end{array}$$

Natural transformations Θ' are in bijective correspondence with bicomodule morphisms $(p(f\otimes f))_+ \to (fp)_+$. These bicomodule morphisms correspond bijectively to functionals $\chi : C\otimes C \to \Bbbk$ satisfying $\sum \chi(c_1 \otimes c_1')f(c_2)f(c_2') = \sum f(c_1c_1')\chi(c_2 \otimes c_2')$. The invertibility of Θ is equivalent to the invertibility of χ.

Similarly, an isomorphism $\Sigma : \Bbbk \to \Bbbk \square_C f_+$ is just an invertible scalar ρ such that $\rho f(1) = \rho 1$. The axioms of a monoidal structure for Θ, Σ translate to the axioms of a monoidal structure on f for χ, ρ.

The relationship between the structures (1) and (3) is as follows. A monoidal structure $\chi : C\otimes C \to \Bbbk, \rho \in \Bbbk$ on f induces a monoidal structure on $(f^+\square_C -)$ given by the D-comodule morphism $m\otimes n \mapsto \sum \chi^{-1}(m_{-1}\otimes n_{-1})m_0 \otimes n_0 : (f^+\square_C M)\otimes (f^+\square_C N) \to f^+\square_C(M\otimes N)$ and by $\lambda \mapsto \rho^{-1}\lambda : \Bbbk \to f^+\square_C \Bbbk$. The proof of the converse, i.e. that when C is finite dimensional every monoidal structure on $(f^+\square_C)$ is of this form for a unique (χ,ρ) is similar to the one presented above for the case of right comodules. $\square$

COROLLARY 1. *In the situation above the functors $u^+\square_{\Bbbk} - = {}_0(-) : \mathcal{M}^D \to {}^C\mathcal{M}^D$, $(-)_0 = -\square_{\Bbbk} u_+ : {}^C\mathcal{M} \to {}^C\mathcal{M}^D$, have canonical structures of monoidal functors.*

2.4. Some useful monoidal functors on comodule categories. In this subsection we describe the functors we use along this work.

DEFINITION 7. If C and D are coalgebras we define the functors

$$(-)^\circ : {}^C\mathcal{M}^D \to {}^{D^{\mathrm{cop}}}\mathcal{M}^{C^{\mathrm{cop}}} \quad (-)^r : (\mathcal{M}_f^C)^{\mathrm{op}} \to {}^C\mathcal{M}_f \quad (-)^\ell : ({}^C\mathcal{M}_f)^{\mathrm{op}} \to \mathcal{M}_f^C$$

The functor $(-)^\circ$ is the identity on arrows, and if $M \in {}^C\mathcal{M}^D$ with coaction $\chi(m) = \sum m_{-1}\otimes m_0 \otimes m_1$, then M° has M as underlying space and coaction $\chi^\circ(m) = \sum m_1 \otimes m_0 \otimes m_{-1}$. In the case when C, D are coquasi bialgebras, $(-)^\circ$ has a canonical structure of a monoidal functor $({}^C\mathcal{M}^D)^{\mathrm{rev}} \to {}^{D^\circ}\mathcal{M}^{C^\circ}$ given by the usual symmetry of vector spaces $\mathrm{sw} : M^\circ \otimes N^\circ \cong (N\otimes M)^\circ$ and the identity $\Bbbk \to \Bbbk^\circ$.

The functor $(-)^r$ is defined as follows. If $M \in \mathcal{M}_f^C$, the underlying space of M^r is $M^\vee$, the linear dual of M. If c and e denote the standard coevaluation and evaluation, the coaction for M^r is:

$$M^\vee \xrightarrow{\mathrm{id}\otimes \mathrm{c}} M^\vee \otimes M \otimes M^\vee \xrightarrow{\mathrm{id}\otimes\chi\otimes\mathrm{id}} M^\vee \otimes M \otimes C \otimes M^\vee \xrightarrow{\mathrm{e}\otimes\mathrm{id}\otimes\mathrm{id}} C \otimes M^\vee. \quad (17)$$

On arrows, $(-)^r$ is given by the usual (linear) duality functor. We call M^r the *right adjoint* of M. When C is a coquasi bialgebra, $(-)^r$ has the following canonical structure of a monoidal functor $(\mathcal{M}_f^C)^{\mathrm{op}} \to {}^C\mathcal{M}_f$. The unit constraint is the canonical isomorphism $\Bbbk \cong \Bbbk^\vee$; if $M, N \in \mathcal{M}_f^C$, then the transformation $M^r \otimes N^r \to (M \otimes N)^r$ is given by the canonical arrows $M^\vee \otimes N^\vee \to (M \otimes N)^\vee$, which are isomorphisms by dimension considerations. We should remark that here we are not thinking $M^\vee$ as a categorical dual of the vector space M but rather as the internal hom $\mathbf{Vect}(M, \Bbbk)$. This is the reason why $(-)^r$ does not reverse the order of the tensor products.

The definition of $(-)^\ell$ is analogous, if $N \in {}^C\mathcal{M}_f$, then:

$$N^\vee \xrightarrow{\mathrm{c}\otimes\mathrm{id}} N^\vee \otimes N \otimes N^\vee \xrightarrow{\mathrm{id}\otimes\chi\otimes\mathrm{id}} N^\vee \otimes C \otimes N \otimes N^\vee \xrightarrow{\mathrm{id}\otimes\mathrm{id}\otimes\mathrm{e}} C \otimes N^\vee. \quad (18)$$

If $N \in {}^C\mathcal{M}_f$, we call N^ℓ the *left adjoint* of N. When C is a coquasi bialgebra we have a monoidal functor $(-)^\ell : ({}^C\mathcal{M}_f)^{\mathrm{op}} \to \mathcal{M}_f^C$.

For future reference we record the following results that can be proved directly.

Lemma 1. *Observe that $(-)^r$ and $(-)^\ell$ are inverse monoidal equivalences and that $(-)^{r\ell} = (-)^{\ell r}$. The monoidal isomorphisms $M^{r\ell} \cong M \cong M^{\ell r}$ are just the canonical linear isomorphisms $M \cong M^{\vee\vee}$.*

Lemma 2. *For any morphism of coalgebras $f : C \to D$, the diagrams below commute.*

$$\begin{array}{ccc} (\mathcal{M}_f^C)^{\mathrm{op}} & \xrightarrow{(-\Box_C f_+)^{\mathrm{op}}} & (\mathcal{M}_f^D)^{\mathrm{op}} \\ {\scriptstyle (-)^r}\downarrow & & \downarrow{\scriptstyle (-)^r} \\ {}^C\mathcal{M}_f & \xrightarrow[f^+\Box_C -]{} & {}^D\mathcal{M}_f \end{array} \qquad \begin{array}{ccc} ({}^C\mathcal{M}_f)^{\mathrm{op}} & \xrightarrow{(f^+\Box_C -)^{\mathrm{op}}} & ({}^D\mathcal{M}_f)^{\mathrm{op}} \\ {\scriptstyle (-)^\ell}\downarrow & & \downarrow{\scriptstyle (-)^\ell} \\ \mathcal{M}_f^C & \xrightarrow[-\Box_C f_+]{} & \mathcal{M}_f^D \end{array}$$

$$\begin{array}{ccc} (\mathcal{M}^C)^{\mathrm{rev}} & \xrightarrow{(-\Box_D f_+)^{\mathrm{rev}}} & (\mathcal{M}^D)^{\mathrm{rev}} \\ {\scriptstyle (-)^\circ}\downarrow & & \downarrow{\scriptstyle (-)^\circ} \\ {}^{C^\circ}\mathcal{M} & \xrightarrow[f^{\mathrm{cop}+}\Box_{C^{\mathrm{cop}}} -]{} & {}^{D^\circ}\mathcal{M} \end{array}$$

If moreover f is a monoidal morphism, the diagrams commute as diagrams of monoidal functors.

Proof. Recall that if f has a monoidal structure $\chi : C \otimes C \to \Bbbk, \rho \in \Bbbk$, the monoidal structures on $(-\Box_C f_+)$ and $(f^+\Box_C -)$ are induced by χ, ρ and χ^{-1}, ρ^{-1}, respectively. Also, it is easy to show that the monoidal structures on $(-\Box_C f_+)^{\mathrm{op}}$ and $(f^+\Box_C -)^{\mathrm{op}}$ are induced by χ^{-1}, ρ^{-1} and χ, ρ respectively. The verification of the Lemma is direct. $\Box$

Lemma 3. *For any coquasi bialgebra C the following two monoidal functors are equal.*

$$({}^{C}\mathcal{M}_f)^{\mathrm{oprev}} \xrightarrow{((-)^{\circ})^{\mathrm{op}}} (\mathcal{M}_f^{C^{\circ}})^{\mathrm{op}} \xrightarrow{(-)^{r}} {}^{C^{\circ}}\mathcal{M}_f$$

$$({}^{C}\mathcal{M}_f)^{\mathrm{oprev}} \xrightarrow{((-)^{\ell})^{\mathrm{rev}}} (\mathcal{M}_f^{C})^{\mathrm{rev}} \xrightarrow{(-)^{\circ}} {}^{C^{\circ}}\mathcal{M}_f$$

LEMMA 4. *The following two monoidal functors are monoidally isomorphic to the identity functor via the canonical maps* $M \mapsto M^{\vee\vee}$

$$\mathcal{M}_f^C \xrightarrow{(-)^r} {}^{C}\mathcal{M}_f^{\mathrm{op}} \xrightarrow{(-)^{\circ}} (\mathcal{M}_f^{C^{\circ}})^{\mathrm{oprev}} \xrightarrow{(-)^r} {}^{C^{\circ}}\mathcal{M}_f^{\mathrm{rev}} \xrightarrow{(-)^{\circ}} \mathcal{M}_f^C$$

$${}^{C}\mathcal{M}_f \xrightarrow{(-)^{\ell}} (\mathcal{M}_f^C)^{\mathrm{op}} \xrightarrow{(-)^{\circ}} ({}^{C^{\circ}}\mathcal{M}_f)^{\mathrm{oprev}} \xrightarrow{(-)^{\ell}} (\mathcal{M}_f^{C^{\circ}})^{\mathrm{rev}} \xrightarrow{(-)^{\circ}} \mathcal{M}_f^C$$

3. Duality

In the case that the map S is invertible –for example if the coquasi Hopf algebra H is finite dimensional– the monoidal categories ${}^{H}\mathcal{M}_f$, $\mathcal{M}_f^H$ and ${}^{H}\mathcal{M}_f^H$ are rigid. In the case of ${}^{H}\mathcal{M}^H$, *e.g.*, we need to construct for every object M a left and a right dual –denoted as ${}^*M$ and M^* respectively– together with the corresponding evaluation and coevaluation maps.

If $(M,\chi) \in {}^{H}\mathcal{M}_f^H$ and $M^{\vee}$ is the dual of the underlying vector space, ${}^*M = (M^{\vee}, {}^*\chi)$ where ${}^*\chi$ is the composition

$$M^{\vee} \xrightarrow{\mathrm{id}\otimes \mathrm{c}} M^{\vee}\otimes M \otimes M^{\vee} \xrightarrow{\mathrm{id}\otimes\chi\otimes\mathrm{id}} M^{\vee}\otimes H\otimes M\otimes H\otimes M^{\vee} \xrightarrow{\mathrm{sw}\otimes\mathrm{id}\otimes\mathrm{sw}}$$

$$H\otimes M^{\vee}\otimes M\otimes M^{\vee}\otimes H \xrightarrow{\mathrm{id}\otimes\mathrm{e}\otimes\mathrm{id}\otimes\mathrm{id}} H\otimes M^{\vee}\otimes H \xrightarrow{\overline{S}\otimes\mathrm{id}\otimes S} H\otimes M^{\vee}\otimes H. \quad (19)$$

The evaluation and coevaluation morphisms are given by

$$\mathrm{ev}^{\ell} : {}^*M\otimes M \xrightarrow{\mathrm{id}\otimes\chi} {}^*M\otimes H\otimes M\otimes H \xrightarrow{\mathrm{id}\otimes\beta\overline{S}\otimes\mathrm{id}\otimes\alpha} {}^*M\otimes M \xrightarrow{\mathrm{e}} \Bbbk$$

$$\mathrm{coev}^{\ell} : \Bbbk \xrightarrow{\mathrm{c}} M\otimes {}^*M \xrightarrow{\chi\otimes\mathrm{id}} H\otimes M\otimes H\otimes {}^*M \xrightarrow{\alpha\overline{S}\otimes\mathrm{id}\otimes\beta\otimes\mathrm{id}} M\otimes {}^*M.$$

It is not hard to check using (9) that ev^{ℓ} and coev^{ℓ} are morphisms in ${}^{H}\mathcal{M}^H$. Moreover, the maps $\mathrm{ev}^{\ell} : {}^*M\otimes M \to \Bbbk \in {}^{H}\mathcal{M}^H$ and $\mathrm{coev}^{\ell} : \Bbbk \to M\otimes {}^*M \in {}^{H}\mathcal{M}^H$ satisfy the following two equations, which are direct consequences of (10) and (11).

$$\mathrm{id}_M = M \xrightarrow{\mathrm{coev}^{\ell}\otimes\mathrm{id}} (M\otimes {}^*M)\otimes M \xrightarrow{\Phi_{M,{}^*M,M}} M\otimes({}^*M\otimes M) \xrightarrow{\mathrm{id}\otimes\mathrm{ev}^{\ell}} M$$

$$\mathrm{id}_{{}^*M} = {}^*M \xrightarrow{\mathrm{id}\otimes\mathrm{coev}^{\ell}} {}^*M\otimes(M\otimes {}^*M) \xrightarrow{\Phi^{-1}_{{}^*M,M,{}^*M}} ({}^*M\otimes M)\otimes {}^*M \xrightarrow{\mathrm{ev}^{\ell}\otimes\mathrm{id}} {}^*M.$$

Analogously, $M^* = (M^{\vee},\chi^*)$, where χ^* is the composition

$$M^{\vee} \xrightarrow{\mathrm{c}\otimes\mathrm{id}} M^{\vee}\otimes M\otimes M^{\vee} \xrightarrow{\mathrm{id}\otimes\chi\otimes\mathrm{id}} M^{\vee}\otimes H\otimes M\otimes H\otimes M^{\vee} \xrightarrow{\mathrm{sw}\otimes\mathrm{id}\otimes\mathrm{sw}}$$

$$H\otimes M^{\vee}\otimes M\otimes M^{\vee}\otimes H \xrightarrow{\mathrm{id}\otimes\mathrm{id}\otimes\mathrm{e}\otimes\mathrm{id}} H\otimes M^{\vee}\otimes H \xrightarrow{S\otimes\mathrm{id}\otimes\overline{S}} H\otimes M^{\vee}\otimes H.$$

The corresponding right evaluation and coevaluation morphisms are:

$$\mathrm{ev}^r : M\otimes M^* \xrightarrow{\chi\otimes\mathrm{id}} H\otimes M\otimes H\otimes M^* \xrightarrow{\beta\otimes\mathrm{id}\otimes\alpha\overline{S}\otimes\mathrm{id}} M\otimes M^* \xrightarrow{\mathrm{e}} \Bbbk$$

$$\mathrm{coev}^r : \Bbbk \xrightarrow{\mathrm{c}} M^*\otimes M \xrightarrow{\mathrm{id}\otimes\chi} M^*\otimes H\otimes M\otimes H \xrightarrow{\mathrm{id}\otimes\alpha\otimes\mathrm{id}\otimes\beta\overline{S}} M\otimes M^*$$

As before one easily verifies that ev^r and coev^r are morphisms in ${}^{H}\mathcal{M}^H$ and also that they define a right duality.

OBSERVATION 11. In explicit terms the comodule structures for the duals are given by the following formulæ. If $(M,\chi) \in {}^H\mathcal{M}^H$ and $f \in {}^*M$ and $m \in M$, then ${}^*\chi(f) = \sum f_{-1} \otimes f_0 \otimes f_1 \in H \otimes {}^*M \otimes H$ if and only if:

$$\sum f_0(m) f_{-1} \otimes f_1 = \sum f(m_0)\overline{S}(m_{-1}) \otimes S(m_1).$$

Similarly if $(M,\chi) \in {}^H\mathcal{M}^H$ and $f \in M^*$ and $m \in M$, then $\chi^*(f) = \sum f_{-1} \otimes f_0 \otimes f_1 \in H \otimes M^* \otimes H$ if and only if:

$$\sum f_0(m) f_{-1} \otimes f_1 = \sum f(m_0) S(m_{-1}) \otimes \overline{S}(m_1).$$

LEMMA 5. *For the category $\mathcal{M}^H$, the duality functors can be expressed in terms of the functors in Definition 7, in the following way.*

$${}^*(-) : (\mathcal{M}^H_f)^{\mathrm{op}} \xrightarrow{(-)^r} {}^H\mathcal{M} \xrightarrow{(-)^\circ} \mathcal{M}^{H^{\mathrm{cop}}}_f \xrightarrow{-\square_{H^{\mathrm{cop}}} S_+} \mathcal{M}^H_f$$

$$(-)^* : (\mathcal{M}^H_f)^{\mathrm{op}} \xrightarrow{((-)^\circ)^{\mathrm{op}}} ({}^{H^{\mathrm{cop}}}\mathcal{M}_f)^{\mathrm{op}} \xrightarrow{(-)^\ell} \mathcal{M}^{H^{\mathrm{cop}}}_f \xrightarrow{-\square_{H^{\mathrm{cop}}} \overline{S}_+} \mathcal{M}^H_f$$

THEOREM 3. *If H is a finite dimensional coquasi Hopf algebra, then its antipode has a canonical structure of a monoidal morphisms of coquasi bialgebras $S : H^\circ \to H$. Moreover, this structure is given by the functional $\chi^S : H \otimes H \to \Bbbk$*

$$\chi^S(x \otimes y) = \sum \phi^{-1}(S(y_3) \otimes S(x_3) \otimes x_5)\alpha(x_4)\phi(S(y_2)S(x_2) \otimes x_6 \otimes y_5)$$
$$\alpha(y_4)\beta(x_8 y_7)\phi(S(y_1)S(x_1) \otimes (x_7 y_6) \otimes S(x_9 y_8)).$$

and corresponds to the usual monoidal structure of the left dual functor ${}^*(-)$.

PROOF. By general categorical principles, the left dual functor has a canonical monoidal structure ${}^*(-) : (\mathcal{M}^H)^{\mathrm{oprev}} \to \mathcal{M}^H$. This can be explicitly computed in terms of the coquasi Hopf algebra structure of H. On the other hand, we know the monoidal structures of the equivalences $(-)^r$ and $(-)^\circ$, hence we can explicitly compute the monoidal structure of $(-\square_{H^{\mathrm{cop}}} S_+)$. The latter is given by a monoidal structure on the coquasi bialgebra morphism $S : H^\circ \to H$ (see Theorem 2), and in fact it is given by the functional χ^S above. □

Note that the formula for χ^S above is written in function of the comultiplication of H not of the domain of S, *i.e.*, H^{cop}.

The functional in the theorem above appeared in [**4**], and in the dual case of quasi Hopf algebras in [**8**].

THEOREM 4. *Let H be a finite dimensional coquasi Hopf algebra and consider the monoidal structure on S introduced in Proposition 3 above. The canonical linear isomorphisms $M \cong M^{\vee\vee}$ provide the components for monoidal natural isomorphisms*

$${}^{**}(-) \cong -\square_H S^2_+ : \mathcal{M}^H_f \to \mathcal{M}^H_f \qquad (-)^{**} \cong -\square_H \overline{S}^2_+ : \mathcal{M}^H_f \to \mathcal{M}^H_f.$$

PROOF. If follows directly from Proposition 3, Lemma 2 applied to $S : H^{\mathrm{cop}} \to H$ and Lemma 4. □

4. The fundamental theorem of Hopf modules

In this section we present a generalization to the case of coquasi–Hopf algebras of the fundamental theorem on Hopf modules introduced by Sweedler in [**28**]. For a modern presentation of this general case see also [**23, 25**].

4.1. The fundamental theorem. In this section, in order to use (formal) duality arguments, we work with a braided monoidal category that we call $\mathcal{V}$ and with unit object $\Bbbk$ (see [**13**] as a general reference). Moreover we assume that $\mathcal{V}$ has equalizers and that the tensor product preserves equalizers in each variable.

Recall that a braiding for a monoidal category $\mathcal{V}$ is a natural isomorphism $\gamma_{X,Y} : X \otimes Y \to Y \otimes X$ satisfying two coherence axioms (see [**13**, Section 1]). The braiding induces a monoidal structure on the functor $\otimes : \mathcal{V} \times \mathcal{V} \to \mathcal{V}$. See [**13**, Section 5].

Using the coherence results in [**13**], we may assume without loss of generality that $\mathcal{V}$ is a strict monoidal category.

In the above set up, one can define coalgebra, comodule, coquasi bialgebra and coquasi Hopf algebra objects in $\mathcal{V}$. The braiding ensures that the tensor product of two coalgebras is a coalgebra, and likewise with comodules. The fact that the tensor product in $\mathcal{V}$ preserves equalizers in each variable, allows us to define the cotensor product of bicomodules in exactly the same manner than in the case that $\mathcal{V}$ is the category of vector spaces.

If C is a coquasi bialgebra we denote as $\mathcal{V}^C$, ${}^C\mathcal{V}$, ${}^C\mathcal{V}^C$, ${}^C_C\mathcal{V}^C$ and ${}^C\mathcal{V}^C_C$ the categories of right, left and bicomodules, and the category of left and right Hopf modules in $\mathcal{V}$ respectively. The first three categories have monoidal structures induced by the tensor product of $\mathcal{V}$. The category of bicomodules has also a monoidal structure given by the cotensor product $\square_C$.

In the same manner than in Definition 4, we can define the bicomodules f^+ and f_+ for a morphism $f : C \to D$ of coalgebra objects in $\mathcal{V}$.

For a coquasi bialgebra C in $\mathcal{V}$, we call $u : \Bbbk \to C$ the unit morphism. In this situation we can consider a pair of adjoint functors $(u^+\square_\Bbbk -) \dashv (u_+\square_C -) : {}^C\mathcal{V}^C \to \mathcal{V}^C$. Explicitly, if M is a bicomodule, $u_+\square_C M$ is the right comodule of left coinvariants ${}^{\mathrm{co}C}M$ and if N is a right comodule, $u^+\square_\Bbbk N$ is the basic object $N \in \mathcal{V}$ with the same right coaction than N and with left coaction given by $u \otimes \mathrm{id}_N : N \to C \otimes N$.

Definition 8. If C is a coquasi bialgebra in $\mathcal{V}$, we define the free module functor $F : {}^C\mathcal{V}^C \to {}^C_C\mathcal{V}^C$ as $F(M) = C \otimes M$ for $M \in {}^C\mathcal{V}^C$. This functor, together with the forgetful functor $U : {}^C_C\mathcal{V}^C \to {}^C\mathcal{V}^C$ constitute a pair of adjoint functors: $F \dashv U : {}^C_C\mathcal{V}^C \to {}^C\mathcal{V}^C$. Define the functor $L : \mathcal{V}^C \to {}^C_C\mathcal{V}^C$ as the composition $L = F \circ (u^+\square_\Bbbk -)$. Clearly L will have a right adjoint given as $(u_+\square_C -)\circ U$.

The monoidal structure $\square_C$, lifts from ${}^C\mathcal{V}^C$ to the category of Hopf modules in such a way that if $U : {}^C_C\mathcal{V}^C \to {}^C\mathcal{V}^C$ is the forgetful functor, then the adjunction $F \dashv U : {}^C_C\mathcal{V}^C \to {}^C\mathcal{V}^C$ is monoidal (*i.e.*, U is lax monoidal, F is strong monoidal and the unit and counit of the adjunction are monoidal natural transformations).

Without imposing further restrictions on the category $\mathcal{V}$, it is not hard to prove the following result, that is analogous to [**23**, Prop. 3.6] the difference being that in the above mentioned paper the result is formulated for the case that $\mathcal{V} = \mathbf{Vect}$.

A general theorem that yields in particular the lemma we present below, appears in [**20**, Prop. 3.4].

In this lemma we will need the following piece of notation. If C, D, C', D' are coalgebras in $\mathcal{V}$ and $U \in {}^C\mathcal{V}^D$ and $V \in {}^{C'}\mathcal{V}^{D'}$, we will denote by $U \bullet V \in {}^{C\otimes C'}\mathcal{V}^{D\otimes D'}$ the object $U \otimes V$ equipped with the obvious bicomodule structure.

LEMMA 6. *The functor* $L : (\mathcal{V}^C, \Bbbk, \otimes) \to ({}^C_C\mathcal{V}^C, C, \Box_C)$ *is fully faithful and monoidal.*

PROOF. It is well known that the functor L is fully faithful if and only if the unit of the adjunction $L \dashv (u_+\Box_C-)U$ is an isomorphism. It follows from the dual of [**12**, Lemma A1.1.1], that it is enough to exhibit a natural isomorphism between $(u_+\Box_C-)UL$ and the identity functor of $\mathcal{V}^C$. The composition $UL : \mathcal{V}^C \to {}^C\mathcal{V}^C$ can be written as $UL(M) = (C \bullet M)\Box_{C^{\otimes 2}}p_+$. We have natural isomorphisms

$$u_+\Box_C(C \bullet M)\Box_{C^{\otimes 2}}p_+ \cong (u_+ \bullet M)\Box_{C^{\otimes 2}}p_+ \cong M\Box_C(u_+ \bullet C)\Box_{C^{\otimes 2}}p_+ \cong M$$

where the last isomorphism is induced by

$$(u_+ \bullet C)\Box_{C^{\otimes 2}}p_+ \cong ((u \otimes \mathrm{id}_C)p)_+ \cong (\mathrm{id}_C)_+ = C.$$

This shows that there is a natural isomorphism $u_+\Box_C UL(M) \cong M$.

We will now exhibit a canonical monoidal structure on L. The basic observation is that, for $M \in \mathcal{M}^C$, $L(M) = C \otimes (u^+\Box_\Bbbk M)$ is isomorphic to $(C \bullet M)\Box_{C^{\otimes 2}}p_+$, where $C \in {}^C\mathcal{M}^C$ is the regular bicomodule. Then, we can form the composition

$$\begin{aligned}
L(M)\Box_C L(N) &\cong ((C \bullet M)\Box_{C^{\otimes 2}}p_+)\Box_C((C \bullet M)\Box_{C^{\otimes 2}}p_+) \\
&\cong (((C \bullet M)\Box_{C^{\otimes 2}}p_+) \bullet N)\Box_{C^{\otimes 2}}p_+ \\
&\cong (C \bullet M \bullet N)\Box_{C^{\otimes 3}}(p_+ \bullet C)\Box_{C^{\otimes 2}}p_+ \qquad (20)\\
&\cong (C \bullet M \bullet N)\Box_{C^{\otimes 3}}(C \bullet p_+)\Box_{C^{\otimes 2}}p_+ \qquad (21)\\
&\cong (C \bullet (M \otimes N))\Box_{C^{\otimes 2}}p_+ \\
&\cong L(M \otimes N).
\end{aligned}$$

All the isomorphisms above follow easily form the definition of the cotensor product except for the isomorphism between (20) and (21), which is induced by the isomorphism $(p_+ \bullet C)\Box_{C^{\otimes 2}}p_+ \cong ((p \otimes \mathrm{id}_C)p)_+ \cong ((\mathrm{id}_C \otimes p)p)_+ \cong (C \bullet p_+)\Box_{C^{\otimes 2}}p_+$ that is induced by the associator ϕ.

The isomorphism described above together with the obvious isomorphism $\Bbbk \cong L(\Bbbk)$ provide a monoidal structure for L. The axioms of a monoidal functor follow easily from the axioms satisfied by the associator ϕ. $\Box$

In the presence of an antipode, one obtains the following strengthening of the above results. This form of the fundamental theorem on Hopf modules for coquasi Hopf algebras is a consequence of [**20**, Theorem 7.2]. It follows easily by a simple adaptation of the arguments of the Section 11 of the same work.

THEOREM 5. *For an arbitrary coquasi Hopf algebra in* $\mathcal{V}$, *the associated functor* L *is a monoidal equivalence.*

Observe that we do not ask $\mathcal{V}$ to be abelian or additive. Neither we assume anything about the existence of duals in $\mathcal{V}$. The only requirements on $\mathcal{V}$ are that it is braided monoidal, with equalizers that are preserved by the tensor product. For a version of Theorem 5 over **Vect**, see for example [**25**].

In order to apply this theorem to our context, we need its right version that will be deduced below.

OBSERVATION 12. Consider the braided monoidal category $\mathcal{V}^{\mathrm{rev}}$, which has the same underlying category as $\mathcal{V}$, the same unit object but the reverse tensor product $X \otimes^{\mathrm{rev}} Y = Y \otimes X$, see Section 1. If $\gamma_{X,Y} : X \otimes Y \to Y \otimes X$ is the braiding in $\mathcal{V}$, then the braiding in $\mathcal{V}^{\mathrm{rev}}$ is $\gamma^{\mathrm{rev}}_{X,Y} = \gamma_{Y,X}$. The symmetry in the definition of coquasi bialgebra implies that if (C,p,u,ϕ) is a coquasi bialgebra in $\mathcal{V}$, then (C,p,u,ϕ) is a coquasi bialgebra in $\mathcal{V}^{\mathrm{rev}}$. Moreover, if (S,α,β) is an antipode for the coquasi bialgebra H in $\mathcal{V}$, then (S,β,α) is an antipode for H in $\mathcal{V}^{\mathrm{rev}}$.

COROLLARY 2. *Suppose H is a coquasi Hopf algebra with invertible antipode in $\mathcal{V}$. Then the functor $R : {}^H\mathcal{V} \to {}^H\mathcal{V}^H_H$ defined as the composition of $-\Box_{\Bbbk} u_+ : {}^H\mathcal{V} \to {}^H\mathcal{V}^H$ with the free right Hopf module functor ${}^H\mathcal{V}^H \to {}^H\mathcal{V}^H_H$ is a monoidal equivalence from ${}^H\mathcal{V}$ to ${}^H\mathcal{V}^H_H$.*

PROOF. Let us denote $\mathcal{V}^{\mathrm{rev}}$ by $\mathcal{W}$. If (H,u,p,ϕ) is a coquasi bialgebra in $\mathcal{V}$, as we saw in Observation 12, (H,u,p,ϕ) is a coquasi bialgebra in $\mathcal{W}$, and if (S,α,β) is an antipode for H in $\mathcal{V}$ then (S,β,α) is an antipode for H in $\mathcal{W}$.

Clearly, as monoidal categories we have that ${}^H\mathcal{V} = (\mathcal{W}^H)^{\mathrm{rev}}$, ${}^H\mathcal{V}^H = ({}^H\mathcal{W}^H)^{\mathrm{rev}}$ and ${}^H\mathcal{V}^H_H = ({}^H_H\mathcal{W}^H)^{\mathrm{rev}}$. The functor $-\Box_{\Bbbk} u_+$ corresponds to $u^+\Box - : \mathcal{W}^H \to {}^H\mathcal{W}^H$ and the free H-module functor ${}^H\mathcal{V}^H \to {}^H\mathcal{V}^H_H$ to the free H-module functor ${}^H\mathcal{W}^H \to {}^H_H\mathcal{W}^H$. Then R is just L^{rev} for the coquasi Hopf algebra H in $\mathcal{W}$, and hence it is a monoidal equivalence as we wanted to guarantee. □

4.2. The case of invertible antipode. In this section we prove that the composition of the functors $u^+\Box_{\Bbbk}- : \mathcal{M}^H \to {}^H\mathcal{M}^H$ and the free right H–module functor $F : {}^H\mathcal{M}^H \to {}^H\mathcal{M}^H_H$ is a monoidal equivalence when H has an invertible antipode.

First we deal with the general case of a coquasi bialgebra.

LEMMA 7. *For a coquasi bialgebra C, the functor $F(u^+\Box_{\Bbbk}-) : \mathcal{M}^C \to {}^C\mathcal{M}^C_C$ is left adjoint to $(u_+\Box_H-)U : {}^C\mathcal{M}^C_C \to \mathcal{M}^C$ where $U : {}^C\mathcal{M}^C_C \to {}^C\mathcal{M}^C$ is the forgetful functor. Moreover, the counit transformation corresponding to the above adjunction is the following: if $M \in {}^C\mathcal{M}^C_C$, then $\varepsilon_M : {}_0({}^{\mathrm{co}C}M) \otimes C \to M$ is the map $\varepsilon_M(m \otimes c) = m \cdot c$.*

PROOF. The assertion about the adjunction is clear. If ε' and ε'' are the counits of $F \dashv U$ and $(u^+\Box_{\Bbbk}-) \dashv (u_+\Box_C-)$ respectively, then the counit of the composition of these functors is

$$\varepsilon : F(u^+\Box_{\Bbbk}-)(u_+\Box_C-)U \xrightarrow{F\varepsilon''U} FU \xrightarrow{\varepsilon'} \mathrm{id}.$$

For $M \in {}^C\mathcal{M}^C$, the transformation ε''_M is the inclusion of ${}_0({}^{\mathrm{co}C}M)$ in M, while for $N \in {}^C\mathcal{M}^C_C$, $\varepsilon'_N : N \otimes C \to N$ is given by the right action of C on N. Therefore ε is indeed given by the above formula. □

DEFINITION 9. Let H be a coquasi Hopf algebra. Define a functor $\mathcal{I} : \mathcal{M}^H \to {}^H\mathcal{M}$ as the composition $\mathcal{M}^H \xrightarrow{(-)^\circ} {}^{H^{\mathrm{cop}}}\mathcal{M} \xrightarrow{S^+\Box_{H^{\mathrm{cop}}}-} {}^H\mathcal{M}$. In other words, on objects $\mathcal{I}(M,\chi) = (M, (S \otimes \mathrm{id})\mathrm{sw}\chi)$, and on arrows $\mathcal{I}$ is the identity.

COROLLARY 3. *In the situation above $\mathcal{I} : \mathcal{M}^{H\,\mathrm{rev}} \to {}^H\mathcal{M}$ has a canonical structure of a monoidal functor.*

PROOF. It follows immediately from: the equality

$$\mathcal{I} = (S^+\Box_{H^{\mathrm{cop}}}-)\circ(-)^\circ : \mathcal{M}^H \xrightarrow{(-)^\circ} {}^{H^{\mathrm{cop}}}\mathcal{M} \xrightarrow{S^+\Box_{H^{\mathrm{cop}}}-} {}^H\mathcal{M},$$

the fact that $(-)^\circ$ is monoidal –see the comments after Definition 7– and Theorem 2. □

The following natural transformation will be crucial in the proof of Radford's formula.

THEOREM 6. *Let H be a coquasi Hopf algebra with invertible antipode. For $M \in \mathcal{M}^H$ the arrows $\tau_M : M \otimes H \to M \otimes H$ defined as $\tau_M(m \otimes h) = \sum m_0 \phi^{-1}(m_1 \otimes S(m_3) \otimes h_2)\beta(m_2) \otimes S(m_4)h_1$ are the components of a natural transformation between the functors $F \circ (-)_0 \circ \mathcal{I}$ and $F \circ {}_0(-) : \mathcal{M}^H \to {}^H\mathcal{M}^H_H$. Moreover, the natural transformation τ is invertible and its inverse is given for all $M \in \mathcal{M}^H$ by the formula: $\tau_M^{-1}(m \otimes h) = \sum \phi(S(m_1) \otimes m_3 \otimes h_1)\alpha(m_2)m_0 \otimes m_4h_2$.*

PROOF. It is convenient for the proof to split the map τ_M as follows. First define the map $\pi_M : (\mathcal{I}M)_0 \to {}_0M \otimes H$ as $\pi_M(m) = \sum m_0 \otimes \beta(m_1)S(m_2)$. An elementary computation shows that τ_M is equals the following composition.

$$(\mathcal{I}M)_0 \otimes H \xrightarrow{\pi_M \otimes \mathrm{id}} ({}_0M \otimes H) \otimes H \xrightarrow{\Phi_{{}_0M,H,H}} {}_0M \otimes (H \otimes H) \xrightarrow{\mathrm{id} \otimes p} {}_0M \otimes H$$

The structure of H–bicomodule on $(F \circ (-)_0 \circ \mathcal{I})(M) = M \otimes H$ is given by the formula $m \otimes h \mapsto \sum S(m_1)h_1 \otimes m_0 \otimes h_2 \otimes h_3$, while the structure on $(F \circ {}_0(-))(M) = M \otimes H$ is $m \otimes h \mapsto \sum h_1 \otimes m_0 \otimes h_2 \otimes m_1 h_3$. A direct verification shows that π_M is a morphism of bicomodules. Hence τ_M is a composition of morphisms of bicomodules. The compatibility of τ_M with the right action of H is deduced directly from the fact that $\pi_M \otimes \mathrm{id}$ and $(\mathrm{id} \otimes p)\Phi_{{}_0M,H,H}$ are morphisms of right H–modules. The H–equivariance of the first morphism is obvious, while the equivariance of the second is a consequence of the following general fact –that we apply in the situation that $\mathcal{C} = {}^H\mathcal{M}^H$ and $A = H$–. If (A, p, u) is an algebra –also called a monoid– in an arbitrary monoidal category $\mathcal{C}$ (that in accordance with [**13**] can be assumed to be strict) then, for any object X the arrow $\mathrm{id} \otimes p : X \otimes A \otimes A \to X \otimes A$ is a morphism of right A–modules.

Finally, the verification of that the maps τ_M and τ_M^{-1} are indeed inverses to each other is a direct computation. □

A version of the above lemma for quasi Hopf algebras appears in [**25**]. Our proof is similar.

The following result is an immediate consequence of Theorem 6 and Corollary 2.

COROLLARY 4. *In the situation above, the functor $F(u^+\square_{\Bbbk}-) : (\mathcal{M}^H)^{\mathrm{rev}} \to {}^H\mathcal{M}^H_H$ has a unique monoidal structure such that τ is a monoidal natural transformation. In particular, with this structure $F(u^+\square_{\Bbbk}-)$ is a monoidal equivalence.*

We end the section with the following observation, that will be used in Section 6.2.

OBSERVATION 13. If $M, N \in \mathcal{M}^H$ and $P, Q \in {}^H\mathcal{M}$, there exist canonical isomorphisms of bicomodules

$$((u^+\square_{\Bbbk}M) \otimes H)\square_H((u^+\square_{\Bbbk}N) \otimes H) \cong (u^+\square_{\Bbbk}N) \otimes ((u^+\square_{\Bbbk}M) \otimes H)$$

$$((P\square_{\Bbbk}u_+) \otimes H)\square_H((Q\square_{\Bbbk}u_+) \otimes H) \cong (P\square_{\Bbbk}u_+) \otimes ((Q\square_{\Bbbk}u_+) \otimes H)$$

If $f : (P\square_{\Bbbk}u_+) \to (u^+\square_{\Bbbk}M)$ and $g : (Q\square_{\Bbbk}u_+) \to (u^+\square_{\Bbbk}N)$ are morphisms of bicomodules, then the diagram in Figure 1 commutes.

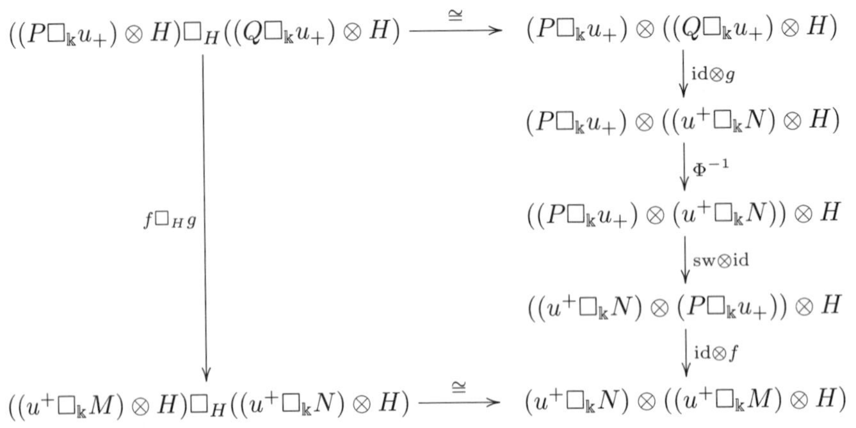

$$
\begin{array}{ccc}
((P\Box_{\Bbbk}u_+)\otimes H)\Box_H((Q\Box_{\Bbbk}u_+)\otimes H) & \xrightarrow{\cong} & (P\Box_{\Bbbk}u_+)\otimes((Q\Box_{\Bbbk}u_+)\otimes H) \\
 & & \Big\downarrow{\scriptstyle \mathrm{id}\otimes g} \\
 & & (P\Box_{\Bbbk}u_+)\otimes((u^+\Box_{\Bbbk}N)\otimes H) \\
 & & \Big\downarrow{\scriptstyle \Phi^{-1}} \\
{\scriptstyle f\Box_H g}\Big\downarrow & & ((P\Box_{\Bbbk}u_+)\otimes(u^+\Box_{\Bbbk}N))\otimes H \\
 & & \Big\downarrow{\scriptstyle \mathrm{sw}\otimes\mathrm{id}} \\
 & & ((u^+\Box_{\Bbbk}N)\otimes(P\Box_{\Bbbk}u_+))\otimes H \\
 & & \Big\downarrow{\scriptstyle \mathrm{id}\otimes f} \\
((u^+\Box_{\Bbbk}M)\otimes H)\Box_H((u^+\Box_{\Bbbk}N)\otimes H) & \xrightarrow{\cong} & (u^+\Box_{\Bbbk}N)\otimes((u^+\Box_{\Bbbk}M)\otimes H)
\end{array}
$$

FIGURE 1

5. The Frobenius isomorphism and the object of cointegrals

Suppose that H is a coquasi Hopf algebra with invertible antipode. If M is a left H–module in the category ${}^H\mathcal{M}_f^H$ its left dual ${}^*M$, is an object in ${}^H\mathcal{M}_H^H$ in a functorial way. If $a_M : H\otimes M \to M$ is a H–module structure of M in ${}^H\mathcal{M}^H$, the corresponding structure $a_{{}^*M} : {}^*M \otimes H \to {}^*M$ is given by

$$
{}^*M\otimes H \xrightarrow{\mathrm{id}\otimes\mathrm{coev}^\ell} ({}^*M\otimes H)\otimes(M\otimes{}^*M) \xrightarrow{\cong} ({}^*M\otimes(H\otimes M))\otimes{}^*M
$$
$$
\xrightarrow{(\mathrm{id}\otimes a_M)\otimes\mathrm{id}} ({}^*M\otimes M)\otimes{}^*M \xrightarrow{\mathrm{ev}^\ell\otimes 1} {}^*M. \quad (22)
$$

From now on we assume that H is a finite dimensional coquasi Hopf algebra. If we take $H\in {}^H\mathcal{M}_f^H$ as a left H–module with respect to the regular action, its right dual ${}^*H$ is canonically an object in ${}^H\mathcal{M}_H^H$. An explicit description of the right H–structure defined above for ${}^*H$ is the following: if $f\in {}^*H$ and $x,y\in H$, $a_{{}^*H}(f\otimes x)(y) = (f\cdot x)(y)$ is

$$
\begin{aligned}
&(f\cdot x)(y) = \\
&\quad \sum \phi^{-1}(\overline{S}(x_5y_7)x_1\otimes y_3\otimes \overline{S}(y_1))\alpha\overline{S}(y_2)\phi(\overline{S}(x_4y_6)\otimes x_2\otimes y_4)\beta\overline{S}(x_3y_5) \\
&\quad f(x_6y_8)\phi(S(x_7y_9)x_{11}\otimes y_{13}\otimes S(y_{15}))\phi^{-1}(S(x_8y_{10})\otimes x_{10}\otimes y_{12})\alpha(x_9y_{11})\beta(y_{14}).
\end{aligned} \tag{23}
$$

It is important to notice that in the above formula –and in the formula for the Frobenius isomorphism– we obtain the expression for $f\cdot x\in {}^*H$ in terms of the *standard* evaluation of vector spaces $H^\vee\otimes H\to\Bbbk$.

THEOREM 7. *If H is a finite dimensional coquasi Hopf algebra, then there exists a unique up to isomorphism one dimensional object $W\in\mathcal{M}^H$ such that there is an isomorphism ${}_0W\otimes H\cong {}^*H\in {}^H\mathcal{M}_H^H$. Moreover, W can be taken as the space of left cointegrals of the Hopf algebra H and the isomorphism –called the Frobenius isomorphism– is the map $\mathcal{F}$ given by*

$$
\mathcal{F}(\varphi\otimes x) = \varphi\cdot x \tag{24}
$$

where the action used is the one defined in (22) *and applied to* $M = {}^*H$ *in accordance to the formula* (23).

PROOF. The existence and uniqueness of W follows immediately from the fundamental theorem on Hopf modules–see more specifically Corollary 2–. The characterization of W as a space of left cointegrals is deduced directly from the explicit description of the inverse functor of $F_{\circ 0}(-)$ as the composition of the forgetful functor $U : {}^H\mathcal{M}_H^H \to \mathcal{M}_H^H$ with the left fixed part functor –see the considerations previous to Lemma 6–. Thus, W is the space of left coinvariants of ${}^*H$ with respect to the coaction described in (19). In explicit terms $W = \{\varphi \in {}^*H : {}^*\chi(\varphi) = 1 \otimes \varphi\}$. Using the description of ${}^*\chi$, appearing in Observation 11, we conclude that $\varphi \in W$ if and only if for all $x \in H$, $\varphi(x)1 = \sum \overline{S}(x_1)\varphi(x_2)$. In other words $\varphi(x)1 = \sum x_1\varphi(x_2)$ and then $\varphi \in {}^*H$ is a left cointegral. The description of the counit of the adjunction as the map given by the action –see Lemma 7– will yield the characterization (24). $\square$

OBSERVATION 14. In the same manner than in the classical case, from the existence of the isomorphism $\mathcal{F}$ we conclude that W, the space of left cointegrals, is one dimensional. Hence, one can prove the existence of a group like element $a \in H$ such that $\sum \varphi(x_1)x_2 = \varphi(x)a$ for all $\varphi \in W$ and $x \in H$. The element $a \in H$ is called the *modular* element.

LEMMA 8. *The coaction* $\chi_W : W \to W \otimes H$ *is of the form* $\chi_W(\varphi) = \varphi \otimes a^{-1}$ *for* $a \in H$ *as above.*

PROOF. As W is one dimensional, the coaction χ_W is of the form $\chi_W(\varphi) = \varphi \otimes b$ for $b \in H$. The definition of the right comodule structure on W (see Observation 11) yields –for $x \in H$– the formula: $\sum \varphi(x_1)S(x_2) = \varphi(x)b$. It follows then that $S(a) = a^{-1} = b$ –see Observation 4–. $\square$

OBSERVATION 15. The comodule W is isomorphic to ${a^{-1}}_+ \in \mathcal{M}^H$, where $a^{-1} : \Bbbk \to H$ is the coalgebra morphism induced by the multiplication by a^{-1}. Similarly, the coaction in ${}^*W$ is given as $\chi_{{}^*W}(t) = t \otimes a$ for $t \in {}^*W$. Hence, ${}^*W$ is isomorphic to $a_+ \in \mathcal{M}^H$.

Recall that we abbreviated the functors $u^+\square_\Bbbk-$ and $-\square_\Bbbk u_+$ by ${}_0(-)$ and $(-)_0$ respectively.

OBSERVATION 16. $W \in \mathcal{M}^H$ as well as ${}_0W \in {}^H\mathcal{M}^H$ are invertible objects in the corresponding monoidal categories. In other words, the functors $-\otimes W : \mathcal{M}^H \to \mathcal{M}^H$ and $-\otimes {}_0W : {}^H\mathcal{M}^H \to {}^H\mathcal{M}^H$ are equivalences and it is clear that the inverse equivalences are obtained by tensoring with the corresponding duals.

For use in the next section we write down the following observation.

OBSERVATION 17. Define $c_W^l, c_W^r : {}^H\mathcal{M}^H \to {}^H\mathcal{M}^H$ as follows: $c_W^l = ({}_0W \otimes -) \otimes {}_0{}^*W$ and $c_W^r = {}_0W \otimes (- \otimes {}_0{}^*W)$

It is clear that c_W^l and c_W^r are monoidal functors that are naturally isomorphic via the natural transformation given by the obvious associator. In the notations of Theorem 7 and using the fact that c_W^l, c_W^r are monoidal functors, we conclude that $({}_0W \otimes H) \otimes {}_0^*W$ and $({}_0W \otimes H) \otimes {}_0^*W$ are algebras in the category ${}^H\mathcal{M}^H$.

6. Radford's formula

In this section we use categorical methods to prove Radford's formula expressing S^4 in terms of conjugation with a functional and a group like element. In the second part of this section we prove the monoidality of the functional.

6.1. Radford's formula. We use the notations of the last section and assume that H is a finite dimensional coquasi Hopf algebra. We will take basis elements $\varphi \in W$, $t \in {}^*W$ normalized in such a way that $t(\varphi) = 1$.

LEMMA 9. *In the notations of Theorem 7 the isomorphism in* ${}^H\mathcal{M}^H$

$$\gamma : ({}_0W \otimes H) \otimes {}_0^*W \xrightarrow{\mathcal{F}\otimes \mathrm{id}} {}^*H \otimes {}_0^*W \cong {}^*({}_0W \otimes H) \xrightarrow{({}^*\mathcal{F})^{-1}} {}^{**}H$$

is a morphism of algebras. Moreover, if we define the Nakayama isomorphism $\mathcal{N} : H \to {}^{**}H$ *by the formula:* $\mathcal{N}(x) = \gamma((\varphi \otimes x) \otimes t)$, *then the commutativity of the diagram below characterizes* $\mathcal{N}$*:*

$$\begin{array}{ccccc} H \otimes (W \otimes H) & & \xrightarrow{\cong} & & (H \otimes W) \otimes H \\ \downarrow{\scriptstyle \mathcal{N}\otimes\mathcal{F}} & & & & \downarrow{\scriptstyle \mathcal{F}\,\mathrm{sw}\otimes \mathrm{id}} \\ {}^{**}H \otimes {}^*H & \xrightarrow[\mathrm{ev}^\ell_{{}^*H}]{} & \Bbbk & \xleftarrow[\mathrm{ev}^\ell_H]{} & {}^*H \otimes H \end{array}$$

PROOF. The multiplicativity of γ follows immediately from the fact that $\mathcal{F}$ is a morphism of H–modules and from the commutativity of the following diagram that is a direct consequence of the definition of the action $a_{{}^*H} : {}^*H \otimes H \to {}^*H$ –see Definition (22)–.

$$\begin{array}{ccccc} ({}^*H \otimes H) \otimes H & & \xrightarrow{\cong} & & {}^*H \otimes (H \otimes H) \\ \downarrow{\scriptstyle a_{{}^*H}\otimes \mathrm{id}} & & & & \downarrow{\scriptstyle \mathrm{id}\otimes p} \\ {}^*H \otimes H & \xrightarrow[\mathrm{ev}^\ell_H]{} & \Bbbk & \xleftarrow[\mathrm{ev}^\ell_H]{} & {}^*H \otimes H \end{array}$$

The assertion concerning $\mathcal{N}$ follows directly from the definitions. □

OBSERVATION 18. (1) The fact that γ is a morphism of algebras is valid in the following general context. Let $\mathcal{C}$ be a rigid monoidal category and let $a, w \in \mathcal{C}$ be respectively an algebra and an arbitrary object. Let $\mathcal{F} : w \otimes a \to {}^*a$ be an invertible morphism of a–modules in $\mathcal{C}$, then the object $(w \otimes a) \otimes {}^*w$ is an algebra in $\mathcal{C}$ and the map

$$\gamma : (w \otimes a) \otimes {}^*w \xrightarrow{\mathcal{F}\otimes \mathrm{id}} {}^*a \otimes {}^*w \cong {}^*(w \otimes a) \xrightarrow{({}^*\mathcal{F})^{-1}} {}^{**}a$$

is a morphism of algebras.
(2) In the case of ordinary Hopf algebras, the commutativity of the diagram that characterizes $\mathcal{N}$ after identifying H with its double dual reads as $\varphi(y\mathcal{N}(x)) = \varphi(xy)$ that is the usual definition of the Nakayama automorphism.

LEMMA 10. *In the notations of Theorem 6 and Theorem 7 if M is an object in* $\mathcal{M}^H$, *then the morphism* $\xi_M : {}^{**}\mathcal{I}(M)_0 \otimes (({}_0W \otimes H) \otimes {}_0^*W) \to {}^{**}{}_0M \otimes (({}_0W \otimes H) \otimes {}_0^*W)$ *defined by the commutativity of the diagram below is a morphism of right*

$({}_0W \otimes H) \otimes {}^*_0W$*-modules.*

$$\begin{array}{ccc} {}^{**}(\mathcal{I}(M)_0 \otimes H) & \xrightarrow{{}^{**}\tau_M} & {}^{**}({}_0M \otimes H) \\ \cong\downarrow & & \downarrow\cong \\ {}^{**}\mathcal{I}(M)_0 \otimes {}^{**}H & \xrightarrow{\omega_M} & {}^{**}_0M \otimes {}^{**}H \\ \mathrm{id}\otimes\gamma^{-1}\downarrow & & \downarrow\mathrm{id}\otimes\gamma^{-1} \\ {}^{**}\mathcal{I}(M)_0 \otimes (({}_0W \otimes H) \otimes {}^*_0W) & \xrightarrow{\xi_M} & {}^{**}_0M \otimes (({}_0W \otimes H) \otimes {}^*_0W) \end{array} \tag{25}$$

PROOF. Being τ_M a morphism of H–modules it is clear that ω_M is a morphism of ${}^{**}H$–modules. Then, from the fact that γ is an algebra morphism and that all the modules involved are free over the corresponding algebra objects, it follows that ξ_M is a morphism of $(W_0 \otimes H) \otimes {}^*W_0$–modules. □

DEFINITION 10. Define $\nu_M : {}^{**}\mathcal{I}(M)_0 \otimes ({}_0W \otimes H) \to {}^{**}_0M \otimes ({}_0W \otimes H)$ as the unique morphism such that the diagram below commutes.

$$\begin{array}{ccc} {}^{**}\mathcal{I}(M)_0 \otimes (({}_0W \otimes H) \otimes {}^*_0W) & \xrightarrow{\xi_M} & {}^{**}_0M \otimes (({}_0W \otimes H) \otimes {}^*_0W) \\ \cong\downarrow & & \downarrow\cong \\ ({}^{**}\mathcal{I}(M)_0 \otimes ({}_0W \otimes H)) \otimes {}^*_0W & \xrightarrow{\nu_M \otimes \mathrm{id}_{{}^*_0W}} & ({}^{**}_0M \otimes ({}_0W \otimes H)) \otimes {}^*_0W \end{array}$$

The existence of ν_M and the fact that it is a morphism in ${}^H\mathcal{M}^H$ follows immediately from the considerations of Observation 16. Moreover from the fact that ξ_M is a morphism in ${}^H\mathcal{M}^H_{({}_0W\otimes H)\otimes {}^*_0W}$, it follows that ν_M is a morphism in ${}^H\mathcal{M}^H_H$.

The monoidality of $\mathcal{I}$ gives canonical isomorphisms $\mathcal{I}(M^{**}) \cong {}^{**}\mathcal{I}(M)$. Composing with ν_M we get an isomorphism $\widehat{\nu}_M : \mathcal{I}(M^{**})_0 \otimes ({}_0W \otimes H) \to {}^{**}_0M \otimes ({}_0W \otimes H)$.

For $M \in \mathcal{M}^H$, consider the following composition or arrows in ${}^H\mathcal{M}^H_H$, with domain ${}_0W^* \otimes (({}^{**}_0M \otimes {}_0W) \otimes H)$.

$$\begin{aligned} \zeta_M : {}_0W^* \otimes {}^{**}_0M \otimes {}_0W \otimes H &\xrightarrow{\mathrm{id}\otimes\widehat{\nu}_M^{-1}} {}_0W^* \otimes \mathcal{I}(M^{**})_0 \otimes {}_0W \otimes H \to \\ &\xrightarrow{\mathrm{id}\otimes\mathrm{sw}\otimes\mathrm{id}} {}_0W^* \otimes {}_0W \otimes \mathcal{I}(M^{**})_0 \otimes H \xrightarrow{\mathrm{ev}\otimes\tau_{M^{**}}} {}_0M^{**} \otimes H \end{aligned} \tag{26}$$

Here we omitted the associativity constraints for simplicity. However this does not introduce any ambiguity as long as we know how to associate the domain and codomain, by the coherence theorem for monoidal categories.

The composition ζ_M is a morphism in ${}^H\mathcal{M}^H_H$. Indeed, $\widehat{\nu}_M$ and $\tau_{M^{**}}$ are morphisms of Hopf modules; the morphism $\mathrm{id} \otimes \mathrm{sw} \otimes \mathrm{id}$ is the image under the free Hopf module functor ${}^H\mathcal{M}^H \to {}^H\mathcal{M}^H_H$ of the morphism of bicomodules $\mathrm{id} \otimes \mathrm{sw} : {}_0W^* \otimes \mathcal{I}(M^{**})_0 \otimes {}_0W \to {}_0W^* \otimes {}_0W \otimes \mathcal{I}(M^{**})_0$. Observe that sw is a morphism of bicomodules because the trivial comodule structures in each tensor factor are added on opposite sides.

DEFINITION 11. Denote by $\mu_M : W^* \otimes ({}^{**}M \otimes W) \to M^{**}$ the unique morphism in $\mathcal{M}^H$ such that $\mu_M \otimes \mathrm{id}_H = \zeta_M$.

It is clear that μ is a natural isomorphism between the functors $W^* \otimes ({}^{**}(-) \otimes W)$ and $(-)^{**} : \mathcal{M}^H \to \mathcal{M}^H$.

COROLLARY 5. *The canonical linear isomorphisms $M \cong M^{\vee\vee}$ together with μ give a natural isomorphism in $\mathcal{M}^H$*

$$M\square_H\, p(a^{-1} \otimes p(\overline{S}^2 \otimes a))_+ \to M\square_H S^2_+. \tag{27}$$

PROOF. First we use Theorem 4, and substitute ${}^{**}M$ by $M\square_H S^2_+$ and M^{**} by $M\square_H \overline{S}^2_+$. In this manner we obtain from μ_M an isomorphism

$$W^* \otimes ((M\square_H \overline{S}^2_+) \otimes W) \to M\square_H S^2_+.$$

Now using the fact that $W \cong a_+^{-1}$ –Observation 15– and the conclusions of Observation 10 part b) we deduce our result. □

THEOREM 8 (Radford's formula). *There exists an invertible functional $\sigma : H \to \Bbbk$ such that for all $x \in H$*

$$a^{-1}(\overline{S}^2(x)a) = S^2(\sigma \rightharpoonup x \leftharpoonup \sigma^{-1}).$$

PROOF. It easily follows form Corollary 5 and Theorem 1. Indeed in the situation of Corollary 5 the theorem guarantees the existence of a functional σ such that –see Observation 8, (13)– $p(a^{-1} \otimes p(\overline{S}^2 \otimes a))(x) = S^2(\sigma \rightharpoonup x \leftharpoonup \sigma^{-1})$. □

The functional σ defined in the theorem above is the analogue for finite dimensional coquasi Hopf algebras of the modular function of a finite dimensional Hopf algebra. See Section 7.

OBSERVATION 19. The above formula can be transformed into another similar to the classical formula:

$$S^4(x) = (a^{-1}(\widehat{\sigma} \rightharpoonup x \leftharpoonup \widehat{\sigma}^{-1}))a \tag{28}$$

where $\widehat{\sigma}$ is another invertible functional that can be computed explicitly in terms of the above information.

6.2. Monoidality. In this section we prove that the natural isomorphism μ of Definition 11 is monoidal. We shall work as if the monoidal category $({}^H\mathcal{M}^H, \Bbbk, \otimes)$ were strict, and hence ignore the associativity and unit constraints. This can be formalized by passing to an monoidally equivalent strict monoidal category. Indeed, our proof does not depend on the fact that we are working with the category of comodules, but only on certain properties satisfied by the various arrows we consider.

The functor $\mathcal{M}^H \to \mathcal{M}^H$ given by $M \mapsto W^* \otimes M \otimes W$ has a canonical monoidal structure given by the constraints

$$W^* \otimes M \otimes W \otimes W^* \otimes N \otimes W \xrightarrow{\mathrm{id}\otimes\mathrm{id}\otimes\mathrm{ev}\otimes\mathrm{id}\otimes\mathrm{id}} W^* \otimes M \otimes N \otimes W$$

$$\Bbbk \xrightarrow{\mathrm{coev}} W^* \otimes W \xrightarrow{\cong} W^* \otimes \Bbbk \otimes W$$

These morphisms are isomorphisms because W is an invertible object –it has dimension one–.

THEOREM 9. *The natural transformation μ in Definition 11 is monoidal.*

The assertion that μ is a monoidal natural transformation is expressed in the commutativity of the diagrams in Figure 2.

$$
\begin{array}{ccc}
W^* \otimes {}^{**}M \otimes W \otimes W^* \otimes {}^{**}N \otimes W & \xrightarrow{\mu_M \otimes \mu_N} & {}^{**}M \otimes {}^{**}N \\
\Big\downarrow{\scriptstyle \mathrm{id}\otimes\mathrm{id}\otimes\mathrm{ev}\otimes\mathrm{id}\otimes\mathrm{id}} & \nearrow{\scriptstyle \mu_{M\otimes N}} & \\
W^* \otimes M \otimes N \otimes W & &
\end{array}
\qquad
\begin{array}{ccc}
 & & \Bbbk \\
 & \swarrow & \Big\downarrow{\scriptstyle \mathrm{id}} \\
W^* \otimes \Bbbk \otimes W & \xrightarrow[\mu_{\Bbbk}]{} & \Bbbk
\end{array}
\tag{29}
$$

FIGURE 2

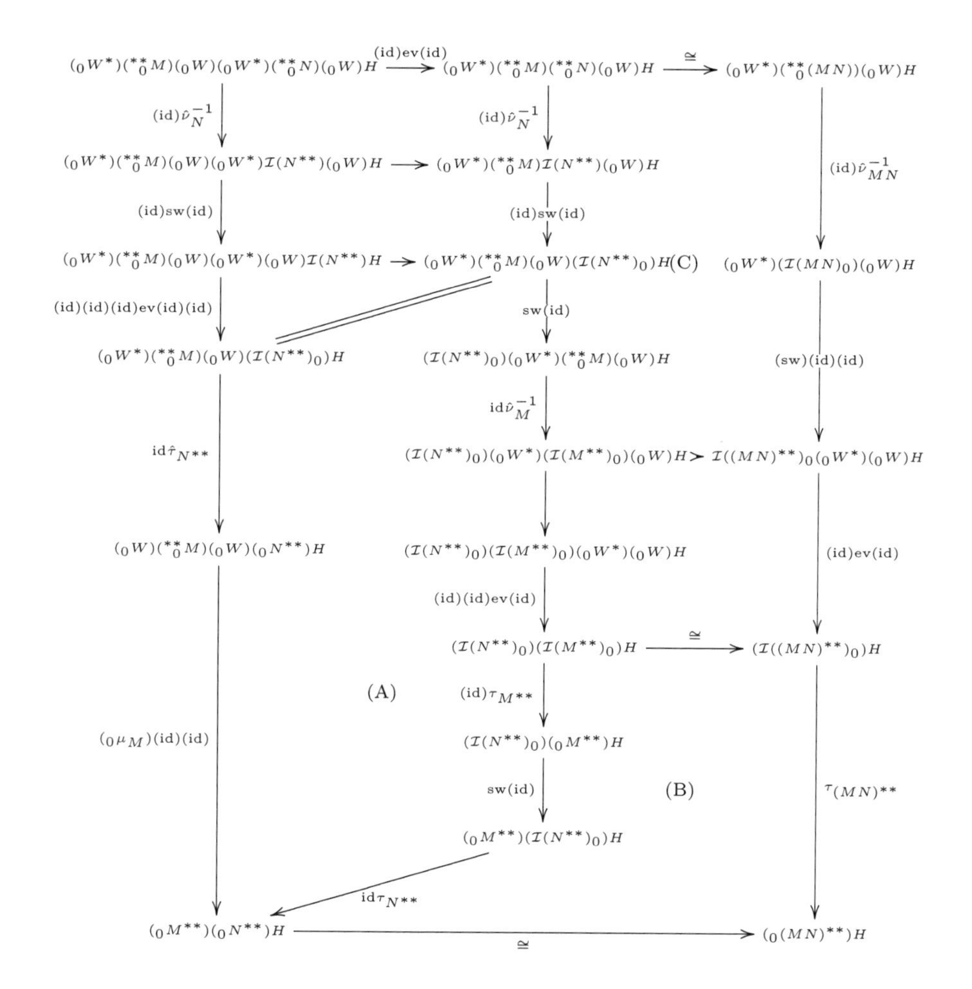

FIGURE 3

PROOF. We divide the proof in two parts. In some diagrams, we omit the symbol $\otimes$ as a saving space measure, adding parenthesis when necessary.

First axiom. The image of the diagram on the left hand side of Figure 2 is the exterior rectangle in Figure 3. So it is enough to show the latter commutes, as the functor $M \mapsto {}_0M \otimes H$ is an equivalence by Corollary 4. The sub diagrams left blank commute trivially.

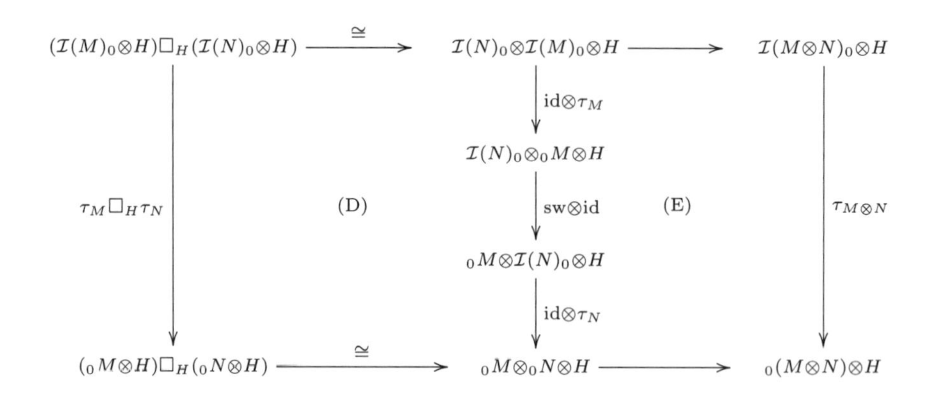

FIGURE 4

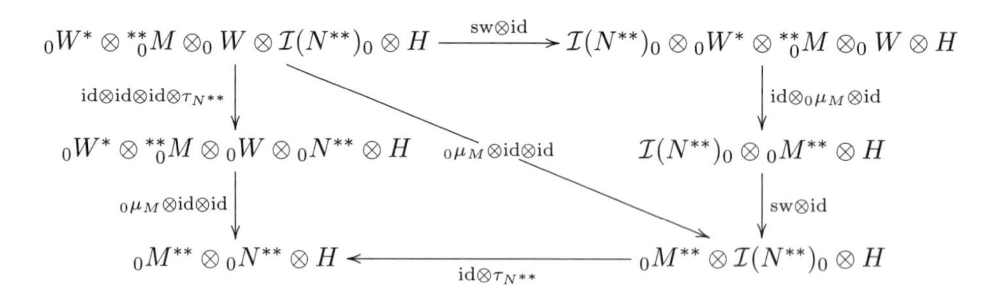

FIGURE 5

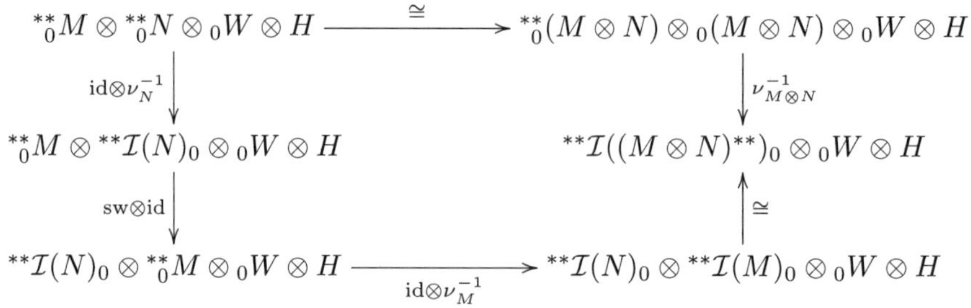

$$
\begin{array}{ccc}
{}^{**}_0M\otimes {}^{**}_0N\otimes {}_0W\otimes H & \xrightarrow{\cong} & {}^{**}_0(M\otimes N)\otimes {}_0(M\otimes N)\otimes {}_0W\otimes H \\
\downarrow{\scriptstyle \mathrm{id}\otimes\nu_N^{-1}} & & \downarrow{\scriptstyle \nu_{M\otimes N}^{-1}} \\
{}^{**}_0M\otimes {}^{**}\mathcal{I}(N)_0\otimes {}_0W\otimes H & & {}^{**}\mathcal{I}((M\otimes N)^{**})_0\otimes {}_0W\otimes H \\
\downarrow{\scriptstyle \mathrm{sw}\otimes\mathrm{id}} & & \uparrow{\scriptstyle \cong} \\
{}^{**}\mathcal{I}(N)_0\otimes {}^{**}_0M\otimes {}_0W\otimes H & \xrightarrow[\mathrm{id}\otimes\nu_M^{-1}]{} & {}^{**}\mathcal{I}(N)_0\otimes {}^{**}\mathcal{I}(M)_0\otimes {}_0W\otimes H
\end{array}
$$

FIGURE 6

The diagram marked by (A) in Figure 3 is just the commutative rectangle in Figure 5. This is easy to show using the naturality of sw and the definition of μ. The diagram (B) in Figure 3 commutes if the diagram marked by (E) in Figure 4 commutes for all M, N. To show this, observe that the exterior rectangle in Figure 4 commutes by monoidality of τ and that the sub diagram (D) commutes by Observation 13.

Finally, the diagram marked by (C) in Figure 4 commutes if and only if the diagram in Figure 6 does. If we tensor this diagram with ${}_0W^*$ on the right, after

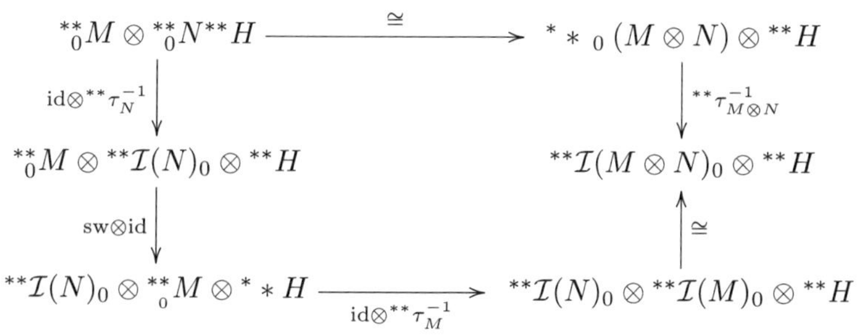

$$
\begin{array}{ccc}
{}^{**}_{0}M \otimes {}^{**}_{0}N^{**}H & \xrightarrow{\cong} & * * {}_0(M \otimes N) \otimes {}^{**}H \\
\Big\downarrow {\scriptstyle \mathrm{id}\otimes {}^{**}\tau_N^{-1}} & & \Big\downarrow {\scriptstyle {}^{**}\tau^{-1}_{M\otimes N}} \\
{}^{**}_{0}M \otimes {}^{**}\mathcal{I}(N)_0 \otimes {}^{**}H & & {}^{**}\mathcal{I}(M \otimes N)_0 \otimes {}^{**}H \\
\Big\downarrow {\scriptstyle \mathrm{sw}\otimes\mathrm{id}} & & \Big\uparrow {\scriptstyle \cong} \\
{}^{**}\mathcal{I}(N)_0 \otimes {}^{**}_{0}M \otimes * * H & \xrightarrow[\mathrm{id}\otimes {}^{**}\tau_M^{-1}]{} & {}^{**}\mathcal{I}(N)_0 \otimes {}^{**}\mathcal{I}(M)_0 \otimes {}^{**}H
\end{array}
$$

FIGURE 7

composing with the isomorphism $\gamma : {}_0W \otimes H \otimes {}_0W^* \cong {}^{**}H$ of Lemma 9, we get the diagram in Figure 7, which commutes as the diagram (E) referred to above does.

Second axiom. Now we prove the commutativity of the diagram involving $\Bbbk$ in Figure 2. For this we will need some notation. The symbol $\Bbbk$ will denote the trivial *left* H-comodule. Let us denote the canonical isomorphisms between $\Bbbk$ and both ${}^{**}\Bbbk$ and $\Bbbk^{**}$ by j. As $\mathcal{I}$ is a monoidal functor –see Corollary 3–, we have canonical isomorphisms $\delta : {}^{**}\mathcal{I}(\Bbbk) \to \mathcal{I}(\Bbbk^{**})$ in $\mathcal{M}^H$ and $\theta : \mathcal{I}(\Bbbk)_0 \to {}_0\Bbbk$ in ${}^H\mathcal{M}^H$. One useful observation is that, as the monoidal structure on $\mathcal{I}$ is the unique one making τ a monoidal natural transformation, we have $\theta \otimes \mathrm{id} = \tau_{\Bbbk} : \mathcal{I}(\Bbbk)_0 \otimes H \to {}_0\Bbbk \otimes H$.

The diagram we want to show that commutes lies in the category $\mathcal{M}^H$; hence, we may equivalently show that its image under the functor $M \mapsto {}_0M \otimes H$ of Corollary 4 commutes. This new diagram is the one outer diagram in Figure 8. Indeed, the image of $\mu_{\Bbbk}$ is the arrow ζ in (26), that, by naturality of sw, is equal to the composition $(\delta \otimes \mathrm{id})(\mathrm{ev} \otimes \mathrm{id})(\mathrm{id} \otimes \mathrm{sw} \otimes \mathrm{id})(\mathrm{id} \otimes \nu_{\Bbbk}^{-1})$ on the right hand side of the diagram in Figure 8. The only sub diagrams whose commutativity is not obvious are the ones marked with (a), (b) and (c).

The commutativity of the diagram (a) follows form the following observation. By definition, $\nu_{\Bbbk}^{-1} \otimes \mathrm{id}_{{}^*_0W}$ corresponds, up to composing with certain canonical isomorphisms, to ${}^{**}\tau_{\Bbbk}$ –see Lemma 10–. On the other hand, ${}^{**}\theta \otimes \mathrm{id} \otimes \mathrm{id} \otimes \mathrm{id}$: ${}^{**}\mathcal{I}(\Bbbk)_0 \otimes {}_0W \otimes H \otimes {}_0W^* \to {}^{**}_{0}\Bbbk \otimes {}_0W \otimes H \otimes {}_0W^*$ also corresponds, up to composing with the same canonical isomorphisms, to ${}^{**}\tau_{\Bbbk}$ (this because $\theta \otimes \mathrm{id}_H = \tau_{\Bbbk}$). It follows that the diagram (a) commutes.

The diagram made by (b) commutes because θ is induced by the monoidal structure of $\mathcal{I}$, and monoidal functors preserve duals. Finally, the diagram (c) commutes by naturality of τ. $\square$

The monoidality of the natural transformation μ just proved translates into properties of the functional σ in Theorem 8. We compute below in an explicit way the monoidal structure of σ.

OBSERVATION 20. The functional σ induces the natural isomorphism of Corollary 5, which is monoidal since μ is. Therefore, if we know the monoidal structures of the morphisms $p(a^{-1} \otimes p(\bar{S}^2 \otimes a))$ and S^2, we can deduce the equations satisfied by σ. More explicitly, if these morphisms have monoidal structures (χ_1, ρ_1) and (χ_2, ρ_2) respectively, then σ satisfies

$$\chi_1 \star \sigma p = (\sigma \otimes \sigma) \star \chi_2 \qquad \rho_1 \sigma(1) = \rho_2. \tag{30}$$

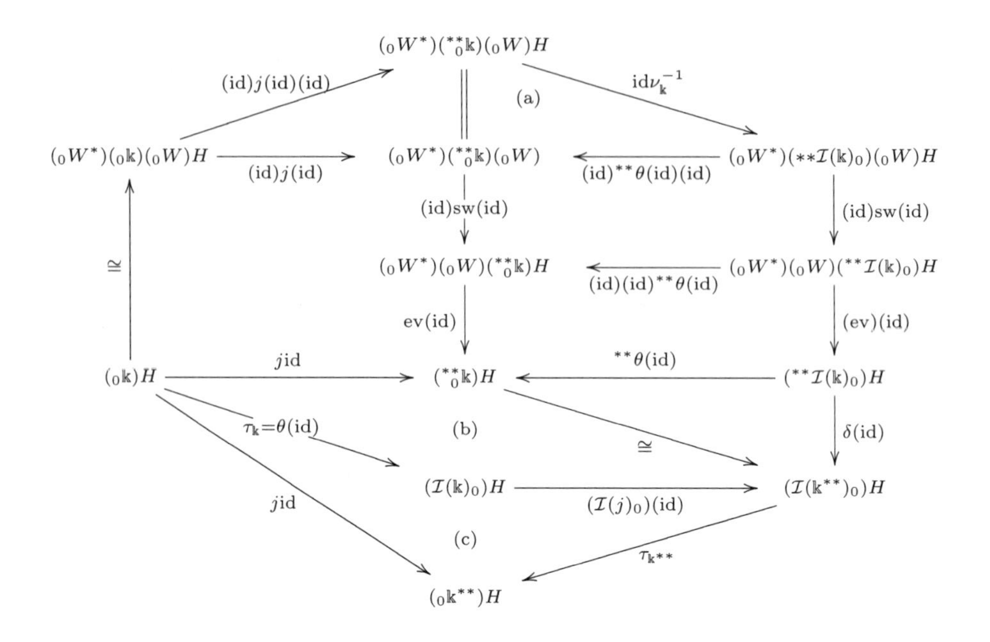

FIGURE 8

The antipode $S : H^\circ \to H$ has a monoidal structure $(\chi^S, 1)$, where χ^S is given explicitly in Proposition 3. By Observation 2 we have that S^2, which is the composition of $S^{\text{cop}} : H \to H^\circ$ and $S : H^\circ \to H$, has $(\chi^S(S \otimes S) \star (\chi^S)^{-1}\text{sw}, 1)$ as monoidal structure. This is because S^{cop} has a monoidal structure $((\chi^S)^{-1}\text{sw}, 1)$. The inverse of the antipode $\bar{S} : H^\circ \to H$ has a canonical monoidal structure given in terms of χ^S by $((\chi^S)^{-1}(\bar{S} \otimes \bar{S}), 1)$. Thus, $\bar{S}^2$, this is, the composition of $\bar{S}$ with $\bar{S}^{\text{cop}} : H^\circ \to H$, has a monoidal structure $(\chi^S(\bar{S}^2 \otimes \bar{S}^2)\text{sw} \star (\chi^S)^{-1}(\bar{S} \otimes \bar{S}), 1)$.

The morphism $p(a^{-1} \otimes p(\text{id} \otimes a)) : H \to H$ is monoidal with a monoidal structure given by $(\chi_0, 1)$ where χ_0 is the following product in $(H \otimes H)^\vee$.

$$\phi^{-1}(a^{-1} \otimes (-)a \otimes a^{-1}((?)a)) \star \phi((-)a \otimes a^{-1} \otimes (?)a) \star \phi^{-1}(- \otimes a \otimes a^{-1}) \star \phi(- \otimes ? \otimes a)$$

Then, the monoidal structure (χ_1, ρ_1) of the composition of $\bar{S}^2$ with $p(a^{-1} \otimes p(\text{id} \otimes a))$ is given by $\chi_1 = \chi_0(\bar{S}^2 \otimes \bar{S}^2) \star \chi^S(\bar{S}^2 \otimes \bar{S}^2)\text{sw} \star (\chi^S)^{-1}(\bar{S} \otimes \bar{S})$ and $\rho_1 = 1$. We deduce that σ satisfies $\sigma(1) = 1$ and

$$\chi_0(\bar{S}^2 \otimes \bar{S}^2) \star \chi^S(\bar{S}^2 \otimes \bar{S}^2)\text{sw} \star (\chi^S)^{-1}(\bar{S} \otimes \bar{S}) \star \sigma p = (\sigma \otimes \sigma) \star \chi^S(S \otimes S) \star (\chi^S)^{-1}\text{sw}. \tag{31}$$

7. The case of a Hopf algebra

We briefly mention the needed adjustments to the proof above in order to get the classical Radford's formula for S^4. We assume that H is a finite dimensional Hopf algebra and will use without defining them explicitly the following well known maps. The modular function of H that will be denoted by ω, the Nakayama automorphism –denoted by $\mathcal{N}$– whose inverse is related to S and ω by the formula: $S^2 x \leftharpoondown \omega^{-1} = \mathcal{N}^{-1}x$. In particular, we have that $\varepsilon\mathcal{N}^{-1} = \omega^{-1}$.

By substituting the associators as well as α and β by ε in Theorem 6, we obtain the formula $\tau_M(m \otimes h) = \sum m_0 \otimes S(m_1)h$ for $\tau_M : M \otimes H \to M \otimes H$. Moreover: $\tau_M^{-1}(m \otimes h) = \sum m_0 \otimes m_1 h$.

Concerning duality, if we write explicitly the formulæ (19) that yields the comodule structure on the dual spaces in the case where we consider $H \in {}^H\mathcal{M}^H$ as a left module with respect to the regular action, we obtain a right H–structure that is simply the following: $f \leftharpoonup h \in {}^*H$, $(f \leftharpoonup h)(x) = f(hx)$.

The Frobenius map $\mathcal{F}$ –see Theorem 7–, is $\mathcal{F}(\varphi \otimes h) = \varphi \leftharpoonup h$ for $\varphi \in W$ and $h \in H$. In particular as we mentioned in Observation 18 part (2), the morphism of algebras γ defined in Lemma 9 is given as: $\gamma(\varphi \otimes x \otimes t) = \mathrm{ev}(- \otimes \mathcal{N}x)$.

Once the above is established it is easy to see that the map ξ_M considered in Lemma 10, can be described explicitly by $\xi_M(\mathrm{ev}_m \otimes \varphi \otimes h \otimes t) = \sum \mathrm{ev}_{m_0} \otimes \varphi \otimes \mathcal{N}^{-1}(S(m_1))h \otimes t$.

Hence the map $\nu_M : {}^{**}\mathcal{I}(M)_0 \otimes {}_0W \otimes H \to {}^{**}_{\ 0}M \otimes {}_0W \otimes H$ introduced in Definition 10 is given by: $\nu_M(\mathrm{ev}_m \otimes \varphi \otimes h) = \sum \mathrm{ev}_{m_0} \otimes \varphi \otimes \mathcal{N}^{-1}(S(m_1))h$. Moreover, the map $\widehat{\nu}_M$ has exactly the same expression than ν_M and it is easy to show that $\widehat{\nu}_M^{-1}(\mathrm{ev}_m \otimes \varphi \otimes h) = \sum \mathrm{ev}_{m_0} \otimes \varphi \otimes \mathcal{N}^{-1}(m_1)h$.

Thus, the morphism ζ_M defined in (26) is given by: $\zeta_M(t \otimes \mathrm{ev}_m \otimes \varphi \otimes h) = \sum \mathrm{ev}_{m_0} \otimes \overline{S}(m_1)\mathcal{N}^{-1}(m_2)h$.

Then, as $\zeta_M = \mu_M \otimes \mathrm{id}_H$, with $\mu_M : W^* \otimes {}^{**}M \otimes W \to M^{**}$, it is clear that μ_M satisfies the following equality: $\mu_M(t \otimes \mathrm{ev}_m \otimes \varphi) \otimes h = \sum \mathrm{ev}_{m_0} \otimes \overline{S}(m_1)\mathcal{N}^{-1}(m_2)h$. If we apply $\mathrm{id} \otimes \mathrm{id} \otimes \varepsilon$ to the equality above we obtain:

$$\mu_M(t \otimes \mathrm{ev}_m \otimes \varphi) = \sum \mathrm{ev}_{m_0}(\varepsilon\mathcal{N}^{-1})(m_1) = \sum \mathrm{ev}_{m_0}\omega^{-1}(m_1) = \mathrm{ev}_{\omega^{-1} \rightharpoonup m}$$

Next, we observe that the natural isomorphism constructed in Corollary 5 is simply the map $m \mapsto (\omega^{-1} \rightharpoonup m)$. Applying the bijections proved in Theorem 1, we find that the map σ appearing in Radford's formula –Theorem 8– is simply $\sigma(h) = \varepsilon(\omega^{-1} \rightharpoonup h) = \omega^{-1}(h)$. Hence, we deduce the classical Radford's formula

$$a^{-1}\overline{S}^2(x)a = \omega^{-1} \rightharpoonup S^2(x) \leftharpoonup \omega \quad \text{or} \quad S^4(x) = \omega \rightharpoonup a^{-1}xa \leftharpoonup \omega^{-1}.$$

This shows that the functional σ appearing in Theorem 8 is indeed the coquasi Hopf algebra analogue of the modular function.

Next we explain how the monoidality of σ proved at the end of the previous section generalizes the multiplicativity of the modular function $\omega \in H^\vee$.

Recall that in the case of a Hopf algebra the associativity of the product and the fact that S is a morphism of algebras are expressed as $\phi = \varepsilon \otimes \varepsilon \otimes \varepsilon$ and $\chi^S = \varepsilon \otimes \varepsilon$. Therefore, the equality (31) simplifies to $\sigma p = \sigma \otimes \sigma$, *i.e.*, the functional σ is multiplicative. Hence $\omega = \sigma^{-1}$ is a morphism of algebras.

An important point is that in the proof of the monoidality of σ we did not use integrals (only cointegrals, presented as the comodule W). This is consistent with the results of [**1**].

8. Appendix: categorical background

In this appendix we put together some of the results on functors between categories of comodules we used along the text. As all the results we present are well known we do not provide proofs. A unified presentation of the subject, which we can not include here due to size constraints, can be achieved by a systematic use of [**17**, Theorem 5.31].

We will work with categories enriched in the category of vector spaces over a field $\Bbbk$, sometimes called $\Bbbk$–linear categories. We denote by $[\mathcal{A},\mathcal{B}]$ the $\Bbbk$–linear category of $\Bbbk$–linear functors $\mathcal{A}\to\mathcal{B}$ and natural transformations between them. We also consider two full subcategories: when $\mathcal{A},\mathcal{B}$ have filtered colimits, $\mathrm{Fin}[\mathcal{A},\mathcal{B}]$ the subcategory of finitary (*i.e.*, filtered colimit preserving) functors; and when $\mathcal{A},\mathcal{B}$ have finite limits, $\mathrm{Lex}[\mathcal{A},\mathcal{B}]$ the full subcategory of left exact functors.

The inclusion functor $\mathcal{M}_f^C\to\mathcal{M}^C$ induces equivalences $\mathrm{Fin}[\mathcal{M}^C,\mathcal{B}]\simeq[\mathcal{M}_f^C,\mathcal{B}]$ for all categories with filtered colimits $\mathcal{B}$. The pseudo inverse is given by taking left Kan extensions.

Given a finite dimensional coalgebra C, denote by $\mathcal{A}$ the full subcategory of $\mathcal{M}_f^C$ determined by the regular comodule C. The inclusion functor $\mathcal{A}\hookrightarrow\mathcal{M}_f^C$ induces equivalences $\mathrm{Lex}[\mathcal{M}_f^C,\mathcal{B}]\simeq[\mathcal{A},\mathcal{B}]$. Recall that there is an isomorphism $[\mathcal{A},\mathbf{Vect}]\cong\mathcal{M}_{C^\vee}$. For, to give a functor $\mathcal{A}\to\mathbf{Vect}$ is to give a vector space with a left action of the algebra $\mathcal{M}^C(C,C)$, which is isomorphic to $(C^\vee)^{\mathrm{op}}$.

Now it is easy to deduce that for a finite dimensional coalgebra C there are equivalences $\mathrm{Lex}[\mathcal{M}_f^C,\mathcal{M}^D]\simeq{}^C\mathcal{M}^D$, sending a right exact functor F to the bicomodule $F(C)$. A pseudoinverse for this equivalence is the functor sending a bicomodule M to $-\square_C M$.

In Section 2 we used the following easy observation.

OBSERVATION 21. Let C be a finite dimensional coalgebra, $K:\mathcal{M}_f^C\to\mathcal{M}^C$ be the inclusion functor and $M,N\in{}^C\mathcal{M}^D$. Then we have a string of canonical isomorphisms

$$\begin{aligned}{}^C\mathcal{M}^D(M,N)&\cong\mathrm{Lex}[\mathcal{M}_f^C,\mathcal{M}^D](K(-)\square_C M,K(-)\square_C N)\\&=[\mathcal{M}_f^C,\mathcal{M}^D](K(-)\square_C M,K(-)\square_C N)\\&\cong\mathrm{Fin}[\mathcal{M}^C,\mathcal{M}^D](-\square_C M,-\square_C N)\\&=[\mathcal{M}^C,\mathcal{M}^D](-\square_C M,-\square_C N)\end{aligned}$$

The last piece of categorical background we will need is the *tensor product* of categories with finite limits. This is closely related to Deligne's tensor product of abelian categories of [**6**].

Let $\mathcal{C},\mathcal{D}$ be two categories with finite limits and recall that the category $\mathcal{C}\otimes\mathcal{D}$ has as objects pairs (c,d) with $c\in\mathcal{C}$ and $d\in\mathcal{D}$, and homs $\mathcal{C}\otimes\mathcal{D}((c,d),(c',d'))=\mathcal{C}(c,c')\otimes_\Bbbk\mathcal{D}(d,d')$. A tensor product of $\mathcal{C}$ with $\mathcal{D}$ as categories with finite limits is a category with finite limits $\mathcal{C}\boxtimes\mathcal{D}$ together with a functor $\mathcal{C}\otimes\mathcal{D}\to\mathcal{C}\boxtimes\mathcal{D}$, that is left exact in each variable and that for every finitely complete category $\mathcal{E}$, induces equivalences $\mathrm{Lex}[\mathcal{C}\boxtimes\mathcal{D},\mathcal{E}]\simeq\mathrm{Lex}[\mathcal{C},\mathcal{D};\mathcal{E}]$. Here, the category on the right hand side of the above equivalence is the category of functors $\mathcal{C}\otimes\mathcal{D}\to\mathcal{E}$ that are left exact in each variable, and with arrows the natural transformations between them.

We are interested in the case of categories of finite dimensional comodules over finite dimensional coalgebras, that is dual to the case considered by Deligne in [**6**]. If C,D are finite dimensional coalgebras, it is not hard to show that the functor $\otimes_\Bbbk:\mathcal{M}_f^C\otimes\mathcal{M}_f^D\to\mathcal{M}_f^{C\otimes D}$ is the tensor product of $\mathcal{M}_f^C$ with $\mathcal{M}_f^D$ as categories with finite limits.

References

[1] Beattie, M., Bulacu, D. and Torrecillas, B. *Radford's S^4 formula for co-Frobenius Hopf algebras.* J. Algebra, **307** , 1, (2007), pp. 330-342

[2] Bénabou, J. *Introduction to bicategories.* Reports of the Midwest Category Seminar. Springer, Berlin. (1967) pp. 1–77.
[3] Bespalov, Y., Kerler, T., Lyubashenko, V. and Turaev, V. *Integrals for braided Hopf algebras.* J. Pure and Applied Algebra, **148**, (2000), pp. 113-164.
[4] Bulacu D., Chiriţă B. *Dual Drinfeld double by diagonal crossed coproduct*, Rev. Roumaine Math. Pures Appl. 47 (3) (2002) 271–294.
[5] Bulacu, D. and Caenepeel, S. *Integrals for (dual) quasi–Hopf algebras, applications.* J. Algebra, **266**, 2, (2003), pp. 552-583.
[6] Deligne, P. *Catégories tannakiennes*, The Grothendieck Festschrift, Vol. II, Progr. Math., vol. 87, Birkhäuser Boston, Boston, MA, 1990, pp. 111–195.
[7] Doi, Y. and Takeuchi, M. *BiFrobenius algebras.* In Andruskiewtisch, N. (ed.) et al. *New trends in Hopf algebra theory.* Proceedings of the colloquium on quantum groups and Hopf algebras, La Falda, Sierras de Córdoba, Argentina, August 9-13, 1999. Providence, RI: American Mathematical Society (AMS) (2000).
[8] Drinfel'd,V. G. *Quasi–Hopf algebras.* Leningrad Math. **1**, (1990), pp. 1419–1457.
[9] Etingof, P., Nikshych, D. and Ostrik, V. *An analogue of Radofrd's S^4 formula for finite tensor categories.* Int. Math. Res. Not. **54**, (2004), pp. 2915-2933.
[10] Ferrer Santos, W. and Haim, M. *Radford's formula for biFrobenius algebras and applications.* To appear in Communications in Algebra.
[11] Hausser, F. and Nill, F. *Integral Theory for Quasi-Hopf Algebras.* arXiv:math.QA/9904164.
[12] Johnstone, P.T. *Sketches of an elephant: a topos theory compendium. Vol. 1*, volume 43 of *Oxford Logic Guides.* The Clarendon Press Oxford University Press, New York, 2002.
[13] Joyal, A. and Street, R. *Braided tensor categories.* Adv. Math, **102**, 1, (1993), pp. 20–78.
[14] Kadison, L. and Stolin, A. A. *An approach to Hopf algebras via Frobenius coordinates I.* Beitr. Algebra Geom. **42**, 2, (2001), pp. 359-384.
[15] Kadison, L. *An approach to Hopf algebras via Frobenius Coordinates.* J. Algebra, **295**, 1, (2006), pp. 27-43.
[16] Kassel, C. *Quantum Groups.* New York: Springer–Verlag. Graduate Texts in Mathematics, **155**, 1995, xii+523 p.
[17] Kelly, G.M. *Basic concepts of enriched category theory*, London Mathematical Society Lecture Note Series, vol. 64, Cambridge University Press, Cambridge, 1982.
[18] Larson, R. *Characters of Hopf algebras.* J. Algebra, **17**, (1971), pp. 352-368.
[19] Majid, S. *Foundations of quantum group theory*, Cambridge University Press, Cambridge, 1995.
[20] Lopez Franco, I. *Hopf modules for autonomous pseudomonoids and the monoidal centre.* (preprint) `http://uk.arxiv.org/abs/0710.3853`.
[21] Nikshych, D. *On the structure of weak Hopf algebras.* Adv. Math, **170**, (2002), pp. 257-286.
[22] Radford, D. *The order of the antipode of a finite dimensional Hopf algebra.* Amer. J. Math. **98**, (1976), pp. 333-355.
[23] Schauenburg, P. *Hopf modules and the double of a quasi-Hopf algebra*, Trans. Amer. Math. Soc. **354** (2002), no. 8, 3349–3378 (electronic).
[24] Schauenburg, P. *Hopf bimodules, coquasibialgebras, and an exact sequence of Kac*, Adv. Math. **165**, 2, (2002), 194–263.
[25] Schauenburg, P. *Two characterizations of finite quasi-Hopf algebras.* J. Algebra, **273**, 2, (2004), pp. 538–550 (2004).
[26] Schneider, H.-J. *Lectures on Hopf Algebras*, notes by S. Natale. Trabajos de Matemática, Vol. 31/95, (1995), FaMAF, Córdoba. Argentina.
[27] Sweedler, M. *Integrals for Hopf algebras.* Ann. of Math. **91**,(1969), pp. 323-335.
[28] Sweedler, M. *Hopf algebras.* New York: W.A. Benjamin, Inc. 1969, 336 p.

Facultad de Ciencias, Universidad de la República, Iguá 4225, 11400 Montevideo, Uruguay,

E-mail address: `wrferrer@cmat.edu.uy`

DPMMS, Centre for Mathematical Sciences, Wilberforce Road, Cambridge CB3 0WB, UK

E-mail address: `I.Lopez-Franco@dpmms.cam.ac.uk`

Contemporary Mathematics
Volume **483**, 2009

Group Identities on Symmetric Units in Alternative Loop Algebras

Edgar G. Goodaire and César Polcino Milies

To our friend, Professor Ivan Shestakov, on the occasion of his sixtieth birthday

Abstract. If $\alpha = \sum \alpha_\ell \ell$ is an element of an alternative loop ring RL, we denote by $\alpha^\sharp$ the element $\sum \alpha_\ell \ell^{-1}$ and call α *symmetric* if $\alpha^\sharp = \alpha$. In previous work, the authors have considered the possibility that the unit loop of RL satisfies a group identity. Here, we assume merely that the symmetric units of RL (with R a field or the ring of rational integers) satisfy a group identity.

1. Introduction

From a commutative, associative ring R with 1 and a loop L, one forms the loop algebra RL precisely as if L were a group. The algebra RL is *alternative* if it satisfies the alternative laws

$$(yx)x = yx^2 \quad \text{and} \quad x(xy) = x^2y.$$

In the early 1980s, the existence of loop algebras that are alternative but not associative was established [**Goo83, CG86**] and since that time there has been quite a bit of research into the problems associated with such algebras and the underlying *RA loops* which produce them. A good place to find a discussion of some of these problems is [**GJM96**], which is also the best reference for RA loops and alternative loop algebras. Of note is the fact that an RA loop is Moufang and hence *diassociative*: subloops generated by two elements are associative. In fact, if three elements of a Moufang loop associate in some order, then the subloop they generate is a group.

Just as with group rings, the set of units (that is, invertible elements) in an alternative loop ring RL is closed under products and inverses and hence forms

2000 *Mathematics Subject Classification.* Primary 17D05; Secondary 20N05.

The first author wishes to thank FAPESP of Brasil for financial support and the Instituto de Matemática e Estatística of the Universidade de São Paulo where he was a guest while the work on this paper was accomplished.

This research was supported by a Discovery Grant from the Natural Sciences and Engineering Research Council of Canada, by FAPESP, Proc. 2005/60411-8 and CNPq., Proc. 300243/79-0(RN) of Brasil.

a (Moufang) loop $\mathcal{U}(RL)$, the *unit loop* of RL, and this loop contains L. This observation makes it natural to ask how many properties of L are inherited by $\mathcal{U}(RL)$. The answer is "not many." For example, L is solvable, nilpotent, FC, has the torsion product property and is torsion over its centre, but $\mathcal{U}(RL)$ rarely satisfies any of these conditions [**GM96b, GM96c, GM97, GM02, GM01**]. Generalizing all these concepts, the authors have recently considered the possibility that $\mathcal{U}(RL)$ satisfies a group identity [**GM**].

One defines the notion of a group identity on a group G like this. Let K denote the free group on a set of variables $x_1, x_2, \ldots$ and let $w(x_1, x_2, \ldots, x_n) \in K$ be a nonempty reduced word in K. Then $w = 1$ is a *group identity* for G if $w(g_1, g_2, \ldots, g_n) = 1$ for all $g_1, g_2, \ldots, g_n \in G$. We extend the notion of group identity to Moufang loops with the following definition.

DEFINITION 1.1. A Moufang loop M satisfies a group identity if and only if there is a nonempty reduced word $w = w(x_1, x_2)$ in the free group on two variables such that $w(\ell_1, \ell_2) = 1$ for all $\ell_1, \ell_2 \in M$.

The restriction to two generators here is actually artificial because of diassociativity and the fact that a free group on n generators can always be embedded in a free group on just two generators [**Rob82**, Theorem 6.1.1]. For example, any nilpotent Moufang loop satisfies a group identity in our sense. To see this, let $f = 1$ be an identity (in a finite number of nonassociating variables) satisfied by a nilpotent Moufang loop L. Now view f as a word in the free **group** on these same variables and express f as a word $w(x_1, x_2)$ in two variables. Suppose a and b are elements of L and we apply w to the pair a, b. Since $\langle a, b\rangle$ is a group and it is nilpotent, $w(a, b) = 1$, so L satisfies the group identity $w = 1$.

Since Moufang loops are *inverse property loops*, that is, $(ab)^{-1} = b^{-1}a^{-1}$ for any a, b, the map $a \mapsto a^{-1}$ is an involution (an antiautomorphism of order 2) and this extends linearly to an involution of an alternative loop ring RL, which we denote $\alpha \mapsto \alpha^\sharp$.[1] Thus, for $\alpha = \sum \alpha_\ell \ell \in RL$, $\alpha^\sharp = \sum \alpha_\ell \ell^{-1}$.

Call a unit α *symmetric* if $\alpha^\sharp = \alpha$ and let $\mathcal{U}^+(RL)$ be the set of symmetric units in RL. (Conditions under which this set is actually a loop are known [**GM06**].) In this paper, we suppose existence of a group identity on $\mathcal{U}^+(RL)$ and begin with some basic results.

Recall that a loop is *Hamiltonian* if it is not an abelian group and every subloop is normal. Throughout, we use $\mathcal{Z}(L)$ and $\mathcal{Z}(RL)$ to denote the centres of L and RL, respectively.

LEMMA 1.2. *If L is a Hamiltonian Moufang 2-loop (possibly associative) and R is any commutative, associative ring with 1, then $\mathcal{U}^+(RL)$ satisfies the group identity $(u, v) = 1$.*

PROOF. Let L be a Hamiltonian Moufang 2-loop. Then $L = L_1 \times E$ is the direct product of an elementary abelian 2-group E and a loop L_1 which is the Cayley loop if L is not associative [**Nor52**], [**GJM96**, Theorem II.4.8] and the quaternion group of order 8 otherwise (see, for example, [**Hal59**, Theorem 12.5.4], for a proof of this classical theorem). Thus, the centre of L consists precisely of the elements of L

[1] In group rings, the map $\alpha \mapsto \alpha^\sharp$ is denoted $\alpha \mapsto \alpha^*$, but since alternative loop rings have a canonical (and, in general, different) involution $\alpha \to \alpha^*$ (see Section 4), we use $\sharp$ instead of $*$ for the involution of central interest here.

that have order 2. With s the unique nonidentity commutator of L, we note also that the inverse of a noncentral element ℓ is $s\ell$. Now the set of symmetric elements of RL is spanned by loop elements of order 2 and ring elements of the form $\ell+\ell^{-1}$, $\ell \notin \mathcal{Z}(L)$. In particular, every symmetric unit is a linear combination of elements central in L and ring elements of the form $\ell+\ell^{-1}=(1+s)\ell$, $\ell \notin \mathcal{Z}(L)$. Such elements are conjugacy class sums in RL, so they span the centre of RL [**GJM96**, Corollary III.1.5]. Also, each such element is symmetric. So $\mathcal{U}^+(RL)=\mathcal{Z}(RL)$ and the result follows. □

2. The Finite Case

The questions we explore in this paper have been considered in the context of group rings by various authors [**GSV98, SV06**]. Here is a theorem of A. Giambruno, S. K. Sehgal and A. Valenti that we will use later.

THEOREM 2.1. [**GSV98**] *Let F be a field of characteristic $p \geq 0$, $p \neq 2$, and let G be a finite group. Then $\mathcal{U}^+(FG)$ satisfies a group identity if and only if the set P of p-elements of G is a normal subgroup of G and G/P is an abelian group or a Hamiltonian 2-group.*

For alternative loop rings that are not associative we have an analogous result. The proof requires the fact that for any prime p, the set L_p of p-elements in an RA loop L is a normal subloop of L and central if p is odd [**CG86**, proof of Theorem 6], [**GJM96**, Proposition V.1.1].

THEOREM 2.2. *Let F be a field of characteristic $p \geq 0$ and let L be a finite RA loop. Then $\mathcal{U}^+(FL)$ satisfies a group identity if and only if*

(1) *$p=2$, or*
(2) *L is a Hamiltonian 2-loop, or*
(3) *p is odd, $L=L_p \times L_2$, and L_2 is a Hamiltonian 2-loop.*

PROOF. If $p=2$, $\mathcal{U}(FL)$ is nilpotent [**GM97**] and hence satisfies a group identity. If L is a Hamiltonian 2-loop, $\mathcal{U}^+(FL)$ satisfies a group identity by Lemma 1.2. Let F be a field of odd characteristic p and suppose $L=L_p \times L_2$ with L_2 Hamiltonian. Then $FL=FL_p[L_2]$ is a loop ring of L_2 over a central coefficient ring and Lemma 1.2 says that $\mathcal{U}^+(FL)$ satisfies a group identity. This gives the theorem in one direction.

Now assume that $\mathcal{U}^+(FL)$ satisfies a group identity and that $\operatorname{char} F=p\neq 2$. Take $a,x \in L$ with $ax \neq xa$ and let z be an element of prime order $q \neq p$. Then $G=\langle a,x,z\rangle$ is a finite nonabelian group and $\mathcal{U}^+(FG)$ satisfies a group identity. If $p=0$, the set P of p-elements is $\{1\}$ and $G=G/P$ is a Hamiltonian 2-group by Theorem 2.1. It follows that L is a Hamiltonian 2-loop. If $p>0$, then P is central (because p is odd), so G/P is nonabelian and hence a 2-group. So $q=2$ and we learn that the only primes dividing $|L|$ are 2 and p. Normality of L_p and L_2 shows that $L=L_p \times L_2$. Applying Theorem 2.1 to the associative subloop of L_2 generated by two noncommuting elements, we see that L_2 is Hamiltonian. □

3. Torsion Loops

In this section, L is a torsion loop that is not necessarily finite. As before, our results both use and extend a theorem about group algebras.

THEOREM 3.1. [**GSV98**] *Let F be a field and G a torsion group with FG semiprime. Then $\mathcal{U}^+(FG)$ satisfies a group identity if and only if G is abelian or a Hamiltonian 2-group.*

In Theorems 4.2.12 and 4.2.13 of Passman's classic text [**Pas77**], one can find necessary and sufficient conditions for a group algebra FG over a field F to be semiprime. In characteristic 0, FG is always semiprime, whereas in positive characteristic p, it is necessary and sufficient that G contain no finite normal subgroups of order divisible by p. Suppose FL is alternative loop algebra over the field F. As in the associative case, FL is semiprime in characteristic 0. In characteristic $p > 0$, FL is semiprime if and only if L contains no central element of order p [**GJM96**, Corollary VI.3.6]. Since elements of odd order in L are central, if $p > 2$ and FL is semiprime, then L contains no elements of order p.

This is a key idea used in the proof of our extension of Theorem 3.1 to alternative loop algebras. Note that our theorem makes no assumption about FL being semiprime.

THEOREM 3.2. *Let F be a field of characteristic $p \geq 0$ and let L be a torsion RA loop. Then $\mathcal{U}^+(FL)$ satisfies a group identity if and only if $p = 2$ or $L = L_0 \times P$ is the direct product of a Hamiltonian 2-loop L_0 and an abelian p-group P.*

PROOF. If $p = 2$, then $\mathcal{U}(FL)$ is nilpotent [**GM97**] and hence satisfies a group identity. If $p \neq 2$ and $L = L_0 \times P$ as in the statement, then we can think of the loop algebra FL as the loop algebra of L_0 with coefficients in the central associative ring FP, so $\mathcal{U}(FL)$ satisfies a group identity by Lemma 1.2. Thus we have the result in one direction.

Conversely, assume that $\mathcal{U}^+(FL)$ satisfies a group identity and $p \neq 2$. We may write $L = L_0 \times P$ where L_0 is an RA loop whose elements have order prime to p and P is a p-group [**GJM96**, Proposition V.1.1] and, since the elements of P have odd order, they are central. So P is an abelian group. Take any $a, x \in L_0$ which do not commute and let $z \in L_0$ be an element central in L_0. Then $G = \langle a, x, z \rangle$ is a nonabelian torsion group and FG is semiprime. By Theorem 3.1, G is a Hamiltonian 2-group, so L_0 is a Hamiltonian 2-loop. □

4. Nontorsion Loops

In this final section, we consider RA loops with no finiteness restrictions whatsoever and present theorems concerning the symmetric units of alternative loop algebras over fields and also over the ring Z of rational integers. Our results make reference to the torsion elements of a loop L. If L is RA, this set forms a locally finite subloop which is finite if L is finitely generated [**GM95**, Lemma 2.1], [**GM96a**, Lemma 1.4], [**GJM96**, Lemma VIII.4.1].

For group rings over the integers, Giambruno, Sehgal and Valenti have shown that if $\mathcal{U}^+(\mathsf{Z}G)$ satisfies a group identity, then any torsion subgroup H of G is either abelian or a Hamiltonian 2-group and every subgroup of H is normal in G [**GSV98**, Theorem 4]. This helps us to characterize RA loops whose symmetric units (in $\mathsf{Z}L$) satisfy a group identity.

THEOREM 4.1. *Let L be an RA loop. Then the following are equivalent:*

(1) *$\mathcal{U}^+(\mathsf{Z}L)$ satisfies a group identity.*
(2) *$\mathcal{U}(\mathsf{Z}L)$ satisfies a group identity.*

(3) *For every finitely generated group $G \subseteq L$, $\mathcal{U}(\mathbb{Z}G)$ satisfies a group identity.*
(4) *The torsion subloop T of L is either an abelian group or a Hamiltonian 2-loop, and every subloop of T is normal in L.*

When any, and hence all, of these conditions is satisfied, $\mathcal{U}(\mathbb{Z}G)$ satisfies the identity $(u^2, v^2) = 1$.

PROOF. The equivalence of (2), (3) and (4) is known and each implies that $(u^2, v^2) = 1$ is a group identity [**GJM96**, Corollary XII.2.9]. Since (1) is an obvious consequence of (2), we have only to establish that (2) follows from (1), so assume that $\mathcal{U}^+(\mathbb{Z}L)$ satisfies a group identity. Let $t \in T$, the torsion subloop of L, and suppose $xt \neq tx$ for some $x \in L$. Let $t_0 \in T$ be any central element. The subloop $G = \langle t_0, t, x \rangle$ is a group because t_0, t and x associate and $\mathcal{U}^+(\mathbb{Z}G)$ satisfies a group identity, so the result of Giambruno et al mentioned earlier tells us that both t and t_0 are 2-elements and $x^{-1}tx \in \langle t \rangle$. In an RA loop, as with groups, the last observation shows that every subloop of T is normal in L [**GJM96**, Corollary IV.1.11]. In particular, every subloop of T is normal in T so, if this is a group, the result for groups shows that T is either abelian or a Hamiltonian 2-group whereas, if T is not a group, then it is a Hamiltonian Moufang 2-loop. □

Now we consider loop algebras over a field. In the associative case, Sehgal and Valenti have a helpful result.

THEOREM 4.2. [**SV06**, Theorem 4] *Let F be an infinite field, G a group whose torsion elements comprise a set $T \neq G$ and suppose FG is semiprime. Suppose $\mathcal{U}^+(FG)$ satisfies a group identity. Then we know the following.*
(1) *If* $\operatorname{char} F = p > 2$, *then T is an abelian p'-group (that is, all elements of T have order relatively prime to p).*
(2) *If* $\operatorname{char} F = 0$, *then T is an abelian group or a Hamiltonian 2-group.*
(3) *All idempotents of FT are central in FG.*

A few preliminary remarks will help the reader with our proof of the alternative analogue. An RA loop L has a unique nonidentity *commutator/associator* s, which is necessarily central and of order 2. The map $\ell \mapsto \ell^*$ where

$$\ell^* = \begin{cases} \ell & \text{if } \ell \text{ is central} \\ s\ell & \text{otherwise} \end{cases}$$

is an involution of L that extends linearly to an involution of any loop ring RL (which we continue to denote $*$). Now L contains a group G of index 2 and so we have $L = G \cup Gu$ for any $u \in L \setminus G$. Thus elements of RL have the form $x + yu$, where x and y are in the group ring RG, and $(x + yu)^* = x^* + sy$. It is known that an element of RL is central if and only if it is invariant under $*$, so we obtain a very useful test for centrality: $x + yu$ is central if and only if $x^* = x$ and $sy = y$. This material is quite basic to the theory of RA loops and can be found, for example, in Section III.4 of [**GJM96**].

THEOREM 4.3. *Let F be an infinite field of characteristic $p \geq 0$ and let L be an RA loop with torsion subloop $T \neq L$. Assume that FL is semiprime or that the set of elements of L of order prime to p is finite. If $p = 0$, assume also that F contains no solutions to $x^2 + y^2 + z^2 + w^2 = -1$. Then the following statements are equivalent.*
(1) *$\mathcal{U}^+(FL)$ satisfies a group identity.*

(2) $p = 2$ *or*
  (a) T *is an abelian group or a (possibly associative) Hamiltonian 2-loop and*
  (b) *if* $G \subseteq L$ *is a group with torsion subgroup* $T(G)$, *then all idempotents of* $FT(G)$ *are central in* FG.

(3) $p = 2$ *or every idempotent of* FT *is central in* FL.

In the case that FL *is semiprime and* $p \neq 2$, $\mathcal{U}^+(FL)$ *satisfies the group identity* $(u, v) = 1$.

PROOF. The last statement follows because when $p \neq 2$, the proof that (2) implies (3) that we present below shows that the symmetric units commute.

Assume (1), that $\mathcal{U}^+(FL)$ satisfies a group identity, and assume also that $p \neq 2$. Let $t_1, t_2 \in T$ be 2-elements. Since g^2 is central for any $g \in L$ and since L contains an element of infinite order, the centre of L contains an element a of infinite order. Let $H = \langle t_1, t_2, a\rangle$ be the group generated by t_1, t_2 and a. Then H contains no p-element so the group algebra FH is semiprime. The symmetric units of FH satisfy a group identity so, if $p > 2$, Theorem 4.2 says that $t_1t_2 = t_2t_1$. Since elements of odd order are central, it follow that T is commutative and hence associative [**GM96b**], [**GJM96**, Corollary IV.2.4]. On the other hand, if $p = 0$ and T is not commutative, let t_1 and t_2 be any elements of T that do not commute and let a be a central element of infinite order. This time Theorem 4.2 says that $H = \langle t_1, t_2, a\rangle$ is a Hamiltonian 2-group, so T is a Hamiltonian 2-loop. Now let $G \subseteq L$ be a group with torsion subgroup $T(G)$. We complete the proof of (2) by showing that all idempotents of $FT(G)$ are central in FG. This follows immediately by Theorem 4.2 if $p = 0$, or if FG is semiprime, so assume $p > 0$ and that G contains p-elements.

Write $G = G_0 \times P$ where P is a central p-group and G_0 contains no p-elements. Let e be an idempotent in $FT(G)$. Then the subloop M generated by the support[2] of e is finitely generated and torsion, so it is finite [**GJM96**, Lemma VIII.4.1]. Write $M = M_0 \times P_0$ where M_0 contains no p-elements and $|P_0| = p^m$ for some positive integer m. Write $e = \sum_{i,j} k_{ij} m_i n_j$ with $m_i \in M_0$, $n_j \in P_0$. Then $e^{p^m} = \sum_{i,j} k_{ij} m_i^{p^m} n_j^{p^m} = \sum_{i,j} k_{ij} m_i^{p^m} = e$, showing that $n_j = 1$ for all indices j. Thus actually $e \in FM_0 \subseteq FG_0$. Since FG_0 is semiprime and $\mathcal{U}^+(FG_0)$ satisfies a group identity, it follows that e commutes with every element in G_0. It also commutes with every element of H because H is central, so e is central in FG. This establishes (2).

Now assume statement (2) and that $p \neq 2$. We first prove that every idempotent of FT is central in FL in the case that FL is semiprime. This is known to be the case if $p > 0$ or T is abelian [**GM**], so we assume that $p = 0$ and T is a Hamiltonian 2-loop. Replacing T by the subloop generated by the support of an idempotent in FT, if necessary, we may assume that T is finitely generated and hence finite. Thus $T = T_1 \times E$ is the direct product of an abelian 2-group and a subloop T_1 which is Q, the quaternion group of order 8, if T is associative, and the Cayley loop $M_{16}(Q)$ otherwise. (See the proof of Lemma 1.2 for appropriate references.) Thinking of FT as the loop algebra of T_1 with coefficients in FE, it is basic that FE is the direct sum of copies of F, so FT is the direct sum of copies of FT_1. In characteristic 0, $FQ \cong 4F \oplus (F, -1, -1)$, where $(F, -1, -1)$ is the quaternion

[2]The *support* of an element $\alpha = \sum_{\ell \in S} \alpha_\ell \ell$ in a loop ring is the set $\{\ell \in S \mid \alpha_\ell \neq 0\}$ of loop elements that actually appear in the representation of α, that is, with nonzero coefficient.

algebra over F, and $F[M_{16}(Q)] = 8F \oplus (F, -1, -1, -1)$ where $(F, -1, -1, -1)$ is the Cayley-Dickson algebra over F [**GJM96**, Corollaries VII.2.3 and VII.2.4]. Since $x^2+y^2+z^2+w^2 = -1$ has no solutions in F, this quaternion and this Cayley-Dickson algebra are division algebras [**GJM96**, Theorem I.3.4]. This already shows that every idempotent of FT is central in FT, but we want centrality in FL. Since every idempotent in FT is the sum of primitive idempotents, to obtain such centrality, it suffices to show that the primitive idempotents of FT are central in FL.

We now show that all elements of FE are central in FL by showing that E is central in L. So take $x \in L$ and $1 \neq e \in E$. Then $G = \langle x, e \rangle$ is a group and, by hypothesis, every idempotent of $FT(G)$ is central in FG. The main theorem of [**CM88**] says that $x^{-1}ex$ is a power of e, so $x^{-1}ex = e$. Thus e is central and we may complete the proof by showing that the primitive idempotents of FT_1 are central in FL. This task is straightforward because these idempotents appear in the literature [**GM96a**]. Presenting $Q = \langle a, b \mid a^4 = 1, b^2 = a^2 \rangle$, the five primitive idempotents of FQ are

$$\begin{aligned}
e_1 &= \frac{\epsilon}{8}(1 + a + a^2 + a^3 + b + ab + a^2 + a^3b) \\
e_2 &= \frac{\epsilon}{8}(1 + a + a^2 + a^3 - b - ab - a^2 - a^3b) \\
e_3 &= \frac{\epsilon}{8}(1 - a + a^2 - a^3 + b - ab + a^2b - a^3b) \\
e_4 &= \frac{\epsilon}{8}(1 - a + a^2 - a^3 - b + ab - a^2b + a^3b) \\
e_5 &= \frac{\epsilon}{8}(1 - a^2)
\end{aligned}$$

where ϵ is the identity of F which, as an idempotent of FE, is central in FL. Since $e_i^* = e_i$ for each i, each e_i is central in FL. The nine primitive idempotents of $F[M_{16}(Q)]$ are e_5 and the eight elements

$$e_{j1} = \frac{\epsilon}{2}(1 + u)e_j \quad \text{and} \quad e_{j2} = \frac{\epsilon}{2}(1 - u)e_j,$$

$j = 1, 2, 3, 4$. As before, ϵ is central in FL. Since $e_{ij} = x + yu$, with $x, y \in FQ$, $x^* = x$ and $sy = y$, each e_{ij} is also central. This establishes (3) in the case that FL is semiprime.

If FL is not semiprime, then $\operatorname{char} F = p > 2$ and L contains p-elements. As before, we write $L = L_0 \times P$ where P is a central p-group and L_0 contains no p-elements. Given an idempotent e in FT, the subloop $M = \langle \operatorname{supp}(e) \rangle$ is finitely generated and torsion, hence finite. Write $M = M_0 \times P_0$ where $|P_0| = p^m$ for some integer m and, once again, write $e = \sum_{i,j} k_{ij} m_i n_j$, with $m_i \in M_0$, $n_j \in P_0$. This time, we consider the image $\overline{e} = \sum_{i,j} k_{ij} \overline{m}_i n_j$ in the group $\overline{M} = M/\langle s \rangle \cong M_0/\langle s \rangle \times P_0$ and compute

$$\overline{e}^{p^m} = \sum_{i,j} k_{ij} \overline{m}_i^{p^m} n_j^{p^m} = \sum_{i,j} k_{ij} \overline{m}_i^{p^m} = \overline{e}.$$

Since $\overline{e}^{p^m} = \overline{e}$, $e \in FM_0 \subseteq FL_0$. Since FL_0 is semiprime, e commutes with every element in L_0 and also with every element in P because P is central. Hence e is central in FL and we have established (3).

Finally we assume statement (3) and prove (1). As noted before, $\mathcal{U}(FL)$ is nilpotent in characteristic 2, so we may also assume that $p \neq 2$ and hence that every idempotent of FT is central in FL. As before, we assume initially that FL is semiprime and show that in this case the symmetric units commute. In doing this, we may assume that L is finitely generated and hence that T is finite. In this case, FT is the sum of simple algebras which are fields and quaternion algebras if T is associative, and fields and Cayley-Dickson algebras otherwise [**GJM96**, Corollary VI.4.8]. Since every idempotent of FT is central in FT, all simple components are division rings, D_i. Using [**GJM96**, Lemma XII.1.1], we conclude that any unit u of FL can be written in the form $\sum d_i\ell_i$, $d_i \in D_i$, $\ell_i \in L$. Thus $u^\sharp = \sum \ell_i^{-1} d_i^\sharp$ so, if u is symmetric, $\ell_i^{-1} d_i^\sharp = d_i\ell_i$ for each i. It follows that ℓ_i^2 has support in FT, so $\ell_i \in T$ and $u \in FT$. If $p > 0$, T is an abelian group because $p \neq 2$ [**GM96a**, Theorem 2.3], while if $p = 0$, T is a Hamiltonian 2-loop [**GM96a**, Theorem 3.3]. In either case, the symmetric units of FT commute (using Lemma 1.2 in the Hamiltonian case).

Now suppose that FL is not semiprime, but that the group P of p-elements is finite. Let u and v be symmetric units. As before, we may assume that L is finite and write $L = L_0 \times P_0$ where L_0 is a subloop consisting of elements of order prime to p and $P_0 \subseteq P$. We think of FL as $F[P]L_0$, the group ring of L_0 with coefficients in $F[P]$, and pass to the semiprime loop ring $(F[P]/N)L_0$, where N denotes the radical of $F[P]$. By the case already considered, the images $\overline{u}$, $\overline{v}$ of u and v, respectively, commute. Thus $(u, v) \in 1 + N$, so $(u, v)^{p^t} = 1$ for some sufficiently high power p^t of p (independent of u and v) because N is nilpotent. This establishes (3) and completes the proof. □

REMARK 4.4. It is interesting to compare the equivalence of (1) and (3) in Theorem 4.3 with a previous result of the authors which states that the full unit loop $\mathcal{U}(FL)$ satisfies a group identity if and only if $p = 2$ or T **is an abelian group and** every idempotent of FT is central in FL [**GM**]. Unlike the situation when the coefficient ring is Z, a group identity on $\mathcal{U}^+(FL)$ is, in general, not enough to force a group identity on $\mathcal{U}(FL)$.

References

[CG86] Orin Chein and Edgar G. Goodaire, *Loops whose loop rings are alternative*, Comm. Algebra **14** (1986), no. 2, 293–310.

[CM88] Sônia P. Coelho and C. Polcino Milies, *A note on central idempotents in group rings II*, Proc. Edinburgh Math. Soc. **31** (1988), 211–215.

[GJM96] E. G. Goodaire, E. Jespers, and C. Polcino Milies, *Alternative loop rings*, North-Holland Math. Studies, vol. 184, Elsevier, Amsterdam, 1996.

[GM] Edgar G. Goodaire and César Polcino Milies, *Polynomial and group identities in alternative loop algebras*, J. Algebra Appl., to appear post 2008.

[GM95] ______, *On the loop of units of an alternative loop ring*, Nova J. Algebra Geom. **3** (1995), no. 3, 199–208.

[GM96a] ______, *Central idempotents in alternative loop algebras*, Nova J. Math. Game Theory Algebra **5** (1996), no. 3, 207–214.

[GM96b] ______, *Finite conjugacy in alternative loop algebras*, Comm. Algebra **24** (1996), no. 3, 881–889.

[GM96c] ______, *The torsion product property in alternative algebras*, J. Algebra **184** (1996), 58–70.

[GM97] ______, *Nilpotent Moufang unit loops*, J. Algebra **190** (1997), 88–99.

[GM01] ______, *Alternative loop rings with solvable unit loops*, J. Algebra **240** (2001), no. 1, 25–39.

[GM02] ______, *Moufang unit loops torsion over their centres*, Quaestiones Math. **25** (2002), 1–12.

[GM06] ———, *Symmetric units in alternative loop rings*, Algebra Colloq. **13** (2006), no. 3, 361–370.

[Goo83] Edgar G. Goodaire, *Alternative loop rings*, Publ. Math. Debrecen **30** (1983), 31–38.

[GSV98] A. Giambruno, S. K. Sehgal, and A. Valenti, *Symmetric units and group identities*, Manuscripta Math. **96** (1998), 443–461.

[Hal59] M. Hall, Jr., *The theory of groups*, MacMillan, New York, 1959.

[Nor52] D. A. Norton, *Hamiltonian loops*, Proc. Amer. Math. Soc. **3** (1952), 56–65.

[Pas77] D. S. Passman, *The algebraic structure of group rings*, Wiley-Interscience, New York, 1977.

[Rob82] D. J. S. Robinson, *A course in the theory of groups*, Springer-Verlag, New York, 1982.

[SV06] S. K. Sehgal and A. Valenti, *Group algebras with symmetric units satisfying a group identity*, Manuscripta Math. **119** (2006), 243–254.

Memorial University of Newfoundland, St. John's, Newfoundland, Canada A1C 5S7

E-mail address: edgar@math.mun.ca

Instituto de Matemática e Estatística, Universidade de São Paulo, Caixa Postal 66.281, CEP 05315-970, São Paulo SP, Brasil

E-mail address: polcino@ime.usp.br

Contemporary Mathematics
Volume **483**, 2009

On Multiplicity Problems for Finite-Dimensional Representations of Hyper Loop Algebras

Dijana Jakelić and Adriano Moura

Abstract. Given a hyper loop algebra over a non-algebraically closed field, we address multiplicity problems in the underlying abelian tensor category of finite-dimensional representations. Namely, we give formulas for the ℓ-characters of the simple objects, the Jordan-Hölder multiplicities of the Weyl modules, and the Clebsch-Gordan coefficients.

Introduction

Let $\mathcal{C}$ be a Jordan-Hölder tensor category. Roughly speaking, this is an abelian category with finite-length objects equipped with a tensor product (for a precise definition of abelian and tensor categories see, for instance, [**9**, Chapter 5]). Given two objects V and W of $\mathcal{C}$ with V simple, let $\mathrm{mult}_V^{\mathcal{C}}(W)$ be the multiplicity of V as an irreducible constituent of W in the category $\mathcal{C}$. We will often refer to $\mathrm{mult}_V^{\mathcal{C}}(W)$ as the Jordan-Hölder multiplicity of V inside W in the category $\mathcal{C}$. Formulas for Jordan-Hölder multiplicities make one of the most exciting subjects in combinatorial representation theory. For instance, if $\mathfrak{g}$ is a Kac-Moody algebra, the Jordan-Hölder multiplicities for an integrable module W may be computed from the character of W. In order to do that one has to know the characters of the irreducible integrable modules. These are, in turn, given by the celebrated Weyl-Kac character formula which, in some special cases, may be used to recover some famous identities including Rogers-Ramanujan's identities (see [**4, 25**] and references therein). An important subclass of multiplicity problems is the one of computing Clebsch-Gordan coefficients, i.e., of computing $\mathrm{mult}_V^{\mathcal{C}}(W)$ in a given category $\mathcal{C}$ when W is a tensor product of two simple objects. In the case of a Kac-Moody algebra these coefficients may, in principle, be found as a consequence of the Weyl-Kac character formula.

2000 *Mathematics Subject Classification.* 17B65; 17B10; 20G42.

Key words and phrases. hyperalgebras, loop groups, affine Kac-Moody algebras, finite-dimensional representations.

The research of A.M. is partially supported by CNPq and FAPESP. A.M also thanks the Department of Mathematics, Statistics, and Computer Science of the University of Illinois at Chicago for its hospitality and support during the period this note was written.

In this note we address some of the aforementioned problems for the category $\mathcal{C}(\tilde{\mathfrak{g}})_{\mathbb{K}}$ of finite-dimensional representations of the hyper loop algebra $U(\tilde{\mathfrak{g}})_{\mathbb{K}}$ over a field $\mathbb{K}$. Here, $\mathfrak{g}$ denotes a fixed simple finite-dimensional Lie algebra over the complex numbers. The finite-dimensional representation theory of hyper loop algebras was initiated in [**20, 21**]. As remarked in the introduction of [**21**], in an appropriate algebraic geometric framework, it should turn out that studying finite-dimensional representations of these algebras is equivalent to studying finite-dimensional representations of certain algebraic loop groups (in the defining characteristic). We remark that the hyperalgebra $U(\mathfrak{g})_{\mathbb{K}}$ of the underlying Chevalley group of universal type is a subalgebra of $U(\tilde{\mathfrak{g}})_{\mathbb{K}}$.

It was shown in [**21**] that the passage from algebraically closed fields to non-algebraically closed ones is much more interesting in the context of hyper loop algebras than in the case of the hyperalgebra $U(\mathfrak{g})_{\mathbb{K}}$. For instance, multiplicity formulas, such as the ones mentioned in the first paragraph, for the category $\mathcal{C}(\mathfrak{g})_{\mathbb{K}}$ of finite-dimensional $U(\mathfrak{g})_{\mathbb{K}}$-modules are invariant under field extensions. However, it was already clear from several results and examples in [**21**] that these problems have a much richer relation with Galois theory in the case of hyper loop algebras.

The purpose of this note is to solve certain multiplicity problems for $\mathcal{C}(\tilde{\mathfrak{g}})_{\mathbb{K}}$. Namely, we find formulas for the following: (1) the ℓ-characters of the simple objects, (2) the Jordan-Hölder multiplicities of the Weyl modules, and (3) the Clebsch-Gordan coefficients. By a solution we mean the following. Suppose $\mathbb{K}$ is a non-algebraically closed field and $\mathbb{F}$ is an algebraic closure of $\mathbb{K}$. Our formulas express the multiplicities for the category $\mathcal{C}(\tilde{\mathfrak{g}})_{\mathbb{K}}$ in terms of the corresponding multiplicities for the category $\mathcal{C}(\tilde{\mathfrak{g}})_{\mathbb{F}}$ via Galois theory. We also review the literature related to these problems in the algebraically closed setting. It turns out that the multiplicities in the category $\mathcal{C}(\tilde{\mathfrak{g}})_{\mathbb{F}}$ may be expressed in terms of the corresponding formulas for the category $\mathcal{C}(\mathfrak{g})_{\mathbb{F}}$. We give bibliographic references for the later and remark that some of these problems are still open in the category $\mathcal{C}(\mathfrak{g})_{\mathbb{F}}$ when $\mathbb{F}$ has positive characteristic.

The paper is organized as follows. In §1, we review several prerequisites such as the construction of the hyperalgebras and hyper loop algebras as well as the relevant results about their finite-dimensional representations. In §2, we present the results about the Clebsch-Gordan problem. In §3, we study the ℓ-characters of the simple objects of $\mathcal{C}(\tilde{\mathfrak{g}})_{\mathbb{K}}$. This is equivalent to computing their Jordan-Hölder multiplicities in the category of finite-dimensional representations of a certain polynomial subalgebra of $U(\tilde{\mathfrak{g}})_{\mathbb{K}}$. We end the section by expressing the results of §2 in terms of ℓ-characters. Finally, §4 is concerned with the computation the Jordan-Hölder multiplicities for the Weyl modules.

1. Preliminaries

1.1. Throughout the paper, let $\mathbb{F}$ be a fixed algebraically closed field and $\mathbb{K}$ a subfield of $\mathbb{F}$ having $\mathbb{F}$ as an algebraic closure. Also, $\mathbb{C}, \mathbb{Z}, \mathbb{Z}_+, \mathbb{N}$ denote the sets of complex numbers, integers, non-negative integers, and positive integers, respectively. Given a ring $\mathbb{A}$, the underlying multiplicative group of units is denoted by $\mathbb{A}^\times$. The dual of a vector space V is denoted by V^*. The symbol $\cong$ means "isomorphic to".

If $\mathbb{L}$ is a field and A an associative unitary $\mathbb{L}$-algebra, the expression "V is an A-module" means V is an $\mathbb{L}$-vector space equipped with an A-action, i.e., a homomorphism of associative unitary $\mathbb{L}$-algebras $A \to \mathrm{End}_{\mathbb{L}}(V)$. The action of $a \in A$ on $v \in V$ is denoted simply by av. If V is an $\mathbb{L}$-vector space and $\mathbb{M}$ is a field extension of $\mathbb{L}$, set $V^{\mathbb{M}} = \mathbb{M} \otimes_{\mathbb{L}} V$ and identify V with $1 \otimes V \subseteq V^{\mathbb{M}}$. If V is an A-module, then $V^{\mathbb{M}}$ is naturally an $A^{\mathbb{M}}$-module.

The next lemma is immediate and will be repeatedly used in the proofs of the main results.

LEMMA 1.1. *Let A be a $\mathbb{K}$-algebra, V an A-module, and $0 \subseteq V_1 \subseteq V_2 \subseteq \cdots \subseteq V_m = V$ a composition series for V. Then $0 \subseteq V_1^{\mathbb{F}} \subseteq V_2^{\mathbb{F}} \subseteq \cdots \subseteq V_m^{\mathbb{F}} = V^{\mathbb{F}}$ is a sequence of inclusions of $A^{\mathbb{F}}$-modules and $V_j^{\mathbb{F}}/V_{j-1}^{\mathbb{F}} \cong (V_j/V_{j-1})^{\mathbb{F}}$.* □

1.2. Given two $\mathbb{K}$-vector spaces V and W, let $\varphi_{V,W}^{\mathbb{F}} : V^{\mathbb{F}} \otimes_{\mathbb{F}} W^{\mathbb{F}} \to (V \otimes_{\mathbb{K}} W)^{\mathbb{F}}$ be the isomorphism of $\mathbb{F}$-vector spaces determined by $(a \otimes v) \otimes (b \otimes w) \mapsto (ab) \otimes (v \otimes w)$.

Recall that if A is a Hopf algebra the tensor product of A-modules can be equipped with a structure of A-module by using the comultiplication. The Hopf algebra structure of A can be extended to one on $A^{\mathbb{F}}$. In particular, the comultiplication is given by $\Delta^{\mathbb{F}}(b \otimes x) = (\varphi_{A,A}^{\mathbb{F}})^{-1}(b \otimes \Delta(x))$, where Δ is the comultiplication on A. One now easily checks the following lemma which will be used in the proof of Theorem 2.1 below.

LEMMA 1.2. *Let A be a Hopf algebra over $\mathbb{K}$. If V, W are A-modules, $\varphi_{V,W}^{\mathbb{F}}$ is an isomorphism of $A^{\mathbb{F}}$-modules.* □

1.3. Let I be the set of vertices of a finite-type connected Dynkin diagram and let $\mathfrak{g}$ be the corresponding simple Lie algebra over $\mathbb{C}$ with a fixed Cartan subalgebra $\mathfrak{h}$. Fix a set of positive roots R^+ and let

$$\mathfrak{n}^{\pm} = \bigoplus_{\alpha \in R^+} \mathfrak{g}_{\pm\alpha} \quad \text{where} \quad \mathfrak{g}_{\pm\alpha} = \{x \in \mathfrak{g} : [h, x] = \pm\alpha(h)x, \ \forall\, h \in \mathfrak{h}\}.$$

Fix also a Chevalley basis $\{x_\alpha^{\pm}, h_i : \alpha \in R^+, i \in I\}$ for $\mathfrak{g}$ so that $\mathfrak{n}^{\pm} = \mathbb{C}x_\alpha^{\pm}$ and $[x_{\alpha_i}^+, x_{\alpha_i}^-] = h_i$. Here, α_i $(i \in I)$ denote the simple roots. The fundamental weights are denoted by ω_i $(i \in I)$, while Q, P, Q^+, and P^+ denote the root and weight lattices with the corresponding positive cones, respectively. We equip $\mathfrak{h}^*$ with the partial order given by $\lambda \leq \mu$ iff $\mu - \lambda \in Q^+$.

Let $\tilde{\mathfrak{g}} = \mathfrak{g} \otimes \mathbb{C}[t, t^{-1}] = \tilde{\mathfrak{n}}^- \oplus \tilde{\mathfrak{h}} \oplus \tilde{\mathfrak{n}}^+$ be the loop algebra over $\mathfrak{g}$ with bracket given by $[x \otimes f(t), y \otimes g(t)] = [x, y] \otimes (f(t)g(t))$. Given $\alpha \in R^+, i \in I$, and $r \in \mathbb{Z}$, let also $x_{\alpha,r}^{\pm} = x_\alpha^{\pm} \otimes t^r, h_{i,r} = h_i \otimes t^r$. Clearly, these elements form a basis of $\tilde{\mathfrak{g}}$. We identify $\mathfrak{g}$ with the subalgebra of $\tilde{\mathfrak{g}}$ given by the subset $\mathfrak{g} \otimes 1$ and identify the elements $x_{\alpha,0}^{\pm}$ and $h_{i,0}$ with the elements $x_\alpha^{\pm}$ and h_i, respectively.

1.4. For a Lie algebra $\mathfrak{a}$, let $U(\mathfrak{a})$ be its universal enveloping algebra. Consider the $\mathbb{Z}$-subalgebra $U(\tilde{\mathfrak{g}})_{\mathbb{Z}}$ (resp. $U(\mathfrak{g})_{\mathbb{Z}}$) of $U(\tilde{\mathfrak{g}})$ generated by the elements $(x_{\alpha,r}^{\pm})^{(k)} := \frac{(x_{\alpha,r}^{\pm})^k}{k!}$ (resp. $(x_\alpha^{\pm})^{(k)} := \frac{(x_\alpha^{\pm})^k}{k!}$) for all $\alpha \in R^+, r, k \in \mathbb{Z}, k \geq 0$. $U(\mathfrak{g})_{\mathbb{Z}}$ is Kostant's integral form of $U(\mathfrak{g})$ [**24**] and $U(\tilde{\mathfrak{g}})_{\mathbb{Z}}$ is Garland's integral form of $U(\tilde{\mathfrak{g}})$ [**17**]. Set $U(\tilde{\mathfrak{n}}^{\pm})_{\mathbb{Z}} = U(\tilde{\mathfrak{n}}^{\pm}) \cap U(\tilde{\mathfrak{g}})_{\mathbb{Z}}$ and define $U(\tilde{\mathfrak{h}})_{\mathbb{Z}}, U(\mathfrak{n}^{\pm})_{\mathbb{Z}}$, and $U(\mathfrak{h})_{\mathbb{Z}}$ similarly. We have $U(\mathfrak{g})_{\mathbb{Z}} = U(\mathfrak{n}^-)_{\mathbb{Z}} U(\mathfrak{h})_{\mathbb{Z}} U(\mathfrak{n}^+)_{\mathbb{Z}}$ and $U(\tilde{\mathfrak{g}})_{\mathbb{Z}} = U(\tilde{\mathfrak{n}}^-)_{\mathbb{Z}} U(\tilde{\mathfrak{h}})_{\mathbb{Z}} U(\tilde{\mathfrak{n}}^+)_{\mathbb{Z}}$. Moreover, given an ordering on $R^+ \times \mathbb{Z}$, the monomials in the elements $(x_{\alpha,r}^{\pm})^{(k)}$ obtained in the obvious way from the corresponding PBW monomials associated to our fixed bases for $\mathfrak{n}^{\pm}$ and $\tilde{\mathfrak{n}}^{\pm}$ form bases for $U(\mathfrak{n}^{\pm})_{\mathbb{Z}}$ and $U(\tilde{\mathfrak{n}}^{\pm})_{\mathbb{Z}}$, respectively. The set $\{\prod_{i \in I} \binom{h_i}{k_i} : k_i \in \mathbb{Z}_+\}$ is a basis for $U(\mathfrak{h})_{\mathbb{Z}}$ where $\binom{h_i}{k} = \frac{1}{k!} h_i(h_i - 1) \cdots (h_i - k + 1)$.

As for $U(\tilde{\mathfrak{h}})_{\mathbb{Z}}$, define elements $\Lambda_{i,r} \in U(\tilde{\mathfrak{h}})$ by the following equality of formal power series in the variable u:

$$\sum_{r=0}^{\infty} \Lambda_{i,\pm r}\ u^r = \exp\left(-\sum_{s=1}^{\infty} \frac{h_{i,\pm s}}{s}\ u^s\right).$$

It turns out that $\Lambda_{i,r} \in U(\tilde{\mathfrak{h}})_{\mathbb{Z}}$ and that the unitary $\mathbb{Z}$-subalgebras generated by the sets $\mathbf{\Lambda}^{\pm} = \{\Lambda_{i,\pm r} : i \in I, r \in \mathbb{N}\}$ are the polynomial algebras $\mathbb{Z}[\mathbf{\Lambda}^{\pm}]$ (cf. [**21**, Proposition 1.7]). Moreover, multiplication defines an isomorphism of $\mathbb{Z}$-algebras

$$U(\tilde{\mathfrak{h}})_{\mathbb{Z}} \cong \mathbb{Z}[\mathbf{\Lambda}^{-}] \otimes_{\mathbb{Z}} U(\mathfrak{h})_{\mathbb{Z}} \otimes_{\mathbb{Z}} \mathbb{Z}[\mathbf{\Lambda}^{+}].$$

Given a field $\mathbb{L}$, let $U(\mathfrak{g})_{\mathbb{L}}$ and $U(\tilde{\mathfrak{g}})_{\mathbb{L}}$ be the hyperalgebra and the hyper loop algebra of $\mathfrak{g}$ over $\mathbb{L}$ which are defined by

$$U(\mathfrak{g})_{\mathbb{L}} = \mathbb{L} \otimes_{\mathbb{Z}} U(\mathfrak{g})_{\mathbb{Z}} \quad \text{and} \quad U(\tilde{\mathfrak{g}})_{\mathbb{L}} = \mathbb{L} \otimes_{\mathbb{Z}} U(\tilde{\mathfrak{g}})_{\mathbb{Z}}.$$

Define also $U(\tilde{\mathfrak{n}}^{\pm})_{\mathbb{L}}$, etc., in the obvious way.

From now on, we set $\mathbf{\Lambda} = \mathbf{\Lambda}^{+}$ and let $\mathbb{L}[\mathbf{\Lambda}] = \mathbb{L} \otimes_{\mathbb{Z}} \mathbb{Z}[\mathbf{\Lambda}] \subseteq U(\tilde{\mathfrak{g}})_{\mathbb{L}}$. We keep the notation $(x^{\pm}_{\alpha,r})^{(k)}$, $\binom{h_i}{k}$, and $\Lambda_{i,r}$ for the images of these elements in $U(\tilde{\mathfrak{g}})_{\mathbb{L}}$.

The Hopf algebra structure on $U(\tilde{\mathfrak{g}})$ induces a Hopf algebra structure on $U(\tilde{\mathfrak{g}})_{\mathbb{L}}$. We denote the corresponding augmentation ideal by $U(\tilde{\mathfrak{g}})^0_{\mathbb{L}}$ (and similarly for the factors of the triangular decompositions above). $U(\mathfrak{g})_{\mathbb{L}}$ is a Hopf subalgebra of $U(\tilde{\mathfrak{g}})_{\mathbb{L}}$.

REMARK. If the characteristic of $\mathbb{L}$ is zero, the algebra $U(\tilde{\mathfrak{g}})_{\mathbb{L}}$ is naturally isomorphic to the enveloping algebra $U(\tilde{\mathfrak{g}}_{\mathbb{L}})$ where $\tilde{\mathfrak{g}}_{\mathbb{L}} = \mathbb{L} \otimes_{\mathbb{Z}} \tilde{\mathfrak{g}}_{\mathbb{Z}}$ and $\tilde{\mathfrak{g}}_{\mathbb{Z}} = \tilde{\mathfrak{g}} \cap U(\tilde{\mathfrak{g}})_{\mathbb{Z}}$ (and similarly for $\mathfrak{g}$ in place of $\tilde{\mathfrak{g}}$).

For further details regarding this subsection see [**20**, §1] and [**21**, §1.3].

1.5. For a field $\mathbb{L}$, let $\mathcal{P}^+_{\mathbb{L}}$ be the monoid of I-tuples of polynomials $\boldsymbol{\omega} = (\boldsymbol{\omega}_i(u))$ such that $\boldsymbol{\omega}_i(0) = 1$ for all $i \in I$ and let $\mathcal{P}_{\mathbb{L}}$ be the corresponding abelian group of rational functions. Let also $\boldsymbol{M}_{\mathbb{L}}$ be the set of unitary $\mathbb{L}$-algebra homomorphisms $\boldsymbol{\varpi} : \mathbb{L}[\mathbf{\Lambda}] \to \mathbb{L}$. Notice that $\boldsymbol{M}_{\mathbb{L}}$ can be identified with a submonoid of the multiplicative monoid of I-tuples of formal power series by the assignment $\boldsymbol{\varpi} \mapsto (\boldsymbol{\varpi}_i(u))$ where $\boldsymbol{\varpi}_i(u) = 1 + \sum_{r \in \mathbb{N}} \varpi_{i,r} u^r$ and $\varpi_{i,r} = \boldsymbol{\varpi}(\Lambda_{i,r})$. If $\mathbb{L}$ is algebraically closed, we can identify $\mathcal{P}_{\mathbb{L}}$ with a subgroup of $\boldsymbol{M}_{\mathbb{L}}$ by expanding the denominators of the elements of $\mathcal{P}_{\mathbb{L}}$ into the corresponding products of geometric power series.

Given $\mu \in P$ and $a \in \mathbb{L}^{\times}$, let $\boldsymbol{\omega}_{\mu,a}$ be the element of $\mathcal{P}_{\mathbb{L}}$ whose i-th coordinate rational function is $(\boldsymbol{\omega}_{\mu,a})_i(u) = (1 - au)^{\mu(h_i)}$. Let $\mathcal{Q}^+_{\mathbb{L}}$ be the sub-monoid of $\mathcal{P}_{\mathbb{L}}$ generated by the elements $\boldsymbol{\omega}_{\alpha,a}$ with $\alpha \in R^+$ and $a \in \mathbb{L}^{\times}$ and define a partial order on $\mathcal{P}_{\mathbb{L}}$ by setting $\boldsymbol{\varpi} \leq \boldsymbol{\omega}$ iff $\boldsymbol{\omega} \in \boldsymbol{\varpi}\mathcal{Q}^+_{\mathbb{L}}$.

1.6. Let $\boldsymbol{\varpi} \in \boldsymbol{M}_{\mathbb{F}}$. Define $\mathcal{F}(\boldsymbol{\varpi})$ to be an $\mathbb{F}[\mathbf{\Lambda}]$-module isomorphic to the quotient of $\mathbb{F}[\mathbf{\Lambda}]$ by the maximal ideal generated by the elements $\Lambda_{i,r} - \varpi_{i,r}$. In particular, $\mathcal{F}(\boldsymbol{\varpi})$ is a one-dimensional $\mathbb{F}$-vector space. Given a nonzero vector $v \in \mathcal{F}(\boldsymbol{\varpi})$, let $\mathcal{K}(\boldsymbol{\varpi}) = \mathbb{K}[\mathbf{\Lambda}]v$ and $\deg(\boldsymbol{\varpi}) = \dim_{\mathbb{K}}(\mathcal{K}(\boldsymbol{\varpi}))$. It is easy to see that the isomorphism class of $\mathcal{K}(\boldsymbol{\varpi})$ as a $\mathbb{K}[\mathbf{\Lambda}]$-module does not depend on the choice of v. Finally, given $g \in \mathrm{Aut}(\mathbb{F}/\mathbb{K})$, let $g(\boldsymbol{\varpi})$ be the I-tuple of power series whose i-th entry is $1 + \sum_{r \in \mathbb{N}} g(\varpi_{i,r})u^r$ and set $[\boldsymbol{\varpi}] = \{g(\boldsymbol{\varpi}) : g \in \mathrm{Aut}(\mathbb{F}/\mathbb{K})\}$. The set $[\boldsymbol{\varpi}]$ is called the conjugacy class of $\boldsymbol{\varpi}$ over $\mathbb{K}$. Let $\boldsymbol{M}_{\mathbb{F},\mathbb{K}}, \mathcal{P}_{\mathbb{F},\mathbb{K}}$, and $\mathcal{P}^+_{\mathbb{F},\mathbb{K}}$ be the corresponding sets of conjugacy classes in $\boldsymbol{M}_{\mathbb{F}}, \mathcal{P}_{\mathbb{F}}$, and $\mathcal{P}_{\mathbb{F}}^+$, respectively . We have (see [**21**, Theorem 1.2]):

THEOREM 1.3.

(a) $\mathcal{K}(\boldsymbol{\varpi})$ is an irreducible $\mathbb{K}[\boldsymbol{\Lambda}]$-module for every $\boldsymbol{\varpi} \in \boldsymbol{M}_{\mathbb{F}}$ and every finite-dimensional irreducible $\mathbb{K}[\boldsymbol{\Lambda}]$-module is of this form for some $\boldsymbol{\varpi} \in \boldsymbol{M}_{\mathbb{F}}$.

(b) $\mathcal{K}(\boldsymbol{\varpi}) \cong \mathcal{K}(\boldsymbol{\varpi}')$ iff $\boldsymbol{\varpi}' \in [\boldsymbol{\varpi}]$.

(c) Suppose $\deg(\boldsymbol{\varpi}) < \infty$. Then $\mathcal{K}(\boldsymbol{\varpi})^{\mathbb{F}} \cong \bigoplus_{\boldsymbol{\varpi}' \in [\boldsymbol{\varpi}]} V_{\boldsymbol{\varpi}'}$ with $V_{\boldsymbol{\varpi}'}$ an indecomposable self-extension of $\mathcal{F}(\boldsymbol{\omega}')$ of length $\deg(\boldsymbol{\omega})/|[\boldsymbol{\omega}]|$, where $|[\boldsymbol{\omega}]|$ is the cardinality of $[\boldsymbol{\omega}]$. □

The number $\deg(\boldsymbol{\omega})/|[\boldsymbol{\omega}]|$ is called the inseparability degree of $\boldsymbol{\omega}$ over $\mathbb{K}$ and will be denoted by $\operatorname{indeg}(\boldsymbol{\omega})$.

1.7. The basic facts about the finite-dimensional representation theory of $U(\mathfrak{g})_{\mathbb{K}}$ are as follows (see [**21**, §1.4] and references therein).

THEOREM 1.4. Let V be a finite-dimensional $U(\mathfrak{g})_{\mathbb{K}}$-module. Then:

(a) $V = \bigoplus_{\mu \in P} V_\mu$, where $V_\mu = \{v \in v : \binom{h_i}{k} v = \binom{\mu(h_i)}{k} v\}$.

(b) If V is irreducible, there exists $\lambda \in P^+$ such that

$$\dim_{\mathbb{K}}(V_\lambda) = 1, \qquad U(\mathfrak{n}^+)^0_{\mathbb{K}} V_\lambda = 0, \quad \text{and} \quad V = U(\mathfrak{n}^-)_{\mathbb{K}} V_\lambda. \tag{1.1}$$

(c) For every $\lambda \in P^+$ there exists a finite-dimensional irreducible $U(\mathfrak{g})_{\mathbb{K}}$-module, denoted by $V_{\mathbb{K}}(\lambda)$, which satisfies (1.1).

(d) If V satisfies (1.1) for some $\lambda \in P^+$ and $V_\mu \neq 0$, then $\mu \leq \lambda$.

(e) The $U(\mathfrak{g})_{\mathbb{F}}$-module $V_{\mathbb{F}}(\lambda), \lambda \in P^+$, is isomorphic to $(V_{\mathbb{K}}(\lambda))^{\mathbb{F}}$. □

We also recall the following basic facts on characters. Let $\mathcal{C}(\mathfrak{g})_{\mathbb{K}}$ be the category of finite-dimensional $U(\mathfrak{g})_{\mathbb{K}}$-modules. If V is an object in $\mathcal{C}(\mathfrak{g})_{\mathbb{K}}$, the character of V is the function $\operatorname{char}(V) : P \to \mathbb{Z}_+$, $\mu \mapsto \operatorname{char}(V)_\mu := \dim(V_\mu)$. Observe that the last part of the above theorem implies that $\operatorname{char}(V_{\mathbb{K}}(\lambda)) = \operatorname{char}(V_{\mathbb{L}}(\lambda))$ for every $\lambda \in P^+$ and every field $\mathbb{L}$ having the same characteristic as $\mathbb{K}$.

It is convenient to also regard $\operatorname{char}(V)$ as an element of the integral group ring $\mathbb{Z}[P]$ of P. It follows that there exists a ring homomorphism char : $\operatorname{Gr}(\mathcal{C}(\mathfrak{g})_{\mathbb{K}}) \to \mathbb{Z}[P]$ where $\operatorname{Gr}(\mathcal{C}(\mathfrak{g})_{\mathbb{K}})$ is the Grothendieck ring of $\mathcal{C}(\mathfrak{g})_{\mathbb{K}}$. In particular, if V and W are objects in $\mathcal{C}(\mathfrak{g})_{\mathbb{K}}$ and M is an extension of V by W we have

$$\operatorname{char}(M) = \operatorname{char}(V) + \operatorname{char}(W) \quad \text{and} \quad \operatorname{char}(V \otimes W) = \operatorname{char}(V)\operatorname{char}(W). \tag{1.2}$$

Using the first equality above, one easily devises an algorithm that proves the following corollary of Theorem 1.4.

COROLLARY 1.5. Let V be an object of $\mathcal{C}(\mathfrak{g})_{\mathbb{K}}$ and $\lambda \in P^+$. The Jordan-Hölder multiplicity $\operatorname{mult}^{\mathcal{C}(\mathfrak{g})_{\mathbb{K}}}_{V_{\mathbb{K}}(\lambda)}(V)$ is completely determined by $\operatorname{char}(V)$. □

In practice, in order to compute $\operatorname{mult}^{\mathcal{C}(\mathfrak{g})_{\mathbb{K}}}_{V_{\mathbb{K}}(\lambda)}(V)$ from a given $\operatorname{char}(V)$, one must know $\operatorname{char}(V_{\mathbb{K}}(\mu))$ for all $\mu \in P^+$ such that $\operatorname{char}(V)_\mu \neq 0$. Character formulas are some of the most interesting results in combinatorial representation theory. In the characteristic zero case, there are several formulas for computing $\operatorname{char}(V_{\mathbb{K}}(\lambda))$ which can be found in [**19**, Chapter VI]. In positive characteristic, the general solution for this problem is still open. We refer to [**1, 11**] for surveys of the related topics. See also [**12, 13**] for the most recent developments.

1.8. In the next two subsections we review some results proved in [**21**]. We begin with results on the classification of the finite-dimensional irreducible $U(\tilde{\mathfrak{g}})_{\mathbb{K}}$-modules.

A finite-dimensional $U(\tilde{\mathfrak{g}})_{\mathbb{K}}$-module V is said to be a highest-quasi-ℓ-weight module if there exists $\boldsymbol{\omega} \in \boldsymbol{M}_{\mathbb{F}}$ and a $\mathbb{K}[\boldsymbol{\Lambda}]$-submodule $\mathcal{V}$ of V such that

$$\mathcal{V} \cong \mathcal{K}(\boldsymbol{\omega}), \qquad U(\tilde{\mathfrak{n}}^+)^0_{\mathbb{K}}\mathcal{V} = 0, \quad \text{and} \quad V = U(\tilde{\mathfrak{n}}^-)_{\mathbb{K}}\mathcal{V}. \tag{1.3}$$

When $\deg(\boldsymbol{\omega}) = 1$, we simply say V is a highest-ℓ-weight module. The conjugacy class $[\boldsymbol{\omega}]$ is said to be the highest (quasi)-ℓ-weight of V and $\mathcal{V}$ is said to be the highest-(quasi)-ℓ-weight space of V.

Let $\mathrm{wt} : \mathcal{P}_{\mathbb{F}} \to P$ be the group homomorphism such that $\mathrm{wt}(\boldsymbol{\omega})(h_i) = \deg(\boldsymbol{\omega}_i(u))$ for all $i \in I$ and all $\boldsymbol{\omega} \in \mathcal{P}^+_{\mathbb{F}}$. Here $\deg(\boldsymbol{\omega}_i(u))$ is the degree of $\boldsymbol{\omega}_i(u)$ as a polynomial.

THEOREM 1.6. Let V be a finite-dimensional $U(\tilde{\mathfrak{g}})_{\mathbb{K}}$-module. Then:

(a) $V = \bigoplus_{[\boldsymbol{\varpi}] \in \mathcal{P}_{\mathbb{F},\mathbb{K}}} V_{\boldsymbol{\varpi}}$ with $V_{\boldsymbol{\varpi}}$ a $\mathbb{K}[\boldsymbol{\Lambda}]$-submodule of V isomorphic to a self-extension of $\mathcal{K}(\boldsymbol{\varpi})$. Moreover, $V_\mu = \bigoplus_{[\boldsymbol{\varpi}] \,:\, \mathrm{wt}(\boldsymbol{\varpi}) = \mu} V_{\boldsymbol{\varpi}}$.

(b) If V is irreducible, V is a highest-quasi-ℓ-weight module with highest quasi-ℓ-weight in $\mathcal{P}^+_{\mathbb{F},\mathbb{K}}$.

(c) Given $\boldsymbol{\omega} \in \mathcal{P}^+_{\mathbb{F}}$, there exists a universal finite-dimensional highest-quasi-ℓ-weight $U(\tilde{\mathfrak{g}})_{\mathbb{K}}$-module of highest quasi-ℓ-weight $[\boldsymbol{\omega}]$, denoted by $W_{\mathbb{K}}(\boldsymbol{\omega})$. Moreover, $W_{\mathbb{K}}(\boldsymbol{\omega})$ has a unique irreducible quotient denoted by $V_{\mathbb{K}}(\boldsymbol{\omega})$.

(d) The isomorphism classes of $W_{\mathbb{K}}(\boldsymbol{\omega})$ and $V_{\mathbb{K}}(\boldsymbol{\omega})$ depend only on $[\boldsymbol{\omega}] \in \mathcal{P}^+_{\mathbb{F},\mathbb{K}}$.

(e) If V is a highest-quasi-ℓ-weight module of highest quasi-ℓ-weight $[\boldsymbol{\omega}] \in \mathcal{P}^+_{\mathbb{F},\mathbb{K}}$, then $V_{\boldsymbol{\varpi}} \neq 0$ only if there exists $\boldsymbol{\varpi}' \in [\boldsymbol{\varpi}]$ such that $\boldsymbol{\varpi}' \leq \boldsymbol{\omega}$. Moreover, $V_{\boldsymbol{\omega}}$ is the highest-quasi-ℓ-weight space of V. □

A conjugacy class $[\boldsymbol{\varpi}]$ for which $V_{\boldsymbol{\varpi}} \neq 0$ is called a quasi-ℓ-weight of V and $V_{\boldsymbol{\varpi}}$ is the corresponding quasi-ℓ-weight space. When $\deg(\boldsymbol{\varpi}) = 1$, i.e., when $\boldsymbol{\varpi} \in \mathcal{P}_{\mathbb{K}} \subseteq \mathcal{P}_{\mathbb{F},\mathbb{K}}$, we simply say $[\boldsymbol{\varpi}]$ is an ℓ-weight of V. The module $W_{\mathbb{K}}(\boldsymbol{\omega})$ is called the Weyl module of highest (quasi)-ℓ-weight $[\boldsymbol{\omega}]$.

1.9. Now we recall results concerning base change. All the results of this note can be regarded as applications of part (b) the following theorem.

THEOREM 1.7. Let $\boldsymbol{\omega} \in \mathcal{P}^+_{\mathbb{F}}$, $V = V_{\mathbb{F}}(\boldsymbol{\omega})$, and $W = W_{\mathbb{F}}(\boldsymbol{\omega})$.

(a) If v is a nonzero vector in $V_{\boldsymbol{\omega}}$, then $U(\tilde{\mathfrak{g}})_{\mathbb{K}}v \cong V_{\mathbb{K}}(\boldsymbol{\omega})$ and $\mathrm{char}(V_{\mathbb{K}}(\boldsymbol{\omega})) = \deg(\boldsymbol{\omega})\mathrm{char}(V_{\mathbb{F}}(\boldsymbol{\omega}))$. Similarly, if v is a nonzero vector in $W_{\boldsymbol{\omega}}$, then $U(\tilde{\mathfrak{g}})_{\mathbb{K}}v \cong W_{\mathbb{K}}(\boldsymbol{\omega})$ and $\mathrm{char}(W_{\mathbb{K}}(\boldsymbol{\omega})) = \deg(\boldsymbol{\omega})\ \mathrm{char}(W_{\mathbb{F}}(\boldsymbol{\omega}))$.

(b) $(V_{\mathbb{K}}(\boldsymbol{\omega}))^{\mathbb{F}} \cong \bigoplus_{\boldsymbol{\omega}' \in [\boldsymbol{\omega}]} V^{\mathbb{F}}(\boldsymbol{\omega}')$ where $V^{\mathbb{F}}(\boldsymbol{\omega}')$ is an indecomposable self-extension of $V_{\mathbb{F}}(\boldsymbol{\omega}')$ of length $\mathrm{indeg}(\boldsymbol{\omega})$. Similarly, $(W_{\mathbb{K}}(\boldsymbol{\omega}))^{\mathbb{F}} \cong \bigoplus_{\boldsymbol{\omega}' \in [\boldsymbol{\omega}]} W^{\mathbb{F}}(\boldsymbol{\omega}')$ where $W^{\mathbb{F}}(\boldsymbol{\omega}')$ is an indecomposable self-extension of $W_{\mathbb{F}}(\boldsymbol{\omega}')$ of length $\mathrm{indeg}(\boldsymbol{\omega})$. □

2. Clebsch-Gordan Problem

2.1. We now show how to solve the Clebsch-Gordan problem for the category $\mathcal{C}(\tilde{\mathfrak{g}})_{\mathbb{K}}$ of finite-dimensional $U(\tilde{\mathfrak{g}})_{\mathbb{K}}$-modules in terms of Clebsch-Gordan coefficients for $\mathcal{C}(\mathfrak{g})_{\mathbb{F}}$. In order to do so, we first reduce the problem to the category $\mathcal{C}(\tilde{\mathfrak{g}})_{\mathbb{F}}$. To simplify notation, given $\boldsymbol{\omega} \in \mathcal{P}_{\mathbb{F}}^+$ and a finite-dimensional $U(\tilde{\mathfrak{g}})_{\mathbb{K}}$-module V, let

$$\text{mult}_{\boldsymbol{\omega}}^{\mathbb{K}}(V) = \text{mult}_{V_{\mathbb{K}}(\boldsymbol{\omega})}^{\mathcal{C}(\tilde{\mathfrak{g}})_{\mathbb{K}}}(V). \tag{2.1}$$

Similarly, given $\boldsymbol{\omega}, \boldsymbol{\varpi}, \boldsymbol{\pi} \in \mathcal{P}_{\mathbb{F}}^+$, let

$$\text{mult}_{\boldsymbol{\omega}}^{\mathbb{K}}(\boldsymbol{\varpi}, \boldsymbol{\pi}) = \text{mult}_{V_{\mathbb{K}}(\boldsymbol{\omega})}^{\mathcal{C}(\tilde{\mathfrak{g}})_{\mathbb{K}}}(V_{\mathbb{K}}(\boldsymbol{\varpi}) \otimes V_{\mathbb{K}}(\boldsymbol{\pi})). \tag{2.2}$$

Notice that, if $\text{wt}(\boldsymbol{\omega}) = \text{wt}(\boldsymbol{\varpi}\boldsymbol{\pi})$, then $\text{mult}_{\boldsymbol{\omega}}^{\mathbb{F}}(\boldsymbol{\varpi}, \boldsymbol{\pi}) \leq 1$ with equality holding if and only if $\boldsymbol{\omega} = \boldsymbol{\varpi}\boldsymbol{\pi}$.

Given $\boldsymbol{\omega}, \boldsymbol{\varpi}, \boldsymbol{\pi} \in \mathcal{P}_{\mathbb{F}}$, define

$$[\boldsymbol{\omega} : \boldsymbol{\varpi}, \boldsymbol{\pi}] = \{(\boldsymbol{\varpi}', \boldsymbol{\pi}') \in [\boldsymbol{\varpi}] \times [\boldsymbol{\pi}] : \boldsymbol{\varpi}'\boldsymbol{\pi}' = \boldsymbol{\omega}\}. \tag{2.3}$$

Quite clearly the cardinality $|[\boldsymbol{\omega} : \boldsymbol{\varpi}, \boldsymbol{\pi}]|$ depends only on the conjugacy classes of $\boldsymbol{\omega}, \boldsymbol{\varpi}, \boldsymbol{\pi}$.

THEOREM 2.1. *Let $\boldsymbol{\omega}, \boldsymbol{\varpi}, \boldsymbol{\pi} \in \mathcal{P}_{\mathbb{F}}^+$. Then,*

$$\text{mult}_{\boldsymbol{\omega}}^{\mathbb{K}}(\boldsymbol{\varpi}, \boldsymbol{\pi}) = \frac{\text{indeg}(\boldsymbol{\varpi})\ \text{indeg}(\boldsymbol{\pi})}{\text{indeg}(\boldsymbol{\omega})} \sum_{\boldsymbol{\varpi}' \in [\boldsymbol{\varpi}]} \sum_{\boldsymbol{\pi}' \in [\boldsymbol{\pi}]} \text{mult}_{\boldsymbol{\omega}}^{\mathbb{F}}(\boldsymbol{\varpi}', \boldsymbol{\pi}').$$

In particular, $\text{mult}_{\boldsymbol{\omega}}^{\mathbb{K}}(\boldsymbol{\varpi}, \boldsymbol{\pi}) = \frac{\text{indeg}(\boldsymbol{\varpi})\ \text{indeg}(\boldsymbol{\pi})}{\text{indeg}(\boldsymbol{\omega})} |[\boldsymbol{\omega} : \boldsymbol{\varpi}, \boldsymbol{\pi}]|$ *when* $\text{wt}(\boldsymbol{\omega}) = \text{wt}(\boldsymbol{\varpi}\boldsymbol{\pi})$.

PROOF. Let $V = V_{\mathbb{K}}(\boldsymbol{\varpi}) \otimes_{\mathbb{K}} V_{\mathbb{K}}(\boldsymbol{\pi})$. It follows from Theorem 1.7(b) together with Lemma 1.1 applied to a composition series of V in the category $\mathcal{C}(\tilde{\mathfrak{g}})_{\mathbb{K}}$ that

$$\text{mult}_{\boldsymbol{\omega}}^{\mathbb{K}}(V)\deg(\boldsymbol{\omega}) = \sum_{\boldsymbol{\omega}' \in [\boldsymbol{\omega}]} \text{mult}_{\boldsymbol{\omega}'}^{\mathbb{F}}(V^{\mathbb{F}}) = |[\boldsymbol{\omega}]|\ \text{mult}_{\boldsymbol{\omega}}^{\mathbb{F}}(V^{\mathbb{F}}).$$

On the other hand, it follows from Lemma 1.2 and Theorem 1.7(b) that

$$\text{mult}_{\boldsymbol{\omega}}^{\mathbb{F}}(V^{\mathbb{F}}) = \text{indeg}(\boldsymbol{\varpi})\ \text{indeg}(\boldsymbol{\pi}) \sum_{\boldsymbol{\varpi}' \in [\boldsymbol{\varpi}]} \sum_{\boldsymbol{\pi}' \in [\boldsymbol{\pi}]} \text{mult}_{\boldsymbol{\omega}}^{\mathbb{F}}(\boldsymbol{\varpi}', \boldsymbol{\pi}'). \qquad \square$$

As a consequence of the previous theorem, we have the following alternate proof of the nontrivial direction of the statement of [**21**, Theorem 2.18].

COROLLARY 2.2. *Suppose $\boldsymbol{\varpi}, \boldsymbol{\pi} \in \mathcal{P}_{\mathbb{F}}^+$ are such that $V_{\mathbb{F}}(\boldsymbol{\varpi}) \otimes V_{\mathbb{F}}(\boldsymbol{\pi}) \cong V_{\mathbb{F}}(\boldsymbol{\varpi}\boldsymbol{\pi})$ and $\deg(\boldsymbol{\varpi}\boldsymbol{\pi}) = \deg(\boldsymbol{\varpi})\deg(\boldsymbol{\pi})$. Then, $V_{\mathbb{K}}(\boldsymbol{\varpi}\boldsymbol{\pi}) \cong V_{\mathbb{K}}(\boldsymbol{\varpi}) \otimes V_{\mathbb{K}}(\boldsymbol{\pi})$.*

PROOF. The second statement of Theorem 2.1 implies that $\text{mult}_{\boldsymbol{\varpi}\boldsymbol{\pi}}^{\mathbb{K}}(\boldsymbol{\varpi}, \boldsymbol{\pi}) > 0$. On the other hand, the hypotheses of the corollary together with Theorem 1.7(a) imply $\dim(V_{\mathbb{K}}(\boldsymbol{\varpi}) \otimes V_{\mathbb{K}}(\boldsymbol{\pi})) = \dim(V_{\mathbb{K}}(\boldsymbol{\varpi}\boldsymbol{\pi}))$. $\square$

2.2. Next, we recall how to rephrase the classification of finite-dimensional irreducible $U(\tilde{\mathfrak{g}})_{\mathbb{F}}$-modules in terms of evaluation representations. We begin with the existence of the so-called evaluation maps (see [**20**, Proposition 3.3]).

PROPOSITION 2.3. *For every $a \in \mathbb{F}^{\times}$, there exists a surjective algebra homomorphism $\text{ev}_a : U(\tilde{\mathfrak{g}})_{\mathbb{F}} \to U(\mathfrak{g})_{\mathbb{F}}$ called the evaluation map at a.* $\square$

If V is a $U(\mathfrak{g})_{\mathbb{F}}$-module, we denote by $V(a)$ the pull-back of V by ev_a. Then, for $\lambda \in P^+$, $V_{\mathbb{F}}(\boldsymbol{\omega}_{\lambda,a}) \cong (V_{\mathbb{F}}(\lambda))(a)$ (see [**20**, §3B]). Observe that every $\boldsymbol{\omega} \in \mathcal{P}_{\mathbb{F}}^+$ can be uniquely written as a product of the form

$$\boldsymbol{\omega} = \prod_{j=1}^{m} \boldsymbol{\omega}_{\lambda_j,a_j} \quad \text{for some} \quad m \in \mathbb{N}, \lambda_j \in P^+, a_j \in \mathbb{F}^\times \text{ with } a_j \neq a_k \text{ if } j \neq k. \tag{2.4}$$

Two elements $\boldsymbol{\varpi}, \boldsymbol{\pi} \in \mathcal{P}_{\mathbb{F}}^+$ are said to be relatively prime if $\boldsymbol{\varpi}_i(u)$ is relatively prime to $\boldsymbol{\pi}_j(u)$ for every $i, j \in I$. The next proposition is an immediate consequence of [**20**, Corollary 3.5 and Equation (3-13)].

PROPOSITION 2.4. *If $\boldsymbol{\varpi}, \boldsymbol{\pi} \in \mathcal{P}_{\mathbb{F}}^+$ are relatively prime, then $V_{\mathbb{F}}(\boldsymbol{\varpi}) \otimes V_{\mathbb{F}}(\boldsymbol{\pi}) \cong V_{\mathbb{F}}(\boldsymbol{\varpi}\boldsymbol{\pi})$. In particular, the finite-dimensional irreducible $U(\tilde{\mathfrak{g}})_{\mathbb{F}}$-modules can be realized as tensor products of evaluations representations, i.e., if $\boldsymbol{\omega} = \prod_{j=1}^{m} \boldsymbol{\omega}_{\lambda_j,a_j}$ as in (2.4), then $V_{\mathbb{F}}(\boldsymbol{\omega}) \cong \bigotimes_{j=1}^{m} V_{\mathbb{F}}(\boldsymbol{\omega}_{\lambda_j,a_j})$.* □

The next proposition is easily established.

PROPOSITION 2.5. *Let $\lambda, \mu \in P^+, V = V_{\mathbb{F}}(\lambda) \otimes V_{\mathbb{F}}(\mu)$, and let $0 \subseteq V_1 \subseteq V_2 \subseteq \cdots \subseteq V_m = V$ be a composition series for V. Then $0 \subseteq V_1(a) \subseteq V_2(a) \subseteq \cdots \subseteq V_m(a) = V(a)$ is a composition series for $V_{\mathbb{F}}(\boldsymbol{\omega}_{\lambda,a}) \otimes V_{\mathbb{F}}(\boldsymbol{\omega}_{\mu,a})$.* □

Using the last two propositions, one easily reduces the Clebsch-Gordan problem for $\mathcal{C}(\tilde{\mathfrak{g}})_{\mathbb{F}}$ to the one for $\mathcal{C}(\mathfrak{g})_{\mathbb{F}}$. By Corollary 1.5, solutions for $\mathcal{C}(\mathfrak{g})_{\mathbb{F}}$ can be obtained from the knowledge of $\mathrm{char}(V_{\mathbb{F}}(\lambda)), \lambda \in P^+$. The later is an open problem in general, as mentioned in subsection 1.7. If $\mathbb{F}$ has characteristic zero there are several formulas for computing Clebsch-Gordan coefficients in $\mathcal{C}(\mathfrak{g})_{\mathbb{F}}$ (see for instance [**16**]). New insights into this problem were introduced by the advent of crystal bases (see [**3, 26, 27**] and references therein).

3. ℓ-Characters

3.1. In this section, we consider the notion of ℓ-characters – the classical analogue of the q-characters defined by Frenkel and Reshetikhin in [**15**]. This is equivalent to studying $\mathrm{mult}_{\mathcal{K}(\boldsymbol{\varpi})}^{\mathcal{C}(\boldsymbol{\Lambda})_{\mathbb{K}}}(V_{\mathbb{K}}(\boldsymbol{\omega}))$ for every $\boldsymbol{\varpi} \in \mathcal{P}_{\mathbb{F}}$ and every $\boldsymbol{\omega} \in \mathcal{P}_{\mathbb{F}}^+$. Here, $\mathcal{C}(\boldsymbol{\Lambda})_{\mathbb{K}}$ is the category of finite-dimensional $\mathbb{K}[\boldsymbol{\Lambda}]$-modules. Evidently, every object in $\mathcal{C}(\tilde{\mathfrak{g}})_{\mathbb{K}}$ is also an object in $\mathcal{C}(\boldsymbol{\Lambda})_{\mathbb{K}}$ (recall also Theorem 1.6(a)).

DEFINITION 3.1. Let V be a finite-dimensional $U(\tilde{\mathfrak{g}})_{\mathbb{K}}$-module. The ℓ-character of V is the function $\mathrm{char}_\ell(V) : \mathcal{P}_{\mathbb{F},\mathbb{K}} \to \mathbb{Z}_+$ given by $[\boldsymbol{\varpi}] \mapsto \mathrm{char}_\ell(V)_{\boldsymbol{\varpi}} := \dim(V_{\boldsymbol{\varpi}})$.

Quite clearly

$$\mathrm{mult}_{\mathcal{K}(\boldsymbol{\varpi})}^{\mathcal{C}(\boldsymbol{\Lambda})_{\mathbb{K}}}(V) = \frac{\mathrm{char}_\ell(V)_{\boldsymbol{\varpi}}}{\deg(\boldsymbol{\varpi})}. \tag{3.1}$$

Hence, studying $\mathrm{char}_\ell(V)$ is equivalent to studying the Jordan-Hölder multiplicities of V in the category $\mathcal{C}(\boldsymbol{\Lambda})_{\mathbb{K}}$.

Using the partial order on $\mathcal{P}_{\mathbb{F}}$ defined in §1.5, one proves the following analogue of Corollary 1.5 as a corollary of Theorem 1.6.

COROLLARY 3.2. *Let V be an object of $\mathcal{C}(\tilde{\mathfrak{g}})_{\mathbb{K}}$ and $\boldsymbol{\omega} \in \mathcal{P}_{\mathbb{F}}^+$. The Jordan-Hölder multiplicity $\mathrm{mult}_{\boldsymbol{\omega}}^{\mathbb{K}}(V)$ is completely determined by $\mathrm{char}_\ell(V)$.* □

A similar coment as the one we made after Corollary 1.5 is in place here, as well. Namely, the knowledge of $\mathrm{char}_\ell(V_{\mathbb{K}}(\boldsymbol{\varpi}))$ for every $\boldsymbol{\varpi} \in \mathcal{P}_{\mathbb{F}}^+$ is required in order to carry out, in practice, the computation of $\mathrm{mult}_{\boldsymbol{\omega}}^{\mathbb{K}}(V)$ from a given $\mathrm{char}_\ell(V)$.

3.2. The results of §2.1 and (1.2) enable us to reduce the problem of computing $\mathrm{char}_\ell(V_{\mathbb{F}}(\boldsymbol{\omega}))$ for $\boldsymbol{\omega} \in \mathcal{P}_{\mathbb{F}}^+$ to those of computing Clebsch-Gordan coefficients for $\mathcal{C}(\mathfrak{g})_{\mathbb{F}}$ as well as $\mathrm{char}(V_{\mathbb{F}}(\lambda))$, $\lambda \in P^+$. The next theorem expresses $\mathrm{char}_\ell(V_{\mathbb{K}}(\boldsymbol{\omega}))$ and $\mathrm{char}_\ell(W_{\mathbb{K}}(\boldsymbol{\omega}))$ in terms of $\mathrm{char}_\ell(V_{\mathbb{F}}(\boldsymbol{\omega}))$ and $\mathrm{char}_\ell(W_{\mathbb{F}}(\boldsymbol{\omega}))$, respectively.

THEOREM 3.3. *For every $\boldsymbol{\omega} \in \mathcal{P}_{\mathbb{F}}^+$ and every $\boldsymbol{\varpi} \in \mathcal{P}_{\mathbb{F}}$ we have*

$$\mathrm{char}_\ell(V_{\mathbb{K}}(\boldsymbol{\omega}))_{\boldsymbol{\varpi}} = \deg(\boldsymbol{\omega}) \sum_{\boldsymbol{\varpi}' \in [\boldsymbol{\varpi}]} \mathrm{char}_\ell(V_{\mathbb{F}}(\boldsymbol{\omega}))_{\boldsymbol{\varpi}'}$$

and

$$\mathrm{char}_\ell(W_{\mathbb{K}}(\boldsymbol{\omega}))_{\boldsymbol{\varpi}} = \deg(\boldsymbol{\omega}) \sum_{\boldsymbol{\varpi}' \in [\boldsymbol{\varpi}]} \mathrm{char}_\ell(W_{\mathbb{F}}(\boldsymbol{\omega}))_{\boldsymbol{\varpi}'}.$$

PROOF. Set $V = V_{\mathbb{K}}(\boldsymbol{\omega})$. By Theorem 1.3(c) and Lemma 1.1 applied to a composition series of V in the category $\mathcal{C}(\boldsymbol{\Lambda})_{\mathbb{K}}$ we have

$$\mathrm{char}_\ell(V)_{\boldsymbol{\varpi}} = \sum_{\boldsymbol{\varpi}' \in [\boldsymbol{\varpi}]} \mathrm{char}_\ell(V^{\mathbb{F}})_{\boldsymbol{\varpi}'}.$$

On the other hand, it is not difficult to see from the proof of Theorem 1.7(b) (see the proof of Theorem 2.12 and §A.3 of [**21**]) that every $g \in \mathrm{Aut}(\mathbb{F}/\mathbb{K})$ induces an isomorphism of $U(\tilde{\mathfrak{g}})_{\mathbb{K}}$-modules $V_{\mathbb{F}}(\boldsymbol{\omega}) \to V_{\mathbb{F}}(g(\boldsymbol{\omega}))$. In particular, for every $\boldsymbol{\omega}' \in [\boldsymbol{\omega}]$ and every $\boldsymbol{\varpi} \in \mathcal{P}_{\mathbb{F}}$, we have

$$\sum_{\boldsymbol{\varpi}' \in [\boldsymbol{\varpi}]} \mathrm{char}_\ell(V_{\mathbb{F}}(\boldsymbol{\omega}))_{\boldsymbol{\varpi}'} = \sum_{\boldsymbol{\varpi}' \in [\boldsymbol{\varpi}]} \mathrm{char}_\ell(V_{\mathbb{F}}(\boldsymbol{\omega}'))_{\boldsymbol{\varpi}'}.$$

Since, by Theorem 1.7(b), $V^{\mathbb{F}}$ is a $U(\tilde{\mathfrak{g}})_{\mathbb{F}}$-module of length $\deg(\boldsymbol{\omega})$ whose irreducible constituents are of the form $V_{\mathbb{F}}(\boldsymbol{\omega}')$ with $\boldsymbol{\omega}' \in [\boldsymbol{\omega}]$, it follows that

$$\sum_{\boldsymbol{\varpi}' \in [\boldsymbol{\varpi}]} \mathrm{char}_\ell(V^{\mathbb{F}})_{\boldsymbol{\varpi}'} = \deg(\boldsymbol{\omega}) \sum_{\boldsymbol{\varpi}' \in [\boldsymbol{\varpi}]} \mathrm{char}_\ell(V_{\mathbb{F}}(\boldsymbol{\omega}))_{\boldsymbol{\varpi}'}.$$

This proves the first formula. For the second one, we observe that all the above steps can be carried out with $W_{\mathbb{K}}(\boldsymbol{\omega})$ in place of $V_{\mathbb{K}}(\boldsymbol{\omega})$ with the corresponding obvious modifications. □

3.3. Due to Corollary 3.2, the formula of Theorem 2.1 can, in principle, be recovered from $\mathrm{char}_\ell(V)$ with $V = V_{\mathbb{K}}(\boldsymbol{\varpi}) \otimes V_{\mathbb{K}}(\boldsymbol{\pi})$. In the algebraically closed case, the notion of ℓ-character can be reinterpreted as a ring homomorphism $\mathrm{char}_\ell : \mathrm{Gr}(\mathcal{C}(\tilde{\mathfrak{g}})_{\mathbb{F}}) \to \mathbb{Z}[\mathcal{P}_{\mathbb{F}}]$ where $\mathrm{Gr}(\mathcal{C}(\tilde{\mathfrak{g}})_{\mathbb{F}})$ is the Grothendieck ring of $\mathcal{C}(\tilde{\mathfrak{g}})_{\mathbb{F}}$ and $\mathbb{Z}[\mathcal{P}_{\mathbb{F}}]$ is the integral group ring of $\mathcal{P}_{\mathbb{F}}$. In particular, if V and W are objects in $\mathcal{C}(\tilde{\mathfrak{g}})_{\mathbb{F}}$ and M is an extension of V by W we have

$$\mathrm{char}_\ell(M) = \mathrm{char}_\ell(V) + \mathrm{char}_\ell(W) \quad \text{and} \quad \mathrm{char}_\ell(V \otimes W) = \mathrm{char}_\ell(V)\,\mathrm{char}_\ell(W). \tag{3.2}$$

The second formula follows easily from the comultiplication of the elements $\Lambda_{i,r}$ (cf. the proof of [**20**, Corollary 3.5]). In the non-algebraically closed case, $\mathcal{P}_{\mathbb{F},\mathbb{K}}$ is not a group, but we can still form the free $\mathbb{Z}$-module $\mathbb{Z}[\mathcal{P}_{\mathbb{F},\mathbb{K}}]$ and reinterpret ℓ-character as a group homomorphism from the Grothendieck group of $\mathcal{C}(\tilde{\mathfrak{g}})_{\mathbb{K}}$ to $\mathbb{Z}[\mathcal{P}_{\mathbb{F},\mathbb{K}}]$. We again have $\mathrm{char}_\ell(M) = \mathrm{char}_\ell(V) + \mathrm{char}_\ell(W)$ whenever M is an extension of V

by W. Although multiplication is not defined on $\mathbb{Z}[\mathcal{P}_{\mathbb{F},\mathbb{K}}]$ and, hence, the second formula of (3.2) does not make sense anymore, we have a replacement formula given by Theorem 3.4 below in the case of irreducible modules.

Given $\eta \in \mathbb{Z}[\mathcal{P}_{\mathbb{F}}]$, say

$$\eta = \sum_{\boldsymbol{\varpi} \in \mathcal{P}_{\mathbb{F}}} \eta(\boldsymbol{\varpi})\, \boldsymbol{\varpi} \qquad \text{with} \quad \eta(\boldsymbol{\varpi}) \in \mathbb{Z},$$

define

$$[\eta] = \sum_{[\boldsymbol{\varpi}] \in \mathcal{P}_{\mathbb{F},\mathbb{K}}} \eta([\boldsymbol{\varpi}])\, [\boldsymbol{\varpi}] \in \mathbb{Z}[\mathcal{P}_{\mathbb{F},\mathbb{K}}] \qquad \text{where} \quad \eta([\boldsymbol{\varpi}]) = \sum_{\boldsymbol{\varpi}' \in [\boldsymbol{\varpi}]} \eta(\boldsymbol{\varpi}'). \tag{3.3}$$

THEOREM 3.4. For every $\boldsymbol{\varpi}, \boldsymbol{\pi} \in \mathcal{P}_{\mathbb{F}}^+$,

$$\mathrm{char}_\ell(V_{\mathbb{K}}(\boldsymbol{\varpi}) \otimes V_{\mathbb{K}}(\boldsymbol{\pi})) =$$

$$\left[\mathrm{indeg}(\boldsymbol{\varpi})\, \mathrm{indeg}(\boldsymbol{\pi}) \sum_{\boldsymbol{\varpi}' \in [\boldsymbol{\varpi}]} \sum_{\boldsymbol{\pi}' \in [\boldsymbol{\pi}]} \mathrm{char}_\ell(V_{\mathbb{F}}(\boldsymbol{\varpi}'))\, \mathrm{char}_\ell(V_{\mathbb{F}}(\boldsymbol{\pi}')) \right].$$

PROOF. Similar to the proofs of Theorems 2.1 and 3.3. We omit the details (see also the proof of Theorem 4.1 below). □

Notice that the formulas of Theorem 3.3 may be rewritten using (3.3) as

$$\mathrm{char}_\ell(V_{\mathbb{K}}(\boldsymbol{\omega})) = \deg(\boldsymbol{\omega})[\mathrm{char}_\ell(V_{\mathbb{F}}(\boldsymbol{\omega}))]$$

and

$$\mathrm{char}_\ell(W_{\mathbb{K}}(\boldsymbol{\omega})) = \deg(\boldsymbol{\omega})[\mathrm{char}_\ell(W_{\mathbb{F}}(\boldsymbol{\omega}))].$$

4. Jordan-Hölder Multiplicities for Weyl Modules

4.1. We now address the problem of computing $\mathrm{mult}^{\mathbb{K}}_{\boldsymbol{\varpi}}(W_{\mathbb{K}}(\boldsymbol{\omega}))$ for every $\boldsymbol{\omega}, \boldsymbol{\varpi} \in \mathcal{P}_{\mathbb{F}}^+$. In light of Corollary 3.2, it is in principle possible to combine the two formulas of Theorem 3.3 to compute $\mathrm{mult}^{\mathbb{K}}_{\boldsymbol{\varpi}}(W_{\mathbb{K}}(\boldsymbol{\omega}))$ in terms of $\mathrm{char}_\ell(W_{\mathbb{F}}(\boldsymbol{\omega}))$. However, it is actually easier to proceed similarly to the proofs of Theorems 2.1 and 3.3 to obtain the following.

THEOREM 4.1. If $\boldsymbol{\omega}, \boldsymbol{\varpi} \in \mathcal{P}_{\mathbb{F}}^+$,

$$\begin{aligned}\mathrm{mult}^{\mathbb{K}}_{\boldsymbol{\varpi}}(W_{\mathbb{K}}(\boldsymbol{\omega})) &= \frac{\mathrm{indeg}(\boldsymbol{\omega})}{\mathrm{indeg}(\boldsymbol{\varpi})} \sum_{\boldsymbol{\omega}' \in [\boldsymbol{\omega}]} \mathrm{mult}^{\mathbb{F}}_{\boldsymbol{\varpi}}(W_{\mathbb{F}}(\boldsymbol{\omega}')) \\ &= \frac{\deg(\boldsymbol{\omega})}{\deg(\boldsymbol{\varpi})} \sum_{\boldsymbol{\varpi}' \in [\boldsymbol{\varpi}]} \mathrm{mult}^{\mathbb{F}}_{\boldsymbol{\varpi}'}(W_{\mathbb{F}}(\boldsymbol{\omega})).\end{aligned}$$

PROOF. Set $V = W_{\mathbb{K}}(\boldsymbol{\omega})$. It follows from the first part of Theorem 1.7(b) together with Lemma 1.1 applied to a composition series of V in the category $\mathcal{C}(\tilde{\mathfrak{g}})_{\mathbb{K}}$ that

$$\mathrm{mult}^{\mathbb{K}}_{\boldsymbol{\varpi}}(V)\deg(\boldsymbol{\varpi}) = \sum_{\boldsymbol{\varpi}' \in [\boldsymbol{\varpi}]} \mathrm{mult}^{\mathbb{F}}_{\boldsymbol{\varpi}'}(V^{\mathbb{F}}) = |[\boldsymbol{\varpi}]|\, \mathrm{mult}^{\mathbb{F}}_{\boldsymbol{\varpi}}(V^{\mathbb{F}}).$$

On the other hand, the second part of Theorem 1.7(b) gives

$$\mathrm{mult}^{\mathbb{F}}_{\boldsymbol{\pi}}(V^{\mathbb{F}}) = \mathrm{indeg}(\boldsymbol{\omega}) \sum_{\boldsymbol{\omega}' \in [\boldsymbol{\omega}]} \mathrm{mult}^{\mathbb{F}}_{\boldsymbol{\pi}}(W_{\mathbb{F}}(\boldsymbol{\omega}')) \qquad \text{for all} \qquad \boldsymbol{\pi} \in \mathcal{P}_{\mathbb{F}}^+.$$

This completes the proof of the first equality.

To prove the second equality, one proceeds similarly to the proof of Theorem 3.3 to get

$$\sum_{\varpi' \in [\varpi]} \mathrm{mult}^{\mathbb{F}}_{\varpi'}(W_{\mathbb{F}}(\omega)) = \sum_{\varpi' \in [\varpi]} \mathrm{mult}^{\mathbb{F}}_{\varpi'}(W_{\mathbb{F}}(\omega')) \qquad \text{for all} \qquad \omega' \in [\omega].$$

Now the second part of Theorem 1.7(b) gives

$$\sum_{\varpi' \in [\varpi]} \mathrm{mult}^{\mathbb{F}}_{\varpi'}(V^{\mathbb{F}}) = \deg(\omega) \sum_{\varpi' \in [\varpi]} \mathrm{mult}^{\mathbb{F}}_{\varpi'}(W_{\mathbb{F}}(\omega)).$$

□

4.2. We now review the results leading to a method for reducing the computation of $\mathrm{mult}^{\mathbb{F}}_{\varpi}(W_{\mathbb{F}}(\omega))$ to the computation of certain multiplicities in the category $\mathcal{C}(\mathfrak{g})_{\mathbb{F}}$.

We observe that all the results of this subsection in the positive characteristic setting depend on the conjecture of [**20**] which says that all Weyl modules for $U(\tilde{\mathfrak{g}})_{\mathbb{F}}$, when $\mathbb{F}$ is of positive characteristic, can be obtained by reduction modulo p from appropriate Weyl modules in characteristic zero. This statement can be regarded as an analogue of a conjecture of Chari and Pressley [**10**] saying that all Weyl modules for $U(\tilde{\mathfrak{g}})_{\mathbb{C}}$ can be obtained as classical limits of appropriate Weyl modules for quantum affine algebras. Chari-Pressley's conjecture has been recently proved for $\mathfrak{g}$ of type A in [**5**] and for simply laced $\mathfrak{g}$ in [**14**]. Moreover, it has been pointed out by Nakajima that the general case follows from the global and crystal basis theory developed in [**2, 22, 23, 28, 29**] (see [**5, 14**] for brief summaries of Nakajima's arguments). We expect the conjecture of [**20**] can be proved using similar arguments.

We start with the following proposition.

Proposition 4.2. *Let $\lambda \in P^+$ and $a \in \mathbb{F}^\times$. If $0 \subseteq V_1 \subseteq V_2 \subseteq \cdots \subseteq V_m = V$ is a composition series for $V = W_{\mathbb{F}}(\omega_{\lambda,a})$ as a $U(\mathfrak{g})_{\mathbb{F}}$-module, then $0 \subseteq V_1(a) \subseteq V_2(a) \subseteq \cdots \subseteq V_m(a)$ is a composition series for $W_{\mathbb{F}}(\omega_{\lambda,a})$ as a $U(\tilde{\mathfrak{g}})_{\mathbb{F}}$-module.*

Proof. In characteristic zero this is proved in [**6**, Proposition 3.3]. The positive characteristic case follows from the characteristic zero case using the conjecture of [**20**] (cf. [**20**, Proposition 4.11]). □

Hence, in the spirit of Corollary 1.5, in order to compute $\mathrm{mult}^{\mathbb{F}}_{\varpi}(W_{\mathbb{F}}(\omega))$ when $\omega = \omega_{\lambda,a}$, it suffices to know $\mathrm{char}(W_{\mathbb{F}}(\omega_{\lambda,a}))$. In characteristic zero, this character may be computed as a by-product of the proofs of Chari-Pressley's conjecture mentioned above. Namely, characters are unchanged by taking classical limits. Moreover, it follows from [**2, 7**] that the Weyl modules for quantum affine algebras may be realized as tensor products of the so-called fundamental representations. Hence, by the multiplicative property of characters with respect to tensor products, it is left to compute the characters of the later representations. For this, and other related problems, we refer to [**8, 18**] and references therein. Finally, the positive characteristic case then follows from the conjecture of [**20**], since the conjecture implies the character of Weyl modules is unchanged by the reduction modulo p process studied in [**20**]. In particular, $\mathrm{char}(W_{\mathbb{F}}(\omega_{\lambda,a}))$ depends only on λ (and hence on $\mathfrak{g}$), but not on $\mathbb{F}$ or a.

A method for computing $\mathrm{mult}^{\mathbb{F}}_{\boldsymbol{\varpi}}(W_{\mathbb{F}}(\boldsymbol{\omega}))$ for general $\boldsymbol{\omega}$ is now obtained by combining the special case $\boldsymbol{\omega} = \boldsymbol{\omega}_{\lambda,a}$ with the solutions of the Clebsch-Gordan problem for the category $\mathcal{C}(\tilde{\mathfrak{g}})_{\mathbb{F}}$ and the next theorem.

THEOREM 4.3. Let $\boldsymbol{\omega} \in \mathcal{P}^{+}_{\mathbb{F}}$ and write $\boldsymbol{\omega} = \prod_{j=1}^{m} \boldsymbol{\omega}_{\lambda_j,a_j}$ as in (2.4). Then $W_{\mathbb{F}}(\boldsymbol{\omega}) \cong \bigotimes_{j=1}^{m} W_{\mathbb{F}}(\boldsymbol{\omega}_{\lambda_j,a_j})$. □

The theorem above was proved in [**10**] for the characteristic zero case and follows again from the conjecture of [**20**] for the positive characteristic setting.

References

1. H. Andersen, *The irreducible characters for semi-simple algebraic groups and for quantum groups*, Proceedings of the International Congress of Mathematicians Zürich 1994, Birkhäuser (1995), 732–743.
2. J. Beck and H. Nakajima, *Crystal bases and two-sided cells of quantum affine algebras*, Duke Math. J. **123** (2004), 335–402.
3. A. Berenstein and A. Zelevinsky, *Tensor product multiplicities, canonical bases and totally positive varieties*, Invent. Math. **143** (2001), 77–128.
4. S. Capparelli, J. Lepowsky, and A. Milas, *The Rogers-Ramanujan recursion and intertwining operators*, Commun. Contemp. Math. **5** (2003), 947–966.
5. V. Chari and S. Loktev, *Weyl, fusion and Demazure modules for the current algebra of* $\mathfrak{sl}_{r+1}$, Adv. Math. **207** (2006), 928–960.
6. V. Chari and A. Moura, *Spectral characters of finite-dimensional representations of affine algebras*, J. Algebra **279** (2004), 820–839.
7. ______, *Characters and blocks for finite-dimensional representations of quantum affine algebras*, Internat. Math Res. Notices (2005), 257–298.
8. ______, *Characters of fundamental representations of quantum affine algebras*, Acta Appl. Math. **90** (2006), 43–63.
9. V. Chari and A. Pressley, A guide to quantum groups, Cambridge University Press (1995).
10. ______, *Weyl modules for classical and quantum affine algebras*, Represent. Theory **5** (2001), 191–223.
11. S. Donkin, *An introduction to the Lusztig conjecture*, in Representations of reductive groups, Publ. Newton Inst., Cambridge Univ. Press, Cambridge (1998), 173–187.
12. P. Fiebig, *Sheaves on affine Schubert varieties, modular representations and Lusztig's conjecture*, arXiv:0711.0871.
13. ______, *Lusztig's conjecture as a moment graph problem*, arXiv:0712.3909.
14. G. Fourier and P. Littelmann, *Weyl modules, Demazure modules, KR modules, crystals, fusion products and limit constructions*, Adv. Math. **211** (2007), 566–593.
15. E. Frenkel and N. Reshetikhin, *The q-characters of representations of quantum affine algebras and deformations of* $\mathcal{W}$*-algebras*, Contemp. Math. **248** (1999), 163–205.
16. W. Fulton and J. Harris, Representation theory - a first course, GTM 129, Springer (1991).
17. H. Garland, *The arithmetic theory of loop algebras*, J. Algebra **53** (1978), 480–551.
18. D. Hernandez, *On minimal affinizations of representations of quantum groups*, Comm. Math. Phys. **276** (2007), 221–259.
19. J. Humphreys, Introduction to Lie algebras and representation theory, GTM 9, Springer (1972).
20. D. Jakelić and A. Moura, *Finite-dimensional representations of hyper loop algebras*, Pacific J. Math. **233** (2007), 371–402.
21. ______, *Finite-dimensional representations of hyper loop algebras over non-algebraically closed fields*, to appear in Algebras and Representation Theory, arXiv:0711.0795.
22. M. Kashiwara, *Crystal bases of modified quantized enveloping algebras*, Duke Math. J. **73** (1994), 383–413.
23. ______, *On level zero representations of quantized affine algebras*, Duke Math. J. **112** (2002), 117–195.

24. B. Kostant, *Groups over* $\mathbb{Z}$, Algebraic groups and discontinuous subgroups, Proc. Symp. Pure Math. IX, Providence, AMS (1966).
25. J. Lepowsky and R. Wilson, *The structure of standard modules. I. Universal algebras and the Rogers-Ramanujan identities*, Invent. Math. **77** (1984), 199–290.
26. P. Littelmann, *A Littlewood-Richardson rule for symmetrizable Kac-Moody algebras*, Invent. Math. **116** (1994), 329–346.
27. ———, *Crystal graphs and Young tableaux*, J. Algebra **175** (1995), 65–87.
28. H. Nakajima, *Quiver varieties and finite-dimensional representations of quantum affine algebras*, J. Amer. Math. Soc. **14** (2001), 145–238.
29. ———, *Extremal weight modules of quantum affine algebras*, Adv. Stud. Pure Math. **40** (2004), 343–369.

Department of Mathematics, Statistics, and Computer Science,
University of Illinois at Chicago, Chicago, IL 60607-7045, USA
E-mail address: dijana@math.uic.edu

UNICAMP - IMECC, Campinas - SP, 13083-970, Brazil.
E-mail address: aamoura@ime.unicamp.br

Contemporary Mathematics
Volume **483**, 2009

Maximal subalgebras of simple alternative superalgebras.

Jesús Laliena and Sara Sacristán

DEDICATED TO PROFESSOR IVAN SHESTAKOV ON HIS 60TH BIRTHDAY.

ABSTRACT. The maximal subalgebras of the finite dimensional simple alternative nonassociative superalgebras are studied.

1. Introduction.

I. Shestakov and E. Zelmanov proved in [**17**] that prime alternative superalgebras of characteristic $\neq 2, 3$ are either associative or trivial, that is, with zero odd part. Also they showed in the same paper that the description is not true in the characteristic 2 and 3 cases. Later I. Shestakov, in [**15**], classified prime alternative superalgebras without any assumption on characteristic, but with nondegenerate even part. As a conclusion he obtained that every finite dimensional simple alternative nonassociative superalgebra over a field F is either trivial (and then it is a Cayley-Dickson algebra) or the characteristic of F is either 2 or 3, and then five types of alternative nonassociative superalgebras appear.

Here we are interested in describing the maximal subalgebras of these types of superalgebras, more precisely , finite dimensional central simple alternative nonassociative superalgebras in characteristic 2 and 3. Precedents of this work are the papers of E. Dynkin in 1952 (see [**3**], [**4**]), where the maximal subgroups of some classical groups and the maximal subalgebras of semisimple Lie algebras are classified, the papers of M. Racine (see [**12**], [**13**]), who classifies the maximal subalgebras of finite dimensional central simple algebras belonging to one of the following classes: associative, associative with involution, alternative and special and exceptional Jordan algebras; and the paper by A. Elduque in 1986 (see [**5**]), solving the same question for finite dimensional central simple Malcev algebras.

In some previous papers, [**7**], [**8**], [**9**], the authors, jointly with A. Elduque, have described the maximal subalgebras of finite dimensional central simple superalgebras which are either associative or associative with superinvolution, and also the maximal subalgebras of finite dimensional simple Jordan superalgebras with semisimple even part over an algebraically closed field of characteristic 0.

1991 *Mathematics Subject Classification.* Primary 17A70.

Key words and phrases. Simple alternative superalgebras, maximal subalgebras.

The authors have been supported by the Spanish Ministerio de Educación y Ciencia and FEDER (MTM 2007 67884-CO4-02,03).

2. Basic concepts.

First of all, let us recall some basic facts. A *superalgebra* over a field F is just a $\mathbb{Z}_2$-graded algebra $A = A_{\bar{0}} \oplus A_{\bar{1}}$ over F (so $A_\alpha A_\beta \subseteq A_{\alpha+\beta}$ for $\alpha, \beta \in \mathbb{Z}_2$). An element a in A_α ($\alpha = \bar{0}, \bar{1}$) is said to be *homogeneous* of degree α and the notation $\bar{a} = \alpha$ is used. A superalgebra is said to be *nontrivial* if $A_{\bar{1}} \neq 0$ and *simple* if $A^2 \neq 0$ and A contains no proper graded ideal.

An *associative superalgebra* is just a superalgebra that is associative as an ordinary algebra. Here are some examples:

i) $A = M_n(F)$, the algebra of $n \times n$ matrices over F, where
$$A_{\bar{0}} = \left\{ \begin{pmatrix} a & 0 \\ 0 & b \end{pmatrix} : a \in M_r(F), b \in M_s(F) \right\},$$
$$A_{\bar{1}} = \left\{ \begin{pmatrix} 0 & c \\ d & 0 \end{pmatrix} : c \in M_{r\times s}(F), d \in M_{s\times r}(F) \right\},$$
with $r + s = n$. This superalgebra is denoted by $M_{r,s}(F)$.

ii) The (generalized) quaternion algebra $\mathbf{H}(\alpha, \beta)$ with basis $\{1, u, v, uv\}$, such that $u^2 = \alpha, v^2 = \beta$, and $uv = -vu$. The choice of u and v to be odd determines a $\mathbf{Z}_2$-grading in $\mathbf{H}(\alpha, \beta)$, in which $\mathbf{H}(\alpha, \beta)_{\bar{0}} = F.1 + F.uv$ and $H(\alpha, \beta)_{\bar{1}} = F.u + F.v$. It is said that $\mathbf{H}(\alpha, \beta) = \mathbf{H}$ is a *quaternion superalgebra.*

We remark that the center of an associative superalgebra A, denoted by $Z(A)$, is a superalgebra: $Z(A) = Z(A)_{\bar{0}} + Z(A)_{\bar{1}}$. Let $Z = Z(A)_{\bar{0}}$, A is said to be a *central superalgebra* over the field F if $F = Z$.

In [**16**], Wall described the structure of finite dimensional simple associative superalgebras (see also [**2, 14**]).

THEOREM 2.1. *Let A be a finite dimensional non trivial central simple associative superalgebra over a field F. Then either:*

(i) *A is central simple as an (ungraded) algebra over F. In this case $Z(A)_{\bar{1}} = 0$ and A is said to be of* even type.

(ii) *$A_{\bar{0}}$ is a central simple algebra over F and $A = A_{\bar{0}} \oplus A_{\bar{0}}u$ with $u \in Z(A)_{\bar{1}}, u^2 = 1$. In this case A is said to be of* odd type.

A superalgebra $A = A_{\bar{0}} \oplus A_{\bar{1}}$ is said to be an *alternative superalgebra* if and only if for any homogeneous elements a, b, c in A and $x \in A_{\bar{0}}$:

$$(a,b,c) + (-1)^{\bar{b}\bar{c}}(a,c,b) = 0$$
$$(a,b,c) + (-1)^{\bar{a}\bar{b}}(b,a,c) = 0 \qquad (2.1)$$
$$(x,x,a) = 0.$$

where $(a, b, c) = (ab)c - a(bc)$.

For an alternative superalgebra, B, the associative center, $N(B) = \{x \in B : (x, a, b) = 0 \text{ for any } a, b \in B\}$, and the center, $Z(B) = \{x \in N(B) : xa = ax \text{ for any } a \in B\}$, are superalgebras. An alternative superalgebra B is said to be *central* over a field F if $Z(B)_{\bar{0}} = F$.

THEOREM 2.2. *([15]) The classification of finite dimensional nontrivial central simple nonassociative alternative superalgebras over a field F of characteristic 2 and 3 is the following:*

1) The superalgebra $B(1,2)$:
   Here $\mathrm{char}F = 3$ *and* $B = B(1,2)$ *is such that* $B_{\bar{0}} = F.1$, $B_{\bar{1}} = F.x + F.y$, *where 1 is a unit element of* B *and* $xy = -yx = 1$.
2) The superalgebra $B(4,2)$:
   Now $\mathrm{char}F = 3$, *and* $B_{\bar{0}} = M_2(F)$, *the algebra of* 2×2 *matrices over* F, *and* $B_{\bar{1}} = F.m_1 + F.m_2$. *The action of* $B_{\bar{0}}$ *on* $B_{\bar{1}}$ *is given by*
   $$e_{ij}.m_k = \delta_{ik}m_j \quad i,j,k \in \{1,2\}, \quad m.a = \bar{a}.m;$$
   where $a \in B_{\bar{0}}$, $m \in B_{\bar{1}}$, $a \longrightarrow \bar{a}$ *is the symplectic involution in* $A = M_2(F)$. *That is,* $B_{\bar{1}}$ *is the 2-dimensional irreducible Cayley bimodule over* $M_2(F)$. *The odd multiplication on* $B_{\bar{1}}$ *is defined by*
   $$m_1^2 = -e_{21}, \; m_2^2 = e_{12}, \; m_1m_2 = e_{11}, \; m_2m_1 = -e_{22}.$$
3) *The twisted superalgebra of vector type* $B(\Gamma, D, \gamma)$. *Let* Γ *be a commutative and associative algebra over* F, D *a nonzero derivation of* Γ, *and* $\gamma \in \Gamma$. *Denote by* $\bar{\Gamma}$ *an isomorphic copy of the* F*-module* Γ, *with the isomorphism mapping* $a \longrightarrow \bar{a}$. *Consider the direct sum of* F*-modules* $B(\Gamma, D, \gamma) = \Gamma + \bar{\Gamma}$ *and define a multiplication on it by the rules*
   $$\begin{aligned} a.b &= ab, \\ a.\bar{b} &= \overline{ab} = \bar{a}.b, \\ \bar{a}.\bar{b} &= \gamma ab + 2D(a)b + aD(b), \end{aligned}$$
   where $a, b \in \Gamma$ *and* ab *is the product in* Γ. *Define a* $\mathbf{Z}_2$*-grading on* $B(\Gamma, D, \gamma)$ *by setting* $B_{\bar{0}} = \Gamma, B_{\bar{1}} = \bar{\Gamma}$; *then* $B(\Gamma, D, \gamma)$ *is an alternative superalgebra if* $\mathrm{char}F = 3$ *and is not associative if* $D(\Gamma)\Gamma^2 \neq 0$. *Moreover* $B(\Gamma, D, \gamma)$ *is simple if and only if* Γ *is* D*-simple.* Γ *is said to be then a differentiably simple commutative associative ring.*
4) *The Cayley-Dickson superalgebra* $O(4,4)$. *Let* $\mathrm{char}F = 2$, $O(4,4) = O = \mathbf{H} + v\mathbf{H}$ *a Cayley-Dickson algebra over* F *with a natural* $\mathbf{Z}_2$*-grading induced by the Cayley-Dickson process applied to a generalized quaternion algebra* $\mathbf{H}$ *(see, for instance, [18]). Then* $O(4,4)$ *is a simple alternative superalgebra.*
5) *The double Cayley-Dickson superalgebra* $O[u]$. *Let* $\mathrm{char}F = 2, F[u] = F + Fu, u^2 = \alpha \neq 0 \in F$, *be a simple 2-dimensional superalgebra with the even part* F *and the odd part* Fu. *Then* $O[u] = F[u] \otimes_F O = O + Ou$ *is a simple alternative superalgebra, with even part* O *and odd part* Ou.

Notice that the last two cases are $\mathbf{Z}_2$*-graded alternative algebras.*

In this note our purpose is to describe maximal F-subalgebras of alternative superalgebras of the types 1), 2), 4) and 5). At this point in time the maximal subalgebras of superalgebras of type 3) have not been determined.

We will need to know in the sequel some important properties about composition algebras. Cayley-Dickson algebras and quaternion algebras are particular cases of composition algebras. We list these properties in the following comments. The proofs and more information can be found in [**1**] or [**18**].

Every composition algebra, C, has an identity element, 1, and it is endowed with a strictly nondegenerate quadratic form, n, which allows composition, that is, $n(xy) = n(x)n(y)$. Strictly nondegenerate means that the symmetric bilinear form corresponding to it, $t(x, y) = n(x + y) - n(x) - n(y)$, is nondegenerate. The mapping $a \longrightarrow \bar{a} = t(1, a) - a$ is an involution of the algebra and is related to the quadratic form by $n(x) = x\bar{x}$, and for any $a \in C$

$$a^2 - t(1, a)a + n(a) = 0. \tag{2.2}$$

Every subalgebra B of C such that $1 \in B$ and on which the restriction of the bilinear form $t(x, y)$ is nondegenerate allows us to represent each element of C as $a + vb$ with $a, b \in B$ and

$$(a + vb)(c + vd) = (ac + \alpha d\bar{b}) + v(\bar{a}d + cb) \tag{2.3}$$

where $v \in B^{\perp}, 0 \neq \alpha = -n(v)$ (this is part of the Cayley-Dickson process mentioned in theorem 2.2, 4). Composition algebras can be of two types: those in which every element is invertible and then $n(a) \neq 0$ for any $a \neq 0$, and those with zero divisors and then there exists an element $a \neq 0$ such that $n(a) = 0$. The latter are called *s*plit . Split quaternion algebras are isomorphic to $M_2(F)$ and in the split Cayley-Dickson algebras we can find a basis $\{x_i, y_i : \ 0 \leq i \leq 3\}$ such that x_0, y_0 are orthogonal idempotents and

$$\begin{array}{lllll} x_0x_i = x_i, & y_0y_i = y_i, & x_iy_0 = x_i, & y_ix_0 = y_i, & 0 \leq i \leq 3; \\ x_iy_i = x_0, & x_ix_{i+1} = y_{i+2}, & y_iy_{i+1} = x_{i+2}, & y_ix_i = y_0, & \end{array} \tag{2.4}$$

with $1 \leq i \leq 3$, where the indices are taken modulo 3; and all other products are zero. Notice that $< x_0, y_0, x_1, y_1 >$, the subspace generated by $\{x_0, y_0, x_1, y_1\}$, is isomorphic to $M_2(F)$, and if $\{e_{11}, e_{12}, e_{21}, e_{22}\}$ is the usual basis of $M_2(F)$ then $x_0 = e_{11}, y_0 = e_{22}, x_1 = e_{12}, y_1 = e_{21}$. In this case if $v \in < x_0, y_0, x_1, y_1 >^{\perp}$ then we can take $x_2 = ve_{22}, y_2 = ve_{11}, x_3 = ve_{21}, y_3 = -ve_{12}$.

M. Racine gave in [**12**] the classification of maximal subalgebras of central Cayley-Dickson algebras. The authors have known from the referee that, recently, S. M. Gagola III have shown in a preprint that M. Racine have missed a case in characteristic 2. One can compare this classification with the one obtained for maximal subalgebras of $O(4, 4)$ in Theorem 3.2.

3. Maximal subalgebras of $B(1, 2), B(4, 2), O(4, 4)$ and $O[u]$.

Note that any maximal subalgebra M in a finite dimensional simple alternative superalgebra B over a field F contains the identity element. Indeed, if $1 \notin M$, the algebra generated by M and 1: $M + F.1$, is the whole B by maximality. So M is a nonzero graded ideal of B, a contradiction with B being simple. Therefore $1 \in M$.

We begin studying the maximal F-subalgebras of the superalgebras $B(1, 2)$ and $B(4, 2)$.

THEOREM 3.1. (i) *Let $B = B(1, 2)$. A subalgebra M of B is maximal if and only if $M = F.1 + F.a$ with $a \in B_{\bar{1}} - \{0\}$.*

(ii) *Let* $B = B(4,2)$. *A subalgebra* M *of* B *is maximal if and only if either* $M = B_{\bar{0}}$ *or* $M = (F.e_{11} + F.e_{22} + F.e_{12}) + (F.m_2)$, *up to an automorphism of* B, *where* $\{e_{ij}\}_{i,j\in\{1,2\}} \cup \{m_i\}_{i\in\{1,2\}}$ *is the basis of* B *described in theorem 2.2, 2).*

PROOF. The proof of (i) is straightforward.

To show (ii) we remark the following facts:

(a) Since B is simple, $B_{\bar{1}}^2 = B_{\bar{0}}$,
(b) $M_2(F).Fa = B_{\bar{1}}$ and $a^2 \in B_{\bar{0}} - F.1$, for any $a \in B_{\bar{1}} - \{0\}$.

Now let M be a maximal subalgebra of B. If $M_{\bar{0}} = B_{\bar{0}}$, it is easy to check that $M = M_{\bar{0}}$ is maximal because of (b). So we can suppose that $M_{\bar{0}} \neq B_{\bar{0}}$, and then by maximality $M_{\bar{1}} \neq 0$ and hence $M = M_{\bar{0}} + F.a$ because of (a). We notice also that $M_{\bar{0}} \neq F.1$ because of (b).

Since $M_{\bar{0}} \neq B_{\bar{0}}$, $M_{\bar{0}}$ is contained in a maximal subalgebra of $M_2(F)$. From theorem 1 in [**12**] we know that the maximal subalgebras of $M_2(F)$ are either $F.e + F.f + eB_{\bar{0}}f$ with e, f nonzero idempotents in $M_2(F)$ such that $e + f = 1$, or the centralizer in $M_2(F)$ of a quadratic extension field of F, K.

If $M_{\bar{0}}$ is contained in the centralizer in $M_2(F)$ of a quadratic extension field, K, we have $M_{\bar{0}} = K$ because of (b) and the double centralizer theorem (see for instance [**10**] page 104), and so $M = K + F.a$. But since $char F \neq 2$ we can suppose that $K = F.1 + F.x$ with $x^2 = \delta \notin F^2$. Now let $xa = \lambda a$ with $\lambda \in F$, then $\delta a = x(xa) = \lambda^2 a$ from (2.1), which contradicts $\delta \notin F^2$.

If $M_{\bar{0}} = F.e + F.f + eB_{\bar{0}}f$ with e, f nonzero idempotents such that $e + f = 1$, from lemma 13 and its proof in [**15**], we can suppose that $M_{\bar{0}} = F.e_{11} + F.e_{22} + F.e_{12}$ and $M_{\bar{1}} = F.a$ where $a = \lambda m_1 + \mu m_2$ with $\lambda, \mu \in F$, for e_{ij}, m_i as in theorem 2.2, 2). But then since $e_{11}a = \lambda m_1$ and $e_{12}m_1 = m_2$, we deduce from (a) that $\lambda = 0$. So $M = (F.e_{11} + F.e_{22} + F.e_{12}) + (F.m_2)$. We notice that if we had put $M_{\bar{0}} = F.e_{11} + F.e_{22} + F.e_{21}$ then M would be $M = (F.e_{11} + F.e_{22} + F.e_{21}) + (F.m_1)$, but clearly there is an automorphism of $B(4,2)$ applying $< e_{11}, e_{22}, e_{21} > + < m_2 >$ in $< e_{11}, e_{22}, e_{12} > + < m_1 >$, concretely the linear map such that

$$\begin{array}{llllll} e_{11} \longrightarrow e_{22}, & e_{22} \longrightarrow e_{11}, & e_{12} \longrightarrow -e_{21}, \\ e_{21} \longrightarrow -e_{12}, & m_1 \longrightarrow m_2, & m_2 \longrightarrow m_1 \end{array}$$

And if $M_{\bar{0}}$ is not a maximal subalgebra of $M_2(F)$, then since $1 \in M_{\bar{0}}$, $M_{\bar{0}}$ is either semisimple or with radical. Notice that from (b) $M_{\bar{0}} \neq F.1$ and we can suppose also that $M_{\bar{0}}$ is not a quadratic extension field. So from lemma 13 and its proof in [**15**], we can suppose that either $M = (F.e_{11} + F.e_{22}) + F.a$ or $M = (F.1 + F.x) + F.a$ with $x = e_{12}$ and $a = \lambda m_1 + \mu m_2$ with $\lambda, \mu \in F$, being e_{ij}, m_i like in theorem 2.2, 2). If $M_{\bar{0}} = F.e_{11} + F.e_{22}$, then $e_{11}a = \lambda m_1, e_{22}a = \mu m_2$, and because of (a) we have either $a = m_1$ or $a = m_2$. In both cases M is contained in one of the subalgebras described before. And if $M_{\bar{0}} = F.1 + F.e_{12}$, then $e_{12}a = \lambda m_2$. Now, because of (a), $\lambda = 0$ and so $M \subseteq (F.e_{11} + F.e_{22} + F.e_{12}) + (F.m_2)$, and is not maximal.

□

Now we will classify the maximal F-subalgebras of $O(4,4)$. We recall that $car F = 2$ and that $O(4,4)$ is also an alternative algebra.

THEOREM 3.2. *Let $B = O(4,4)$, a subalgebra M of B is maximal if it is one of the following:*

(a) *M is a division quaternion algebra and either $M = B_{\bar{0}}$ or M is also a quaternion superalgebra.*
(b) *$M \cong M_2(F)$ with $M_{\bar{0}}$ a separable quadratic extension field of F.*
(c) *$M = K + vK$ with K a purely inseparable quadratic extension field of F.*
(d) *M is either $< x_0, y_0, x_1, y_1 > + < x_2, y_3 >$ or $< x_0, y_0, y_3 > + < x_1, y_1, x_2 >$ where $\{x_i\}_{i \in \{0,1,2,3\}}, \{y_i\}_{i \in \{0,1,2,3\}}$ is the basis given in (1.7).*

PROOF. Let $O(4,4) = \mathbf{H} + v\mathbf{H}$, and notice that to describe the odd part we can choose any $v \in O(4,4)_{\bar{1}}$ such that $n(v) \neq 0$. Let $M = M_{\bar{0}} + M_{\bar{1}}$ be a maximal subalgebra of $O(4,4)$. We distinguish two cases: i) If $M_{\bar{1}} = 0$ and ii) if $M_{\bar{1}} \neq 0$.

If $M_{\bar{1}} = 0$, since M is maximal, $M = O(4,4)_{\bar{0}} = \mathbf{H}$. Notice that if $M = \mathbf{H}$ is a division algebra from (2.3) we can check that M is a maximal subalgebra. But if $M = \mathbf{H}$ is split, that is, $M \cong M_2(F)$, from the introduction we can suppose that $M =< x_0, y_0, x_1, y_1 >$ and then $M \subsetneq < x_0, y_0, x_1, y_1 > + < x_2, y_3 >$ where $< x_0, y_0, x_1, y_1 > + < x_2, y_3 >$ is a subalgebra, a contradiction. So if $M_{\bar{1}} = 0$, then M is maximal if and only if M is a division quaternion algebra.

If $M_{\bar{1}} \neq 0$, we suppose first that there exist an element $u \in M_{\bar{1}}$ such that $0 \neq n(u) = u\bar{u} \in F$. Notice that $O(4,4) = \mathbf{H} + u\mathbf{H}$. In this case u is invertible and so $M_{\bar{1}} = uM_{\bar{0}}$ and $M_{\bar{0}} = uM_{\bar{1}}$. We observe that then M is maximal if and only if $M_{\bar{0}}$ is maximal in $\mathbf{H}$. In theorem 1 in [**12**], M. Racine described the maximal subalgebras of a central simple finite dimensional associative algebra, and as a consequence we know that now we have three possibilities: either a) M_0 is a separable quadratic extension field of F and then M is either a division quaternion algebra or is isomorphic to $M_2(F)$, or b) $M_{\bar{0}}$ is a purely inseparable quadratic extension field, $F(a)$, and $M = < 1, a > + < u, ua >= (F.1 + f.a) \otimes (F.1 + F.u)$ with the grading given by $F.1 + F.u$, or c) $M_{\bar{0}} = e\mathbf{H}e + f\mathbf{H}f + e\mathbf{H}f$ with e, f orthogonal idempotents in $\mathbf{H}$, and then $dim_F M = 6$ and by (2.4) and the comments before (2.4) we can suppose that $M =< x_0, y_0, y_3 > + < x_1, y_1, x_2 >$. So if $M_{\bar{1}} \neq 0$ and there exists an element $u \in M_{\bar{1}}$ such that $n(u) \neq 0$ then either M is a division quaternion algebra and also a quaternion superalgebra, or $M \cong M_2(F)$ with $M_{\bar{0}}$ a separable quadratic extension field of F, or $M = K + uK$ with K a purely inseparable quadratic extension field of F.

If $M_{\bar{1}} \neq 0$ and $n(x) = 0$ for any $x \in M_{\bar{1}}$, then since $t(1,x) = 0$ ($x \in v\mathbf{H} = \mathbf{H}^{\perp}$), it follows from (2.2) that $x^2 = 0$ for any $x \in M_{\bar{1}}$. So $M_{\bar{1}}$ is a totally isotropic subspace with respect to the quadratic form n. But n is regular, so $dim_F M_{\bar{1}} = 1$ or 2. Notice that for any $0 \neq x \in M_{\bar{1}}$ we can check using (2.3) that $O(4,4)_{\bar{0}} + O(4,4)_{\bar{0}} x$ is a subalgebra such that $M \subseteq O(4,4)_{\bar{0}} + O(4,4)_{\bar{0}} x$. By maximality $O(4,4)_{\bar{0}} = M_{\bar{0}}$ and $O(4,4)_{\bar{0}} x = M_{\bar{1}}$. Now suppose that $dim_F M_{\bar{1}} = 1$, then $M_{\bar{1}} = F.x$. We consider the linear map $f \colon O(4,4)_{\bar{0}} \to O(4,4)_{\bar{1}}$ such that $z \longrightarrow zx$. Then $Imf = M_{\bar{1}}$ and if $z \in Kerf$ then $0 = zx = \bar{z}(zx) = (\bar{z}z)x$ because of $\bar{z} = t(1,z) - z$ and (2.1), and so $n(z) = \bar{z}z = 0$ for any $z \in Kerf$, that is, $Kerf$ is a totally isotropic space with respect to n. Therefore, since n is regular, $dim_F(Kerf) \leq 2$, a contradiction because $dim_F(Imf) + dim_F(Kerf) = dim_F(O(4,4)_{\bar{0}}) = 4$. In consequence $dim_F(M_{\bar{1}}) = 2$ and since $M = O(4,4)_0 + O(4,4)_0 x$, by (2.4) we can suppose that $M =< x_0, y_0, x_1, y_1 > + < x_2, y_3 >$.

□

We finish this section describing the maximal subalgebras of $O[u]$, the double Cayley-Dickson superalgebra.

THEOREM 3.3. *Let $O[u]$ be the double Cayley-Dickson superalgebra, and let M be a subalgebra of A. Then M is a maximal subalgebra of A if and only if either:*

(i) *$M = M_{\bar{0}} \oplus M_{\bar{0}}u$ with $M_{\bar{0}}$ a maximal subalgebra of $O[u]_{\bar{0}}$.*
(ii) *$M = O[u]_{\bar{0}}$.*
(iii) *$O[u]_{\bar{0}} = O$ is a graded algebra: $O = \mathbf{H} \oplus v\mathbf{H}$, and $M = \mathbf{H} \oplus (v\mathbf{H})u$.*

These conditions are mutually exclusive.

PROOF. Consider $\bar{Z} = \bar{Z}_{\bar{0}} + \bar{Z}_{\bar{1}}$ where $\bar{Z}_{\bar{i}} = \{x \in N_{\bar{i}} : xa = (-1)^{ij}ax$ for any $a \in O[u]_j\}$. We know that $\bar{Z} = F + Fu$. Let M be a maximal subalgebra of $O[u]$. Then $M \subseteq \bar{Z}M \subseteq O[u]$ and by maximality either $\bar{Z}M = M$ or $\bar{Z}M = O[u]$.

If $M = \bar{Z}M$, $u \in M$ because $1 \in M$. This implies that $M_{\bar{1}} = M_{\bar{0}}u$ and $M = M_{\bar{0}} + M_{\bar{0}}u$. Since M is a maximal subalgebra of $O[u]$ it follows that $M_{\bar{0}}$ is maximal subalgebra of $O[u]_{\bar{0}} = O$. The converse is clear.

If $\bar{Z}M = O[u]$, then $O[u]_{\bar{0}} = M_{\bar{0}} + M_{\bar{1}}u$ and $O[u]_{\bar{1}} = M_{\bar{1}} + M_{\bar{0}}u$. Since $M_{\bar{0}} \cap M_{\bar{1}}u$ is an ideal of $M_{\bar{0}}$ because of $M_{\bar{0}}(M_{\bar{1}}u) \subseteq M_{\bar{1}}u$ and $(M_{\bar{1}}u)^2 \subseteq M_{\bar{0}}$, and $O[u]_{\bar{0}} = O$ is simple, it follows that $O[u]_{\bar{0}} = M_{\bar{0}} + M_{\bar{1}}u$ is a graded algebra. If the grading is trivial, that is, $M_{\bar{1}}u = 0 = M_{\bar{1}}$, then $M = O[u]_{\bar{0}}$ and M is a maximal subalgebra of $O[u]$. Otherwise, $O[u]_{\bar{0}} = O$ is a graded algebra, but $\mathbf{Z}_2-$gradings of octonions were described in proposition 6 of [**6**], and all of them are obtained like in theorem 2.2 4), that is, $O[u]_{\bar{0}} = \mathbf{H} + v\mathbf{H}$, $O[u] = (\mathbf{H} + v\mathbf{H})) + (\mathbf{H} + v\mathbf{H})u$ and so $M = \mathbf{H} + (v\mathbf{H})u$.

Conversely, if $M = \mathbf{H} + (v\mathbf{H})u$ with $\mathbf{H}$ a quaternion subalgebra of O and $v \in \mathbf{H}^{\perp}$ such that $0 \neq \alpha = n(v)$, then let $x \in O[u]_{\bar{0}} - M_{\bar{0}}$. We can suppose that $x \in v\mathbf{H} - \{0\}$, but then since $(v\mathbf{H})u \subseteq M$ we have $u \in alg < M, x >$ (the algebra generated by the elements of M and by x) and so $\mathbf{H}u \subseteq alg < M, x >$ because $\mathbf{H} \subseteq M$ and also $((v\mathbf{H})u)u = v\mathbf{H} \subseteq alg < M, x >$, that is, $alg < M, x >= O[u]$. And if $x \in O[u]_{\bar{1}} - M_{\bar{1}}$, we can suppose that $x \in \mathbf{H}u$, but then since $\mathbf{H} \subseteq M$ we have $u \in alg < M, x >$ and so $v\mathbf{H} \subseteq alg < M, x >$ because $((v\mathbf{H})u)u = v\mathbf{H}$. Finally notice that also $\mathbf{H}u \subseteq alg < M, x >$, that is, $alg < M, x >= O[u]$. □

References

[1] F. van der Blij, T. A. Springer, The arithmetics of octaves and of the group G_2, *Nederl. Akad. Wetensch. Proc. Ser. A* **21** (1959), 406-418.

[2] C. Draper, A. Elduque, Division superalgebras, in A. Castellón Serrano, J.A. Cuenca Mira, A. Fernández López, C. Martín González (Eds.), Proc. Int. Conf. on Jordan Systems, Málaga 1997, 1999, pp. 77-83.

[3] E. Dynkin, Semi-simple subalgebras of semi-simple Lie algebras, *Mat. Sbornik* **30** (1952), 249-462; *Amer. Math. Soc. Transl.* **6** (1957), 111-244.

[4] E. Dynkin, Maximal subgroups of the classical groups, *Trudy Moskov. Mat. Obsc.* (1952), 39-166; *Amer. Math. Soc. Transl.* **6** (1957), 245-378.

[5] A. Elduque, On maximal subalgebras of central simple Malcev algebras, *J. Algebra* **103** no.1 (1986), 216-227.

[6] A. Elduque, Gradings on Octonions, *J. Algebra***207** (1998), 342-354.

[7] A. Elduque, J. Laliena, S. Sacristán, Maximal subalgebras of associative superalgebras, *J. Algebra* **275** (2004), no. 1, 40-58.

[8] A. Elduque, J. Laliena, S. Sacristán, Maximal subalgebras of Jordan superalgebras, *J. Pure Appl. Algebra* **212** no. 11 (2008), 2461-2478.

[9] A. Elduque, J. Laliena, S. Sacristán, The Kac Jordan superalgebra: automorphisms and maximal subalgebras, *P. Amer. Math. Soc.* **135** (2007), 3805-3813.

[10] I. N. Herstein, Noncommutative Rings, Carus Mathematics Monograph 15, Mathematics Association America, 1968.
[11] M.C. López-Díaz, I. P. Shestakov, Alternative superalgebras with DCC on two-sided ideals, *Comm. Algebra* **33** (2005), 3479-3487.
[12] M. Racine, On Maximal subalgebras, *J. Algebra* **30** (1974), 155-180.
[13] M. Racine, Maximal subalgebras of exceptional Jordan algebras, *J. Algebra* **46** (1977), no. 1, 12-21.
[14] M. Racine, Primitive superalgebras, *J. Algebra* **206** (1998), 588-614.
[15] I. P. Shestakov, Prime alternative superalgebras of arbitrary characteristic, *Algebra i Logika* **36, no.6** (1997), 675-716.
[16] C. T. C. Wall, Graded Brauer Groups, *Jour. Reine Angew. Math.* **213** (1964), 187-199.
[17] E. I. Zelmanov and I.P. Shestakov, Prime alternative superalgebras and nilpotence of the radical of a free alternative algebra, *Izv. Aka. Nauk SSSR Ser. Mat.* **54** (1990), 676-693; English transl. in *Math. USSR Izv.* **37** (1991), no11, 19-36.
[18] K. A. Zhevlakov, A. M. Slinko, I. P. Shestakov, A. I. Shirshov, Rings that are nearly associative, Academic Press, 1982.

Departamento de Matemáticas y Computación, Universidad de La Rioja, 26004, Logroño. Spain
E-mail address: jesus.laliena@unirioja.es

Departamento de Matemáticas y Computación, Universidad de La Rioja, 26004, Logroño. Spain
E-mail address: sara.sacristan@unirioja.es

Contemporary Mathematics
Volume **483**, 2009

Automorphisms of elliptic Poisson algebras

Leonid Makar-Limanov, Umut Turusbekova, and Ualbai Umirbaev

This paper is dedicated to Professor Ivan Shestakov on the occasion of his 60th birthday.

Abstract. We describe the automorphism groups of elliptic Poisson algebras on polynomial algebras in three variables and give an explicit set of generators and defining relations for this group.

1. Introduction

It is well known [**2, 8, 9, 11**] that the automorphisms of polynomial algebras and free associative algebras in two variables are tame. It is also known [**17, 18**] that polynomial algebras and free associative algebras in three variables in the case of characteristic zero have wild automorphisms. It was recently proved [**13**] that the automorphisms of free Poisson algebras in two variables over a field of characteristic 0 are tame. Note that the Nagata automorphism [**14, 17**] gives an example of a wild automorphism of a free Poisson algebra in three variables.

One of the main problems of affine algebraic geometry (see, for example [**5**]) is a description of automorphism groups of polynomial algebras in $n \geq 3$ variables. Unfortunately this problem is still open even when $n = 3$ and there isn't any plausible conjecture about the generators. There was a conjecture that the group of automorphisms of polynomial algebras in three variables is generated by all affine and exponential automorphisms (see [**5**]). This seems to be not true and the first author and D. Wright (oral communication) independently constructed potential counterexamples.

In order to find a plausible conjecture it is necessary to consider many different types of automorphisms. In this paper we describe the groups of automorphisms of the polynomial algebra $k[x, y, z]$ over a field k of characteristic 0 endowed with additional structure, namely, with Poisson brackets. A study of automorphisms of Poisson structures on polynomial algebras is also interesting in view of M. Kontsevich's Conjecture about the existence of an isomorphism between automorphism groups of symplectic algebras and Weyl algebras [**1**].

A complete description of quadratic Poisson brackets on the polynomial algebra $k[x, y, z]$ over a field k of characteristic 0 is given in [**3**], [**4**], and [**10**]. Among

---

1991 *Mathematics Subject Classification.* Primary 17B63, 17B40; Secondary 14H37, 17A36, 16W20.

Key words and phrases. Poisson algebras, automorphisms, derivations.

The first author was supported by an NSA grant and by grant FAPESP, processo 06/59114-1.

corresponding Poisson algebras the most interesting are elliptic Poisson algebras E_α. By definition (see, for example [**15**]), the elliptic Poisson algebra E_α is the polynomial algebra $k[x, y, z]$ endowed with the Poisson bracket defined by

$$\{x, y\} = -\alpha xy + z^2, \quad \{y, z\} = -\alpha yz + x^2, \quad \{z, x\} = -\alpha zx + y^2,$$

where $\alpha \in k$.

We describe the automorphism groups of the elliptic Poisson algebras E_α over a field k of characteristic 0. We also show that E_α doesn't have any nonzero locally nilpotent derivations.

This paper is organized as follows. In Section 2 we give some definitions and examples. In Section 3 we describe Casimir elements of elliptic algebras and prove a lemma related to the non-rationality of elliptic curves. In Section 4 we study the automorphism group of E_α.

2. Definitions and examples

A vector space P over a field k endowed with two bilinear operations $x \cdot y$ (a multiplication) and $\{x, y\}$ (a Poisson bracket) is called *a Poisson algebra* if P is a commutative associative algebra under $x \cdot y$, P is a Lie algebra under $\{x, y\}$, and P satisfies the following identity:

$$\{x, y \cdot z\} = \{x, y\} \cdot z + y \cdot \{x, z\}.$$

Let us call a linear map $\phi : P \longrightarrow P$ an automorphism of P as a Poisson algebra if

$$\phi(xy) = \phi(x)\phi(y), \quad \phi(\{x, y\}) = \{\phi(x), \phi(y)\}$$

for all $x, y \in P$

Similarly, a linear map $D : P \longrightarrow P$ is a derivation of P if

$$D(xy) = D(x)y + xD(y), \quad D(\{x, y\}) = \{D(x), y\} + \{x, D(y)\}$$

for all $x, y \in P$. In other words, D is simultaneously a derivation of P as an associative algebra and as a Lie algebra. It follows that for every $x \in P$ the map

$$\mathrm{ad}_x : P \longrightarrow P, \quad (y \mapsto \{x, y\}),$$

is a derivation of P. It is natural to call these derivations *inner*.

There are two important classes of Poisson algebras.

1) Symplectic algebras S_n. For each n algebra S_n is a polynomial algebra $k[x_1, y_1, \ldots, x_n, y_n]$ endowed with the Poisson bracket defined by

$$\{x_i, y_j\} = \delta_{ij}, \quad \{x_i, x_j\} = 0, \quad \{y_i, y_j\} = 0,$$

where δ_{ij} is the Kronecker symbol and $1 \leq i, j \leq n$.

2) Symmetric Poisson algebras $PS(\mathfrak{g})$. Let $\mathfrak{g}$ be a Lie algebra with a linear basis $e_1, e_2, \ldots, e_k, \ldots$. Then $PS(\mathfrak{g})$ is the usual polynomial algebra $k[e_1, e_2, \ldots, e_k, \ldots]$ endowed with the Poisson bracket defined by

$$\{e_i, e_j\} = [e_i, e_j]$$

for all i, j, where $[x, y]$ is the multiplication of the Lie algebra $\mathfrak{g}$.

Let $k\{x_1, x_2, \ldots, x_n\}$ be a free Poisson algebra in the variables $x_1, x_2, \ldots, x_n$. Recall that $k\{x_1, x_2, \ldots, x_n\}$ is the symmetric Poisson algebra $PS(\mathfrak{g})$, where $\mathfrak{g}$ is the free Lie algebra in the variables $x_1, x_2, \ldots, x_n$. Choose a linear basis

$$e_1, e_2, \ldots, e_m, \ldots$$

of $\mathfrak{g}$ such that $e_1 = x_1, e_2 = x_2, \ldots, e_n = x_n$. The algebra $k\{x_1, x_2, \ldots, x_n\}$ is generated by these elements as a polynomial algebra. Consequently,

$$k[x_1, x_2, \ldots, x_n] \subset k\{x_1, x_2, \ldots, x_n\}.$$

This inclusion was successfully used in [**16**] to determine algebraic dependence of two elements of the polynomial algebra $k[x_1, x_2, \ldots, x_n]$. Namely, two elements f and g of $k[x_1, x_2, \ldots, x_n]$ are algebraically dependent if and only if $\{f, g\} = 0$. It is also proved in [**12**] that if elements f and g of the free Poisson algebra $k\{x_1, x_2, \ldots, x_n\}$ satisfy the equation $\{f, g\} = 0$ then f and g are algebraically dependent. There is a conjecture [**12**] that if $\{f, g\} \neq 0$ then the Poisson subalgebra of $k\{x_1, x_2, \ldots, x_n\}$ generated by f and g is a free Poisson algebra in the variables f and g, i.e., the elements f and g are *free*.

3. Elliptic Poisson algebras

Let

$$C = C(x, y, z) = \frac{1}{3}(x^3 + y^3 + z^3) - \alpha xyz, \qquad \alpha \in k,$$

then the equation $C(x, y, z) = 0$ defines an elliptic curve $\mathfrak{E}_\alpha$ in kP^2. The elliptic Poisson algebra E_α (see, for example [**15**]) is the polynomial algebra $k[x, y, z]$ endowed with the Poisson bracket

$$\{x, y\} = \frac{\partial C}{\partial z}, \quad \{y, z\} = \frac{\partial C}{\partial x}, \quad \{z, x\} = \frac{\partial C}{\partial y}.$$

Consequently,

$$(3.1) \quad \{x, y\} = -\alpha xy + z^2, \quad \{y, z\} = -\alpha yz + x^2, \quad \{z, x\} = -\alpha zx + y^2.$$

This bracket can be written as $\{u, v\} = \mathrm{J}(u, v, C(x, y, z))$ where $\mathrm{J}(u, v, C(x, y, z))$ is the Jacobian, i. e. the determinant of the corresponding Jacobi matrix.

If $\alpha^3 = 1$ and k contains a root ϵ of the equation $\lambda^2 + \lambda + 1 = 0$ then

$$(3.2) \quad C = C(x, y, z) = \frac{1}{3}(x + y + \alpha z)(\epsilon x + \epsilon^2 y + \alpha z)(\epsilon^2 x + \epsilon y + \alpha z),$$

It is well known (see, for example [**7**]) that C is an irreducible polynomial if $\alpha^3 \neq 1$.

The center $Z(A)$ of any Poisson algebra A is defined as a set of all elements $f \in A$ such that $\{f, g\} = 0$ for every $g \in A$. The elements of $Z(A)$ are called *Casimir* elements. The statement of the following lemma is well known (see, for example [**15**]).

LEMMA 3.1. $Z(E_\alpha) = k[C]$.

Proof. Since $\{C, h\} = \mathrm{J}(C, h, C) = 0$ for any $h \in E_\alpha$ we see that $C \in Z(E_\alpha)$. Assume that $Z(E_\alpha) \ni f$ which is algebraically independent with C. Take an element $g \in E_\alpha$ which is algebraically independent over $k[C, f]$. Then for any element $h \in E_\alpha$ there exists a polynomial dependence $H(C, f, g, h) = 0$. Of course, we can assume that $H(C, f, g, h)$ has minimal possible degree in h. So $0 = \{g, H(C, f, g, h)\} = \{g, h\}\frac{\partial H}{\partial h}$ and $\{g, h\} = 0$ since $\frac{\partial H}{\partial h} \neq 0$. Therefore g is also in the center. Take now an element $p \in E_\alpha$. Then $0 = \{p, H(C, f, g, h)\} = \{p, h\}\frac{\partial H}{\partial h}$. Therefore $\{p, h\} = 0$ and E_α is a commutative Poisson algebra, which is not the case. So if $f \in Z(E_\alpha)$ then f is algebraically dependent with C. Since the brackets are homogeneous we can assume that f is homogeneous. But then $f = \mu C^n$ where $\mu \in k$. Indeed, f and C satisfy a polynomial relation $P(C, f) = 0$ which can be assumed homogeneous. So over an algebraic closure of k we will have $f^m = \mu C^n$ where n and m are relatively prime integers and $\mu \in k$ since $f, C \in k[x,y,z]$. If $\alpha^3 \neq 1$ then C is irreducible and so $m = 1$; if $\alpha^3 = 1$ then C is either irreducible or a product of three linear factors and again $m = 1$. □

LEMMA 3.2. *Suppose a, b, $c \in k[x_1, x_2, \ldots, x_n]$ are homogeneous elements of the same degree. If $C(a,b,c) = 0$ and $\alpha^3 \neq 1$ then a, b, and c are proportional to each other.*

Proof. The statement of the lemma is an easy corollary of non-rationality of elliptic curves (see, for example [**7**]). But we prefer to give here an independent proof for the convenience of the reader. So,

$$a^3 + b^3 + c^3 - 3\alpha abc = 0. \tag{3.3}$$

If a and b are divisible by an irreducible polynomial p then c is also divisible by p and we can cancel p^3 out of (3.3). So we can assume that a, b, and c are pair-wise relatively prime. If $\deg(a) \neq 0$ we can also assume that $\frac{\partial a}{\partial x_1} \neq 0$. For any f we put $f' = \frac{\partial f}{\partial x_1}$. Differentiating (3.3), we get

$$a'a^2 + b'b^2 + c'c^2 - \alpha(a'bc + ab'c + abc') = 0.$$

Consequently,

$$\begin{aligned} a'(a^3 + b^3 + c^3 - 3\alpha abc) - a[a'a^2 + b'b^2 + c'c^2 - \alpha(a'bc + ab'c + abc')] \\ = (b^2 - \alpha ac)(a'b - ab') + (c^2 - \alpha ab)(a'c - ac') = 0, \end{aligned}$$

and

$$(b^2 - \alpha ac)(a'b - ab') = (c^2 - \alpha ab)(ac' - a'c). \tag{3.4}$$

Note that $b^2 - \alpha ac \neq 0$ and $c^2 - \alpha ab \neq 0$, otherwise a, b, and c are not pair-wise relatively prime. If $b^2 - \alpha ac$ and $c^2 - \alpha ab$ are relatively prime then (4) gives that $b^2 - \alpha ac$ divides $ac' - a'c$. But $\deg(ac' - a'c) < \deg(b^2 - \alpha ac)$, hence $ac' - a'c = 0$. Then $(a/c)' = 0$ and $a = cr$ where $r \in k(x_2, \ldots, x_n)$. But then a and c again are not relatively prime: any irreducible factor of a which contains x_1 should divide c.

Consequently, $b^2 - \alpha ac$ and $c^2 - \alpha ab$ are not relatively prime. Let p be an irreducible polynomial that divides $b^2 - \alpha ac$ and $c^2 - \alpha ab$. Denote by I the ideal of $k[x_1, x_2, \ldots, x_n]$ generated by p. For $f \in k[x_1, x_2, \ldots, x_n]$ denote by $\bar{f}$ the homomorphic image of f in $k[x_1, x_2, \ldots, x_n]/I$. Then, $\bar{b}^2 = \alpha\bar{a}\bar{c}$ and $\bar{c}^2 = \alpha\bar{a}\bar{b}$. Since a, b, and c are pair-wise relatively prime, $\bar{a}, \bar{b}, \bar{c} \neq 0$. Note that $k[x_1, x_2, \ldots, x_n]/I$ is a domain. So $\bar{a}^2 = \alpha\bar{b}\bar{c}$ since $a^3 + b^3 + c^3 - 3\alpha abc = 0$ and $\bar{b}^2\bar{c}^2 = \alpha^2\bar{a}^2\bar{b}\bar{c} = \alpha^3\bar{b}^2\bar{c}^2$. This gives $\alpha^3 = 1$ contrary to our assumptions. □

4. Automorphisms

Let us denote by φ_γ, where $\gamma \in k^*$, an automorphism of E_α such that

$$\varphi_\gamma(x) = \gamma x, \quad \varphi_\gamma(y) = \gamma y, \quad \varphi_\gamma(z) = \gamma z.$$

Note that $\langle\varphi_\gamma\rangle \cong k^*$, where k^* is the multiplicative group of the field k. Algebra E_α has also automorphisms

$$\tau : E_\alpha \longrightarrow E_\alpha, \quad x \mapsto y, \quad y \mapsto z, \quad z \mapsto x.$$

and

$$\sigma : E_\alpha \longrightarrow E_\alpha, \quad x \mapsto x, \quad y \mapsto \epsilon y, \quad z \mapsto \epsilon^2 z,$$

if a solution ϵ of the equation $\lambda^2 + \lambda + 1 = 0$ is in k.

Denote by $\operatorname{Aut} E_\alpha$ the group of automorphisms of E_α. Since the center of E_α is $k[C]$ the restriction of $\psi \in \operatorname{Aut} E_\alpha$ to $k[C]$ is an automorphism of $k[C]$. Let G be the group of those automorphisms of $k[x, y, z]$ which preserve $k[C]$.

LEMMA 4.1. *The group G consists of linear automorphisms.*

Proof. Let $\psi \in G$. Then by the chain rule

$$\mathrm{J}(\psi(x), \psi(y), \psi(C)) = \mathrm{J}(\psi(x), \psi(y), \psi(z))\frac{\partial\,\psi(C)}{\partial\,\psi(z)} = \delta_1(\psi(z^2) - \alpha\psi(x)\psi(y))$$

where $\delta_1 = \mathrm{J}(\psi(x), \psi(y), \psi(z))$. Also

$$\mathrm{J}(\psi(x), \psi(y), \psi(C)) = \mathrm{J}(\psi(x), \psi(y), k_2 C + k_3) = k_2\{\psi(x), \psi(y)\}$$

since $\psi(C) = \delta_2 C + \delta_3$ where $\delta_2,\ \delta_3 \in k$ and $\delta_2 \neq 0$.

So $\delta_2\{\psi(x), \psi(y)\} = \delta_1(\psi(z^2) - \alpha\psi(x)\psi(y))$ where $\delta_1,\ \delta_2 \in k$ are non-zero constants. Similarly, $\delta_2\{\psi(y), \psi(z)\} = \delta_1(\psi(x^2) - \alpha\psi(y)\psi(z))$ and $\delta_2\{\psi(z), \psi(x)\} = \delta_1(\psi(y^2) - \alpha\psi(z)\psi(x))$. Denote $\psi(x) = a,\ \psi(y) = b,\ \psi(z) = c$. So

$$(4.1)\ \delta_4\{a, b\} = -\alpha ab + c^2, \quad \delta_4\{b, c\} = -\alpha bc + a^2, \quad \delta_4\{c, a\} = -\alpha ca + b^2.$$

for some non-zero $\delta_4 \in k$.

We can assume without loss of generality that $\deg(a) \leq \deg(c)$ and $\deg(b) \leq \deg(c)$. Then (4.1) show that $\deg(a) = \deg(b) = \deg(c)$. Denote by $\bar{f}$ the highest homogeneous part of $f \in k[x, y, z]$.

$$(4.2)\ \delta_4\{\bar{a}, \bar{b}\} = -\alpha\bar{a}\bar{b} + \bar{c}^2, \quad \delta_4\{\bar{b}, \bar{c}\} = -\alpha\bar{b}\bar{c} + \bar{a}^2, \quad \delta_4\{\bar{c}, \bar{a}\} = -\alpha\bar{c}\bar{a} + \bar{b}^2,$$

since these terms are the only terms in the left and right sides of the equalities which may have the highest possible degree.

If $\deg c > 1$ then $C(\bar{a}, \bar{b}, \bar{c}) = 0$ since otherwise $\deg C(a, b, c) > 3$. So $\bar{a}$, $\bar{b}$, and $\bar{c}$ are pair-wise proportional by Lemma 3.2. Therefore

$$\{\bar{a}, \bar{b}\} = \{\bar{b}, \bar{c}\} = \{\bar{c}, \bar{a}\} = 0.$$

Then (4.2) gives

$$\alpha\bar{a}\bar{b} = \bar{c}^2, \quad \alpha\bar{b}\bar{c} = \bar{a}^2, \quad \alpha\bar{c}\bar{a} = \bar{b}^2$$

and

$$\bar{b}^2\bar{c}^2 = \alpha^2\bar{a}^2\bar{b}\bar{c} = \alpha^3\bar{b}^2\bar{c}^2.$$

This is impossible if $\alpha^3 \neq 1$ and hence in this case $\deg(a) = \deg(b) = \deg(c) = 1$.

If $\alpha^3 = 1$ then using the decomposition (3.2) we have $3 = \deg(\psi(C)) = \deg(a + b + \alpha c) + \deg(\epsilon a + \epsilon^2 b + \alpha c) + \deg(\epsilon^2 a + \epsilon b + \alpha c)$. Therefore $\deg(a + b + \alpha c) = \deg(\epsilon a + \epsilon^2 b + \alpha c) = \deg(\epsilon^2 a + \epsilon b + \alpha c) = 1$ and the elements a, b, and c also have degree 1 since $a,\ b,\ c, \alpha + b + \alpha c,\ \epsilon a + \epsilon^2 b + \alpha c,\ \epsilon^2 a + \epsilon b + \alpha c \notin k$.

So, we proved that ψ is an affine automorphism. If $\alpha^3 \neq 1$ then (4.1) gives the linearity of ψ (just look at the lowest homogeneous parts of a, b, and c). If $\alpha^3 = 1$ then non of the $\epsilon^i a + \epsilon^{2i} b + \alpha c$ can contain a constant term since otherwise $\psi(C)$ contains a term of degree two. So ψ is a linear automorphism in this case as well. $\square$

THEOREM 4.2. *Let k be an arbitrary field of characteristic 0 such that the equation $\lambda^2+\lambda+1=0$ has a solution in k. If $\alpha^3\neq 1$ then the group of automorphisms $\operatorname{Aut} E_\alpha$ of the algebra E_α is generated by $\varphi_\gamma(\gamma\in k^*)$, σ, and τ.*

Proof. Let ψ be an arbitrary automorphism of E_α. By Lemma 4.1, ψ is a linear automorphism. Let

$$\begin{aligned}\psi(x)&=a=\beta_{11}x+\beta_{12}y+\beta_{13}z,\\ \psi(y)&=b=\beta_{21}x+\beta_{22}y+\beta_{23}z,\\ \psi(z)&=c=\beta_{31}x+\beta_{32}y+\beta_{33}z.\end{aligned}$$

Denote by B the matrix $(\beta_{ij})_{3\times 3}$.

We have

$$\begin{aligned}\{a,b\}&=(\beta_{11}\beta_{22}-\beta_{12}\beta_{21})\{x,y\}\\ &\quad+(\beta_{12}\beta_{23}-\beta_{13}\beta_{22})\{y,z\}+(\beta_{13}\beta_{21}-\beta_{11}\beta_{23})\{z,x\}\\ &=-\alpha(\beta_{11}\beta_{22}-\beta_{12}\beta_{21})xy+(\beta_{11}\beta_{22}-\beta_{12}\beta_{21})z^2-\alpha(\beta_{12}\beta_{23}-\beta_{13}\beta_{22})yz\\ &\quad+(\beta_{12}\beta_{23}-\beta_{13}\beta_{22})x^2-\alpha(\beta_{13}\beta_{21}-\beta_{11}\beta_{23})xz+(\beta_{13}\beta_{21}-\beta_{11}\beta_{23})y^2\end{aligned}$$

and

$$\begin{aligned}-\alpha ab+c^2&=(\beta_{31}^2-\alpha\beta_{11}\beta_{21})x^2+(\beta_{32}^2-\alpha\beta_{12}\beta_{22})y^2\\ &\quad+(\beta_{33}^2-\alpha\beta_{13}\beta_{23})z^2+[2\beta_{31}\beta_{32}-\alpha(\beta_{11}\beta_{22}+\beta_{12}\beta_{21})]xy\\ &\quad+[2\beta_{32}\beta_{33}-\alpha(\beta_{12}\beta_{23}+\beta_{13}\beta_{22})]yz+[2\beta_{31}\beta_{33}-\alpha(\beta_{11}\beta_{23}+\beta_{13}\beta_{21})]zx.\end{aligned}$$

Since

$$\{a,b\}=-\alpha ab+c^2,\quad \{b,c\}=-\alpha bc+a^2,\quad \{c,a\}=-\alpha ca+b^2. \tag{4.3}$$

we can write 18 relations between the coefficients of the matrix B:

$$\beta_{ij}^2=\alpha\beta_{i+1j}\beta_{i+2j}+(\beta_{i+1j+1}\beta_{i+2j+2}-\beta_{i+1j+2}\beta_{i+2j+1}) \tag{4.4}$$

and

$$\beta_{ij}\beta_{ij+1}=\alpha\beta_{i+1j+1}\beta_{i+2j} \tag{4.5}$$

for all i and j modulo 3.

Denote by b^j the product of all elements of the jth column of B. Then (4.5) gives

$$b^jb^{j+1}=\alpha^3b^{j+1}b^j$$

So $b^jb^{j+1}=0$ since $\alpha^3\neq 1$ and at least two columns contain zero elements. Therefore the matrix B contains zero entries.

If $\alpha=0$ then $\beta_{ij}\beta_{ij+1}=0$ and every row contains two zero entries. Of course, a row should contain a non-zero element since the determinant of B is not zero.

Assume now that $\alpha\neq 0$. If in some row we have three non-zero elements we can check using (4.5) that all entries of B are not equal to zero. Indeed, if $\beta_{ij}\beta_{ij+1}\neq 0$ then $\beta_{i+1j+1}\beta_{i+2j}\neq 0$ and using this inequality for $j=1,2,3$ we fill all the matrix with non-zero entries. So each row has at least one zero element.

Suppose that every row contains just one zero. Up to renumbering we can assume that $\beta_{11}\beta_{12}\neq 0$ and $\beta_{13}=0$. Then $\beta_{22}\beta_{31}\neq 0$. Now, $\beta_{22}\beta_{23}=\alpha\beta_{33}\beta_{12}$, and $\beta_{21}\beta_{22}=\alpha\beta_{32}\beta_{11}$ in force of (4.5). If $\beta_{23}=0$ then $\beta_{33}=0$ which is impossible since B has a non-zero determinant. Hence $\beta_{23}\neq 0$, $\beta_{33}\neq 0$ and $\beta_{21}=0$, $\beta_{32}=0$.

But $\beta_{11}\beta_{12} = \alpha\beta_{22}\beta_{31}$, $\beta_{22}\beta_{23} = \alpha\beta_{33}\beta_{12}$, and $\beta_{33}\beta_{31} = \alpha\beta_{11}\beta_{23}$. If we multiply these equalities we will get $\alpha^3 = 1$. So there exists a row with two zeros.

If there is a row with two non-zero entries we can assume that $\beta_{11}\beta_{12} \neq 0$. Then $\beta_{22}\beta_{31} \neq 0$ and it is easy to check that any additional non-zero element in the second or third row will lead to two non-zero entries in these rows. But then the determinant of B is zero which is impossible. So every row (and every column) contains just one non-zero element.

Let us assume that $\beta_{11} \neq 0$. Then the relations (4.4) give $\beta_{11}^2 = (\beta_{22}\beta_{33} - \beta_{23}\beta_{32})$ and either $\beta_{22}\beta_{33} = 0$ or $\beta_{23}\beta_{32} = 0$. If $\beta_{11}^2 = -\beta_{23}\beta_{32}$, then $\beta_{23}^2 = -\beta_{32}\beta_{11}$ and $\beta_{32}^2 = -\beta_{11}\beta_{23}$ and we obtain a contradiction by multiplying these equalities. So $\beta_{11}^2 = \beta_{22}\beta_{33}$, $\beta_{22}^2 = \beta_{33}\beta_{11}$, and $\beta_{33}^2 = \beta_{11}\beta_{22}$. We have three solutions to these equations: $\beta_{11} = \beta_{22} = \beta_{33}$, $\beta_{22} = \epsilon\beta_{11}, \beta_{33} = \epsilon^2\beta_{11}$, and $\beta_{22} = \epsilon^2\beta_{11}, \beta_{33} = \epsilon\beta_{11}$. They correspond to the cases $\psi = \varphi_{\beta_{11}}$, $\psi = \sigma\varphi_{\beta_{11}}$, and $\psi = \sigma^2\varphi_{\beta_{11}}$, respectively. Remaining two solutions with $\beta_{12}\beta_{23}\beta_{31} \neq 0$ and $\beta_{13}\beta_{23}\beta_{31} \neq 0$ correspond to additional actions by τ. So, this gives the statement of the theorem. □

Recall that a derivation D of an algebra R is called *locally nilpotent* if for every $a \in R$ there exists a natural number $m = m(a)$ such that $D^m(a) = 0$. If R is generated by a finite set of elements $a_1, a_2, \dots, a_n$ then it is well known (see, for example [**5**]) that a derivation D is locally nilpotent if and only if there exist positive integers m_i, where $1 \leq i \leq n$, such that $D^{m_i}(a_i) = 0$.

COROLLARY 4.3. *Algebra E_α does not have any nonzero locally nilpotent derivations.*

Proof. Let D be a nonzero locally nilpotent derivation of E_α. For every $c \in Z(E_\alpha)$ and $f \in E_\alpha$ we have

$$0 = D\{c, f\} = \{D(c), f\} + \{c, D(f)\} = \{D(c), f\}.$$

So $D(c) \in Z(E_\alpha)$, i.e., $Z(E_\alpha)$ is invariant under the action of D. Therefore D induces a locally nilpotent derivation of $Z(E_\alpha) = k[C]$ and $D(C) = \beta \in k$ (see, for example [**6**]). Since

$$D(C) = D(x)\frac{\partial C}{\partial x} + D(y)\frac{\partial C}{\partial y} + D(z)\frac{\partial C}{\partial z} = \beta$$

and C is homogeneous of degree three, $\beta = 0$.

So $D(C) = 0$ and CD is also a locally nilpotent derivation. For any locally nilpotent derivation its exponent is an automorphism (see, for example [**6**]). So $\exp(CD)$ is an automorphism which is certainly not linear. This contradicts Lemma 4.1. □

Usually it is much easier to describe the locally nilpotent derivations than the automorphisms. But at the moment we do not know any direct proof of Corollary 4.3.

By Theorem 4.2, the automorphism group $\operatorname{Aut} E_\alpha$ of the algebra E_α is generated by $\varphi_\gamma (\gamma \in k^*)$, σ, and τ when $\alpha^3 \neq 1$ and $\epsilon \in k$. Note that $\langle \varphi_\gamma \rangle = \{\varphi_\gamma | \gamma \in k^*\} \cong k^*$, $\langle \tau \rangle \cong Z_3$, and $\langle \sigma \rangle \cong Z_3$, where Z_3 is the cyclic group of order 3. The automorphisms φ_γ, where $\gamma \in k^*$, are related by

$$\varphi_{\gamma_1}\varphi_{\gamma_2} = \varphi_{\gamma_1\gamma_2}. \tag{4.6}$$

Moreover, it is easy to check that

$$\sigma\varphi_\gamma = \varphi_\gamma\sigma, \tau\varphi_\gamma = \varphi_\gamma\tau, \tag{4.7}$$

i.e., φ_γ belongs to the center of $\operatorname{Aut} E_\alpha$.

In addition,

$$\sigma^3 = \mathrm{id}, \tau^3 = \mathrm{id}, \sigma\tau = \varphi_\epsilon \tau\sigma. \tag{4.8}$$

THEOREM 4.4. *Let k be an arbitrary field of characteristic 0 such that the equation $\lambda^2 + \lambda + 1 = 0$ has a solution in k. Then,*

$$\operatorname{Aut} E_\alpha = (\langle\varphi_\gamma\rangle \times \langle\tau\rangle) \rtimes \langle\sigma\rangle = (\langle\varphi_\gamma\rangle \times \langle\sigma\rangle) \rtimes \langle\tau\rangle \cong (k^* \times Z_3) \rtimes Z_3.$$

if $\alpha^3 \neq 1$

Proof. Using (4.6), (4.7), and (4.8), any automorphism $\psi \in \operatorname{Aut} E_\alpha$ can be written as

$$\psi = \varphi_\gamma \tau^i \sigma^j$$

where $\gamma \in k^*$ and $0 \leq i, j \leq 2$. It is easy to check that this representation is unique. The relations (4.6)-(4.8) show that $\langle\varphi_\gamma\rangle \times \langle\tau\rangle$ and $\langle\varphi_\gamma\rangle \times \langle\sigma\rangle$ are normal subgroups of $\operatorname{Aut} E_\alpha$. □

COROLLARY 4.5. *The relations (4.6)-(4.8) are defining relations of the group $\operatorname{Aut} E_\alpha$ with respect to the generators φ_γ, σ, and τ if $\alpha^3 \neq 1$.*

COROLLARY 4.6. *Let k be an arbitrary field of characteristic 0 such that the equation $\lambda^2 + \lambda + 1 = 0$ has no solution in k. If $\alpha^3 \neq 1$ (i.e. $\alpha \neq 1$) then*

$$\operatorname{Aut} E_\alpha \cong k^* \times Z_3.$$

When $\alpha^3 = 1$ the curve C is not elliptic and degenerates into a product of three linear factors if $\epsilon \in k$ (see (3.2)). Let us introduce new variables $u = x + y + \alpha z$, $v = \epsilon x + \epsilon^2 y + \alpha z$, $w = \epsilon^2 x + \epsilon y + \alpha z$. Then E_α is the polynomial algebra $k[u, v, w]$ endowed with the Poisson bracket defined by

$$\{u, v\} = \mu uv, \ \{v, w\} = \mu vw, \ \{w, u\} = \mu wu,$$

where $\mu = \alpha(\epsilon^2 - \epsilon)$. It is easy to check that the maps ψ_γ $(\gamma \in k^*)$ and σ' defined by

$$\psi_\gamma(u) = \gamma u, \ \ \psi_\gamma(v) = v, \ \ \psi_\gamma(w) = w$$

and

$$\sigma'(u) = v, \ \ \sigma'(v) = w, \ \ \sigma'(w) = u$$

are automorphisms of E_α. Note that $\sigma = \epsilon^2\sigma'$. Immediate calculations as in the proofs of Theorems 4.2 and 4.4 give

THEOREM 4.7. *If $\alpha^3 = 1$ and $\epsilon \in k$ is a root of the equation $\lambda^2 + \lambda + 1 = 0$ then ψ_γ $(\gamma \in k^*)$ and σ' generate $\operatorname{Aut} E_\alpha$ and*

$$\operatorname{Aut} E_\alpha \cong (k^*)^3 \rtimes Z_3.$$

If $\alpha^3 = 1$ and $\epsilon \notin k$ then $\alpha = 1$. In this case we cannot use variables u, v, w. But we can consider the algebra $E'_\alpha = E_\alpha \otimes_k k'$ where $k' = k + \epsilon k$ is an extension of k. Every automorphism of E_α can be extended to an automorphism of E'_α and Theorem 4.7 gives the next result.

COROLLARY 4.8. *If $\alpha^3 = 1$ and $\epsilon \notin k$ then the group of automorphisms of E_α consist of the automorphisms given by*

$$x \mapsto \mu_1 x + \mu_2 y + \mu_3 z, \quad y \mapsto \mu_1 y + \mu_2 z + \mu_3 x, \quad z \mapsto \mu_1 z + \mu_2 x + \mu_3 y,$$

where $\mu_1^3 + \mu_2^3 + \mu_3^3 - 3\mu_1\mu_2\mu_3 \neq 0$.

The crucial difference of the case when $\epsilon \notin k$ is the absence of the automorphism σ, just as in the Corollary 4.6.

References

[1] A. Belov-Kanel, M. Kontsevich, Automorphisms of the Weyl Algebra, Letters in Mathematical Physics, 74 (2005), 181–199.

[2] A. J. Czerniakiewicz, Automorphisms of a free associative algebra of rank 2, I, II, Trans. Amer. Math. Soc., 160 (1971), 393–401; 171 (1972), 309–315.

[3] J. Donin and L. Makar-Limanov, Quantization of quadratic Poisson brackets on a polynomial algebra of three variables, J. Pure Appl. Algebra, 129 (1998), no. 3, 247–261.

[4] J. Dufour et A. Haraki, Rotationnels et structures de Poisson quadratiques, C. R. Acad. Sci. Paris, V. 312 I (1991), 137-140.

[5] A. van den Essen, Polynomial automorphisms and the Jacobian conjecture, Progress in Mathematics, 190, Birkhauser verlag, Basel, 2000.

[6] G. Freudenburg, Algebraic theory of locally nilpotent derivations, Springer-Verlag, Berlin, 2006.

[7] D. Husemöller, Elliptic Curves, 2nd edition, Springer-Verlag, New York, 2004.

[8] H. W. E. Jung, Über ganze birationale Transformationen der Ebene, J. reine angew. Math., 184 (1942), 161–174.

[9] W. van der Kulk, On polynomial rings in two variables, Nieuw Archief voor Wiskunde, (3)1 (1953), 33–41.

[10] Z. Liu and P. Xu, On quadratic Poisson structures, Letters in Math. Phys., 26 (1992), 33-42.

[11] L. Makar-Limanov, The automorphisms of the free algebra with two generators, Funksional. Anal. i Prilozhen. 4(1970), no.3, 107-108; English translation: in Functional Anal. Appl. 4 (1970), 262–263.

[12] L. Makar-Limanov, U. U. Umirbaev, Centralizers in free Poisson algebras, Proc. Amer. Math. Soc. 135 (2007), no. 7, 1969–1975.

[13] L. Makar-Limanov, U. Turusbekova, U. Umirbaev, Automorphisms and derivations of free Poisson algebras in two variables, J. Algebra (2008), doi:10.1016/j.jalgebra.2008.01.005

[14] M. Nagata, On the automorphism group of $k[x, y]$, Lect. in Math., Kyoto Univ., Kinokuniya, Tokio, 1972.

[15] A. V. Odesskii, Elliptic algebras, Russian Math. Surveys, 57 (2002), no. 6, 1127–1162.

[16] I. P. Shestakov and U. U. Umirbaev, Poisson brackets and two generated subalgebras of rings of polynomials, Journal of the American Mathematical Society, 17 (2004), 181–196.

[17] I. P. Shestakov and U. U. Umirbaev, Tame and wild automorphisms of rings of polynomials in three variables, Journal of the American Mathematical Society, 17 (2004), 197–227.

[18] U. U. Umirbaev, The Anick automorphism of free associative algebras, J. Reine Angew. Math. 605 (2007), 165–178.

DEPARTMENT OF MATHEMATICS & COMPUTER SCIENCE, BAR-ILAN UNIVERSITY, 52900 RAMAT-GAN, ISRAEL AND DEPARTMENT OF MATHEMATICS, WAYNE STATE UNIVERSITY, DETROIT, MI 48202, USA

E-mail address: `lml@math.wayne.edu`

DEPARTMENT OF MATHEMATICS, EURASIAN NATIONAL UNIVERSITY, ASTANA, 010008, KAZAKHSTAN

E-mail address: `umut.math@mail.ru`

DEPARTMENT OF MATHEMATICS, EURASIAN NATIONAL UNIVERSITY, ASTANA, 010008, KAZAKHSTAN AND DEPARTMENT OF MATHEMATICS, WAYNE STATE UNIVERSITY, DETROIT, MI 48202, USA

E-mail address: `umirbaev@math.wayne.edu`

Contemporary Mathematics
Volume **483**, 2009

JORDAN SUPERALGEBRAS AND THEIR REPRESENTATIONS

Consuelo Martínez and Efim Zelmanov

This paper is dedicated to Ivan Shestakov in his 60th. birthday.

ABSTRACT. This is a survey of representation theory of finite dimensional simple Jordan superalgebras over an algebraically closed field of zero characteristic.

INTRODUCTION

The theory of bimodules over semisimple finite dimensional Jordan algebras was developed by N. Jacobson [**J**]. Here we survey the efforts to extend it to Jordan superalgebras.

Sections 2, 3 contain basic definitions, examples and constructions. In the section 4 we discuss superalgebras of rank (maximal number of pairwise orthogonal idempotents in the even part) ≥ 3. This case is similar to that of Jordan algebras. Much more interesting is the case of rank 2, since it leads to new phenomena and new parametric families of bimodules. These superalgebras are discussed, case by case, in the sections 4, 5 and 6. In particular, the parametric family of irreducible bimodules over the superalgebra $JP(2)$ is related to its embedding into the Cheng-Kac superalgebra, which is a Jordan counterpart of a certain exceptional superconformal algebra (see [**ChK**], [**GLS**]).

Some cases, $JP(n)$ and $Q(n)^{(+)}$, $n \geq 3$ are done by an extension of Jacobson's coordinatization theorem and classification of related alternative bimodules. All other cases are done by lifting Jordan bimodules to Lie modules over the Tits-Kantor-Koecher Lie superalgebra of J and a careful analysis of possible weights.

1. Basic Definitions and Examples

Throughout this paper all algebras are considered over a ground field F that is algebraically closed and has zero characteristic.

1991 *Mathematics Subject Classification.* Primary 17C70 ; Secondary 17C55, 17B10, 17B60.
Key words and phrases. Jordan superalgebras, representations.
The first author was partially supported by MTM 2007-67884-C04-01 and FICYT IB05-186.
The second author was partially supported by the NSF.

A (linear) Jordan algebra is a vector space J with a binary bilinear operation $(x,y) \to xy$ satisfying the following identities:

$xy = yx$

$(x^2y)x = x^2(yx)$

If V is an F-vector space of countable dimension, then $G = G(V)$ denotes the Grassmann (or exterior) algebra over V, that is, the quotient of the tensor algebra over the ideal generated by the symmetric tensors. This algebra $G(V)$ is $\mathbb{Z}/2\mathbb{Z}$-graded, $G(V) = G(V)_{\bar{0}} + G(V)_{\bar{1}}$. Its even part $G(V)_{\bar{0}}$ is the linear span of all tensors of even length and the odd part $G(V)_{\bar{1}}$ is the linear span of all tensors of odd length.

If $\mathcal{V}$ is a variety of algebras defined by homogeneous identities (see [**J**], [**ZSSS**]), a superalgebra $A = A_{\bar{0}} + A_{\bar{1}}$ is a $\mathcal{V}$- superalgebra if its *Grassmann enveloping algebra* $G(A) = A_{\bar{0}} \otimes G(V)_{\bar{0}} + A_{\bar{1}} \otimes G(V)_{\bar{1}}$ lies in $\mathcal{V}$.

Given an element $a \in A_{\bar{0}} \cup A_{\bar{1}}$, $|a|$ will denote its parity (0 or 1).

Thus, a Jordan superalgebra is a $\mathbb{Z}/2\mathbb{Z}$-graded algebra $J = J_{\bar{0}} + J_{\bar{1}}$ satisfying the graded identities

$$xy = (-1)^{|x||y|}yx,$$

and

$$((xy)z)t + (-1)^{|y||z|+|y||t|+|z||t|}((xt)z)y + (-1)^{|x||y|+|x||z|+|x||t|+|z||t|}((yt)z)x = (xy)(zt) + (-1)^{|y||z|}(xz)(yt) + (-1)^{|t|(|y|+|z|)}(xt)(yz).$$

We will denote by $\{x,y,z\} = (xy)z + x(yz) - (-1)^{|x||y|}y(xz)$ the Jordan triple product.

Examples

I) $A = M_{m+n}(F)$, $A_{\bar{0}} = \begin{pmatrix} \star & 0 \\ 0 & \star \end{pmatrix}$, $A_{\bar{1}} = \begin{pmatrix} 0 & \star \\ \star & 0 \end{pmatrix}$ and

II)

$$A = Q(n) = \begin{pmatrix} a & b \\ b & a \end{pmatrix} \mid a,b \in M_n(F) \tag{1.1}$$

are associative superalgebras.

C.T.C. Wall [**W**] proved that every associative simple finite-dimensional superalgebra over the algebraically closed field F is isomorphic to one of them.

III) Let A be an associative (super)algebra. The new operation $a \cdot b = \frac{1}{2}(ab + (-1)^{|a||b|}ba)$ defines a structure of a Jordan (super)algebra on A. We will denote this Jordan (super)algebra as $A^{(+)}$. Those Jordan superalgebras that can be obtained as subalgebras of a superalgebra $A^{(+)}$, with A an associative superalgebra, are called special and exceptional otherwise. The Lie superalgebra $(A, [a,b] = ab - (-1)^{|a||b|}ba)$ is denoted as $A^{(-)}$.

In this way we get the first examples of Jordan simple finite dimensional superalgebras, applying III) to the associative superalgebras I) and II).

1) $M_{m+n}^{(+)}(F)$, $m \geq 1$, $n \geq 1$;

2) $Q(n)^{(+)}$, $n \geq 2$;

If A is an associative superalgebra and $\star : A \to A$ is a superinvolution, that is, $(a^\star)^\star = a$, $(ab)^\star = (-1)^{|a||b|} b^\star a^\star$, then the set of symmetric elements $H(A, \star)$ is a (Jordan) subsuperalgebra of $A^{(+)}$. Similarly the set of skewsymmetric elements $Skew(A, \star)$ is a Lie subsuperalgebra of $A^{(-)}$.

The following two subalgebras of $M^{(+)}_{m+n}$ are of this type

3) Let I_n, I_m be the identity matrices, t the transposition and $U = -U^t = -U^{-1} = \begin{pmatrix} 0 & -I_m \\ I_m & 0 \end{pmatrix}$. Then $\star : M_{n+2m}(F) \to M_{n+2m}(F)$ given by

$$\begin{pmatrix} a & b \\ c & d \end{pmatrix}^\star = \begin{pmatrix} I_n & 0 \\ 0 & U \end{pmatrix} \begin{pmatrix} a^t & -c^t \\ b^t & d^t \end{pmatrix} \begin{pmatrix} I_n & 0 \\ 0 & U^{-1} \end{pmatrix}$$

is a superinvolution.

We will refer to $osp_{n,2m}(F) = Skew(M_{n+2m}(F), \star)$ and $Josp_{n,2m}(F) = H(M_{n+2m}(F), \star)$ as the Lie and the Jordan orthosymplectic superalgebras respectively.

4) The associative superalgebra $M_{n+n}(F)$ has another superinvolution:

$$\begin{pmatrix} a & b \\ c & d \end{pmatrix}^\sigma = \begin{pmatrix} d^t & -b^t \\ c^t & a^t \end{pmatrix}.$$

The Lie superalgebra of skewsymmetric elements and the Jordan superalgebra of symmetric elements are denoted by $P_n(F)$ and $JP_n(F)$ respectively.

5) The 3-dimensional Kaplansky superalgebra, $K_3 = Fe + (Fx + Fy)$, with the multiplication $e^2 = e$, $\quad ex = \frac{1}{2}x, ey = \frac{1}{2}y, \quad [x, y] = e$ is not unital.

6) The 1-parametric family of 4-dimensional superalgebras D_t is defined as $D_t = (Fe_1 + Fe_2) + (Fx + Fy)$ with the product: $e_i^2 = e_i, e_1 e_2 = 0, e_i x = \frac{1}{2}x, e_i y = \frac{1}{2}y, xy = e_1 + te_2$, $t \in F$, $i = 1, 2$.

The superalgebra D_t is simple if $t \neq 0$. In the case $t = -1$, the superalgebra D_{-1} is isomorphic to $M_{1+1}(F)^{(+)}$.

7) Let $V = V_{\bar{0}} + V_{\bar{1}}$ be a $\mathbb{Z}/2\mathbb{Z}$-graded vector space with a superform $(\,|\,) : V \times V \to F$ which is symmetric on $V_{\bar{0}}$, skewsymmetric in $V_{\bar{1}}$ and $(V_{\bar{0}}|V_{\bar{1}}) = (0) = (V_{\bar{1}}|V_{\bar{0}})$.

The superalgebra $J = F1 + V = (F1 + V_{\bar{0}}) + V_{\bar{1}}$ is Jordan.

8) V. Kac introduced the 10-dimensional superalgebra K_{10} that is related (via the Tits-Kantor- Koecher construction) to the exceptional 40-dimensional Lie superalgebra. It was proved in [**MeZ**] that this superalgebra is not a homomorphic image of a special Jordan superalgebra.

9) Let R be an associative commutative superalgebra equipped with a bilinear map $[,] : R \times R \to R$, $[R_{\bar{i}}, R_{\bar{j}}] \subseteq R_{\overline{i+j}}$. A Kantor double is a direct sum of vector spaces $J(R, [,]) = R + vR$, $|v| = 1$, with the product

$$a(vb) = (-1)^{|a|} vab, \ (vb)a = v(ba), \ va.vb = (-1)^{|a|}[a, b].$$

We say that $[,]$ is a Jordan bracket if J is a Jordan superalgebra (see [**KM**]). I. L. Kantor [**Kn**] showed that every Poisson bracket is Jordan. In particular, let V be an n-dimensional vector space with a basis $\xi_1, \ldots, \xi_n; n \geq 2$. Consider the Poisson bracket on the Grassmann superalgebra $G_n = G(V)$, $[f, g] = \sum_{i=1}^n (-1)^{|f|} \frac{\partial f}{\partial \xi_i} \frac{\partial g}{\partial \xi_i}$.

The 2^{n+1}-dimensional Jordan superalgebra $Kan(n) = J(G(V), [,])$ is simple and has a nonsemisimple even part.

V. Kac [**K2**] (see also I. L. Kantor [**Kn**]) proved that every simple finite dimensional Jordan superalgebra over F is isomorphic to one of the superalgebras $M_{n+m}(F)^{(+)}$, $Q_n(F)^{(+)}$, $Josp_{n,2m}(F)$, $JP_n(F)$, a superalgebra of a superform, K_3, D_t, K_{10} or a Kantor superalgebra $Kan(n)$.

Let R be an associative commutative superalgebra with an even derivation $d : R \to R$. In [**MZ1**] we introduced a family of Jordan superalgebras $JKC(R, d)$, which are free R-modules of rank 8. The superalgebras $JCK(F[t^{-1}, t], d/dt)$ are related via the Tits-Kantor-Koecher construction to the exceptional superconformal algebras CK_6 (see [**ChK**], [**GLS**]).

A classification of simple finite dimensional Jordan superalgebras with semisimple even parts over an algebraically closed field of characteristic $p > 2$ was done in [**RZ**]: every such superalgebra is an analog of a simple finite dimensional Jordan superalgebra over a field of zero characteristic plus some new examples that appear in characteristic 3.

In [**MZ1**] it is shown that if the even part is not semisimple, then the only new examples are Kantor doubles $J(O_n \otimes G_m, [,])$ and Cheng-Kac superalgebras $JKC(O_n \otimes G_m, d)$, where $O_n = F[t_1, \ldots, t_n | t_i^p = 0, 1 \leq i \leq n]$ is the algebra of truncated polynomials; $[,]$ is a Jordan bracket, d is an even derivation.

A semisimple finite dimensional Jordan superalgebra is not necessarily a direct sum of simple ones. For the structure of semisimple Jordan superalgebras see [**Z**].

Nicoleta Cantarini and V. Kac [**CK**] classified infinite dimensional irreducible locally compact Jordan superalgebras.

2. Generalities on Jordan bimodules

Let $\mathcal{V}$ be a variety of F-algebras and let J be a $\mathcal{V}$-superalgebra. A $\mathbb{Z}/2\mathbb{Z}$-graded vector space V with operations $V \times J \to V$, $J \times V \to V$ is said to be a $\mathcal{V}$- bimodule over J if the split null extension $V + J$ is a $\mathcal{V}$- superalgebra. Recall that the split null extension is the direct sum of vector spaces $V + J$ with the operation that extends the multiplication of J and the action of J on V while the product of two arbitrary elements in V is zero.

From now on let J be a Jordan superalgebra.

Given an arbitrary set X, there is a unique free J-bimodule $V(X)$ over the set of free generators X. If V' is a J-bimodule, then an arbitrary map $X \to V'$ uniquely extends to a homomorphism of bimodules $V(X) \to V'$.

Let X be a set consisting of one element. For an element $a \in J$ let $R_{V(X)}(a) : V(X) \to V(X)$, $v \to va$.

The subalgebra $U(J)$ of the algebra of all linear transformations of $V(X)$ generated by the operators $R_{V(X)}(a)$, $a \in J$, is called the multiplicative enveloping algebra of J.

Every Jordan bimodule over J is a $U(J)$-right module and viceversa.

A bimodule V over J is said to be one-sided if $\{J, V, J\} = (0)$. In this case, the mapping $a \to R_V(a) \in End_F(V)$ is a homomorphism of J into $End_F(V)^{(+)}$.

Let e be the identity element of the superalgebra J. The subbimodule $W(X) = \{e, V(X), 1-e\} + \{1-e, V(X), 1-e\}$ is the universal one-sided bimodule over J. The associative subalgebra $S(J)$ of $End_F(W(X))$ generated b $R_{W(X)}(J)$ is the universal (associative) enveloping algebra of J (see [**J**]).

The subbimodule $\{e, V(X), e\}$ is the universal unital J-bimodule. The subalgebra $U_1(J)$ of $End_F(\{e, V(X), e\})$ generated by multiplications by J is called the universal unital multiplicative enveloping algebra of J (see [**J**]).

As in [**J**], [Th. 15, p. 103], we have $U(J) \simeq U_1(J) \oplus S(J)$.

If $J = J' \oplus J''$ then $U_1(J) \simeq U_1(J') \oplus U_1(J'') \oplus (U_1(J') \otimes U_1(J''))$ (see [J]).

N. Jacobson [**J**] showed that for a finite dimensional Jordan algebra J the universal multiplicative enveloping algebra $U(J)$ is also finite dimensional. If J is semisimple, then $U(J)$ is also semisimple. Hence, every bimodule over J is completely reducible.

If $V = V_{\bar{0}} + V_{\bar{1}}$ is a $\mathcal{V}$-bimodule over a $\mathcal{V}$-superalgebra A, then the bimodule $V^{op} = V_{\bar{1}}^{op} + V_{\bar{0}}^{op}$, where the parity of the subspace $V_{\bar{i}}^{op}$ is different from $\bar{i}$ and the action of A is defined via

$$av^{op} = (-1)^{|a|}(av)^{op}, \quad v^{op}a = (va)^{op}$$

for arbitrary $a \in A$, $v \in V$ is also a $\mathcal{V}$-bimodule, which is denoted by V^{op} and is called the opposite of the bimodule V.

Finally, let's recall some constructions relating Lie and Jordan algebras.

DEFINITION 2.1. ([**L**]) A Jordan (super)pair $P = (P^-, P^+)$ is a pair of vector (super)spaces with a pair of trilinear operations

$$\{\ ,\ ,\}: P^- \times P^+ \times P^- \to P^-, \qquad \{\ ,\ ,\}: P^+ \times P^- \times P^+ \to P^+$$

that satisfies the following identities:

(P.1) $\{x^\sigma, y^{-\sigma}, \{x^\sigma, z^{-\sigma}, x^\sigma\}\} = \{x^\sigma, \{y^{-\sigma}, x^\sigma, z^{-\sigma}\}, x^\sigma\}$,

(P.2) $\{\{x^\sigma, y^{-\sigma}, x^\sigma\}, y^{-\sigma}, u^\sigma\} = \{x^\sigma, \{y^{-\sigma}, x^\sigma, y^{-\sigma}\}, u^\sigma\}$,

(P.3) $\{\{x^\sigma, y^{-\sigma}, x^\sigma\}, z^{-\sigma}, \{x^\sigma, y^{-\sigma}, x^\sigma\}\} =$
$\{x^\sigma, \{y^{-\sigma}, \{x^\sigma, z^{-\sigma}, x^\sigma\}, y^{-\sigma}\}, x^\sigma\}$,

for every $x^\sigma, u^\sigma \in P^\sigma$, $y^{-\sigma}, z^{-\sigma} \in P^{-\sigma}$, $\sigma = \pm$.

Let $L = L_{-1} + L_0 + L_1$ be a $\mathbb{Z}$-graded Lie (super)-algebra. Then (L_{-1}, L_1) is a Jordan (super)pair with respect to the trilinear operations $\{x^\sigma, y^{-\sigma}, z^\sigma\} = [[x^\sigma, y^{-\sigma}], z^\sigma]$; $x^\sigma, z^\sigma \in L_{\sigma 1}$, $y^{-\sigma} \in L_{-\sigma 1}$, $\sigma = \pm$.

For an arbitrary Jordan (super) pair $P = (P^-, P^+)$, there exists a unique Z-graded Lie (super) algebra $K = K_{-1} + K_0 + K_1$ such that $(K_{-1}, K_1) \simeq P$, $K_0 = [K_{-1}, K_1]$ and for every 3-graded Lie (super)algebra $L = L_{-1} + L_0 + L_1$, an arbitrary homomorphism of the Jordan pairs $P \to (L_{-1}, L_1)$ uniquely extends to a homomorphism of Lie (super)algebras $K \to L$.

We will refer to $K = K(P)$ as the Tits-Kantor-Koecher (in short **TKK) construction** of the pair P

If J is a Jordan superalgebra, let's consider J^- and J^+ two copies of J. Then (J^-, J^+) is a Jordan superpair with the trilinear operations defined, in a natural way, via the triple product of J. That is

$$\{x^\sigma, y^{-\sigma}, z^\sigma\} = \{x, y, z\}^\sigma, \ \sigma = \pm.$$

The Lie superalgebra $K = K(J^-, J^+)$ is called the TKK-construction of J.

Let $J = J_{\bar{0}} + J_{\bar{1}}$ be a simple finite- dimensional Jordan superalgebra. Let's consider $L = K(J)$ its TKK-construction.

If V is a Jordan bimodule over J, then the null extension $V + J$ is a Jordan superalgebra, so we can consider its TKK Lie superalgebra $K(V + J) = (V^- + J^-) + [V^- + J^-, V^+ + J^+] + (V^+ + J^+)$.

Denote $K(V) = V^- + [V^-, J^+] + [J^-, V^+] + V^+ \leq K(V + J)$. Then $K(V)$ is a Lie module over the superalgebra $J^- + [J^-, J^+] + J^+$ which is isomorphic to $K(J)$.

Let W be the maximal $K(J)$-submodule, which is contained in $K(V)_0 = [V^-, J^+] + [J^-, V^+]$. Let $\bar{K}(V) = K(V)/W$.

The following two lemmas were proved in [**MZ6**]

Lemma 2.2. *([**MZ6**]) Let J be a unital Jordan (super)algebra and let V_1, V_2 be two unital Jordan J-bimodules. The following assertions are equivalent:*

(1) $V_1 \simeq V_2$,
(2) $K(V_1) \simeq K(V_2)$,
(3) $\bar{K}(V_1) \simeq \bar{K}(V_2)$.

Lemma 2.3. *([**MZ6**]) For a unital Jordan bimodule V over a unital Jordan (super)algebra J, the following assertions are equivalent:*

(1) *V is an irreducible J-bimodule,*
(2) *$\bar{K}(V)$ is an irreducible $K(J)$-module.*

3. Superalgebras of rank ≥ 3

In this section we will discuss Jordan bimodules over finite dimensional simple Jordan superalgebras whose even part contains 3 pairwise orthogonal idempotents. These superalgebras include the Kac superalgebra K_{10} and the superalgebras of the types $JP(n), n \geq 3$; $Q(n)^{(+)}, n \geq 3$; $Josp(n, 2m), n + m \geq 3$; $M_{m+n}^{(+)}, m + n \geq 3$.

This part of the theory is similar to Jacobson's theory for representations of Jordan algebras.

Theorem 3.1. *([**MZ6**]) Let J be a finite dimensional simple Jordan superalgebra, whose even part contains 3 pairwise orthogonal idempotents. Then its universal multiplicative enveloping algebra $U(J)$ is finite dimensional and semisimple.*

Hence every Jordan J-bimodule is completely reducible.

Let's review the constructions of standard bimodules over classical Jordan algebras.

Let R be an associative algebra with 1 equipped with an involution $\star : R \to R$, $J = H(R, \star)$. An arbitrary right module over R is a one-sided bimodule over J. The Jordan algebra J has also two unital bimodules: the regular bimodule $H(R, \star)$ and the space of skewsymmetric elements $K(R, \star) = \{a \in R | a^\star = -a\}$;

$v.a = \frac{1}{2}(av + va)$, $a \in J$, $v \in H(R, \star)$ or $K(R, \star)$. We will refer to these one-sided and unital bimodules as the standard bimodules over $H(R, \star)$.

Now let $J = R^{(+)}$. Then $a \to a \oplus a^\star$ is an embedding of the Jordan algebras $R^{(+)} \to (R \oplus R)^{(+)}$. Hence, every right module over $R \oplus R$ is a one-sided J-bimodule.

Consider another embedding $R^{(+)} \to M_2(R)^{(+)}$, $a \to \begin{pmatrix} a & 0 \\ 0 & a^\star \end{pmatrix}$. The subspaces $\begin{pmatrix} 0 & H(R,\star) \\ 0 & 0 \end{pmatrix}$, $\begin{pmatrix} 0 & K(R,\star) \\ 0 & 0 \end{pmatrix}$, $\begin{pmatrix} 0 & 0 \\ H(R,\star) & 0 \end{pmatrix}$, $\begin{pmatrix} 0 & 0 \\ K(R,\star) & 0 \end{pmatrix}$ are invariant with respect to Jordan multiplications by $\begin{pmatrix} a & 0 \\ 0 & a^\star \end{pmatrix}$, $a \in R$ and therefore are unital J-bimodules. These bimodules and the regular one will be refered to as standard J-bimodules.

As we have remarked in the introduction the space of symmetric elements of an associative superalgebra with respect to a superinvolution is a Jordan superalgebra.

Two series of the simple finite dimensional Jordan superalgebras are of the type: $JP(n)$ and $Josp(m, 2n)$.

If a Jordan superalgebra J is of the type $H(R, \star)$ or $R^{(+)}$, where R is an associative superalgebra, then the constructions of standard bimodules above carry over to similar constructions of Jordan bimodules over J. These analogs and their opposites will be refered to as standard J-bimodules.

THEOREM 3.2. *([**MZ6**], [**MSZ2**]) An arbitrary irreducible Jordan bimodule over $J = JP(n)$, $n \geq 3$ or $Josp(m, 2n)$, $m + n \geq 3$ is one of the standard types.*

The remaining simple finite dimensional Jordan superalgebras of rank ≥ 3 are of the type $R^{(+)}$, where $R = Q(n)$ or $M_{m+n}(F)$. Of them, only $M_{m+n}(F)$, where at least one of m, n is even, is equipped with a superinvolution ([**R**]).

DEFINITION 3.3. A graded linear mapping $\star : R \to R$ on an associative superalgebra is called *pseudoinvolution* if $(ab)^\star = (-1)^{|a||b|} b^\star a^\star$, $(a^\star)^\star = (-1)^{|a|} a$ for arbitrary elements $a, b \in R_{\bar{0}} \cup R_{\bar{1}}$.

EXAMPLE 3.4. The mapping $\begin{pmatrix} a & b \\ b & a \end{pmatrix} \to \begin{pmatrix} a^t & \sqrt{-1}b^t \\ \sqrt{-1}b^t & a^t \end{pmatrix}$ is a pseudoinvolution on $Q(n)$.

EXAMPLE 3.5. The mapping $\begin{pmatrix} a & b \\ c & d \end{pmatrix} \to \begin{pmatrix} a^t & -c^t \\ b^t & d^t \end{pmatrix}$ is a pseudoinvolution on $M_{m+n}(F)$.

Replacing superinvolutions by pseudoinvolutions in the constructions of unital standard bimodules we still get Jordan bimodules, which we will refer to as standard.

THEOREM 3.6. *([**MZ6**]) An arbitrary irreducible Jordan bimodule over $J = Q(n)^{(+)}$, $n \geq 3$ or $M_{m+n}(F)^{(+)}$, $m + n \geq 3$, is of one of the standard types.*

The proof of the similar results for Jordan algebras of rank ≥ 3 is based on Jacobson Coordinatization Theorem.

Recall that orthogonal idempotents e, f of a Jordan algebra J are said to be strongly connected if there exists an element $a \in J$ such that $\{e, a, f\}^2 = e + f$. According to the Coordinatization Theorem [**J**] if $1 = \sum_{i=1}^n e_i$, where $e_1, \dots e_n$ are pairwise orthogonal strongly connected idempotents and $n \geq 3$ then J is isomorphic to the Jordan algebra of Hermitian $n \times n$-matrices over an alternative algebra A with an involution $\star : A \to A$ such that $H(A, \star)$ lies in the associative center of A. Such involutive algebras are called nuclear.

If $n \geq 4$ then the algebra A is associative. Applying this theorem to split extensions, we see that an arbitrary unital Jordan bimodule over $H(A, \star)$ is isomorphic to the bimodule of Hermitian $n \times n$ matrices over a nuclear involutive alternative A-bimodule.

The Coordinatization Theorem can be extended to superalgebras.

THEOREM 3.7. *([**MSZ2**]) Let $J = J_{\bar{0}} + J_{\bar{1}}$ be a unital Jordan superalgebra, $1 = \sum_{i=1}^n e_i$, where $e_1, \dots e_n$ are pairwise orthogonal strongly connected in $J_{\bar{0}}$ idempotents, $n \geq 3$. Then J is isomorphic to the superalgebra of Hermitian $n \times n$ matrices over an alternative superalgebra A with a nuclear superinvolution. If $n > 3$ then the superalgebra A is associative.*

The category of unital Jordan bimodules over $J = H_n(A)$ is equivalent to the category of alternative A-bimodules with a nuclear involution (if $n = 3$) or to the category of involutive associative bimodules (if $n \geq 4$).

This theorem is applicable to two types of simple finite dimensional Jordan superalgebras: $JP(n)$ and $Q(n)^{(+)}$, $n \geq 3$.

The $JP(n)$ is the Jordan superalgebra of Hermitian $n \times n$ matrices over the associative superalgebra $M_{1+1}(F)$ with the superinvolution $\star : \begin{pmatrix} a & b \\ c & d \end{pmatrix} \to \begin{pmatrix} d & -b \\ c & a \end{pmatrix}$.

N. A. Pisarenko [**P**] proved that every alternative unital bimodule over $M_{1+1}(F)$ is associative and completely reducible and the only irreducible $M_{1+1}(F)$-bimodules are the regular bimodule and its opposite. The regular bimodule has two (up to isomorphism) superinvolutions: $\star$ and $-\star$.

The superalgebra $Q(n)^{(+)}$ is the Jordan superalgebra of Hermitian $n \times n$ matrices over the associative superalgebra $A = (Fe + Fu) \oplus (Ff + Fv)$, $e^2 = e$, $eu = ue = u$, $u^2 = e$, $f^2 = f$, $fv = vf = v$, $v^2 = -f$, with the superinvolution $e^\star = f$, $f^\star = e$, $u^2 = v$, $v^2 = u$.

It is not difficult to show that every alternative nuclear involutive bimodule over A is completely reducible and associative and that there are four irreducible involutive A-bimodules.

THEOREM 3.8. *([**MSZ2**]). Let $J = JP(n)$ or $Q(n)^{(+)}$, $n \geq 3$, $\mathrm{char} F > 2$. Then*

(1) Every Jordan bimodule over J is completely reducible.
(2) Every irreducible Jordan J-bimodule is standard.

The Coordinatization Theorem is not applicable to $Josp(m, 2n)$ or to $M_{m+n}(F)^{(+)}$, $m + n \geq 3$.

In order to estimate the number of irreducible unital Jordan bimodules over the superalgebra J of one of the above types, we lift the bimodule to a module over the Lie superalgebra $TKK(J)$ and estimate the number of possible highest weights.

The exceptional Jordan superalgebra K_{10} is also of rank 3. The bimodules over K_{10} were studied by A.S. Shtern.

THEOREM 3.9. *([**Sh1**]) (1) all Jordan bimodules over K_{10} are completely reducible,*

(2) the only irreducible Jordan bimodules over K_{10} are the regular bimodule and its opposite.

This theorem was generalized in [**MZ3**] to a Kronecker factorization: let J be a Jordan superalgebra which contains K_{10}, let e be the identity element of K_{10}. Then $J = \{e, J, e\} \oplus \{1 - e, J, 1 - e\}$. There exists an associative commutative superalgebra S such that $\{e, J, e\} \simeq K_{10} \otimes S$.

4. $Kan(n)$ and $Q(2)^{(+)}$

Recall that the Kantor superalgebras $Kan(n)$ is the Kantor double of the Grassmann superalgebra G_n on n Grassmann variables $\xi_1, \dots, \xi_n$ with the Poisson bracket $[f, g] = \sum_{i=1}^{n} (-1)^{|f|} \frac{\partial f}{\partial \xi_i} \frac{\partial g}{\partial \xi_i}$, $Kan(n) = G_n + vG_n$.

The Tits-Kantor-Koecher Lie superalgebra L of $Kan(n)$ has a one dimensional center spanned by the element $z = [v^-, v^+]$. If V is a finite dimensional irreducible Jordan bimodule over $Kan(n)$ then $\bar{K}(V)$ is an irreducible module over L. The element z acts on $\bar{K}(V)$ as a scalar multiplication by $\alpha \in F$ or equivalently, $R_V(v)^2 = \alpha Id_V$. In this case we say that V is a bimodule of level α.

EXAMPLE 4.1. Consider the associative commutative superalgebra A generated by one even variable t and n Grassmann variables $\xi_1, \dots, \xi_n$. Consider the Jordan bracket on A defined by $[t, \xi_i] = 0$, $[\xi_i, \xi_j] = -\delta_{ij}$, $[\xi_i, 1] = 0$, $[t, 1] = \alpha t$, $\alpha \in F$ (see [**KMZ**]).

In the Kantor double $K(A, [,]) = A + vA$ the subsuperalgebra $G_n + vG_n$ is isomorphic to $Kan(n)$, whereas the subspace $V(\alpha) = tG_n + vtG_n$ is an irreducible unital Jordan bimodule over it of level α. The bimodule $V(0)$ is isomorphic to the regular bimodule.

In [**Sh2**] Shtern studied finite dimensional irreducible Jordan bimodules over $Kan(n)$, $n > 4$,of level 0. He showed that the only ones are the regular bimodule and its opposite.

In fact, every finite dimensional irreducible Jordan bimodule over $Kan(n)$, $n > 4$ is isomorphic to $V(\alpha)$ or $V(\alpha)^{op}$ for some $\alpha \in F$.

Nothing is known about irreducible bimodules over $Kan(n)$, $n = 2, 3, 4$ or about indecomposable $Kan(n)$-bimodules.

Representation theory of $Q(2)^{(+)}$ is similar to that of $Q(n)^{(+)}$, $n \geq 3$, though the case of $Q(2)^{(+)}$ needs a special proof (see [**MZ2**],[**MSZ2**]). Let the characteristic of the ground field be > 3. Then

(i) the universal multiplicative enveloping algebra $S(Q(2)^{(+)})$ is finite dimensional and semisimple,

(ii) Every irreducible Jordan $Q(2)^{(+)}$-bimodule is standard.

5. D(t)-bimodules and K_3-bimodules

Recall that both, the even and the odd parts of the Jordan superalgebra $D(t)$, $t \in F$, are two-dimensional, $D(t)_{\bar{0}} = Fe_1 + Fe_2$, $D(t)_{\bar{1}} = Fx + Fy$, $e_1^2 = e_1$, $e_2^2 = e_2$, $e_1e_2 = 0$, $e_ix = \frac{1}{2}x$, $e_iy = \frac{1}{2}y$, $1 \le i \le 2$, $[x, y] = e_1 + te_2$.

Clearly, $D(-1) \simeq M_{1+1}(F)^{(+)}$, $D(0) \simeq K_3 \oplus F$, $D(1)$ is a Jordan superalgebra of a superform. Therefore, the superalgebras $D(-1), D(1)$ will be considered in different sections.

The universal associative enveloping algebras for K_3 and for $D(t)$, $t \neq -1, 1$ were determined by I. Shestakov in terms of generators and relators.

Let $osp(1,2)$ denote the Lie subsuperalgebra of $M_{1+2}(F)$, which consists of skewsymmetric elements with respect to the orthosymplectic superinvolution. Let x, y be the standard basis of the odd part of $osp(1,2)$. As always $U(osp(1,2))$ denotes the universal associative enveloping algebra of $osp(1,2)$. Let $U^*(osp(1,2))$ be the ideal (of codimension one) of $U(osp(1,2))$ generated by $osp(1,2)$.

THEOREM 5.1. *(I. P. Shestakov* [**S**]*) The universal enveloping algebra $S(K_3)$ is isomorphic to $U^*(osp(1,2))/id([x,y]^2 - [x,y])$.*

REMARK 5.2. The ideal U^* above appeared because we do not assume an identity in the enveloping algebra $S(K_3)$.

If $charF = 0$ then K_3 does not have any nonzero one-sided Jordan bimodules. If $charF = p > 0$ then such bimodules exist (see [**MZ2**]).

THEOREM 5.3. *(I.P. Shestakov* [**S**]*) Let $t \neq -1, 1$. Then the universal associative enveloping algebra of $D(t)$ is isomorphic to $U(osp(1,2))/id([x,y]^2 - (1 + t)[x,y] + t)$.*

COROLLARY 5.4. If $\text{char}F = 0$, then all finite dimensional one-sided Jordan bimodules over $D(t)$, $t \neq -1, 1$ are completely reducible.

Indeed, finite dimensional representations of the Lie superalgebra $osp(1,2)$ are known to be completely reducible (see [**K1**]).

Now we have to determine all one-sided irreducible Jordan bimodules over $D(t)$.

THEOREM 5.5. *(*[**MZ2**]*)*

(1) *If t can not be represented as $\frac{-m}{m+1}$ or $-\frac{m+1}{m}$, where $m \in Z$, $m \ge 1$, then $D(t)$ does not have any nonzero one-sided Jordan bimodule;*
(2) *If $t = \frac{-m}{m+1}$, $m \in Z$, $m \ge 1$, then $D(t)$ has two finite dimensional irreducible one-sided Jordan bimodules, $V_1(t)$ and $V_1(t)^{op}$, dim $V_1(t)_{\bar{0}} = m+1$, dim$V_1(t)_{\bar{1}} = m + 2$;*
(3) *if $t = -\frac{m+1}{m}$, $m \in Z$, $m \ge 1$, then $D(t)$ has two irreducible finite dimensional one sided bimodules, $V_2(t)$ and $V_2(t)^{op}$, dim $V_2(t)_{\bar{0}} = m + 1$, dim$V_2(t)_{\bar{1}} = m$.*

For the structure of the bimodules $V_1(t)$, $V_2(t)$ see [**MZ2**].

Now we turn to unital Jordan D(t)-bimodules.

Assuming $t \neq -1$, the elements

$$E = \frac{2}{t+1}R(x)^2,\ F = -\frac{2}{t+1}R(y)^2,\ H = \frac{-2}{t+1}(R(x)R(y) + R(y)R(x))$$

of the universal multiplicative algebra of $D(t)$ form an sl_2-triple, $[F, H] = 2F$, $[E, H] = -2E$, $[E, F] = H$.

DEFINITION 5.6. For $\sigma \in \{\bar{0}, \bar{1}\}$, $i \in \{0, 1, \frac{1}{2}\}$, $\lambda \in F$ the Verma module $V(\sigma, i, \lambda)$ is defined as a unital Jordan $D(t)$-bimodule, presented by one generator v of parity σ and the relations $ve_1 = iv$, $vy = 0$, $vH = \lambda v$.

REMARK 5.7. $V(\sigma, i, \lambda)^{op} = V(1 - \sigma, i, \lambda)$.

PROPOSITION 5.8. ([**MZ5**])

(1) $V(\sigma, \frac{1}{2}, \lambda) \neq (0)$ for arbitrary $\lambda \in F$;
(2) $V(\sigma, 1, \lambda) \neq (0)$ if and only if $\lambda = \frac{-2}{t+1}$;
(3) $V(\sigma, 0, \lambda) \neq (0)$ if and only if $\lambda = \frac{-2}{t+1}$.

Standard arguments show that every nonzero Verma bimodule $V(\sigma, i, \lambda)$ contains a largest proper subbimodule $M(\sigma, i, \lambda)$. Hence $V(\sigma, i, \lambda)$ has a unique irreducible homomorphic image $Irr(\sigma, i, \lambda) = V(\sigma, i, \lambda)/M(\sigma, i, \lambda)$.

THEOREM 5.9. *([**MZ5**]) Let $t \neq \pm 1$.*

(1) *If t is not of the type $-\frac{m}{m+2}$, $m \geq 0$ or $-\frac{m+2}{m}$, $m \geq 1$, then the only unital finite dimensional irreducible Jordan $D(t)$-bimodules are $Irr(\sigma, \frac{1}{2}, m)$, $m \geq 1$;* (*)
(2) *if $t = -\frac{m+2}{m}$, $m \geq 1$, then add the bimodules $V(\sigma, 1, m)$, $\sigma = \bar{0}$ or $\bar{1}$, to (*);*
(3) *If $t = -\frac{m}{m+2}$, $m \geq 0$, then add the bimodules $V(\sigma, 0, m)$, $\sigma = \bar{0}$ or $\bar{1}$, to (*).*

REMARK 5.10. The bimodule $V(\sigma, \frac{1}{2}, 0)$ is infinite dimensional and irreducible.

Since the case (iii) of the theorem above includes $t = 0$ and $D(0) \simeq K_3 \oplus F$ we get the following:

COROLLARY 5.11. ([**MZ5**]) The only finite dimensional irreducible Jordan bimodules over the (nonunital) Kaplansky superalgebra K_3 are $Irr(\sigma, \frac{1}{2}, m)$, $m \geq 1$ and $Irr(\sigma, 0, 0)$.

Irreducible unital Jordan $D(t)$-bimodules over fields of positive characteristics were treated in [**T1**] and [**T2**].

Let $m \in Z$, $m > 0$. Let V' denote the subbimodule of $V(\sigma, i, m)$, generated by $vR(x)^{2m+1}$. The quotient bimodule $W(\sigma, i, m) = V(\sigma, i, m)/V'$ is finite dimensional. In fact, it is the largest finite dimensional homomorphic image of $V(\sigma, i, m)$.

THEOREM 5.12. *([**MZ5**])*

(1) *Suppose that $t \neq 0, \neq \pm 1$ and t is not of the types $-\frac{m}{m+2}$, $-\frac{m+2}{m}$, $m \in Z$, $m > 0$. Then every finite dimensional unital Jordan bimodule over $D(t)$ is completely reducible;*

(2) *if* $t = -\frac{m+1}{m-1}$ *or* $-\frac{m-1}{m+1}$, $m \geq 2$, *then* $W(\sigma, \frac{1}{2}, m)$, $\sigma = \bar{0}$ *or* $\bar{1}$, *are the only finite dimensional indecomposable Jordan* $D(t)$*-bimodules, which are not irreducible.*

We have an exact sequence

$$(0) \to Irr(1-\sigma, 1, m-1) \to W(\sigma, \frac{1}{2}, m) \to Irr(\sigma, \frac{1}{2}, m) \to (0),$$

which does not split.

6. Bimodules over $M_{1+1}(F)^{(+)}$

Let $F[z_1, z_2]$ be the polynomial algebra and let $A = F[z_1, z_2][a]/id(a^2 + a - z_1 z_2) = F[z_1, z_2] + F[z_1, z_2]a$. Let K be the field of fractions of A. Consider the F-subspaces $M_{12} = F[z_1, z_2] + F[z_1, z_2]a^{-1}z_2$, $M_{21} = F[z_1, z_2]z_1 + F[z_1, z_2]a$ of K.

Then $\begin{pmatrix} A & M_{12} \\ M_{21} & A \end{pmatrix}$ is an F-subsuperalgebra of $M_{1+1}(K)$. The mapping

$$u : \begin{pmatrix} \alpha_{11} & \alpha_{12} \\ \alpha_{21} & \alpha_{22} \end{pmatrix} \to \begin{pmatrix} \alpha_{11} & \alpha_{12} + \alpha_{21}a^{-1}z_2 \\ \alpha_{12}z_1 + \alpha_{21}a & \alpha_{22} \end{pmatrix}$$

is an embedding of Jordan superalgebras $M_{1+1}(F)^{(+)} \to \begin{pmatrix} A & M_{12} \\ M_{21} & A \end{pmatrix}^{(+)}$. The image of u generates $\begin{pmatrix} A & M_{12} \\ M_{21} & A \end{pmatrix}$. Moreover,

THEOREM 6.1. ([**MZ2**]) *The homomorphism* u *is the universal specialization,*

$$S(M_{1+1}(F)^{(+)}) \simeq \begin{pmatrix} A & M_{12} \\ M_{21} & A \end{pmatrix}^{(+)}.$$

Thus, one-sided Jordan bimodules of $M_{1+1}(F)^{(+)}$ *are in 1-1 correspondence with right modules over* $\begin{pmatrix} A & M_{12} \\ M_{21} & A \end{pmatrix}^{(+)}$.

Remark that these bimodules don't need to be completely reducible. Indeed, let I be an ideal of $F[z_1, z_2]$. Then $\tilde{I} = \begin{pmatrix} I + Ia & I + Ia^{-1}z_2 \\ Iz_1 + Ia & I + Ia \end{pmatrix}^{(+)}$ is an ideal of $\begin{pmatrix} A & M_{12} \\ M_{21} & A \end{pmatrix}^{(+)}$. If $F[z_1, z_2]/I$ is finite dimensional and not semisimple, then so is $\begin{pmatrix} A & M_{12} \\ M_{21} & A \end{pmatrix}^{(+)} / \tilde{I}$.

Irreducible modules over the algebra $\begin{pmatrix} A & M_{12} \\ M_{21} & A \end{pmatrix}^{(+)}$ and hence one-sided irreducible Jordan bimodules over $M_{1+1}(F)^{(+)}$, are in 1-1 correspondence with irreducible modules over A.

Now we will describe irreducible unital finite dimensional Jordan bimodules over $J = M_{1+1}(F)^{(+)}$ following [MS].

Let's fix the standard notation: $e_1 = \begin{pmatrix} 1 & 0 \\ 0 & 0 \end{pmatrix}$, $e_2 = \begin{pmatrix} 0 & 0 \\ 0 & 1 \end{pmatrix}$, $x = \begin{pmatrix} 0 & 1 \\ 0 & 0 \end{pmatrix}$, $y = \begin{pmatrix} 0 & 0 \\ 1 & 0 \end{pmatrix}$, $[x, y] = e_1 - e_2$. For arbitrary scalars $\alpha, \beta, \gamma \in F$ we define a 4-dimensional J-bimodule $V = V(\alpha, \beta, \gamma)$ with a base $v, w \in V_{\bar{0}}$, $z, t \in V_{\bar{1}}$ via

$$ve_1 = v, we_1 = 0, ze_1 = \frac{1}{2}z, te_1 = \frac{1}{2}t,$$

$$ve_2 = 0, we_2 = w, ze_2 = \frac{1}{2}z, te_2 = \frac{1}{2}t,$$

$$vx = z, wx = (\gamma - 1)z - 2\alpha t, zx = \alpha v, tx = \frac{1}{2}((\gamma - 1)v - u),$$

$$vy = t, wy = 2\beta z - (\gamma + 1)t, zy = \frac{1}{2}(\gamma + 1)v + \frac{1}{2}w, ty = \beta v.$$

PROPOSITION 6.2. ([**MS**])

(1) $V(\alpha, \beta, \gamma)$ is a unital Jordan bimodule ,

(2) if $\gamma^2 - 1 - 4\alpha\beta \neq 0$, then $V(\alpha, \beta, \gamma)$ is irreducible.

If $\gamma^2 - 1 - 4\alpha\beta = 0$ then $Fw + Fwx$ is the only proper subbimodule of $V(\alpha, \beta, \gamma)$.

In the case (ii) of the Proposition above denote $V_2(\alpha, \beta, \gamma) = Fw + FwJ_{\bar{1}}$, $V_1(\alpha, \beta, \gamma) = V(\alpha, \beta, \gamma)/V_2(\alpha, \beta, \gamma)$.

THEOREM 6.3. *([**MS**]) Every irreducible unital finite dimensional Jordan J-bimodule is isomorphic to $V(\alpha, \beta, \gamma)$ if $\gamma^2 - 1 - 4\alpha\beta \neq 0$ or $V_1(\alpha, \beta, \gamma)$ or $V_2(\alpha, \beta, \gamma)$, $\gamma^2 - 1 - 4\alpha\beta = 0$, or one of the opposites.*

7. Jordan superalgebras of a superform

Let $V = V_{\bar{0}} + V_{\bar{1}}$ be a $Z/2Z$-graded vector space with a nondegenerate supersymmetric bilinear form $<,>: V \times V \to F$. Let $\dim_F V_{\bar{0}} = m$, $\dim_F V_{\bar{1}} = 2n$. Choose a basis $e_1, \ldots, e_m$ in $V_{\bar{0}}$ such that $< e_i, e_j >= \delta_{ij}$.

Then the universal associative enveloping algebra of the Jordan algebra $F1 + V_{\bar{0}}$ is the Clifford algebra $Cl(m) =< 1, e_1, \ldots, e_m | e_i e_j + e_j e_i = 0, \imath \neq j, e_i^2 = 1 >$ (see [J]). The algebra $Cl(m)$ has a natural $Z/2Z$-grading $Cl(m) = Cl(m)_{\bar{0}} + Cl(m)_{\bar{1}}$, $Cl(m)_{\bar{k}} = span(e_{i_1} \cdots e_{i_q}, q \equiv k \bmod 2)$.

In $V_{\bar{1}}$ we can find a basis $v_1, w_1, \ldots, v_n, w_n$ such that $< v_i, w_j >= \delta_{ij}$, $< v_i, v_j >=< w_i, w_j >= 0$. Consider the Weyl algebra $W_n =< 1, x_i, y_i, 1 \leq i \leq n \mid [x_i, y_j] = \delta_{ij}, [x_i, x_j] = [y_i, y_j] = 0 >$. Assuming $x_i, y_i, 1 \leq i \leq n$, to be odd, we make W_n a superalgebra.

It is easy to see that the universal associative enveloping algebra of $F1 + V$ is the Clifford algebra C of V, $C =< 1, V >$, which is isomorphic to the (super)tensor product $Cl(m) \otimes_F W_n$.

If $char F = 0$ then $J = F1 + V$ does not have nonzero one-sided Jordan bimodules unless $n = 0$.

Unital irreducible finite dimensional Jordan bimodules over $J = F1 + V$ have the same description as in the case of Jordan algebras (see [**J**]).

In the Clifford algebra C of V consider the subspace $C_r = \sum_{i\leq r} \underbrace{V\cdots V}_{i}$, $C_0 = F1 \subseteq C_1 \subseteq C_2 \subseteq \cdots; \cap_{r\geq 0} C_r = C$.

If $r < 0$, then we let $C_r = (0)$.

PROPOSITION 7.1. ([**MZ5**]) For every odd $r \geq 1$, C_r/C_{r-2} is a unital irreducible Jordan J-bimodule.

Let u be an even vector, $V' = V + Fu$. We will extend the superform to V' via $< u, u >= 1, < u, V >= (0)$.

PROPOSITION 7.2. ([**MZ5**]) For every even $r \geq 0$, uC_r/uC_{r-2} is a unital irreducible Jordan J-bimodule.

THEOREM 7.3. ([**MZ5**]) *Every unital irreducible finite dimensioanl J-bimo-dule is isomorphic to C_r/C_{r-2}, r is odd or uC_r/uC_{r-2}, r is even.*

8. JP(2)-bimodules

THEOREM 8.1. ([**MZ2**]) *The universal associative enveloping algebra $S(JP(2))$ is isomorphic to $M_{2+2}(F[t])$, where $F[t]$ is the polynomial algebra in one variable.*

Hence indecomposable one-sided finite dimensional Jordan bimodules over $JP(2)$ correspond to Jordan blocks, whereas irreducible one-sided finite dimensional Jordan bimodules are parametrized by scalars $\alpha \in F$ and all have dimension 4.

Let us turn to unital finite-dimensional Jordan bimodules over $J = JP(2)$. The Tits-Kantor-Koecher Lie superalgebra $L = K(J)$ is the universal central cover of the Lie superalgebra $P(3)$. This universal central cover has a 1-dimensional center [MZ4]. Let z be a nonzero central element of L. If V is a unital finite dimensional Jordan bimodule over J, then $\bar{K}(V)$ is an irreducible finite dimensional module over L, hence z acts on $\bar{K}(V)$ as a scalar multiplication by $\alpha \in F$. In this case, we say that V is of the level α.

Fix a nonzero central element z.

THEOREM 8.2. ([**MZ6**]) *For an arbitrary scalar $\alpha \in F$ there are exactly two (up to opposites) nonisomorphic unital irreducible finite dimensional Jordan bimodules over $JP(2)$ of level α.*

We will now present the explicit realizations of these bimodules.

Consider the associative commutative algebra $\Phi = F1 + Ft$, $t^2 = 0$ and its derivation $d : \Phi \to \Phi$, such that $d(t) = \alpha t$.

Let W be the Weyl algebra of the differential algebra (Φ, d),

$$W = \sum_{i\geq 0} \Phi d^i = F[d] + tF[d], dt - td = \alpha t$$

For an arbitrary $k \geq 0$ the subspace $td^kF[d]$ is an ideal of W.

The following embedding of $JP(2)$ into $M_{2+2}(W)^{(+)}$ was described in [**MZ2**] (see also [**MSZ1**]). The Jordan subsuperalgebra of $M_{2+2}(W)^{(+)}$

$$J = \{\begin{pmatrix} a & 0 \\ h & a^t \end{pmatrix} | a, h \in M_2(F), h^t = h\} + F \begin{pmatrix} 0 & 0 & 0 & -1 \\ 0 & 0 & 1 & 0 \\ 0 & d & 0 & 0 \\ -d & 0 & 0 & 0 \end{pmatrix} \tag{8.1}$$

is isomorphic to $JP(2)$.

Denote $x = \begin{pmatrix} 0 & 0 & 0 & -1 \\ 0 & 0 & 1 & 0 \\ 0 & d & 0 & 0 \\ -d & 0 & 0 & 0 \end{pmatrix}$. Then $z = [x^-, x^+]$ is a nonzero central element of $L = K(J)$.

The inner derivation $R(x)^2$ acts on $M_{2+2}(tF[d])$ as the multiplication by α. Hence $M_{2+2}(tF[d])$ is a J-bimodule of level α.

Denote $e = \begin{pmatrix} t & 0 & 0 & 0 \\ 0 & t & 0 & 0 \\ 0 & 0 & t & 0 \\ 0 & 0 & 0 & t \end{pmatrix}$.

Proposition 8.3. $R(\alpha) = e.J$ is an irreducible subbimodule of $M_{2+2}(tF[d])$.

Remark 8.4. $R(0)$ is the regular bimodule.

Consider the J-bimodule $V = M_{2+2}(tF[d]/tdF[d])$, $\dim_F V = 16$. We will identify $R(\alpha)$ with the subbimodule of V generated by $\bar{e} = e + M_{2+2}(tdF[d])$.

Proposition 8.5. The J-bimodule $S(\alpha) = V/R(\alpha)$ is irreducible.

The 8-dimensional J-bimodules $R(\alpha)$ and $S(\alpha)$ are not isomorphic because the L-modules $\bar{K}(R(\alpha))$, $\bar{K}(S(\alpha))$ have different sets of weights with respect to the Cartan subalgebra of L.

Theorem 8.6. *(**[MZ6]**) The only unital irreducible finite dimensional Jordan J-bimodules of level α are $R(\alpha)$, $S(\alpha)$ and their opposites.*

Aknowledgement: The authors are grateful to the referee for the valuable comments.

References

[CK] N. Cantarini and V. G. Kac, *Classification of linearly compact simple Jordan and generalized Poisson superalgebras* J. Algebra **313** (2007), no. 1, 100–124.

[ChK] S.J. Cheng and V.G. Kac, *A new $N = 6$ superconformal algebra*, *Comm. Math. Phys* **186** (1997), n. 1, 219–231.

[GLS] P. Grozman, D. Leites and I. Shchepochkina, *Lie superalgebras of string theories*, Acta Math. Vietnam **26** (2001), n. 1, 27–63.

[J] N. Jacobson, *Structure and Representation of Jordan algebras*, Amer. Math. Soc. Providence, R.I., 1969.

[K1] V. G. Kac, *Lie Superalgebras*, Advances in Math.**26** (1977), no.1, 8–96.

[K2] V.G. Kac, *Classification of simple Z-graded Lie superalgebras and simple Jordan superalgebras*, Comm. in Algebra **5** (1997), no. 13, 1375–1400.

[KMZ] V.G. Kac, C. Martínez and E. Zelmanov, *Graded simple Jordan superalgebras of growth one*, Memoirs of the AMS **150** (2001), 1–140.

[Kn] I.L. Kantor, *Jordan and Lie superalgebras defined by Poisson brackets*, Algebra and Analysis (1989), 55–79. Amer. Math. Soc. Transl. Ser. (2) **151** (1992), 55–79.

[Kp1] I. Kaplansky, *Superalgebras*, Pacific J. of Math. **86** (1980), 93–98.

[Kp2] I. Kaplansky, *Graded Jordan Algebras I*, Preprint.

[KM] D. King and K. McCrimmon, *The Kantor construction of Jordan superalgebras*, Comm. Algebra **20** (1992), no. 1, 109–126.

[L] O. Loos, *Jordan pairs*, Lecture Notes in Mathematics **460**, Springer-Verlag, Berlin-New York, 1975.

[LS] M.C. López-Diaz and I. P. Shestakov, *Representations of exceptional simple Jordan superalgebras of characteristic 3*, Comm. in Algebra **33** (2005), 331–337.

[MS] C. Martínez and I. Shestakov, *Unital irreducible bimodules over* M_{1+1}, Preprint.

[MSZ1] C. Martínez, I. Shestakov and E. Zelmanov, *Jordan algebras defined by brackets*, J. London Math. Soc. (2) **64** (2001), no.2, 357–368.

[MSZ2] C. Martínez, I. Shestakov and E. Zelmanov, *Jordan bimodules over the superalgebras* $Q(n)$, Preprint.

[MZ1] C. Martínez and E. Zelmanov, *Simple finite-dimensional Jordan Superalgebras in prime characteristic* , Journal of Algebra **236** (2001), no.2, 575–629.

[MZ2] C. Martínez and E. Zelmanov, *Specializations of Simple Jordan Superalgebras*, Canad. Math. Bull. **45** (2002), no.4, 653–671.

[MZ3] C. Martínez and E. Zelmanov, *A Kronecker factorizarion for the exceptional Jordan superalgebra*, J. Pure Appl. Algebra **177** (2003), n.1, 71–78.

[MZ4] C. Martínez and E. Zelmanov, *Lie superalgebras graded by* $P(n)$ *and* $Q(n)$, Proc. Natl. Acad. Sci. USA **100** (2003), no.14, 8130–8137.

[MZ5] C. Martínez and E. Zelmanov, *Unital bimodules over the simple Jordan superalgebras* $D(t)$, Trans. Amer. Math. Soc. **358** (2006), no.8, 3637–3649.

[MZ6] C. Martínez and E. Zelmanov, *Representation theory of Jordan superalgebras II*, Preprint.

[MeZ] Y. Medvedev and E. Zelmanov, *Some counterexamples in the theory of Jordan Algebras*, *Nonassociative Algebraic Models*, Nova Science Publish., S. González and H.C. Myung eds., 1992, pp. 1–16.

[P] N.A. Pisarenko, *The structure of alternative superbimodules*, Algebra and Logic **33** (1995), no.6, 386–397.

[R] M. Racine, *Associative superalgebras with superinvolution*, Proceedings of the International Conference on Jordan Structures, Málaga, 1999, pp. 163–168.

[RZ] M. Racine and E. Zelmanov, *Simple Jordan superalgebras with semisimple even part*, J. of Algebra **270** (2003), no.2, 374–444.

[S] I. Shestakov, *Universal enveloping algebras of some Jordan superalgebras*, Personal communication.

[Sh1] A.S. Shtern, *Representation of an exceptional Jordan superalgebra*, Funktzional Anal. i Prilozhen **21** (1987), 93–94.

[Sh2] A.S. Shtern, *Representation of finite-dimensional Jordan superalgebras of Poisson brackets*, Comm. in Algebra **23** (1995), no.5, 1815–1823.

[T1] M. N. Trushina, *Irreducible representations of a certain Jordan superalgebra, J. Algebra Appl.* **4**, no. 1, 1-14, (2005).

[T2] M. N. Trushina, *Modular representations of the Jordan superalgebras* $D(t)$ *and* K_3, to appear in J. of Algebra.

[W] C.T.C. Wall, *Graded Brauer groups*, J. Reine Angew Math. **213** (1964), 187–199.

[Z] E. Zelmanov, *Semisimple finite dimensional Jordan superalgebras*, Lie algebras, Rings and Related Topics; Fong, Mikhalev, Zelmanov (Eds), Springer, HK, 2000, pp. 227–243.

[ZSSS] K. A. Zhevlakov, A. M. Slinko, I. P. Shestakov and A. I. Shirshov, *Rings that are rearly associative*, Academic Press, New York, 1982.

DEPARTAMENTO DE MATEMÁTICAS, UNIVERSIDAD DE OVIEDO, C/ CALVO SOTELO, S/N, 33007 OVIEDO SPAIN

Current address: Departamento de Matemáticas, Universidad de Oviedo, C/ Calvo Sotelo, s/n, 33007 Oviedo SPAIN

E-mail address: `cmartinez@uniovi.es`

DEPARTMENT OF MATHEMATICS, UNIVERSITY OF CALIFORNIA AT SAN DIEGO, 9500 GILMAN DRIVE, LA JOLLA, CA 92093-0112 USA

E-mail address: `ezelmano@math.ucsd.edu`

Contemporary Mathematics
Volume **483**, 2009

A new proof of Itô's theorem*

Kurt Meyberg

To Ivan Shestakov on the occasion of his 60th birthday.

Abstract

In this note we use the Casimir element in finite group algebras to give a direct short proof of Itô's theorem on character degrees.

Keywords: finite group, abelian normal subgroup, irreducible representation, character degree, Casimir element.

1 Introduction and notations

Let G be a finite group, $A \trianglelefteq G$ an abelian normal subgroup, Irr$G$, Irr$A$ the sets of irreducible characters of G and A; and let e_χ, ε_Θ be the central idempotents in the group algebra $\mathbb{C}\,G$, resp. $\mathbb{C}\,A$, belonging to $\chi \in \operatorname{Irr} G$, $\Theta \in \operatorname{Irr} A$,

$$e_\chi = \frac{\chi(e)}{|G|} \sum_{\chi \in \operatorname{Irr} G} \chi(g^{-1})\, g\,.$$

Itô's theorem states that all degrees $\chi(e)$, $\chi \in \operatorname{Irr} G$, divide the index $[G : A]$.

$$\chi(e) \mid [G : A] \quad \text{for all } \chi \in \operatorname{Irr} G\,.$$

There exist several different proofs for this basic result. It is either shown to be a special case of a much more general result (see for example [1], [4]) or it is deduced step by step, first showing $\chi(e) \mid |G|$, then $\chi(e) \mid [G : Z]$ where Z is the center of G, and then (using basics for induced representations) $\chi(e) \mid [G : A]$ (see for example [2]). We will also use the Casimir element as in [4], but combine it with another interesting formula in $\mathbb{C}\,G$ and we will use another integrality argument to give a very short, direct and transparent proof.

*MSC-Nr.: 20C15, 20C05

2 Integrality

An element x in a unital ring R is called *integral*, if there exists a $k \in \mathbb{N}$ and integers $a_i \in \mathbb{Z}$, such that $x^k + a_{k-1}\, x^{k-1} + \cdots + a_1\, x + a_o\, e = 0$. Very useful is the following observation: $x \in R$ is integral, iff the subring $\mathbb{Z}[x]$, generated by x and $e \in R$ is as $\mathbb{Z}$-module finitely generated. Then, using the fact that submodules of finitely generated $\mathbb{Z}$-modules are again finitely generated, one gets the following basic integrality theorem (see for example [6]):

If the additive group $(R, +)$ of a unital ring R is finitely generated, then every element $x \in R$ is integral. As an immediate application of this theorem we have that every element $\sum_{g\in G} \alpha_g\, g \in \mathbb{C}\, G$ with integral coefficients $\alpha_g \in \mathbb{Z}$ is integral.

We will need a slightly more general result: Since $x\, \varepsilon_\Theta\, x^{-1} = \varepsilon_{\Theta^x}$, where Θ^x is a conjugate of Θ, we have $x\, \varepsilon_\Theta = \varepsilon_{\Theta^x}\, x$, which shows that the subring $\mathbb{Z}[\varepsilon_\Theta\, (\Theta \in \operatorname{Irr} A), G]$ of $\mathbb{C}\, G$ generated by all ε_Θ, $\Theta \in \operatorname{Irr} A$, and all $g \in G$ is a finitely generated $\mathbb{Z}$-module (generated by all $\varepsilon_\Theta\, g$ $(\Theta \in \operatorname{Irr} A,\ g \in G)$) and consequently every element in $\mathbb{Z}[\varepsilon_\Theta\, (\Theta \in \operatorname{Irr} A), G]$ is integral.

3 The proof

We will need the well known formula for the Casimir element ([3], [4], [5])

$$(1)\ C_G := \sum_{g,\, h\in G} g\, h\, g^{-1}\, h^{-1} = \sum_{\chi\in\operatorname{Irr}\, G} \left(\frac{|G|}{\chi(e)}\right)^2 e_\chi$$

and the following formula for the sum of A-conjugates:

$$(2)\ \sum_{a\in A} a\, x\, a^{-1} = |A| \sum_{\Theta\in\operatorname{Irr}\, A} \varepsilon_\Theta\, x\, \varepsilon_\Theta, \qquad x \in G.$$

Proof. Since A is abelian, we have $\Theta(e) = 1$ for all $\Theta \in \operatorname{Irr} A$ and the $\varepsilon_\Theta = \frac{1}{|A|} \sum_{a\in A} \Theta(a^{-1})\, a$ form a basis of $\mathbb{C}\, A$, $a = \sum \Theta(a)\, \varepsilon_\Theta$ for all $a \in A$.

Then
$$\sum_{a\in A} a\, x\, a^{-1} = \sum_{a\in A} \sum_{\Theta\in\operatorname{Irr}\, A} \Theta(a^{-1})\, a\, x\, \varepsilon_\Theta = |A| \sum_{\Theta} \varepsilon_\Theta\, x\, \varepsilon_\Theta\, .$$

□

Remarks.

1. One may verify (2) also by simply using the definition of ε_Θ and the second orthogonality relation for *abelian* A.

2. With a similar argument one gets for an *arbitrary* normal subgroup $H \trianglelefteq G$

$$\sum_{a,b\in H} a\, x\, b\, a^{-1}\, b^{-1} = c_H \sum_{\Theta\in\operatorname{Irr}\, H} \varepsilon_\Theta\, x\, \varepsilon_\Theta \quad (x \in G)$$

where $c_H = \sum a\, b\, a^{-1}\, b^{-1}$ is the Casimir element in $\mathbb{C}\, H$.

Proof of Itô's theorem.

Using formulas (1), (2) and a coset decomposition $G = \bigcup A\,x$ (x from a set V of coset representatives) we get

$$\begin{aligned}\sum \left(\frac{|G|}{\chi(e)}\right)^2 e_\chi &= \sum ghg^{-1}h^{-1} = \sum axbyx^{-1}a^{-1}y^{-1}b^{-1} \;\; (a, b \in A,\, x, y \in V) \\ &= |A|^2 \sum \varepsilon_\Theta\, x\, \varepsilon_\psi\, y\, x^{-1}\, \varepsilon_\Theta\, y^{-1}\, \varepsilon_\psi \quad (\Theta,\, \psi \in \operatorname{Irr} A,\;\; x,\, y \in V)\end{aligned}$$

showing that $\sum \left(\frac{[G:A]}{\chi(e)}\right)^2 e_\chi$ is in $\mathbb{Z}[\varepsilon_\Theta\, (\Theta \in \operatorname{Irr} A), G]$, hence an integral element.

But obviously, an element of the form $\sum \alpha_\chi\, e_\chi$ ($\alpha_\chi \in \mathbb{C}$) is integral, iff all α_χ are algebraic integers. Therefore all rationals $\frac{[G:A]}{\chi(e)}$, $\chi \in \operatorname{Irr} G$, are algebraic integers and consequently rational integers, $\chi(e) \mid [G : A]$. □

References

[1] Carlson, J.F., Roggenkamp, K.W., Itô's theorem and character degrees revisited. Arch. Math. 50, 1988

[2] Isaac, I.M., Character Theory of Finite Groups. Academic Press, New York 1976

[3] Kellersch, P., Meyberg, K., On a Casimir Element of a Finite Group. Comm. in Alg. 25(6), 1997

[4] Leitz, M., Kommutatoren und Itô's Satz über Charaktergrade. Arch. Math. 67, 1996

[5] Meyberg, K., On the Sum of all N-commutators in a Finite Group. Resenhas IME-USP, Vol. 5, 2001

[6] Müller, W., Darstellungstheorie von endlichen Gruppen. Teubner, Stuttgart 1980

Kurt Meyberg
Zentrum Mathematik, Technische Universität München
Boltzmannstr.3
D-85748 Garching b. München
E-mail: meyberg@ma.tum.de

Contemporary Mathematics
Volume **483**, 2009

The ideal of the Lesieur-Croisot elements of a Jordan algebra. II

Fernando Montaner and Maribel Tocón

ABSTRACT. We show that the set of the Lesieur-Croisot elements of a nondegenerate Jordan algebra is an ideal as conjectured in the Guarujá 2004 meeting.

1. Introduction

Goldie's theorem for associative rings asserts that a ring R is a left order in a semiprime artinian ring Q iff R is semiprime and left Goldie; Q being simple iff R is prime. Semiprime left Goldie rings can also be characterized as those for which *essential left (resp. right) ideals are the ones containing regular elements.*

For Jordan algebras, the so-called *Goldie's theorem* is due to Zelmanov for the linear case [**9**] and to Fernández López, García Rus and Montaner [**1**] for the general quadratic case. It characterizes those nondegenerate (respectively strongly prime) Jordan algebras that have a nondegenerate (respectively simple) classical artinian algebra of quotients; such algebras are called *Goldie.* However, a nondegenerate Jordan algebra that satisfies the natural jordanification of the above associative italicized property, i.e. *an inner ideal is essential if and only if it contains injective elements (x such that U_x is injective)*, is not in general Goldie. In fact, it turns out that these algebras are precisely the orders in nondegenerate unital Jordan algebras of finite capacity. We have chosen the denomination *Lesieur-Croisot* to designate these Jordan algebras, in honour of the work of Lesieur and Croisot on prime noetherian rings [**4**].

Inspired by the local PI-theory for Jordan systems ([**6**], [**7**]), in the Guarujá 2004 meeting we claimed that the set of the Lesieur-Croisot elements (i.e. at which local algebras are Lesieur-Croisot) of a nondegenerate Jordan algebra is an ideal and proved the strongly prime case. The aim of this note is to provide an outline of the extension of this proof to the nondegenerate case. The details can be found in [**8**].

2000 *Mathematics Subject Classification.* 17C10.

Key words and phrases. Jordan algebra, Local algebra, Order, Uniform dimension.

The first author was partially supported by the Ministerio de Educación y Ciencia and FEDER (MTM2004-081159-C04-02) and by the Diputación General de Aragón (Grupo de Investigación de Álgebra).

The second author was supported in part by Ministerio de Educación y Ciencia and FEDER (MTM2004-03845) and Universidad de Córdoba.

For unexplained associative notation we refer the reader to [**2**] and for Jordan notation to [**1**], [**3**] and [**5**]. The algebras considered in this paper are defined over an arbitrary ring of scalars Φ.

2. Lesieur-Croisot elements of a Jordan algebra

As mentioned in the introduction, a nondegenerate Jordan algebra satisfying the property, that *an inner ideal is essential if and only if it contains injective elements*, is in general not Goldie. A counterexample can be obtained by considering the Jordan algebra of Clifford type defined on a vector space having an infinite dimensional totally isotropic subspace. They are indeed characterized as follows:

THEOREM 2.1. [**1**, (10.2)] *A Jordan algebra J is a classical order in a nondegenerate unital Jordan algebra Q with finite capacity if and only if it is nondegenerate and satisfies the property, that an inner ideal K of J is essential if and only if K contains an injective element. Moreover Q is simple if and only if J is prime.*

A Jordan algebra is called *Lesieur-Croisot*, *LC* for short, if it satisfies the equivalent conditions of Theorem 2.1.

It follows from Goldie's theorem for Jordan algebras and the fact that nondegenerate artinian Jordan algebras have finite capacity that nondegenerate Goldie Jordan algebras are Lesieur-Croisot. Also, every strongly prime Jordan algebra satisfying a polynomial identity is Lesieur-Croisot since its central closure is simple with finite capacity [**7**, (0.13)].

Following the standard *local* terminology, an element a of a Jordan algebra J is called *Lesieur-Croisot* if the local algebra of J at a is LC, where recall that the local algebra of J at a, denoted J_a, is defined as the quotient algebra

$$J_a := J^{(a)}/ker_J(a)$$

where the homotope $J^{(a)}$ is the Jordan algebra on the Φ-module J endowed with the new Jordan products $x^{2(a)} = U_x a$, $U_x^{(a)} = U_x U_a$, and $ker_J(a) := \{x \in J : U_a x = U_a U_x a = 0\}$. In most cases (e.g. if $1/2 \in \Phi$, or if J is nondegenerate or special) $ker_J(a) = \{x \in J : U_a x = 0\}$. We denote the set of the Lesieur-Croisot elements by $\mathrm{LC}(J)$, that is,

$$\mathrm{LC}(J) := \{x \in J : J_x \text{ is LC}\}$$

For an associative algebra R and $a \in R$, the *local algebra* of R at a is defined as

$$R_a := R^{(a)}/ker_R(a)$$

where $R^{(a)}$ is the a-homotope of R and $ker_R(a) := \{x \in R : axa = 0\}$.

In general, for a particular property P of a Jordan or associative algebra A, an element $a \in A$ is said to be P if the local algebra A_a satisfies the property P. Thus, if the local algebra A_a is PI, i.e. satisfies a polynomial identity, we just say that the element a is PI; or if A is associative and A_a has finite left uniform dimension, we say that a has finite left uniform dimension. The set of the PI-elements of A will be denoted by $\mathrm{PI}(A)$ and the set of the finite left uniform dimension elements by $\mathrm{F}_l(A)$.

3. The Lesieur-Croisot ideal of a strongly prime Jordan algebra

Let J be a strongly prime Jordan algebra.

In this section we recall the proof of the fact that LC(J) is an ideal of J. The strategy relays on the local PI-dichotomy, that is, we distinguish between Jordan algebras having nonzero PI-elements and those without nonzero PI-elements. This dichotomy, as opposed to the one distinguishing between PI Jordan algebras and non-PI Jordan algebras, has proven to be the adequate one when dealing with problems of Goldie theory.

It is not hard to prove that if J has nonzero PI-elements, then LC(J) actually coincides with PI(J) [**8**, (3.3)] and therefore LC(J) is an ideal of J [**6**, (5.4)]. The local PI-less case is more involved: note that in this case J is special of hermitian type and we can then translate our problem to a purely associative one. More precisely:

THEOREM 3.1. [**8**, (4.2) and (4.4)] *If PI(J) $= 0$ and LC(J) $\neq 0$. Then:*

(i) $\Theta(J) = 0$, *where* $\Theta(J) := \{z \in J : ann_J(z) \text{ is essential in } J\}$.
(ii) *For a $*$-tight associative envelope R of J we have that*

$$LC(J) = J \cap (F_l(R) \cap F_l(R)^*)$$

where $F_l(R) = \{x \in R : udim_l(R_x) < \infty\}$.

Thus, in the case PI(J) $= 0$, the problem of showing that LC(J) is an ideal of J reduces to showing that the subset $\mathrm{F}_l(R) \cap \mathrm{F}_l(R)^*$ of a $*$-tight associative envelope R of J is an ideal of R when LC(J) $\neq 0$.

It is well known that if an associative ring R is left nonsingular, i.e. $\mathrm{Z}_l(R) = 0$, where

$$Z_l(R) = \{z \in R : \mathrm{lann}_R(z) \text{ is an essential left ideal of } R\}$$

then $\mathrm{F}_l(R)$ is an ideal of R. This suggests that the set $\mathrm{F}_l(R) \cap \mathrm{F}_l(R)^*$ should be an ideal of R whenever $\mathrm{Z}_l(R) \cap \mathrm{Z}_l(R)^*$ vanishes.

PROPOSITION 3.2. [**8**, (4.6)] *Let R be a semiprime associative algebra. If $Z_l(R) \cap Z_r(R) = 0$, then $F_l(R) \cap F_r(R)$ is an ideal of R contained in $ann_R(Z_l(R) + Z_r(R))$, where $Z_r(R) = \{z \in R : rann_R(z) \text{ is an essential right ideal of } R\}$.*

The proof of the above proposition is based on local algebra techniques and all together gives us the following corollary.

COROLLARY 3.3. *LC(J) is an ideal of J.*

PROOF. Assume LC(J) $\neq 0$, otherwise the result is trivial. By (i) of Theorem 3.1 and the fact that $\Theta(J) = J \cap (Z_l(R) \cap Z_l(R)^*)$ [**1**, (6.14)], we have that

$$\Theta(J) = J \cap (Z_l(R) \cap Z_l(R)^*) = 0$$

which implies that $\mathrm{Z}_l(R) \cap \mathrm{Z}_l(R)^* = 0$ by $*$-tightness of R. Then $\mathrm{F}_l(R) \cap \mathrm{F}_l(R)^*$ is an ideal of R by Proposition 3.2, and then LC(J) is an ideal of J again by $*$-tightness of R and Theorem 3.1. □

4. The Lesieur-Croisot ideal of a nondegenerate Jordan algebra

Let J be a nondegenerate Jordan algebra.

In this section we show that LC(J) is an ideal of J by reducing the nondegenerate case to the strongly prime one, using the notion of *semi-uniform* Jordan algebra.

DEFINITION 4.1. J is called *semi-uniform* if there exists a minimal set of prime ideals $\mathcal{P} = \{P_1, \ldots, P_n\}$ of J such that $P_1 \cap \cdots \cap P_n = 0$. It is shown in [**8**, (5.3)] that the set $\mathcal{P}$ is necessarily unique.

Recall that a *subdirect product* of a collection of Jordan algebras $\{J_\alpha\}$ is any subalgebra of the full direct product of the J_α such that the canonical projections $\pi_\alpha : J \to J_\alpha$ are onto. An *essential subdirect product* is a subdirect product which contains an essential ideal of the full direct product. If J is actually contained in the direct sum of the J_α, then J is called an *essential subdirect sum.*

If J is semi-uniform with associated set of prime ideals $\{P_1, \ldots, P_n\}$, then J is an essential subdirect sum of $J_i := J/P_i$, where each J_i is a strongly prime Jordan algebra, denoted:

$$J \leq_{\text{ess}} J_1 \oplus \cdots \oplus J_n$$

In this case, J is LC if and only if each J_i is LC [**8**, (5.12)].

THEOREM 4.2. *LC(J) is an ideal of J.*

PROOF. It suffices to show that for any $a, b \in \mathrm{LC}(J)$, any $\alpha \in \Phi$ and any $x \in J'$, the elements $\alpha a, a+b, U_x a$ and $U_a x$ lie in LC(J). To see this, set $c := \alpha a, a+b, U_x a$ or $U_a x$.

Let $L := id_J(a) + id_J(b)$. One can prove that J/I, for $I = \mathrm{ann}_J(L)$, is semi-uniform. Hence an essential subdirect sum of finitely many strongly prime Jordan algebras $J_1, \cdots, J_n$:

$$J/I \leq_{\text{ess}} J_1 \oplus \cdots \oplus J_n$$

and has $(J/I)_{c+I} \cong J_c$, $(J/I)_{a+I} \cong J_a$ and $(J/I)_{b+I} \cong J_b$ since $a, b, c \in L$ implies $I \subseteq ker_J(a) \cap ker_J(b) \cap ker_J(c)$. In particular:

$$J_c \leq_{\text{ess}} (J_1)_{c_1} \oplus \cdots \oplus (J_n)_{c_n}$$

for $c_i \in J_i$ being the projection of $c + I$ in J_i (also, if $c = 1$ we set $c_i = 1 \in J_i'$). Analogously

$$J_a \leq_{\text{ess}} (J_1)_{a_1} \oplus \cdots \oplus (J_n)_{a_n} \quad J_b \leq_{\text{ess}} (J_1)_{b_1} \oplus \cdots \oplus (J_n)_{b_n}$$

Since $a, b \in \mathrm{LC}(J)$, i.e. J_a and J_b are LC, we thus have that $(J_i)_{a_i}$ and $(J_i)_{b_i}$ are LC, i.e. $a_i, b_i \in \mathrm{LC}(J_i)$ for all $i = 1, \ldots, n$. But J_i are strongly prime and then by the previous section, we have that LC(J_i) is an ideal of J_i, therefore $a_i + b_i \in \mathrm{LC}(J_i)$. In general we have that $c_i \in \mathrm{LC}(J_i)$, i.e. $(J_i)_{c_i}$ is LC for all $i = 1, \ldots, n$, which implies that J_c is LC, i.e. $c \in \mathrm{LC}(J)$ as desired. □

References

[1] A. Fernández López, E. García Rus and F. Montaner, *Goldie theory for Jordan algebras*, J. Algebra, **248** (2002), 397-471.

[2] K.R. Goodearl, *Ring Theory: Nonsingular Rings and Modules*, Pure and Applied Mathematics, Marcel Dekker, New York, 1976.

[3] N. Jacobson, *Structure theory of Jordan Algebras*, Lectures notes in Mathematics, University of Arkansas **5**, Wiley, New York, 1981.

[4] L. Lesieur and R. Croisot , *Sur les anneaux premiers Noethériens à gauche*, Ann. Sci. Ecole Norm. Sup. **76** (1959), 161-183.

[5] O. Loos, *Jordan pairs*, Lectures Notes in Mathematics **406** Springer-Verlag, Berlin Heidelberg, 1975.

[6] F. Montaner, *Local PI theory of Jordan systems*, J. Algebra **216** (1999), 302-327.

[7] F. Montaner, *Local PI theory of Jordan systems II*, J. Algebra **241** (2001), 473-514.

[8] F. Montaner and M. Tocón, *Local Lesieur-Croisot theory of Jordan algebras*, J. Algebra, **301** (2006), 256-273.

[9] E. Zelmanov, *Goldie's theorem for Jordan algebras; II*, Siberian Math. J.,**28** (1987),44-52; **29** (1988), 68-74.

Departamento de Matemáticas, Universidad de Zaragoza, 50009 Zaragoza, Spain
E-mail address: `fmontane@posta.unizar.es`

Departamento de Estadística, Econometría, Inv. Operativa y Org. Empresa, Universidad de Córdoba, 14071 Córdoba, Spain
E-mail address: `td1tobam@uco.es`

Contemporary Mathematics
Volume **483**, 2009

Unital Algebras, Ternary Derivations, and Local Triality

José M. Pérez–Izquierdo

Dedicated to I. P. Shestakov on the occasion of his 60th birthday.

Abstract. In this paper we classify some unital algebras that, like the octonions, provide a Principle of Local Triality. Alternative and assosymmetric algebras appear naturally in this context.

1. Introduction

Throughout this paper the base field F will be algebraically closed of characteristic zero, and all algebras will be finite dimensional.

A (generalized) octonion algebra over a field F is an eight dimensional unital algebra C equipped with a nondegenerate symmetric bilinear form $(\ ,\ )\colon C\times C \to F$ satisfying

$$q(xy) = q(x)q(y) \tag{1.1}$$

for any $x, y \in C$, where $q(x) = (x, x)$. The Principle of Local Triality establishes that for any $d_1 \in \mathfrak{o}(C, q) = \{d \in \operatorname{End}_F(C) \mid (d(x), y) + (x, d(y)) = 0\}$ there exist unique $d_2, d_3 \in \mathfrak{o}(C, q)$ such that $d_1(xy) = d_2(x)y + xd_3(y)\ \forall x, y \in C$ [**Sch95**, Theorem 3.31]. For any algebra A we may consider the Lie algebra

$$\operatorname{Tder}(A) = \{(d_1, d_2, d_3) \in \operatorname{End}_F(A)^3 \mid d_1(xy) = d_2(x)y + xd_3(y)\ \forall x, y \in A\} \tag{1.2}$$

of ternary derivations of A with the componentwise commutator product. This algebra can be regarded as a generalization of $\operatorname{Der}(A)$, the derivation algebra of A. However, in general the projections $\mathrm{T}_i(A) = \{d_i \in \operatorname{End}_F(A) \mid \exists (d_1, d_2, d_3) \in \operatorname{Tder}(A)\}$ $i = 1, 2, 3$ of $\operatorname{Tder}(A)$ may not agree, and therefore we should not expect an analog of the Local Principle of Triality from A.

In this paper we are concerned with algebras that, like the generalized octonions, induce a Principle of Local Triality so we will make the following assumption

---

2000 *Mathematics Subject Classification.* Primary 17A36; Secondary 17D10.

Key words and phrases. Ternary derivations, Generalized alternative nucleus, Nonassociative algebra.

The author thanks the following for support: MEC (MTM2007–67884–C04–03), FEDER, CONACyT (U44100–F) and Gobierno de La Rioja (ANGI ANGI2005/05,6).

about A:

$$\text{(LT)} \qquad \mathrm{T}_1(A) = \mathrm{T}_2(A) = \mathrm{T}_3(A)$$

and we will write $\mathrm{T}(A)$ instead of $\mathrm{T}_i(A)$. Observe that $\mathrm{T}(A)$ is a Lie algebra with the product defined by the commutator of linear maps. The dependence among d_1, d_2 and d_3 in $(d_1, d_2, d_3) \in \mathrm{Tder}(A)$ becomes specially interesting for unital algebras. Setting $x = 1$ (resp. $y = 1$) in (1.2) we obtain

$$\begin{array}{ccc} d_1 = d_2 + R_{a_3} & & a_3 = d_3(1) \\ & \text{with} & \\ d_1 = d_3 + L_{a_2} & & a_2 = d_2(1) \end{array} \tag{1.3}$$

where L_a and R_a denote the left and right multiplication operators by a respectively.

Since $d \in \mathrm{Der}(A)$ if and only if $(d, d, d) \in \mathrm{Tder}(A)$, any ternary derivation (d_1, d_2, d_3) with $d_1 = d_2 = d_3$ is said to represent a derivation. For unital algebras, (d_1, d_2, d_3) represents a derivation if and only if $d_2(1) = 0 = d_3(1)$. If in addition A satisfies (LT) then the subspace

$$\mathrm{Mal}(A) = \{d(1) \in A \,|\, d \in \mathrm{T}(A)\}$$

measures the difference between $\mathrm{Tder}(A)$ and $\mathrm{Der}(A)$. Under the condition (N) below, we will prove that $\mathrm{Mal}(A)$ is a Malcev algebra with the commutator product $[a, b] = ab - ba$.

The purpose of this paper[1] is to classify finite dimensional simple unital algebras over algebraically closed fields of characteristic zero that satisfy (LT). Our approach relies on the study of $\mathrm{Mal}(A)$. However, in general $\mathrm{Mal}(A)$ might be too small so we need the following two extra conditions:

$$\text{(N)} \qquad (a, \mathrm{Mal}(A), \mathrm{Mal}(A)) = 0 \text{ with } a \in \mathrm{Mal}(A) \text{ implies } a \in \mathrm{N}(A)$$

and

$$\text{(G)} \qquad A \text{ is generated by } \mathrm{Mal}(A)$$

where $\mathrm{N}(A) = \{x \in A \,|\, (x, A, A) = (A, x, A) = (A, A, x) = 0\}$ is the associative nucleus of A and $(x, y, z) = (xy)z - x(yz)$ denotes the usual associator. We will prove that any simple unital algebra that satisfies (LT), (N) and (G) is isomorphic to the tensor product of a simple associative algebra and certain algebras $T_n(C)$ (see Example 3 below).

Algebras related with the Principle of Local Triality have attracted attention [**EO01, Oku05, Oku06**] since they can be used to obtain in a beautiful and symmetric way Lie algebras and the magic square. Our work does not aim at constructing Lie algebras but at classifying a family of nonassociative algebras which admit a nice set of operators acting on them.

Let us show some examples of algebras that satisfy conditions (LT), (N) and (G).

Example 1. Let A be a unital alternative algebra, so

$$(a, x, y) = -(x, a, y) \quad \text{and } (x, y, a) = -(x, a, y) \;\; \forall a, x, y \in A.$$

In terms of the left and right multiplication operators by a, these identities are written as

$$L_a(xy) = T_a(x)y - xL_a(y) \quad \text{and} \quad R_a(xy) = -R_a(x)y + xT_a(y)$$

[1] I thank the referee for careful reading of the paper and many suggestions.

with $T_a = L_a + R_a$. Adding these identities we also have

$$T_a(xy) = L_a(x)y + xR_a(y)$$

and therefore

$$(L_a, T_a, -L_a), (R_a, -R_a, T_a) \text{ and } (T_a, L_a, R_a) \in \mathrm{Tder}(A).$$

Given $(d_1, d_2, d_3) \in \mathrm{Tder}(A)$, set $a_2 = d_2(1)$ and $a_3 = d_3(1)$. Then

$$(d_1, d_2, d_3) - \frac{1}{3}(L_{2a_2+a_3}, T_{2a_2+a_3}, -L_{2a_2+a_3}) \\ - \frac{1}{3}(R_{2a_3+a_2}, -R_{2a_3+a_2}, T_{2a_3+a_2}) \in \mathrm{Tder}(A)$$

represents a derivation (the second and third components kill the unit) and thus

$$\mathrm{Tder}(A) = \{(d,d,d) \mid d \in \mathrm{Der}(A)\} + \langle (L_a, T_a, -L_a), (R_a, -R_a, T_a) \mid a \in A\rangle$$

where $\langle S\rangle$ denotes the subspace spanned by S. Hence, any unital alternative algebra satisfies (LT) and

$$\mathrm{T}(A) = \mathrm{Der}(A) + \langle L_a, R_a \mid a \in A\rangle.$$

In this case $\mathrm{Mal}(A) = A$, which is a Malcev algebra with the commutator product, so (N) and (G) are satisfied too.

Example 2. *Assosymmetric algebras* were studied by Kleinfeld in [**Kle57**] and recently in [**KK08, Bre02, SR00a, SR00b, HJP96, Boe94, SS87, PR77**]. These are algebras that satisfy the identities:

$$(a, x, y) = (x, a, y) = (x, y, a).$$

Surprisingly, a basis is known for free assosymmetric algebras over fields of characteristic $\neq 2, 3$ ([**HJP96**]). As in Example 1, one easily proves that if A is a unital assosymmetric algebra then A satisfies (LT) and

$$\begin{aligned} \mathrm{Tder}(A) &= \{(d,d,d) \mid d \in \mathrm{Der}(A)\} + \langle (L_a, \mathrm{ad}_a, L_a), (R_a, R_a, -\mathrm{ad}_a) \mid a \in A\rangle \\ \mathrm{T}(A) &= \mathrm{Der}(A) + \langle L_a, R_a \mid a \in A\rangle \\ \mathrm{Mal}(A) &= A \end{aligned}$$

where $\mathrm{ad}_a \colon x \mapsto [a, x]$. Since $(x, y, z) = (y, x, z)$ can be written as $L_{[x,y]} = [L_x, L_y]$ then $(\mathrm{Mal}(A), [\ ,\ ])$ is a Lie algebra and (N), (G) hold.

Example 3. The generalized alternative nucleus of an algebra A is defined as

$$\mathrm{N}_{\mathrm{alt}}(A) = \{a \in A \mid (a, x, y) = -(x, a, y) = (x, y, a)\, \forall x, y \in A\}.$$

This set is a Malcev algebra with the product $[a, b] = ab - ba$ and has been used in [**PIS04, PI07**] to construct universal enveloping algebras for Malcev algebras that extends the classical construction for Lie algebras. In [**MPIP01**] finite dimensional simple algebras generated by their generalized alternative nucleus over algebraically closed fields of characteristic zero were classified. Any of these algebras is isomorphic to the tensor product of a finite dimensional simple associative algebra and a finite number of algebras $T_n(C)$ where C denotes the split octonion algebra, the only generalized octonion algebra up to isomorphism over algebraically closed fields. $T_n(C)$ is defined as follows: $T_0(C) = F$, $T_1(C) = C$ and $T_n(C)$ $(n \geq 2)$ is the kernel of the map $\mathrm{Sym}^n(C) \to \mathrm{Sym}^{n-2}(C)$ induced by $x \otimes \cdots \otimes x \mapsto q(x)x \otimes \cdots \otimes x$ where $q(\,)$ denotes the quadratic form on C. Equation (1.1) implies that this map is a homomorphism of algebras and $T_n(C)$ is an ideal of $\mathrm{Sym}^n(C)$. In fact, $T_n(C) = \langle x \otimes \cdots \otimes x \,|\, q(x) = 0\rangle$. The bilinear form induced on $\mathrm{Sym}^n(C)$ by that of C restricts to a nondegenerate symmetric bilinear form on $T_n(C)$. In Section 5 we will prove that these algebras satisfy conditions (LT), (N) and (G). For any $n \geq 1$ we will prove that $\mathrm{N_{alt}}(T_n(C))$ is isomorphic to the Malcev algebra $(C, [\,,\,])$ of octonions with the commutator product and

$$\begin{aligned}
\mathrm{Tder}(T_n(C)) &= \{(d,d,d) \,|\, d \in \mathrm{Der}(T_n(C))\} \\
&\qquad + \langle (L_a, T_a, -L_a), (R_a, -R_a, T_a) \,|\, a \in \mathrm{N_{alt}}(T_n(C))\rangle \\
\mathrm{T}(T_n(C)) &= \mathrm{Der}(T_n(C)) + \langle L_a, R_a \,|\, a \in \mathrm{N_{alt}}(T_n(C))\rangle \\
\mathrm{Mal}(T_n(C)) &= \mathrm{N_{alt}}(T_n(C)).
\end{aligned}$$

2. Connection with Malcev algebras

The following characterization of $\mathrm{Mal}(A)$ will be useful.

LEMMA 2.1. *Let A be a unital algebra that satisfies* (LT). *Then*

$$\mathrm{Mal}(A) = \{a \in A \,|\, L_a, R_a \in \mathrm{T}(A)\}.$$

PROOF. We only have to prove that $\mathrm{Mal}(A) \subseteq \{a \in A \,|\, L_a, R_a \in \mathrm{T}(A)\}$. Given $a \in \mathrm{Mal}(A)$, by (LT) we may choose $(d_1, d_2, d_3) \in \mathrm{Tder}(A)$ with $a = d_2(1)$. Now, (1.3) and (LT) imply that $L_a = d_1 - d_3 \in \mathrm{T}(A)$. Similar arguments show that $R_a \in \mathrm{T}(A)$. □

Let $A^{(-)}$ denote the vector space A with the new product $[x, y] = xy - yx$.

LEMMA 2.2. *Let A be a unital algebra that satisfies* (LT). *Then* $(\mathrm{Mal}(A), [\,,\,])$ *is a subalgebra of $A^{(-)}$.*

PROOF. Given $a, b \in \mathrm{Mal}(A)$, by Lemma 2.1 we have that $L_a, L_b \in \mathrm{T}(A)$. Since $\mathrm{T}(A)$ is a Lie algebra then $[L_a, L_b] \in \mathrm{T}(A)$ and $[a, b] = [L_a, L_b](1) \in \mathrm{Mal}(A)$. □

PROPOSITION 2.3. *Let A be a unital algebra that satisfies* (LT). *Given $a \in \mathrm{Mal}(A)$ there exist, not necessarily unique, $\alpha(a), \beta(a), \alpha'(a)$ and $\beta'(a) \in \mathrm{Mal}(A)$ such that*

$$\begin{array}{lll}
(a, x, y) = (x, \alpha(a), y), & & (x, a, y) = (\alpha'(a), x, y), \\
(x, a, y) = (x, y, \beta(a)) & \text{and} & (x, y, a) = (x, \beta'(a), y)
\end{array} \tag{2.1}$$

for any $x, y \in A$.

PROOF. Given $a \in \mathrm{Mal}(A)$, by Lemma 2.1 we have that $L_a \in \mathrm{T}(A)$ so we may choose $(d_1, d_2, d_3) \in \mathrm{Tder}(A)$ with $d_1 = L_a$. Due to (1.3) $d_3 = L_{a_3}$ and $d_2 = L_a - R_{a_3}$ so $(L_a, L_a - R_{a_3}, L_{a_3}) \in \mathrm{Tder}(A)$ and $a_3 \in \mathrm{T}(A)$. Then the relation $a(xy) = (ax)y - (xa_3)y + x(a_3 y)$ holds in A. With $\alpha(a) = a_3$ we get

$$(a, x, y) = (x, \alpha(a), y).$$

The existence of $\beta(a), \alpha'(a)$ and $\beta'(a)$ is deduced using similar arguments. □

Consider

$$\begin{aligned}
\mathrm{N}_* &= \mathrm{N}_*(\mathrm{Mal}(A)) = \{a \in \mathrm{Mal}(A) \,|\, (a, \mathrm{Mal}(A), \mathrm{Mal}(A)) = 0\}, \\
\mathrm{N_l} &= \{a \in A \,|\, (a, A, A) = 0\}, \ \ \mathrm{N_m} = \{a \in A \,|\, (A, a, A) = 0\} \text{ and} \\
\mathrm{N_r} &= \{a \in A \,|\, (A, A, a) = 0\}.
\end{aligned}$$

It is clear that $a \in \mathrm{N_l} \Leftrightarrow (L_a, L_a, 0) \in \mathrm{Tder}(A)$, so $\mathrm{N_l} \subseteq \mathrm{Mal}(A)$. Similarly $\mathrm{N_m}, \mathrm{N_r} \subseteq \mathrm{Mal}(A)$.

PROPOSITION 2.4. *Let A be a unital algebra that satisfies* (LT) *and* (N), *then*

$$\mathrm{N}(A) = \mathrm{N}_{\mathrm{l}} = \mathrm{N}_{\mathrm{m}} = \mathrm{N}_{\mathrm{r}} = \mathrm{N}_*(\mathrm{Mal}(A)).$$

PROOF. Given $a \in \mathrm{N}_{\mathrm{m}}$ (resp. $a \in \mathrm{N}_{\mathrm{r}}$) and $b, c \in \mathrm{Mal}(A)$, Proposition 2.3 shows that $(a, b, c) = (\alpha'(b), a, c) = 0$ (resp. $(a, b, c) = (\alpha'(b), a, c) = (\alpha'(b), \beta'(c), a) = 0$) so $\mathrm{N}_{\mathrm{l}}, \mathrm{N}_{\mathrm{m}}, \mathrm{N}_{\mathrm{r}} \subseteq \mathrm{N}_* \subseteq \mathrm{N}(A)$ where the latter inclusion follows from condition (N). Since $\mathrm{N}(A)$ is clearly contained in N_{l}, N_{m} and N_{r} the result follows. □

Proposition 2.4 and (1.3) imply

PROPOSITION 2.5. *Let A be a unital algebra that satisfies* (LT) *and* (N), *then:*

(i) $(d_1, d_2, 0) \in \mathrm{Tder}(A)$ *if and only if* $d_1 = d_2 = L_a$ *with* $a \in \mathrm{N}(A)$.
(ii) $(d_1, 0, d_3) \in \mathrm{Tder}(A)$ *if and only if* $d_1 = d_3 = R_a$ *with* $a \in \mathrm{N}(A)$.
(iii) $(0, d_2, d_3) \in \mathrm{Tder}(A)$ *if and only if* $d_2 = R_a, d_3 = -L_a$ *with* $a \in \mathrm{N}(A)$.

PROPOSITION 2.6. *Let A be a unital algebra that satisfies* (LT) *and* (N). *Then for any $a \in \mathrm{Mal}(A)$ there exists $\alpha(a) \in \mathrm{Mal}(A)$ such that*

$$(a, x, y) = (x, \alpha(a), y) = (x, y, a) \quad \forall x, y \in A.$$

Moreover, the correspondence

$$\begin{array}{rccc} \overline{\alpha}\colon & \mathrm{Mal}(A)/\,\mathrm{N}(A) & \to & \mathrm{Mal}(A)/\,\mathrm{N}(A) \\ & [a] & \mapsto & [\alpha(a)] \end{array}$$

is an isomorphism of vector spaces with $\overline{\alpha}^2 = \mathrm{id}$.

PROOF. Given $a, b \in \mathrm{Mal}(A)$ and $x \in A$ we have

$$\begin{aligned} (b, x, \beta(a)) &= (b, a, x) = (a, \alpha(b), x) = (a, x, \beta\alpha(b)) = (x, \alpha(a), \beta\alpha(b)) \\ &= (b, x, \alpha(a)) \end{aligned}$$

thus $\alpha(a) - \beta(a) \in \mathrm{N}_* = \mathrm{N}(A)$. Similar manipulations lead to

$$\alpha(a) - \beta(a), \alpha'(a) - \beta'(a) \text{ and } \alpha'\alpha(a) - a \in \mathrm{N}(A). \tag{2.2}$$

Using (2.2) we get

$$\begin{aligned} (a, b, x) &= (a, x, \alpha(b)) = (x, \alpha(a), \alpha(b)) \\ &= (x, \beta'\alpha(b), \alpha(a)) = (x, b, \alpha(a)) = (x, a, b). \end{aligned}$$

Given $c \in \mathrm{Mal}(A)$,

$$(a, b, c) = (b, \alpha(a), c) = (b, c, \alpha^2(a)) = (\alpha^2(a), b, c)$$

implies that $\alpha^2(a) - a \in \mathrm{N}_* = \mathrm{N}(A)$. Therefore, $(a, x, y) = (x, \alpha(a), y) = (x, y, a)$. Finally, observe that the element $\alpha(a)$ may not be unique. Two such elements will differ by an element of $\mathrm{N}(A)$, thus there exists a well defined linear map $\overline{\alpha}\colon \mathrm{Mal}(A)/\,\mathrm{N}(A) \to \mathrm{Mal}(A)/\,\mathrm{N}(A)$ with $\overline{\alpha}([a]) = [\alpha(a)]$ and $\overline{\alpha}^2 = \mathrm{id}$ (recall that $\alpha^2(a) - a \in \mathrm{N}(A)$). □

Proposition 2.6 allows us to decompose $\mathrm{Mal}(A)$ as $\mathrm{Mal}(A) = M_1 + M_{-1}$ with

$$M_\epsilon = \{a \in \mathrm{Mal}(A) \mid (a, x, y) = \epsilon(x, a, y) = (x, y, a)\} \quad (\epsilon = \pm 1).$$

The following Lemma shows that M_1, M_{-1} are ideals of $(\mathrm{Mal}(A), [\,,\,])$. Clearly the intersection of these ideals is $M_1 \cap M_{-1} = \mathrm{N}(A)$.

LEMMA 2.7. *Let A be a unital algebra that satisfies* (LT) *and* (N). *We have*

(i) $(a_\epsilon, a_{-\epsilon}, A) = (a_\epsilon, A, a_{-\epsilon}) = (A, a_\epsilon, a_{-\epsilon}) = 0$.

(ii) $[M_\epsilon, M_\epsilon] \subseteq M_\epsilon$.
(iii) $[M_\epsilon, M_{-\epsilon}] \subseteq \mathrm{N}(A)$.
(iv) $[(M_\epsilon, M_\epsilon, M_\epsilon), M_{-\epsilon}] = 0$.

PROOF. Part (i) is obvious. To prove (ii) we first observe that the identity

$$a(x, y, z) + (a, x, y)z = (ax, y, z) - (a, xy, z) + (a, x, yz) \tag{2.3}$$

holds in any algebra [**ZSSS82**, p.136]. Now, with the convention that the subindex ϵ in z_ϵ means that $z_\epsilon \in M_\epsilon$, identity (2.3) with $a = a_{-\epsilon}$, $x = x_\epsilon$, $y = y_\epsilon$ and $z = c_{-\epsilon}$ leads to $(a_{-\epsilon}, x_\epsilon y_\epsilon, c_{-\epsilon}) = 0$ so $(a_{-\epsilon}, [x_\epsilon, y_\epsilon], c_{-\epsilon}) = 0$. Given $a = a_\epsilon + a_{-\epsilon}, b = b_\epsilon + b_{-\epsilon} \in \mathrm{Mal}(A)$, by (i) $([x_\epsilon, y_\epsilon], a, b) = ([x_\epsilon, y_\epsilon], a_\epsilon, b_\epsilon) + ([x_\epsilon, y_\epsilon], a_{-\epsilon}, b_{-\epsilon}) = \epsilon(a_\epsilon, [x_\epsilon, y_\epsilon], b_\epsilon) - \epsilon(a_{-\epsilon}, [x_\epsilon, y_\epsilon], b_{-\epsilon}) = \epsilon(a_\epsilon, [x_\epsilon, y_\epsilon], b_\epsilon) = \epsilon(a, [x_\epsilon, y_\epsilon], b)$. Hence $[x_\epsilon, y_\epsilon] \in M_\epsilon$.

Using (2.3) and (i) we get that

$$(a_\epsilon, x_\epsilon, b_\epsilon)y_{-\epsilon} = a_\epsilon(x_\epsilon, b_\epsilon, y_{-\epsilon}) + (a_\epsilon, x_\epsilon, b_\epsilon)y_{-\epsilon} = (a_\epsilon, x_\epsilon, b_\epsilon y_{-\epsilon})$$

so $(a_\epsilon, x_\epsilon, b_\epsilon)y_{-\epsilon} = \epsilon(a_\epsilon, b_\epsilon, x_\epsilon)y_{-\epsilon} = \epsilon(a_\epsilon, b_\epsilon, x_\epsilon y_{-\epsilon}) = (a_\epsilon, x_\epsilon y_{-\epsilon}, b_\epsilon)$. In a similar way we obtain

$$(a_\epsilon, y_{-\epsilon}x_\epsilon, b_\epsilon) = (a_\epsilon y_{-\epsilon}, x_\epsilon, b_\epsilon) = (a_\epsilon, x_\epsilon, b_\epsilon)y_{-\epsilon},$$

therefore $(a_\epsilon, [x_\epsilon, y_{-\epsilon}], b_\epsilon) = 0$. This implies that $[M_\epsilon, M_{-\epsilon}] \subseteq M_\epsilon \cap M_{-\epsilon} = \mathrm{N}(A)$, and we get (iii).

Again with (2.3) we have

$$y_{-\epsilon}(a_\epsilon, x_\epsilon, b_\epsilon) = (y_{-\epsilon}a_\epsilon, x_\epsilon, b_\epsilon) = (a_\epsilon y_{-\epsilon}, x_\epsilon, b_\epsilon) = (a_\epsilon, x_\epsilon, b_\epsilon)y_{-\epsilon}$$

where the second equality follows from (iii). This proves (iv) and concludes the proof. □

Recall that an algebra M with product denoted by $[x, y]$ is called a *Malcev algebra* if $[x, y] = -[y, x]$ and $J(x, y, [x, z]) = [J(x, y, z), x]$ where $J(x, y, z) = [[x, y], z] + [[y, z], x] + [[z, x], y]$ stands for the jacobian of x, y, z. It is obvious that any Lie algebra is a Malcev algebra, although there exist non–Lie Malcev algebras such as the traceless elements of generalized octonions with product given by the commutator.

PROPOSITION 2.8. *Let A be a unital algebra that satisfies* (LT) *and* (N), *then* $\mathrm{Mal}(A)$ *is a Malcev algebra.*

PROOF. The jacobian of three elements can be expressed in terms of associators in A as

$$J(x, y, z) = \sum_{alt}(x, y, z)$$

where $\sum_{alt}(x_1, x_2, x_3) = \sum_\sigma \mathrm{sig}(\sigma)(x_{\sigma(1)}, x_{\sigma(2)}, x_{\sigma(3)})$ with σ running the symmetric group of order three and $\mathrm{sig}(\sigma)$ denotes the signature of σ [**ZSSS82**, p.136]. Thus the Malcev identity on $\mathrm{Mal}(A)$ is equivalent to $\sum_{alt}(a, b, [a, c]) = \sum_{alt}[(a, b, c), a]$ for any $a, b, c \in \mathrm{Mal}(A)$.

Given $a, b, c \in \mathrm{Mal}(A)$, in order to check that $\sum_{alt}(a, b, [a, c]) = \sum_{alt}[(a, b, c), a]$ observe that by Lemma 2.7 and linearity we can assume that $b, c \in M_\epsilon$. Although the dependence on a is quadratic, we can assume that $a \in M_\epsilon$ too because of Lemma 2.7 (iv). If $\epsilon = 1$ then the associators are symmetric in their arguments and the result follows trivially. In case that $\epsilon = -1$ the result follows as for alternative algebras. □

Now we will focus on the structure of A as a $\mathrm{T}(A)$–module to prove that if (LT) and (N) hold and A is simple and unital then it is an irreducible $\mathrm{T}(A)$–module. Since the arguments are similar to those in [**MPIP01**, Lemma 4.4, Proposition 4.5] we will omit the proofs. We will denote $[\mathrm{T}(A), \mathrm{T}(A)]$ by $\mathrm{T}(A)'$. We say that the degree on $M \subseteq A$ of x is $\delta(x)$ if x can be written as $x = p(a_1, \ldots, a_m)$ where $a_1, \ldots, a_m \in M$ and $p(x_1, \ldots, x_m)$ is a non-associative polynomial (constant polynomials are allowed) of degree $\delta(x)$, and there is no other such expression with a polynomial of degree $< \delta(x)$. By convention the degree of 0 is set to $-\infty$.

From (2.1) we get

$$\begin{aligned} L_{ax} &= L_a L_x + [R_{\gamma(a)}, L_x] \quad & R_{ax} &= R_x R_a - [R_x, L_{\gamma'\beta(a)}] \\ L_{xa} &= L_x L_a + [R_{\beta(a)}, L_x] \quad & R_{xa} &= R_a R_x - [R_x, L_{\gamma'(a)}] \end{aligned} \tag{2.4}$$

where $\gamma = \beta\alpha$ and $\gamma' = \alpha'\beta'$ (compare with [**MPIP01**, Lemma 4.2]).

Lemma 2.9. *Let A be a unital algebra that satisfies* (LT) *and* (N). *Given $S \subseteq \mathrm{Mal}(A)$ closed under α, α', β and β' then the degree of $x \in A$ on S is the same as the degree of L_x, R_x on $\langle L_a, R_a \,|\, a \in S\rangle$.*

Proposition 2.10. *Let A be a unital algebra that satisfies* (LT), (N) *and* (G). *If A is simple then A is an irreducible $\mathrm{T}(A)$–module and $\mathrm{T}(A) = \mathrm{T}(A)' \oplus F\,\mathrm{id}$ with $\mathrm{T}(A)' = [\mathrm{T}(A), \mathrm{T}(A)]$ a semisimple Lie algebra.*

Remark 2.11. Proposition 2.10 remains true if instead of $\mathrm{T}(A)$ we consider $\mathrm{InnT}(A)$, the Lie subalgebra of $\mathrm{T}(A)$ generated by the multiplication operators $\{L_a, R_a \,|\, a \in \mathrm{Mal}(A)\}$, and $\mathrm{InnT}(A)' = [\mathrm{InnT}(A), \mathrm{InnT}(A)]$.

3. The Principle of Local Triality

Given $d_1 \in \mathrm{T}(A)'$ there exist $d_2, d_3 \in \mathrm{T}(A)'$ with $(d_1, d_2, d_3) \in \mathrm{Tder}(A)'$. Our goal is to define automorphisms $\zeta\colon d_1 \mapsto d_2$ and $\eta\colon d_1 \mapsto d_3$ of $\mathrm{T}(A)'$ that imitate those of $\mathfrak{o}(C, q)$. However, the associative nucleus becomes an obstruction since for any $a, b \in \mathrm{N}(A)$ we have that $(L_a, L_a, 0)$ and $(L_a, L_a + R_b, -L_b) \in \mathrm{Tder}(A)$, which allows multiple choices for d_2 or d_3. We first remove this nucleus.

Lemma 3.1. *Let A be a unital algebra that satisfies* (LT). *If there exist algebras A_1, A_2 such that A is isomorphic to the algebra $A_1 \otimes A_2$ then A_1, A_2 are unital algebras that satisfy* (LT).

Proof. It is easy to show that A_1, A_2 are unital so we will focus on the condition (LT). Given $(d_1, d_2, d_3) \in \mathrm{Tder}(A_1)$, then $(d_1 \otimes \mathrm{id}, d_2 \otimes \mathrm{id}, d_3 \otimes \mathrm{id}) \in \mathrm{Tder}(A_1 \otimes A_2)$ and $d_2(1) \otimes 1, d_3(1) \otimes 1 \in \mathrm{Mal}(A_1 \otimes A_2)$. For any of these elements, say $a \otimes 1 \in \mathrm{Mal}(A_1 \otimes A_2)$, Proposition 2.3 applied to $A_1 \otimes A_2$ implies that

$$(a, x, y) \otimes 1 = (a \otimes 1, x \otimes 1, y \otimes 1) = (x \otimes 1, \alpha(a \otimes 1), y \otimes 1)$$

so there exists $\alpha(a) \in A_1$ such that $(a, x, y) = (x, \alpha(a), y)$. In particular $(L_a, L_a - R_{\alpha(a)}, L_{\alpha(a)})$ is a ternary derivation of A_1 and $L_a \in \mathrm{T}_1(A_1)$. Similar reasoning with α', β, β' gives that $L_a, R_a \in \mathrm{T}_i(A_1)$ $i = 1, 2, 3$. Since $d_1 = d_2 + R_{d_3(1)}$ and $d_1 = d_3 + L_{d_2(1)}$ then we can conclude that $d_1, d_2, d_3 \in \mathrm{T}_i(A_1)$ $i = 1, 2, 3$. □

Lemma 3.2. *Let A be a simple unital algebra that satisfies* (LT), (N) *and* (G). *Then $\mathrm{N}(A)$ is a simple associative subalgebra of A.*

PROOF. Since $(d_1, d_2, 0) \in \mathrm{Tder}(A)$ if and only if $d_1 = d_2 = L_a$ with $a \in \mathrm{N}(A)$ then $L_{\mathrm{N}(A)} = \{L_a \mid a \in \mathrm{N}(A)\}$ is an ideal of $\mathrm{T}(A)$. In fact, $[L_a, L_b] = L_{[a,b]}$ $\forall a, b \in \mathrm{N}(A)$ implies that $(L_{\mathrm{N}(A)}, [\,,\,])$ is isomorphic to $(\mathrm{N}(A), [\,,\,])$. The Jacobson radical of $\mathrm{N}(A)$ induces a nilpotent ideal in $L_{\mathrm{N}(A)}$, that by Proposition 2.10 must be contained in $F\,\mathrm{id}$, so it vanishes. Therefore, $\mathrm{N}(A)$ is a semisimple associative algebra. Finally, the center of $\mathrm{N}(A)$ induces an abelian ideal on $L_{\mathrm{N}(A)}$, that is again contained in $F\,\mathrm{id}$, so $\mathrm{N}(A)$ is simple. □

Given an algebra A and $S \subseteq A$, the subalgebra generated by S will be denoted by $\mathrm{alg}\langle S \rangle$.

PROPOSITION 3.3. *Let A be a simple unital algebra that satisfies* (LT), (N) *and* (G). *Then* $\mathrm{N}(A)$ *is a simple associative algebra and there exists a subalgebra A_1 of A such that*

(i) *A_1 is a simple unital algebra that satisfies* (LT), (N) *and* (G),
(ii) $\mathrm{N}(A_1) = F1$ *and*
(iii) *A is isomorphic to the algebra* $A_1 \otimes \mathrm{N}(A)$.

Moreover, $\mathrm{T}(A) \cong \mathrm{T}(A_1) \otimes \mathrm{id} + \mathrm{id} \otimes \mathrm{T}(\mathrm{N}(A))$ *and* $\mathrm{Mal}(A) \cong \mathrm{Mal}(A_1) \otimes F + F \otimes \mathrm{N}(A)$.

PROOF. With the actions given by left and right multiplication A is an $\mathrm{N}(A)$–bimodule. Since $\mathrm{N}(A)$ is simple then A decomposes as the direct sum of bimodules isomorphic to $\mathrm{N}(A)$, the only irreducible bimodule, up to isomorphism, of $\mathrm{N}(A)$. Let $A = \oplus N_i$ such a decomposition and $\varphi_i \colon \mathrm{N}(A) \to N_i$ an isomorphism of bimodules. Let $x_i = \varphi_i(1)$. The condition of φ_i being an isomorphism of bimodules is equivalent to $x_i \in A_1$, the centralizer of $\mathrm{N}(A)$ in A. This shows that the map $\varphi \colon A_1 \otimes \mathrm{N}(A) \to A$ given by $x \otimes a \mapsto xa$ is an epimorphism of algebras. Since $\mathrm{N}(A)$ is simple, a usual argument involving the Jacobson density theorem shows that φ is an isomorphism too, so $A \cong A_1 \otimes \mathrm{N}(A)$ and A_1 must be simple. Lemma 3.1 implies that A_1 satisfies (LT).

Since $\mathrm{N}(A)$ is associative, $\mathrm{id} \otimes \mathrm{T}(\mathrm{N}(A)) = \langle \mathrm{id} \otimes L_a, \mathrm{id} \otimes R_a \mid a \in \mathrm{N}(A) \rangle$. As in the proof of Lemma 3.2 this set is an ideal of $\mathrm{T}(A_1 \otimes \mathrm{N}(A))$. By Proposition 2.10 there exists an ideal I of $\mathrm{T}(A_1 \otimes \mathrm{N}(A))$ that complements $\mathrm{id} \otimes \mathrm{T}(\mathrm{N}(A))$:

$$\mathrm{T}(A_1 \otimes \mathrm{N}(A)) = I \oplus (\mathrm{id} \otimes \mathrm{T}(\mathrm{N}(A))). \tag{3.1}$$

To see that $I \subseteq \mathrm{T}(A_1) \otimes \mathrm{id}$ we start with $d \in I$ and d_2, d_3 such that $(d, d_2, d_3) \in \mathrm{Tder}(A_1 \otimes \mathrm{N}(A))$. By commuting (d, d_2, d_3) and $(\mathrm{id} \otimes L_a, \mathrm{id} \otimes L_a, 0)$ with $a \in \mathrm{N}(A)$ we obtain that $[d_2, \mathrm{id} \otimes L_a] = 0$. Similar computations with $(\mathrm{id} \otimes R_a, 0, \mathrm{id} \otimes R_a)$ give $[d_3, \mathrm{id} \otimes R_a] = 0$. By (3.1) it follows that $d_2 = d_2' + \mathrm{id} \otimes R_b$ with $d_2' \in I$ and $b \in \mathrm{N}(A)$. We can change (d, d_2, d_3) by (d, d_2', d_3') with $d_3' = d_3 + \mathrm{id} \otimes L_b$. That is, we may assume that $d_2 \in I$. By commuting (d, d_2, d_3) with $(0, \mathrm{id} \otimes R_a, -\mathrm{id} \otimes L_a)$ we have that $[d_3, \mathrm{id} \otimes L_a] = 0$. Therefore, $d_3 \in I$. Since I commutes with $\mathrm{id} \otimes \mathrm{T}(\mathrm{N}(A))$ any element of I maps A_1, the centralizer of $\mathrm{N}(A)$, into itself, thus (d, d_2, d_3) induces a ternary derivation on A_1. This proves that $I \subseteq \mathrm{T}(A_1) \otimes \mathrm{id}$. Therefore $\mathrm{T}(A) \cong \mathrm{T}(A_1) \otimes \mathrm{id} + \mathrm{id} \otimes \mathrm{T}(\mathrm{N}(A))$. From this it is clear that $\mathrm{Mal}(A) \cong \mathrm{Mal}(A_1) \otimes F + F \otimes \mathrm{N}(A)$.

Under the isomorphism $A \cong A_1 \otimes \mathrm{N}(A)$, $A = \mathrm{alg}\langle \mathrm{Mal}(A) \rangle$ goes to $\mathrm{alg}\langle \mathrm{Mal}(A_1) \otimes 1 + 1 \otimes \mathrm{N}(A) \rangle = \mathrm{alg}\langle \mathrm{Mal}(A_1) \rangle \otimes \mathrm{N}(A)$ so $A_1 = \mathrm{alg}\langle \mathrm{Mal}(A_1) \rangle$ and A_1 satisfies (G). Clearly A_1 also satisfies (N). □

In the view of Proposition 3.3 we may assume that $\mathrm{N}(A) = F1$. Now we are prepared to state the analog of the Principle of Local Triality in our context.

PROPOSITION 3.4. *Let A be a simple unital algebra that satisfies* (LT) *and* (N). *If $\mathrm{N}(A) = F1$ then given $d_1 \in \mathrm{T}(A)'$ there exist unique $d_2, d_3 \in \mathrm{T}(A)'$ such that $(d_1, d_2, d_3) \in \mathrm{Tder}(A)$. The maps*

$$\zeta\colon d_1 \mapsto d_2 \quad \text{and} \quad \eta\colon d_1 \mapsto d_3$$

are automorphisms of the Lie algebra $\mathrm{T}(A)'$. Moreover, $d \in \mathrm{T}(A)'$ is a derivation if and only if $\zeta(d) = d$ and $\eta(d) = d$.

PROOF. The existence of d_2 and d_3 is clear. To prove the uniqueness we observe that if $(d_1, d_2', d_3') \in \mathrm{Tder}(A)$ then $(0, d_1 - d_1', d_2 - d_2') \in \mathrm{Tder}(A)$. By Proposition 2.5 $d_1 - d_1' = R_a$ and $d_2 - d_2' = -L_a$ with $a \in \mathrm{N}(A) = F1$. Thus $d_1 - d_1', d_2 - d_2' \in \mathrm{T}(A)' \cap F\,\mathrm{id} = \{0\}$. Similar arguments prove that ζ and η are monomorphisms. Condition (LT) ensures that they are automorphisms. The last claim in the statement is obvious. □

The assumption $\mathrm{N}(A) = F1$ also allows a natural choice of $\alpha(a)$ in the statement of Proposition 2.6. Consider

$$\mathrm{Mal}(A)' = \{a \in \mathrm{Mal}(A) \mid L_a \in \mathrm{T}(A)'\}.$$

Since $\mathrm{T}(A) = \mathrm{T}(A)' \oplus F\,\mathrm{id}$ then $\mathrm{Mal}(A) = \mathrm{Mal}(A)' \oplus F1$. We may choose $\alpha(a)$ such that

$$\alpha(1) = 1, \ \ \alpha(a) \in \mathrm{Mal}(A)' \text{ if } a \in \mathrm{Mal}(A)', \text{ and } (a, x, y) = (x, \alpha(a), y) = (x, y, a).$$

The uniqueness of $\alpha(a) \in \mathrm{Mal}(A)'$ in case that $a \in \mathrm{Mal}(A)'$ is clear since the difference of two such elements will belong to $\mathrm{Mal}(A)' \cap \mathrm{N}(A) = \{0\}$. By the proof of Proposition 2.6, $\alpha^2(a) - a \in \mathrm{Mal}(A)' \cap \mathrm{N}(A) = \mathrm{Mal}(A)' \cap F1 = 0$. Therefore $\alpha^2 = \mathrm{id}$.

PROPOSITION 3.5. *Let A be a simple unital algebra with $\mathrm{N}(A) = F1$ that satisfies* (LT), (N) *and* (G). *Then*

$$\begin{aligned}\mathrm{Mal}(A)' &= \{a \in \mathrm{Mal}(A) \mid L_a \in \mathrm{T}(A)'\} \\ &= \{a \in \mathrm{Mal}(A) \mid R_a \in \mathrm{T}(A)'\} = \{d(1) \mid d \in \mathrm{T}(A)'\}\end{aligned}$$

PROOF. Take $a \in \mathrm{Mal}(A)'$. On the one hand we can find $d_1, d_2 \in \mathrm{T}(A)'$ with $(d_1, d_2, L_a) \in \mathrm{Tder}(A)$. On the other hand $(\alpha(a), x, y) = (x, a, y)$ implies that $(L_{\alpha(a)}, L_{\alpha(a)} - R_a, L_a) \in \mathrm{Tder}(A)$ too. Subtracting both ternary derivations we get that $(d_1 - L_{\alpha(a)}, d_2 - L_{\alpha(a)} + R_a, 0)$ is a ternary derivation. Under our hypothesis $\mathrm{N}_l = F1$ so there exists an scalar $\epsilon \in F$ such that $d_1 - L_{\alpha(a)} = \epsilon\,\mathrm{id}$ and $d_2 - L_{\alpha(a)} + R_a = -\epsilon\,\mathrm{id}$. Since $\alpha(a) \in \mathrm{Mal}(A)'$ then $L_a \in \mathrm{T}(A)'$ so $\epsilon\,\mathrm{id} = d_1 - L_{\alpha(a)}$ is a linear combination of commutators of linear maps. This is only possible for $\epsilon = 0$. Hence $R_a = -d_2 + L_{\alpha(a)} \in \mathrm{T}(A)'$. This shows that $\mathrm{Mal}(A)' \subseteq \{a \in \mathrm{Mal}(A) \mid R_a \in \mathrm{T}(A)'\}$. To show that this inclusion is an equality we only have to compare the codimensions of both spaces inside $\mathrm{Mal}(A)$ (notice that $\mathrm{id} \notin \mathrm{T}(A)'$ implies that $1 \notin \{a \in \mathrm{Mal}(A) \mid R_a \in \mathrm{T}(A)'\}$).

Given $d \in \mathrm{T}(A)'$ choose $d_2, d_3 \in \mathrm{T}(A)'$ such that $(d, d_2, d_3) \in \mathrm{Tder}(A)$. By (1.3), $d = d_2 + R_{a_3} = d_3 + L_{a_2}$ for some a_2, a_3, so $R_{a_3}, L_{a_2} \in \mathrm{T}(A)'$ and $a_2, a_3 \in \mathrm{Mal}(A)'$. In particular $d(1) = a_2 + a_3 \in \mathrm{Mal}(A)'$. This proves the inclusion $\{d(1) \mid d \in \mathrm{T}(A)'\} \subseteq \mathrm{Mal}(A)'$. The equality is clear from the definition of $\mathrm{Mal}(A)'$. □

4. Classification

Once we have dealt with the associative nucleus $\mathrm{N}(A)$ we return to the study of $\mathrm{Mal}(A) = M_1 + M_{-1}$. Our goal is to show that $M_1 = F1$ so we can invoke Theorem 4.1 in [**MPIP01**] to obtain the classification:

THEOREM 4.1 ([**MPIP01**]). *Any simple finite dimensional unital algebra, over an algebraically closed field of characteristic zero which is generated by its generalized alternative nucleus is isomorphic to the tensor product of a simple associative algebra and $\mathrm{T}_n(C)$ for some n.*

Since any element $a \in M_1$ satisfies $(a, x, y) = (x, a, y) = (x, y, a)$, this part of $\mathrm{Mal}(A)$ is very much related with the assosymmetric algebras studied by Kleinfeld in [**Kle57**] where he proves that any assosymmetric ring in which the only ideal of square zero is 0 must be associative. This result suggests that $M_1 = F1$. However, we do not even know that the identity $(x, y, z) = (y, x, z) = (y, z, x)$ holds in $\mathrm{alg}\langle M_1\rangle$. It only holds when x is taken from M_1, so we need another approach.

THEOREM 4.1. *Let A be a simple unital algebra with $\mathrm{N}(A) = F1$ that satisfies* (LT), (N) *and* (G). *Then* $\mathrm{Mal}(A) = M_{-1}$.

PROOF. We split the proof in several steps.

Step 1. We have that $[M_1, M_{-1}] = 0$.

Given $a \in M_1, b \in M_{-1}$, on the one hand, by Lemma 2.7 $[M_1, M_{-1}] \subseteq \mathrm{N}(A) = F1$ so $[a, b] \in F1$. On the other hand, (2.4) implies that the trace of $L_{[a,b]} \in F\,\mathrm{id}$ is zero. Therefore a, b commute.

Step 2. $\mathrm{InnT}(A)' = I_1' \oplus I_{-1}'$ *where $I_\epsilon' = \mathrm{alg}\langle L_a, R_a \mid a \in M_\epsilon \cap \mathrm{Mal}'(A)\rangle$ are semisimple Lie algebras ($\epsilon = \pm 1$).*

Lemma 2.7 and the previous step imply that the multiplication operators by elements in M_1 commute with those by elements in M_{-1}. Hence $\mathrm{InnT}(A)$ is the sum of two commuting ideals I_1 and I_{-1} with $I_\epsilon = \mathrm{alg}\langle L_a, R_a \mid a \in M_\epsilon\rangle$. The intersection $I_1 \cap I_{-1}$ is the center of $\mathrm{InnT}(A)$, i.e. $I_1 \cap I_{-1} = F\,\mathrm{id}$ (see Remark 2.11). The derived algebra of $\mathrm{InnT}(A)$ is $\mathrm{InnT}(A)' = I_1' \oplus I_{-1}'$ with $I_\epsilon' = [I_\epsilon, I_\epsilon] = \mathrm{alg}\langle L_a, R_a \mid a \in M_\epsilon \cap \mathrm{Mal}(A)'\rangle$. The ideals I_ϵ' are direct summands of a semisimple Lie algebra, so they are semisimple Lie algebras.

Step 3. $\zeta|_{I_1'}$ and $\eta|_{I_1'}$ are automorphisms of I_1' and $(\zeta|_{I_1'} - \mathrm{id})^2 = 0 = (\eta|_{I_1'} - \mathrm{id})^2$.

The action of ζ and η on the generators of I_1' is given by

$$\zeta(L_a) = L_a - R_a,\ \ \zeta(R_a) = R_a,\ \ \eta(L_a) = L_a \text{ and } \eta(R_a) = R_a - L_a.$$

Thus I_1' is stable under the automorphisms ζ and η.

For any $a, b \in M_1 \cap \mathrm{Mal}(A)'$, $(a, b, x) = (b, a, x)$ and $(x, a, b) = (x, b, a)$ are equivalent to $[L_a, L_b] = L_{[a,b]}$ and $[R_a, R_b] = -R_{[a,b]}$. The image of $[L_a, R_b] + \mathrm{ad}_{[a,b]}$ by ζ is $\zeta([L_a, R_b] + \mathrm{ad}_{[a,b]}) = [L_a - R_a, R_b] + \mathrm{ad}_{[a,b]} - R_{[a,b]} = [L_a, R_b] + \mathrm{ad}_{[a,b]} - [R_a, R_b] - R_{[a,b]} = [L_a, R_b] + \mathrm{ad}_{[a,b]}$. Similarly $[L_a, R_b] + \mathrm{ad}_{[a,b]}$ is fixed by η, therefore $[L_a, R_b] + \mathrm{ad}_{[a,b]}$ is a derivation. It follows that $I_1' \subseteq \mathrm{Der}(A) + L_{M_1'} + R_{M_1'}$ with $M_1' = M_1 \cap \mathrm{Mal}(A)'$. $\mathrm{Der}(A)$ is killed by $\zeta - \mathrm{id}$ and $\eta - \mathrm{id}$ while $L_{M_1'}$ and $R_{M_1'}$ are killed by $(\zeta - \mathrm{id})^2$ and $(\eta - \mathrm{id})^2$, so $(\zeta - \mathrm{id})^2$ and $(\eta - \mathrm{id})^2$ vanish on I_1'.

Step 4. $\chi = \zeta|_{I_1'} - \mathrm{id}$ is a derivation of I_1'.

Given $d, d' \in I_1'$, since ζ is an automorphism then

$$\chi([d, d']) = [\chi(d), d'] + [d, \chi(d')] + [\chi(d), \chi(d')].$$

If we apply χ to both sides of this formula we obtain $[\chi(d), \chi(d')] = 0$. Therefore χ is a derivation of I_1' that squares to zero.

Step 5. $M_1 = F1$.

Since I_1' is semisimple then χ is inner, thus there exists $d \in I_1'$ such that $\mathrm{ad}_d^2 = 0$. Jacobson–Morozov theorem [**Jac51, Mor42**] (or a direct computation with the Killing form) implies that $d = 0$. Therefore, $\zeta|_{I_1'} = \mathrm{id}$. In particular $L_a = \zeta(L_a) = L_a - R_a$ for all $a \in M_1'$, so $M_1' = 0$ and $M_1 = F1$. □

THEOREM 4.2. *Let A be a simple unital algebra that satisfies* (LT), (N) *and* (G). *Then A is isomorphic to the tensor product of algebras $T_n(C)$ and a simple associative algebra.*

PROOF. By Proposition 3.3 $A \cong A_1 \otimes \mathrm{N}(A)$ with $\mathrm{N}(A)$ a simple associative algebra (see Lemma 3.2) so it suffices to classify the subalgebra A_1 which in addition satisfies that $\mathrm{N}(A_1) = F1$. By Theorem 4.1 $\mathrm{Mal}(A_1) = M_{-1}$ so A_1 is a simple finite dimensional unital algebra generated by its generalized alternative nucleus. By [**MPIP01**, Theorem 4.1] we obtain the desired result. □

REMARK 4.3. The only assosymmetric algebra $T_n(C)$ is $T_0(C) = F$ so associative algebras are the only assosymmetric algebras in Theorem 4.2.

5. Appendix: Bilinear Forms

The presence of a nondegenerate symmetric bilinear form on A leads to some interesting constructions that we collect in this section. We will also prove that the algebras $T_n(C)$ verify (LT), (N) and (G). Let us first introduce the objects that we will study.

In this section A will be a unital algebra with an involution $x \mapsto \bar{x}$ and a nondegenerate symmetric bilinear form $(\ ,\ )$ such that

$$(xy, z) = (x, z\bar{y}) = (y, \bar{x}z). \tag{5.1}$$

We will also assume that

$$\mathrm{N}(A) = F1 \text{ and } \mathrm{K}(A) = \{x \in A \mid [x, A] = 0\} = F1. \tag{5.2}$$

Octonions C are examples of these algebras. The involution $x \mapsto \bar{x} = 2(1, x)1 - x$ as well as the bilinear form of C extend to $\mathrm{Sym}^n(C)$ and induce a corresponding involution and a nondegenerate symmetric bilinear form on $T_n(C)$ that satisfy (5.1). Property (5.2) follows from [**MPIP01**, Remark 4.11].

The proof of the following lemma is left to the reader.

LEMMA 5.1. $\mathrm{N_l}(A) = \mathrm{N_m}(A) = \mathrm{N_r}(A) = \mathrm{N}(A) = F1$.

Since the bilinear form is nondegenerate then for any $d \in \mathrm{End}_F(A)$ there exists $d^* \in \mathrm{End}_F(A)$, the adjoint of d, such that $(d(x), y) = (x, d^*(y))$. The map $x \mapsto \overline{d(\bar{x})}$ will be denoted by $\bar{d}$.

LEMMA 5.2. *We have*

(i) $(x, y) = (\bar{x}, \bar{y})$ *for any* $x, y \in A$, *and*
(ii) $\overline{d^*} = (\bar{d})^*$ *for any* $d \in \mathrm{End}_F(A)$.

PROOF. Part (i) follows from the existence of unit in A. Part (ii) follows from (i). □

Given $(d_1, d_2, d_3) \in \mathrm{Tder}(A)$ we have

$$(d_1(xy), z) = \begin{cases} (xy, d_1^*(z)) = (x, d_1^*(z)\bar{y}) \\ (d_2(x)y + xd_3(y), z) = (x, d_2^*(z\bar{y}) + \overline{z\bar{d}_3(y)}) \end{cases}$$

and therefore $(-d_2^*, -d_1^*, \bar{d}_3) \in \mathrm{Tder}(A)$. This defines an automorphism

$$\begin{aligned} \epsilon_{(12)} \colon \mathrm{Tder}(A) &\to \mathrm{Tder}(A) \\ (d_1, d_2, d_3) &\mapsto (-d_2^*, -d_1^*, \bar{d}_3). \end{aligned}$$

Similarly one defines

$$\begin{aligned} \epsilon_{(13)} \colon \mathrm{Tder}(A) &\to \mathrm{Tder}(A) \\ (d_1, d_2, d_3) &\mapsto (-d_3^*, \bar{d}_2, -d_1^*). \end{aligned}$$

Since $\epsilon_{(12)}^2 = \mathrm{id} = \epsilon_{(13)}^2$ and $\epsilon_{(12)}\epsilon_{(13)}\epsilon_{(12)} = \epsilon_{(13)}\epsilon_{(12)}\epsilon_{(13)}$ we obtain a representation of the symmetric group in three letters as automorphisms of $\mathrm{Tder}(A)$ sending (12) to $\epsilon_{(12)}$ and (13) to $\epsilon_{(13)}$. The image of a permutation σ will be denoted by ϵ_σ. For instance

$$\epsilon_{(123)} \colon (d_1, d_2, d_3) \mapsto (-\bar{d}_3^*, -\bar{d}_1^*, d_2). \tag{5.3}$$

The proof of the following lemma is also left to the reader.

LEMMA 5.3. *We have that*

(i) $\mathrm{T}_2(A) = \mathrm{T}_3(A) = \{d^* \mid d \in \mathrm{T}_1(A)\}$.
(ii) $\mathrm{T}_i(A)$ *(*$i = 1, 2, 3$*) is closed under the automorphism* $d \mapsto \bar{d}$.

Let $\mathrm{T}_i(A)^0$ be the elements of $\mathrm{T}_i(A)$ of zero trace. Since the right associative nucleus is $F1$ then for any $d_2 \in \mathrm{T}_2(A)^0$ there exist unique $d_3 \in \mathrm{T}_3(A)^0$ (recall that $(\lambda\,\mathrm{id}, 0, \lambda\,\mathrm{id})$ is a ternary derivation for any $\lambda \in F$) and $d_1 \in \mathrm{T}_1(A)$ such that $(d_1, d_2, d_3) \in \mathrm{Tder}(A)$.

LEMMA 5.4. *The automorphism*

$$\begin{aligned} \Theta \colon \mathrm{T}_2(A)^0 &\to \mathrm{T}_2(A)^0 \\ d_2 &\mapsto d_3 \end{aligned}$$

satisfies $\Theta^3 = \mathrm{id}$.

PROOF. Since $(-\bar{d}_2^*, d_3, -\bar{d}_1^*) \in \mathrm{Tder}(A)$ then there exists $\lambda \in F$ such that the trace of $-\bar{d}_1^* + \lambda\,\mathrm{id}$ is zero and $(-\bar{d}_2^* + \lambda\,\mathrm{id}, d_3, -\bar{d}_1^* + \lambda\,\mathrm{id}) \in \mathrm{Tder}(A))$, thus $\Theta^2(d_2) = -\bar{d}_1^* + \lambda\,\mathrm{id}$. By (5.3), $(-\bar{d}_3^* + \lambda\,\mathrm{id}, -\bar{d}_1^* + \lambda\,\mathrm{id}, d_2) \in \mathrm{Tder}(A)$ implies that $\Theta^3(d_2) = d_2$. □

The Lie algebra $\mathrm{T}_2(A)^0$ decomposes as the direct sum of eigenspaces relative to Θ

$$\mathrm{T}_2(A)^0 = \mathrm{T}_2(A)_1^0 \oplus \mathrm{T}_2(A)_\mu^0 \oplus \mathrm{T}_2(A)_{\mu^2}^0 \tag{5.4}$$

with $\mu^2 + \mu + 1 = 0$ and $\mathrm{T}_2(A)_{\mu^i}^0 = \{d \in \mathrm{T}_2(A)^0 \mid \Theta(d) = \mu^i d\}$.

PROPOSITION 5.5. *Let* $J \colon \mathrm{T}_2(A)^0 \to \mathrm{T}_2(A)^0$ *be the automorphism defined by* $d \mapsto \bar{d}$. *Then*

(i) $\Theta^2 = J\Theta J$,
(ii) $\mathrm{T}_2(A)_{\mu^2}^0 = J(\mathrm{T}_2(A)_\mu^0)$.

PROOF. It is easy to see that if $(d_1, d_2, d_3) \in \mathrm{Tder}(A)$ then $(\bar{d}_1, \bar{d}_3, \bar{d}_2) \in \mathrm{Tder}(A)$. In case that $d_2, d_3 \in \mathrm{T}_2(A)^0$ then $J\Theta J\Theta(d_2) = J\Theta(\bar{d}_3) = J(\bar{d}_2) = d_2$. Thus $J\Theta J\Theta = \mathrm{id}$. This proves (i). Part (ii) follows from (i). □

We can get an explicit description of the spaces $\mathrm{T}_2(A)^0_{\mu^i}$. The trace of $d \in \mathrm{End}_F(A)$ is denoted by $\mathrm{tr}(d)$.

PROPOSITION 5.6.

(i) *For any* $d \in \mathrm{T}_2(A)^0_1$ *we have that* $d(1) \in F1$ *and* $d - d(1)\,\mathrm{id} \in \mathrm{Der}(A)$.
(ii) *We have that* $\mathrm{T}_2(A)^0_\mu = \langle R_{\mu a} - L_a \mid \mathrm{tr}(R_{\mu a} - L_a) = 0$ *and* $\mu^2(x,y,a) + (a,x,y) - \mu(x,a,y) = 0\,\forall x,y \in A\rangle$.
(iii) *We have that* $\mathrm{T}_2(A)^0_{\mu^2} = \langle R_{\mu^2 b} - L_b \mid \mathrm{tr}(R_{\mu^2 b} - L_b) = 0$ *and* $\mu(x,y,b) + (b,x,y) - \mu^2(x,b,y) = 0\,\forall x,y \in A\rangle$.

PROOF. If $\Theta(d) = d$ then there exists d_1 such that $(d_1, d, d) \in \mathrm{Tder}(A)$. By (1.3) we have that $d_1 = d + R_{d(1)} = d + L_{d(1)}$, so $L_{d(1)} = R_{d(1)}$ and $d(1) \in \mathrm{K}(A) = F1$. Since $(d + d(1)\,\mathrm{id}, d, d)$, $(2d(1)\,\mathrm{id}, d(1)\,\mathrm{id}, d(1)\,\mathrm{id}) \in \mathrm{Tder}(A)$ it follows that $(d - d(1)\,\mathrm{id}, d - d(1)\,\mathrm{id}, d - d(1)\,\mathrm{id}) \in \mathrm{Tder}(A)$, so $d - d(1)\,\mathrm{id}$ is a derivation.

Given $d \in \mathrm{T}_2(A)^0_\mu$, there exists d_1 such that $(d_1, d, \mu d) \in \mathrm{Tder}(A)$. By (1.3), $d_1 = d + R_{\mu a} = \mu d + L_a$ with $a = d(1)$. Therefore, $d = (\mu - 1)^{-1}(R_{\mu a} - L_a)$ with

$$(L_a - R_{\mu^2 a}, L_a - R_{\mu a}, L_{\mu a} - R_{\mu^2 a}) \in \mathrm{Tder}(A). \tag{5.5}$$

Since (5.5) is equivalent to the identity

$$\mu^2(x,y,a) + (a,x,y) - \mu(x,a,y) = 0 \tag{5.6}$$

then we obtain (ii). Similar computations give (iii). □

Given an arbitrary algebra B, consider the sets

$$\begin{aligned} \mathrm{N}_\mu &= \{a \in B \mid \mu^2(x,y,a) + (a,x,y) - \mu(x,a,y) = 0\,\forall x,y \in B\}, \\ \mathrm{N}_{\mu^2} &= \{b \in B \mid \mu(x,y,b) + (b,x,y) - \mu^2(x,b,y) = 0\,\forall x,y \in B\} \end{aligned} \tag{5.7}$$

where $\mu^2 + \mu + 1 = 0$ as above. Since the roles of N_μ and N_{μ^2} interchange when we choose μ^2 instead of μ, the properties of N_μ and N_{μ^2} run parallel. We will invoke this duality in the following. From the definition, it is clear that

$$a \in \mathrm{N}_\mu \Leftrightarrow D_a = (L_a - R_{\mu^2 a}, L_a - R_{\mu a}, L_{\mu a} - R_{\mu^2 a}) \in \mathrm{Tder}(B)$$

and by duality

$$b \in \mathrm{N}_{\mu^2} \Leftrightarrow D'_b = (L_b - R_{\mu b}, L_b - R_{\mu^2 b}, L_{\mu^2 b} - R_{\mu b}) \in \mathrm{Tder}(B).$$

It is also clear that $\mathrm{N}_{\mathrm{alt}}(B) = \mathrm{N}_\mu \cap \mathrm{N}_{\mu^2}$ and that N_μ and N_{μ^2} are stable under $\mathrm{Der}(B)$.

Given $a \in \mathrm{N}_\mu$ and $b \in \mathrm{N}_{\mu^2}$ we define the operators

$$D_{a,b} = -D_{b,a} = [L_a, L_b] + [L_a, R_b] + [R_a, R_b].$$

PROPOSITION 5.7. *For any* $a \in \mathrm{N}_\mu$ *and* $b \in \mathrm{N}_{\mu^2}$ *we have that*

$$[D_a, D'_b] = (D_{a,b}, D_{a,b}, D_{a,b}).$$

Therefore $D_{a,b} \in \mathrm{Der}(B)$.

PROOF. Since $\mu(\mu^2(b,y,a)+(a,b,y)-\mu(b,a,y))=0=\mu(a,y,b)+(b,a,y)-\mu^2(a,b,y)$ then, adding both equalities, we obtain that $(b,y,a)+(a,y,b)=0$, that is,

$$[L_a,R_b]=[R_a,L_b].$$

From this relation it is easy to check that the components of the ternary derivation $[D_a,D'_b]$ are equal to $D_{a,b}$. □

PROPOSITION 5.8. *Given $a,a'\in \mathrm{N}_\mu$ and $b,b'\in \mathrm{N}_{\mu^2}$ then*

$$[D_a,D_{a'}]=D'_{[a,a']},\quad [D'_b,D'_{b'}]=D_{[b,b']}.$$

In particular, $[\mathrm{N}_\mu,\mathrm{N}_\mu]\subseteq \mathrm{N}_{\mu^2}$ and $[\mathrm{N}_{\mu^2},\mathrm{N}_{\mu^2}]\subseteq \mathrm{N}_\mu$.

PROOF. From the relation $\mu^2(x,a',a)+(a,x,a')-\mu(x,a,a')=0$ we deduce that

$$\mu^2R_aR_{a'}-\mu^2R_{a'a}-[L_a,R_{a'}]-\mu R_{a'}R_a-\mu R_{aa'}=0.$$

Changing a to a' and subtracting the new relation from the previous one we obtain

$$[R_a,R_{a'}]=-R_{[a,a']}-[L_a,R_{a'}]-[R_a,L_{a'}]. \tag{5.8}$$

The same argument starting with $\mu^2(a',y,a)+(a,a',y)-\mu(a',a,y)=0$ gives

$$[L_a,L_{a'}]=L_{[a,a']}-[R_a,L_{a'}]-[L_a,R_{a'}]. \tag{5.9}$$

From (5.8) and (5.9) we obtain that $[D_a,D_{a'}]=D'_{[a,a']}$. The other formula in the statement follows by duality. □

COROLLARY 5.9. *Let $\mathrm{N}_{\mathrm{alt}}(B)'=[\mathrm{N}_{\mathrm{alt}}(B),\mathrm{N}_{\mathrm{alt}}(B)]$, then*

$$[\mathrm{N}_{\mathrm{alt}}(B)',\mathrm{N}_\mu+\mathrm{N}_{\mu^2}]\subseteq \mathrm{N}_{\mathrm{alt}}(B).$$

PROOF. Given $c,c'\in \mathrm{N}_{\mathrm{alt}}(B)=\mathrm{N}_\mu\cap \mathrm{N}_{\mu^2}$, on the one hand we have that $\mathrm{ad}_{[c,c']}$ interchanges N_μ and N_{μ^2}. On the other hand, $\mathrm{ad}_{[c,c']}=2D_{c,c'}-[\mathrm{ad}_c,\mathrm{ad}_{c'}]$ implies that N_μ and N_{μ^2} are invariant by $\mathrm{ad}_{[c,c']}$. Therefore, $\mathrm{ad}_{[c,c']}(\mathrm{N}_\mu+\mathrm{N}_{\mu^2})\subseteq \mathrm{N}_\mu\cap\mathrm{N}_{\mu^2}=\mathrm{N}_{\mathrm{alt}}(B)$. □

PROPOSITION 5.10. *$T_n(C)$ satisfies* (LT), (N) *and* (G).

PROOF. Recall that $\mathrm{Sym}^n(C)=\mathrm{alg}\langle a\otimes 1\cdots\otimes 1+\cdots+1\otimes\cdots\otimes 1\otimes a\mid a\in C\rangle$. The generators $a\otimes 1\cdots\otimes 1+\cdots+1\otimes\cdots\otimes 1\otimes a$ belong to $\mathrm{N}_{\mathrm{alt}}(\mathrm{Sym}^n(C))$ and the Lie algebra generated by the corresponding left and right multiplication operators by these generators is isomorphic to the Lie algebra generated by $\{L_a,R_a\colon C\to C\mid a\in C\}$, i.e. $\mathfrak{o}(C,q)\oplus F\,\mathrm{id}$. In fact, the left (resp. right) multiplication operator by $a\otimes 1\cdots\otimes 1+\cdots+1\otimes\cdots\otimes 1\otimes a$ corresponds to the action of $L_a\in\mathfrak{o}(C,q)$ (resp. R_a) on $T_n(C)$ when considered as an $\mathfrak{o}(C,q)$–submodule of $C^{\otimes n}$. By [**MPIP01**, Lemma 4.4], the $\mathfrak{o}(C,q)$–submodules of $\mathrm{Sym}^n(C)$ are exactly the ideals, so $\mathrm{Sym}^n(C)$ is a direct sum of ideals that are simple algebras (this decomposition is well–known: $\mathrm{Sym}^n(C)\cong V(n\lambda_1)\oplus V((n-2)\lambda_1)\oplus\cdots$ where λ_1 is the first fundamental weight of $\mathfrak{o}(C,q)$, a simple Lie algebra of type D_4). $T_n(C)$ is one of these ideals (namely $T_n(C)\cong V(n\lambda_1)$), so the projection of $\langle a\otimes 1\cdots\otimes 1+\cdots+1\otimes\cdots\otimes 1\otimes a\mid a\in C\rangle$ on $T_n(C)$ lies inside $\mathrm{N}_{\mathrm{alt}}(T_n(C))$ and generates $T_n(C)$, which shows that $T_n(C)$ satisfies (G). In fact, in [**MPIP01**, Remark 4.11] it was proved that this projection is surjective ($n\geq 1$) and that $(\mathrm{N}_{\mathrm{alt}}(T_n(C)),[\,,\,])\cong(C,[\,,\,])$.

The associator of $a,b,c\in\mathrm{N}_{\mathrm{alt}}(T_n(C))$ can be expressed in terms of the commutator as $J(a,b,c)=6(a,b,c)$. Under the isomorphism $(\mathrm{N}_{\mathrm{alt}}(T_n(C)),[\,,\,])\cong(C,[\,,\,])$

any $a \in \mathrm{N}_{\mathrm{alt}}(T_n(C))$ with $(a, \mathrm{N}_{\mathrm{alt}}(T_n(C)), \mathrm{N}_{\mathrm{alt}}(T_n(C))) = 0$ will correspond to $a \in C$ with $(a, C, C) = 0$. Since $\mathrm{N}(C) = F1$, this implies that $a \in F1$, which proves (N).

The Lie algebra T_+ generated by $\langle \mathrm{ad}_a \mid a \in \mathrm{N}_{\mathrm{alt}}(T_n(C))' \rangle$ is isomorphic to the Lie algebra $\{d \in \mathfrak{o}(C, q) \mid d(1) = 0\}$, a simple Lie algebra of type B_3. The action of this algebra on $T_n(C)$ is completely reducible and multiplicity free, so the only trivial submodule is $F1$. Since, by Corollary 5.9, $\mathrm{N}_\mu, \mathrm{N}_{\mu^2}$ and $\mathrm{N}_{\mathrm{alt}}(T_n(C))$ are T_+–submodules and $[\mathrm{N}_{\mathrm{alt}}(T_n(C))', \mathrm{N}_{\mathrm{alt}}(T_n(C)) + \mathrm{N}_\mu + \mathrm{N}_{\mu^2}] \subseteq \mathrm{N}_{\mathrm{alt}}(T_n(C))$ it follows that $\mathrm{N}_\mu = \mathrm{N}_{\mathrm{alt}}(T_n(C)) = \mathrm{N}_{\mu^2}$. By Proposition 5.6 we get that $\mathrm{T}_2(T_n(C)) = \mathrm{Der}(T_n(C)) + \langle L_a, R_a \mid a \in \mathrm{N}_{\mathrm{alt}}(T_n(C)) \rangle$. Any derivation of $T_n(C)$ is determined by its restriction to $(\mathrm{N}_{\mathrm{alt}}(T_n(C)), [\,,\,]) \cong (C, [\,,\,])$. The derivations of $(C, [\,,\,])$ are linear combinations of restrictions of derivations $D_{a,b}$ with $a, b \in C$ so $\mathrm{Der}(T_n(C)) = \langle D_{a,b} \mid a, b \in \mathrm{Mal}(T_n(C))' \rangle$. In particular all derivations of $T_n(C)$ are is skew–symmetric with respect to the bilinear form, so $\mathrm{T}_2(T_n(C))^* = \mathrm{T}_2(T_n(C))$. Since $d \mapsto -d^*$ interchanges $\mathrm{T}_1(T_n(C))$ and $\mathrm{T}_2(T_n(C))$ it follows that $T_n(C)$ verifies (LT) and that $\mathrm{Mal}(T_n(C)) = \mathrm{N}_{\mathrm{alt}}(T_n(C))$. □

COROLLARY 5.11. *For any simple associative algebra A and $r, n_1, \ldots, n_r \in \mathbb{N}$, the tensor product $A \otimes T_{n_1}(C) \otimes \cdots \otimes T_{n_r}(C)$ satisfies (LT), (N) and (G).*

PROOF. Conditions (N) and (G) are easily checked so we only have to deal with condition (LT). Let $B = T_{n_1}(C) \otimes \cdots \otimes T_{n_r}(C)$. In case that $r = 0$ or $n_1 = \cdots = n_r = 0$ (LT) holds trivially so we may assume that $n_1, \ldots, n_r \geq 1$. By Proposition 3.3. in [**MPIP01**] $\mathrm{N}_{\mathrm{alt}}(B) = \mathrm{N}_{\mathrm{alt}}(T_{n_1}(C)) \otimes 1 \cdots \otimes 1 + \cdots + 1 \otimes \cdots \otimes 1 \otimes \mathrm{N}_{\mathrm{alt}}(T_{n_r}(C))$. The proof of Proposition 4.7. in [**MPIP01**] shows that the multiplication operators by elements in different summands $1 \otimes \cdots \otimes \mathrm{N}_{\mathrm{alt}}(T_{n_i}(C)) \otimes \cdots \otimes 1$ commute. Thus, the Lie algebra T_+ generated by $\langle \mathrm{ad}_a \mid a \in \mathrm{N}_{\mathrm{alt}}(B)' \rangle$ decomposes as the direct sum of ideals isomorphic to simple Lie algebras of B_3. We can proceed as in the proof of Proposition 5.10 to obtain that B satisfies (LT). □

References

[Boe94] A. H. Boers, *On assosymmetric rings*, Indag. Math. (N.S.) **5** (1994), no. 1, 9–27.

[Bre02] Murray Bremner, *Additive structure of free left-symmetric and assosymmetric rings*, Int. J. Math. Game Theory Algebra **12** (2002), no. 1, 23–37.

[EO01] Alberto Elduque and Susumu Okubo, *Local triality and some related algebras*, J. Algebra **244** (2001), no. 2, 828–844.

[HJP96] Irvin Roy Hentzel, David Pokrass Jacobs, and Luiz Antonio Peresi, *A basis for free assosymmetric algebras*, J. Algebra **183** (1996), no. 2, 306–318.

[Jac51] Nathan Jacobson, *Completely reducible Lie algebras of linear transformations*, Proc. Amer. Math. Soc. **2** (1951), 105–113.

[KK08] Hyuk Kim and Kyunghee Kim, *The structure of assosymmetric algebras*, J. Algebra **319** (2008), no. 6, 2243–2258.

[Kle57] Erwin Kleinfeld, *Assosymmetric rings*, Proc. Amer. Math. Soc. **8** (1957), 983–986.

[Mor42] V. V. Morozov, *On a nilpotent element in a semi-simple Lie algebra*, C. R. (Doklady) Acad. Sci. URSS (N.S.) **36** (1942), 83–86.

[MPIP01] Patrick J. Morandi, José M. Pérez-Izquierdo, and S. Pumplün, *On the tensor product of composition algebras*, J. Algebra **243** (2001), no. 1, 41–68.

[Oku05] Susumu Okubo, *Symmetric triality relations and structurable algebras*, Linear Algebra Appl. **396** (2005), 189–222.

[Oku06] ———, *Algebras satisfying symmetric triality relations*, Non-associative algebra and its applications, Lect. Notes Pure Appl. Math., vol. 246, Chapman & Hall/CRC, Boca Raton, FL, 2006, pp. 313–321.

[PI07] José M. Pérez-Izquierdo, *Algebras, hyperalgebras, nonassociative bialgebras and loops*, Adv. Math. **208** (2007), no. 2, 834–876.

[PIS04] José M. Pérez-Izquierdo and Ivan P. Shestakov, *An envelope for Malcev algebras*, J. Algebra **272** (2004), no. 1, 379–393.

[PR77] David Pokrass and David Rodabaugh, *Solvable assosymmetric rings are nilpotent*, Proc. Amer. Math. Soc. **64** (1977), no. 1, 30–34.

[Sch95] Richard D. Schafer, *An introduction to nonassociative algebras*, Dover Publications Inc., New York, 1995, Corrected reprint of the 1966 original.

[SR00a] K. Suvarna and G. Rama Bhupal Reddy, *On prime assosymmetric rings*, Acta Cienc. Indica Math. **26** (2000), no. 4, 393–394.

[SR00b] K. Suvarna and G. Ramabhupal Reddy, *On flexibility of prime assosymmetric rings*, Jñānābha **30** (2000), 37–38.

[SS87] K. Sitaram and K. Suvarna, *Some prime assosymmetric rings*, Indian J. Pure Appl. Math. **18** (1987), no. 12, 1103–1106.

[ZSSS82] K. A. Zhevlakov, A. M. Slin′ko, I. P. Shestakov, and A. I. Shirshov, *Rings that are nearly associative*, Pure and Applied Mathematics, vol. 104, Academic Press Inc. [Harcourt Brace Jovanovich Publishers], New York, 1982, Translated from the Russian by Harry F. Smith.

Dpto. Matemáticas y Computación, Universidad de La Rioja, 26004 Logroño, Spain
E-mail address: jm.perez@unirioja.es

Contemporary Mathematics
Volume **483**, 2009

A decomposition of LF-quasigroups

Peter Plaumann, Liudmila Sabinina, and Larissa Sbitneva

ABSTRACT. In this note we proof a decomposition theorem (as semi–direct product) for suitable isotopes of LF-quasigroups which are smooth or finite. The two factors arising in this product correspond to fundamental types of transsymmetric spaces.

Introduction

A quasigroup $(Q, \cdot, /, \backslash)$ is called an LF-quasigroup if the identity

$$(LF) \qquad x \cdot (y \cdot z) = (x \cdot y)(x \backslash x \cdot z)$$

holds. LF-quasigroups have first been considered by Murdoch in [**12**] and have been studied intensively by Belousov and his school (see [**1**], [**2**]). Considering the map $e : Q \longrightarrow Q$ definded by $e(x) = e_x = x \backslash x$ the identity (LF) takes the form

$$x \cdot (y \cdot z) = (x \cdot y) \cdot (e_x \cdot z).$$

We say that a quasigroup Q is smooth if Q is a $\mathcal{C}^\infty$-manifold and all operations are $\mathcal{C}^\infty$-maps. A *regular transsymmetric structure* on a $\mathcal{C}^\infty$-manifold M consists of a family $\{\mathcal{S}_x\}_{x \in M}$ of (local) diffeomorphisms of M and a $\mathcal{C}^\infty$-map e from M to M such that

$$(TS) \qquad \mathcal{S}_x \mathcal{S}_y = \mathcal{S}_{\mathcal{S}_x y} \mathcal{S}_{e(x)}, \qquad e(x) = S_x^{-1} x \text{ (locally)}$$

holds (see [**15**], [**14**, (A.4.3), p. 206]). A special case of this definition is the class of *generalized symmetric spaces* which are characterized by

$$(GS) \qquad \mathcal{S}_x \mathcal{S}_y = \mathcal{S}_{\mathcal{S}_x y} \mathcal{S}_x, \qquad S_x y = y \implies y = x \text{ (locally)}$$

(see [**10**], [**9**]).

If in addition in a generalized symmetric space one has

$$(S) \qquad \mathcal{S}_x^2 = \mathrm{id}_M,$$

then M is a *symmetric space* (see [**11**]).

Given a smooth LF-quasigroup Q, the manifold Q becomes a transsymmetric space choosing for $\mathcal{S}_x$ the left translations L_x given by $L_x y = x \cdot y$. If Q is left distributive, that is if $e = \mathrm{id}_Q$, then one obtains this way a generalized symmetric

---

1991 *Mathematics Subject Classification.* Primary 20N05; Secondary 22F30.
Key words and phrases. Quasigroups, transsymmetric spaces, Fitting decomposition.
This work has been supported partially by PROMEP (SEP, México). The first author thanks the Programm "Cátedras Especiales. SRE" of the Mexican Government for financial support.

space. This generalized symmetric space is a symmetric space if in Q the so called key identity $x(xy) = y$ holds.

In this note we treat the algebraic structure of an LF-quasigroup Q up to isotopism, making systematic use of the fact that $e : Q \longrightarrow Q$ is an endomorphism. After applying a suitable isotopy we always can assume that Q has a left identity (Proposition 1.3). Under this assumption, the most general splitting theorem we obtain says that a finite LF-quasigroup with a left identity is a semi-direct product of a left nuclearly nilpotent normal subquasigroup with a subquasigroup isotopic to a left distributive quasigroup (Corollary 3.4). For smooth connected LF-quasigroups with a left identity we prove a similar result (Theorem 3.6). Our investigations follow suggestions of Lev Sabinin.

If an LF-quasigroup Q satisfies the additional identity

$$(RF) \qquad xy \cdot z = (x \cdot z/z)(yz),$$

it is called an F-quasigroup. Answering a question posed by V. D. Belousov in 1967, it is shown in [**6**] that every F-quasigroup is isotopic to a Moufang loop which is the product of its nucleus N and its Moufang center C. Belousov has shown that a distributive quasigroup is isotopic to a commutative Moufang loop. It seems remarkable that under much weaker assumptions we obtain a similar factorisation.

1. Algebraic Preliminaries

Let $(Q, \cdot, /, \backslash)$ be a quasigroup. For $a \in Q$ one puts $e_a = a \backslash a$. If the identity

$$x(yz) = (xy)(e_x z)$$

holds in Q, then Q is called an LF-quasigroup (see [**1**], [**2**]). Special cases of this notion are the left distributive quasigroups (LD-quasigroups), where $e_x = x$ and the groups where $e_x z = z$ for all $x, z \in Q$.

Proposition 1.1. *If Q is an LF-quasigroup, then the mapping e is an endomorphism.*

Proof. (cf. [**2**, p. 108], [**16**]). Substituting z by e_y in the identity

$$x(yz) = (xy)(e_x z)$$

and using $ye_y = y$ we obtain

$$xy = (xy)e_{xy} = (xy)(e_x e_y).$$

Hence $e_{xy} = e_x e_y$. □

An *isotopy* of a quasigroup $(Q, \circ, /, \backslash)$ is a triple $T = (\alpha, \beta, \gamma)$ of bijections on Q. Defining an operation $\circ_T$ on Q by

$$x \circ_T y = \gamma^{-1}(\alpha(x) \circ \beta(y))$$

one obtains a quasigroup $Q^T = (Q, \circ_T, /_T, \backslash_T)$ (see [**3**], Chapter III). Isotopism is an equivalence relation between quasigroups. If $\gamma = \mathrm{id}_Q$, then T is called a *principal isotopy* of Q. The following lemma is well known (see [**1**, p. 15f.]).

Lemma 1.1. *Let T be an isotopy of a quasigroup $(Q, \cdot)$ such that Q^T is a loop. Then there are elements $a, b \in Q$ for which the isotopy $T_{a,b} = (R_a^{-1}, L_b^{-1}, \mathrm{id}_Q)$ of Q defines a quasigroup $Q^{T_{a,b}}$ isomorphic to Q^T. The neutral element of $Q^{T_{a,b}}$ is $b \cdot a$.*

PROPOSITION 1.2. *If Q is an LF-quasigroup and the mapping e is an automorphism of Q, then Q is isotopic to an LD-quasigroup.*

PROOF. For the isotopy $T = (\mathrm{id}_Q, e, \mathrm{id}_Q)$ the quasigroup Q^T is an LD-quasigroup (see [**2**, p. 108]). □

PROPOSITION 1.3. *For every LF-quasigroup Q there exists an LF-quasigroup Q_1 with a left identity which is isotopic to Q.*

PROOF. Consider the isotopy $T = (\mathrm{id}_Q, L_a^{-1}, \mathrm{id}_Q)$. Then Q^T is an LF-quasigroup in which the element a is a left identity (see [**1**], [**16**]). □

A loop $G = (G, *, 1_G)$ is called an LM-loop if there is a mapping $\phi : G \longrightarrow G$, called the parameter, such that the identity

$$x * (y * z) = (x * (y * \mathcal{J}\phi x)) * (\phi x * z)$$

holds, where $a * \mathcal{J}a = 1_G$ for all $a \in G$.

PROPOSITION 1.4. *Every loop isotopic to an LF-quasigroup $(Q, \cdot)$ is an LM-loop. If Q has a left identity 1_l then for the isotopy $T_0 = (R_{1_l}^{-1}, \mathrm{id}, \mathrm{id})$ the following hold:*

(1) *The parameter of the LM-loop $(Q, \cdot_{T_0})$ is the mapping e.*
(2) *The mapping e is an endomorphism of the loop $(Q, \cdot_{T_0})$.*
(3) *The kernel of e is a subgroup of $(Q, \cdot_{T_0})$.*

PROOF. In [**2**, p. 108] it is shown that, under the isotopy $T = (R_a^{-1}, L_b^{-1}, \mathrm{id}_Q)$ with arbitrary elements $a, b \in Q$, one obtains an LM-loop Q^T with the parameter $\phi_{a,b} = R_a^{-1} L_b^{-1} e R_a$. We specialize to $a = b = 1_l$. Then (1) is obvious. Furthermore

$$e(x \cdot_{T_0} y) = e(x/1_l \cdot y) = e(x)/e(1_l) \cdot e(y) = e(x)/1_l \cdot e(y) = e(x) \cdot_{T_0} e(y).$$

Finally (3) follows from the fact that a loop which is isotopic to group is a group. □

One observes that the parameter $\phi_{a,b}$ is bijective if and only if the map e is bijective.

PROPOSITION 1.5. *Every loop isotopic to an LM-loop is an LM–loop.*

PROOF. Let L be an LM-loop with the parameter ϕ and let $T = (R_a, L_b, \mathrm{id}_L)$ be an isotopy of L. Defining $\beta_1(y) = b((y/a)(\mathcal{J}\phi b))$ we see that L^T is an LM-loop with the parameter $\beta_1 R_a \phi R_a^{-1}$ (see [**2**, Theorem 3.18, p. 109]). □

A loop L has been called an RCC-loop if the identity

$$x(yz) = ((xy)/x)(xz)$$

holds (see [**4**], [**8**]). If an RCC-loop L has the inverse property (see [**3**]), one obtains

$$L_x L_y L_x^{-1} = L_{x \cdot y \mathcal{J} x}.$$

Hence L is an LM-loop with the parameter $\phi = \mathrm{id}_L$. Suppose that such a loop L were an isotope of some LF-quasigroup L_1. Then

$$\mathrm{id}_L = R_a^{-1} L_b^{-1} e R_a,$$

which implies $L_b = e$. But from this it follows that L_1 is a group. There are, however, RCC-loops with the inverse property which are not groups (see [**5**]). Thus we have shown

PROPOSITION 1.6. *There are LM-loops which are not isotopic to an LF-quasigroup.*

2. LF-quasigroups with a left identity

Justified by Proposition 1.3, in this section we shall consider LF-quasigroups Q with a left neutral element denoted by 1_l .

In the variety $\mathcal{Q}_l$ of quasigroups with a left identity element 1_l we define the *left kernel* of an homorphism $\gamma : Q \longrightarrow Q'$ by $\ker_l \gamma = \{x \in Q \mid \gamma(x) = 1_l\}$. Since $1_l \in \ker_l \gamma$ it follows from [**7**, Section 3, p. 104] that $\ker_l \gamma$ is a normal subquasigroup of Q. The homomorhism theory for $\mathcal{Q}_l$ is described in [**7**], [**1**, p. 55f], [**3**, Chapter IV.9]. For $Q \in \mathcal{Q}_l$ and a homomorphism η defined on Q with $\ker_l \eta = K$ the quasigroup $\eta(Q)$ is isomorphic to the factor quasigroup Q/K consisting of the cosets Kx, $x \in Q$, where the multiplication in Q/K is given by $(Kx)(Ky) = K(xy)$.

For a quasigroup Q the set

$$N_\lambda(Q) = \{u \in Q \mid u(xy) = (ux)y, \text{ for all } x, y \in N\}$$

is called the *left nucleus* of Q.

REMARK 2.1. Observe for $Q \in \mathcal{Q}_l$ that $N_\lambda(Q)$ is a subgroup of Q, that 1_Q is the neutral element of $N_\lambda(Q)$ and that $N_\lambda(Q)$ has not to be normal in Q.

PROPOSITION 2.1. *Let Q be an LF-quasigroup with a left identity 1_l. Then the left kernel of the endomorphism e coincides with the left nucleus $N_\lambda(Q)$ of Q. Furthermore $N_\lambda(Q)$ is normal in Q and is the greatest subgroup of Q.*

PROOF. An element $x \in Q$ lies in $N_\lambda(Q)$ if and only if for all $y, z \in Q$ the identity

$$(xy)(e_x z) = x(yz) = (xy)z$$

holds. But this is equivalent to $e_x = 1_l$.

Assume that H is a subgroup of Q. For $h \in H$ one has $e_h = h\backslash h$ and $h/h = 1_l$. Since 1_l is the neutral element of H, it follows that $e_h = 1_l$. Hence $h \in \ker_l e = N_\lambda(Q)$. □

If the image C of the endomorphism e of an LF-quasigroup Q with a left identity is a complement of the left nucleus $N_\lambda(Q)$ of Q, we say that Q is *nuclearly decomposable* and call C the *nuclear complement of* Q.

THEOREM 2.2. *Let Q be an LF-quasigroup with a left identity 1_l.*

(1) *Q is nuclearly decomposable if and only if $e|_{e(Q)}$ is a bijection,*
(2) *If C is a nuclear complement in Q, then C is isotopic to an LD-quasigroup. In particular $N_\lambda(C) = 1_l$.*

PROOF. The theorem follows from Proposition 1.2 and Proposition 2.1. □

3. LF-quasigroups of finite degree

The homomorphism theory in the variety $\mathcal{Q}_l$, as described in Section 2, allows to use the concept of a semidirect product analogously to its use in the variety of groups. One says that a quasigroup Q is the semidirect product of a normal subquasigroup N and a subquasigroup H if $Q = NH$ and $\{1_l\} = N \cap H$ hold. In this case we write $Q = N \rtimes H$.

PROPOSITION 3.1. *Let Q be a finite quasigroup with a left identity and let η be an endomorphism of Q for which $\ker_l \eta \cap \eta(Q) = \{1_l\}$. Then Q is the semidirect product of $\ker_l \eta$ and $\eta(Q)$.*

PROOF. As mentioned above, the quasigroup $\eta(Q)$ is isomorphic to $Q/(\ker_l \eta)$. Hence $|Q| = |\ker_l \eta| \cdot |\eta(Q)|$, and the proposition follows. □

We say that an endomorphism η of an arbitrary quasigroup Q has *degree d* if

$$\eta^{d+1}(Q) = \eta^d(Q) \supsetneq \eta^{d-1}(Q).$$

If in an LF-quasigroup Q the endomorphism e has degree d, we say that Q has degree d. The following lemma is a consequence of Proposition 3.1.

LEMMA 3.1. *a) If a quasigroup $Q \in \mathcal{Q}_l$ satisfies the minimal condition for subquasigroups containing 1_Q then every endomorphism of Q has finite degree.*

b) For an endomorphism η of a finite quasigroup $Q \in \mathcal{Q}_l$ there is a natural number d such that $Q = \ker_l(\eta^d) \rtimes \eta^d(G)$.

Justified by the geometric considerations in the introduction we call an LF-quasigroup with a left identity 1_l a *GS-quasigroup* if its left nucleus is equal to $\{1_l\}$. As a consequence of Proposition 2.1 and Proposition 1.2 one obtains

PROPOSITION 3.2. *If Q is a GS-quasigroup, then Q is isotopic to an LD-quasigroup.*

Let $\mathcal{Q}$ be a class of quasigroups with a left identity satisfying

(Q1) $\mathcal{Q}$ is closed under subquasigroups and epimorphic images,

(Q2) For $Q \in \mathcal{Q}$ the left nucleus $N_\lambda(Q)$ is a normal subquasigroup of Q.

Following the approach of [**3**, Chapter VI.1] for a quasigroup $Q \in \mathcal{Q}$ we define the *ascending left nuclear series* of Q by

$$N_\lambda^1(Q) = N_\lambda(Q), \qquad N_\lambda^{i+1}(Q)/N_\lambda^i(Q) = N_\lambda\big(Q/N_\lambda^i(Q)\big)$$

for all $i \in \mathbb{N}$.

If the ascending left nuclear series of Q becomes stationary, its final term $N_\lambda^\infty(Q)$ is called the *left hypernucleus* of Q. The quasigroup Q is called *left nuclearly nilpotent* if $Q = N_\lambda^\infty(Q)$.

LEMMA 3.2. *Let Q be an LF-quasigroup. Then for all $i \in \mathbb{N}$*

(1) $\ker_l(e^i) = N_\lambda^i(Q)$,

(2) $N_\lambda(e^i(Q)) \subseteq N_\lambda^i(Q) \cap e^i(Q)$.

PROOF. (1) By induction it follows from Proposition 2.1 that $\ker_l(e^i) = N_\lambda^i(Q)$ for all $i \in \mathbb{N}$.

(2) Since $N_\lambda(e^i(Q))$ is a group, by Proposition 2.1 we have

$$N_\lambda(e^i(Q)) \subseteq N_\lambda(Q) \cap e^i(Q) \subseteq N_\lambda^i(Q) \cap e^i(Q).$$

□

THEOREM 3.3. *Let Q be a finite LF-quasigroup with a left identity. Then there is a natural number d such that $Q = \ker_l(e^d) \rtimes e^d(Q)$. Furthermore,*

(a) $\ker_l(e^d)$ *is the left hypernucleus of Q,*

(b) $e^d(Q)$ *is a GS-quasigroup.*

PROOF. The existence of the claimed decomposition follows from Lemma 3.1, and (a) is a consequence of Lemma 3.2(1).

From (a) and Lemma 3.2(2) it follows that

$$N_\lambda^d(Q) \subseteq \ker_l(e^d) \cap e^d(Q) = \{1_l\}.$$

Thus (b) is shown, too. □

Now Proposition 3.2 gives us the following

COROLLARY 3.4. A finite LF-quasigroup with a left identity is a semi-direct product of its hypernucleus and the isotope of an LD-quasigroup.

We say that a smooth connected quasigroup Q in the variety $\mathcal{Q}_l$ is an *almost semi-direct product* of a closed normal subquasigroup N and a closed subquasigroup H if $Q = NH$ and if $N \cap H$ is discrete. We denote this by $Q = N \rtimes_a H$.

PROPOSITION 3.3. *Let Q be a smooth connected quasigroup with a left identity and let η be a continuous endomorphism of Q for which $\ker_l \eta \cap \eta(Q)$ is discrete. Then Q is an almost semi-direct product of $\ker_l \eta$ and $\eta(Q)$.*

PROOF. It is well known that $\dim Q = \dim \ker_l \eta + \dim \eta(Q)$. Hence it follows that $(\ker_l \eta)\eta(Q)$ is a subquasigroup of Q which has the same dimension as Q. From this we conclude that $(\ker_l \eta)\eta(Q) = Q$. □

LEMMA 3.5. *Every continuous endomorphism η of a smooth connected quasigroup Q with a left identity has finite degree $d \in \mathbb{N}$, and Q is an almost semi-direct product of $\ker_l \eta^d$ and $\eta^d(G)$.*

PROOF. Since the subquasigroups $\eta^i(Q)$ are connected, it follows from

$$Q \supseteq \eta(Q) \supseteq \cdots \supseteq \eta^i(Q) \supseteq \ldots.$$

that η has degree $d \leq \dim Q$. But then the restriction of all powers η^i, $i \in \mathbb{N}$ to $\eta^d(Q = \eta^{d+1}(Q)$ are surjective, and it follows that the subquasigroups $\ker_l(\eta^i) \cap \eta^d(Q)$ are discrete for all $i \in \mathbb{N}$. Hence by Proposition 3.3 the quasigroup Q is the almost semi-direct product of $\ker_l \eta^d$ and $\eta^d(Q)$. □

As an immediate consequence of Lemma 3.5 one obtains

PROPOSITION 3.4. *If η is a continuous endomorphism of a smooth connected quasigroup Q with a left identity, then there is a natural number d such that Q is the almost semi-direct product of $\ker_l \eta^d$ with $\eta^d(Q)$.*

We say that a smooth connected quasigroup Q with a left identity is *locally* a GS-quasigroup if $N_\lambda(Q)$ is discrete.

THEOREM 3.6. *Let Q be a smooth connected LF-quasigroup with a left identity. Then there is a natural number d such that $Q = \ker_l(e^d) \rtimes_a e^d(G)$. Furthermore,*

(a) *$\ker_l(e^d)$ is the left hypernucleus of Q,*
(b) *$e^d(Q)$ is locally a GS-quasigroup.*

PROOF. (a) The existence of the claimed decomposition follows from Lemma 3.5, and (a) is a consequence of Lemma 3.2 (1).

(b) From (a) and Lemma 3.2(2) it follows that $N_\lambda^d(Q) \subseteq \ker_l(e^d) \cap e^d(Q)$. Thus (b) follows from the discreteness of $\ker_l(e^d) \cap e^d(Q)$. □

References

[1] V. D. Belousov, *Foundations of the theory of quasigroups and loops*, Izdat. Nauka, Moscow, 1967 (In Russian).

[2] V. D. Belousov, *Elements of Quasigroup Theory: A Special Course*, Kishinev State University Press, Kishinev, 1981 (Russian).

[3] R. B. Bruck, *A survey of binary systems*. Springer-Verlag (Ergebnisse der Mathematik und ihrer Grenzgebiete. Neue Folge. Heft 20.) 1958.

[4] E. G. Goodaire, D. A. Robinson, *Semi-direct products and Bol loops*, Demonstratio Math. **27** (1994), 573–588.

[5] M. K. Kinyon, K. Kunen, J. D. Phillips, *Diassociativity in conjugacy closed loops*, Comm. Algebra **32** (2004), 767–786.

[6] T. Kepka, M. K. Kinyon, J. D. Phillips, *On the structure of F-quasigroups*, J. of Algebra, **317** no. 2 (2007), 435–461.

[7] F. Kiokemeister, *A Theory of Normality for Quasigroups*, Amer. J. of Math. **70** no. 1 (1948), 99-106.

[8] K. Kunen, *The structure of conjugacy closed loops*, Trans. Amer. Math. Soc. **352** (2000), 2889–2911.

[9] O. Kowalski, *Generalized symmetric spaces*, Springer-Verlag, Lecture Notes in Mathematics, vol. 805, 1980.

[10] A. J. Ledger, *Espaces de Riemann symétriques généralisés*, C. R. Acad. Sci. Paris Sr. A-B 264, 1967.

[11] O. Loos, *Symmetric spaces. I: General theory*, W. A. Benjamin, 1969.

[12] D. C. Murdoch, *Quasigroups which satisfy certain generalized associative laws*, Amer. J. Math., **61** (1939), 509-522.

[13] H. O. Pflugfelder, *Quasigroups and loops: introduction*, Heldermann Verlag, Sigma Series in Pure Mathematics, vol. 7, 1990.

[14] L. Sabinin, *Smooth quasigroups and loops*, Kluwer Academic Publisher, Mathematics and its Applications, vol. 492, 1999.

[15] L. V. Sabinin, L. L. Sabinina, *On geometry of transsymmetric spaces*, Webs and quasigroups (Moscow, 1989), Tver. Gos. Univ. Press, (1991), 117–122.

[16] L. V Sabinin, L. L. Sabinina, *On the theory of left F-quasigroups*, Algebras Groups Geom. **12** (1995), 127–137.

[17] M. F. Smiley, *Notes on Left Division Systems with Left Unity*, Amer. J. Math., **74** (1953), 679-682.

Mathematisches Institut Universität Erlangen-Nürnberg Bismarckstrasse 1 1/2 D-91054 Erlangen GERMANY

E-mail address: `peter.plaumann@mi.uni-erlangen.de`

Facultad de Ciencias, Universidad Autónoma del Estado de Morelos, Avenida Universidad 1001, 62209, Cuernavaca, Morelos, MÉXICO

E-mail address: `liudmila@buzon.uaem.mx`

Facultad de Ciencias, Universidad Autónoma del Estado de Morelos, Avenida Universidad 1001, 62209, Cuernavaca, Morelos, MÉXICO

E-mail address: `larissasbitneva@hotmail.com`

Contemporary Mathematics
Volume **483**, 2009

Braided and coboundary monoidal categories

Alistair Savage

Dedicated to Ivan Shestakov on his sixtieth birthday

ABSTRACT. We discuss and compare the notions of braided and coboundary monoidal categories. Coboundary monoidal categories are analogues of braided monoidal categories in which the role of the braid group is replaced by the cactus group. We focus on the categories of representations of quantum groups and crystals and explain how while the former is a braided monoidal category, this structure does not pass to the crystal limit. However, the categories of representations of quantum groups of finite type also possess the structure of a coboundary category which does behave well in the crystal limit. We explain this construction and also a recent interpretation of the coboundary structure using quiver varieties. This geometric viewpoint allows one to show that the category of crystals is in fact a coboundary monoidal category for arbitrary symmetrizable Kac-Moody type.

Introduction

In this expository paper we discuss and contrast two types of categories – braided monoidal categories and coboundary monoidal categories – paying special attention to how the categories of representations of quantum groups and crystals fit into this framework. Monoidal categories are essentially categories with a tensor product, such as the categories of vector spaces, abelian groups, sets and topological spaces. Braided monoidal categories are well-studied in the literature. They are monoidal categories with an action of the braid group on multiple tensor products. The example that interests us the most is the category of representations of a quantum group $U_q(\mathfrak{g})$. Coboundary monoidal categories are perhaps less well known than their braided cousins. The concept is similar, the difference being that the role of the braid group is now played by the so-called *cactus group*. A key component in the definition of a coboundary monoidal category is the *cactus commutor*, which assumes the role of the braiding.

The theory of crystals can be thought of as the $q \to \infty$ (or $q \to 0$) limit of the theory of quantum groups. In this limit, representations are replaced by

2000 *Mathematics Subject Classification.* Primary: 17B37, 18D10; Secondary: 16G20.
This research was supported by the Natural Sciences and Engineering Research Council (NSERC) of Canada.

combinatorial objects called crystal graphs. These are edge-colored directed graphs encoding important information about the representations from which they come. Developing concrete realizations of crystals is an active area of research and there exist many different models.

It is interesting to ask if the structure of a braided monoidal category passes to the crystal limit. That is, does one have an induced structure of a braided monoidal category on the category of crystals. The answer is no. In fact, one can prove that it is impossible to give the category of crystals the structure of a braided monoidal category (see Proposition 5.6). However, the situation is more hopeful if one instead considers coboundary monoidal categories. For quantum groups of finite type, there is a way – a unitarization procedure introduced by Drinfel'd [**5**] – to use the braiding on the category of representations to define a cactus commutor on this category. This structure passes to the crystal limit and one can define a coboundary structure on the category of crystals in finite type (see [**7, 10**]). Kamnitzer and Tingley [**9**] gave an alternative definition of the crystal commutor which makes sense for quantum groups of arbitrary symmetrizable Kac-Moody type. However, while this definition agrees with the previous one in finite type, it is not obvious that it satisfies the desired properties, giving the category of crystals the structure of a coboundary category, in other types.

In [**24**], the author gave a geometric realization of the cactus commutor using quiver varieties. In this setting, the commutor turns out to have a very simple interpretation – it corresponds to simply taking adjoints of quiver representations. Equipped with this geometric description, one is able to show that the crystal commutor satisfies the requisite properties and thus the category of crystals, in arbitrary symmetrizable Kac-Moody type, is a coboundary category.

In the current paper, when discussing the topics of quantum groups and crystals, we will often restrict our attention to the quantum group $U_q(\mathfrak{sl}_2)$ and its crystals. This allows us to perform explicit computations illustrating the key concepts involved. The reader interested in the more general case can find the definitions in the references given throughout the paper.

The organization of this paper is as follows. In Section 1, we introduce the braid and cactus groups that play an important role in the categories in which we are interested. In Sections 2 through 4, we define monoidal, braided monoidal, and coboundary monoidal categories. We review the theory of quantum groups and crystals in Section 5. In Section 6 we recall the various definitions of cactus commutors in the categories of representations of quantum groups and crystals. Finally, in Section 7, we give the geometric interpretation of the commutor.

The author would like to thank J. Kamnitzer and P. Tingley for very useful discussions during the writing of this paper and for helpful comments on an earlier draft.

1. The braid and cactus groups

1.1. The braid group.

DEFINITION 1.1 (Braid group). *For n a positive integer, the n-*strand Braid group *$\mathcal{B}_n$ is the group with generators $\sigma_1, \dots, \sigma_{n-1}$ and relations*

(1) $\sigma_i \sigma_j = \sigma_j \sigma_i$ *for* $|i-j| \geq 2$, *and*
(2) $\sigma_i \sigma_{i+1} \sigma_i = \sigma_{i+1} \sigma_i \sigma_{i+1}$ *for* $1 \leq i \leq n-2$.

These relations are known as the braid relations *and the second is often called the* Yang-Baxter equation.

Recall that the symmetric group S_n is the group on generators $s_1, \ldots, s_{n-1}$ satisfying the same relations as for the σ_i above in addition to the relations $s_i^2 = 1$ for all $1 \le i \le n-1$. We thus have a surjective group homomorphism $\mathcal{B}_n \twoheadrightarrow S_n$. The kernel of this map is called the *pure braid group.*

The braid group has several geometric interpretations. The one from which its name is derived is the realization in terms of braids. An *n-strand braid* is an isotopy class of a union of n non-intersecting smooth curve segments (strands) in $\mathbb{R}^3$ with end points $\{1, 2, \ldots, n\} \times \{0\} \times \{0, 1\}$, such that the third coordinate is strictly increasing from 0 to 1 in each strand. The set of all braids with multiplication giving by placing one braid on top of another (and rescaling so that the third coordinate ranges from 0 to 1) is isomorphic to $\mathcal{B}_n$ as defined algebraically above.

The braid group is also isomorphic to the mapping class group of the n-punctured disk – the group of self-homeomorphisms of the punctured disk with n-punctures modulo the subgroup consisting of those homeomorphisms isotopic to the identity map. One can picture the isomorphism by thinking of each puncture being connected to the boundary of the disk by a string. Each homeomorphism of the n-punctured disk can then be seen to yield a braiding of these strings. The pure braid group corresponds to the classes of homeomorphisms that map each puncture to itself.

A similar geometric realization of the braid group is as the fundamental group of the configuration space of n points in the unit disk D. A loop from one configuration to itself in this space defines an n-strand braid where each strand is the trajectory in $D \times [0, 1]$ traced out by one of the n points. If the points are labeled, then we require each point to end where it started and the corresponding fundamental group is isomorphic to the pure braid group.

1.2. The cactus group. Fix a positive integer n. For $1 \le p < q \le n$, let

$$\hat{s}_{p,q} = \begin{pmatrix} 1 & \cdots & p-1 & p & p+1 & \cdots & q & q+1 & \cdots & n \\ 1 & \cdots & p-1 & q & q-1 & \cdots & p & q+1 & \cdots & n \end{pmatrix} \in S_n.$$

Since $s_{i,i+1} = s_i$, these elements generate S_n. If $1 \le p < q \le n$ and $1 \le k < l \le n$, we say that $p < q$ and $k < l$ are *disjoint* if $q < k$ or $l < p$. We say that $p < q$ *contains* $k < l$ if $p \le k < l \le q$.

Definition 1.2 (Cactus group). *For n a positive integer, the n-*fruit cactus *group J_n is the group with generators $s_{p,q}$ for $1 \le p < q \le n$ and relations*

(1) $s_{p,q}^2 = 1$,
(2) $s_{p,q} s_{k,l} = s_{k,l} s_{p,q}$ *if $p < q$ and $k < l$ are disjoint, and*
(3) $s_{p,q} s_{k,l} = s_{r,t} s_{p,q}$ *if $p < q$ contains $k < l$, where $r = \hat{s}_{p,q}(l)$ and $t = \hat{s}_{p,q}(k)$.*

It is easily checked that the elements $\hat{s}_{p,q}$ of the symmetric group satisfy the relations defining the cactus group and thus the map $s_{p,q} \mapsto \hat{s}_{p,q}$ extends to a surjective group homomorphism $J_n \twoheadrightarrow S_n$. The kernel of this map is called the *pure cactus group.*

The cactus group also has a geometric interpretation. In particular, the kernel of the surjection $J_n \twoheadrightarrow S_n$ is isomorphic to the fundamental group of the Deligne-Mumford compactification $\overline{M}_0^{n+1}(\mathbb{R})$ of the moduli space of real genus zero curves with $n+1$ marked points. The generator $s_{p,q}$ of the cactus group corresponds

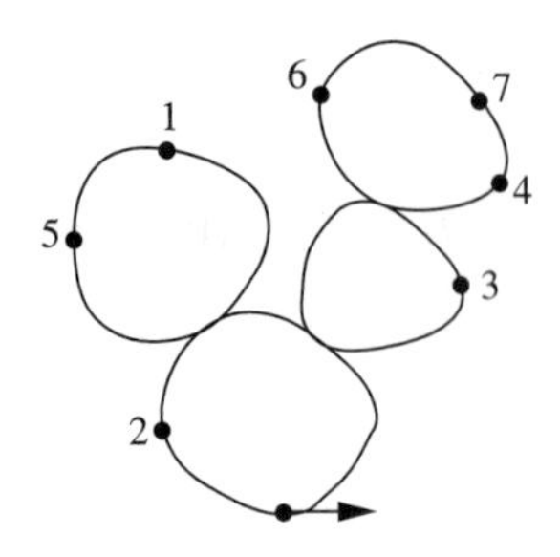

FIGURE 1. A 7-fruited cactus

to a path in $\overline{M}_0^{n+1}(\mathbb{R})$ in which the marked points $p, \ldots, q$ balloon off into a new component, this components flips, and then the component collapses, the points returning in reversed order. Elements of $\overline{M}_0^{n+1}(\mathbb{R})$ look similar to cacti of the genus *Opuntia* (the marked points being flowers) which justifies the name *cactus group* (see Figure 1). Note the similarity with the last geometric realization of the braid group mentioned in Section 1.1. Both the pure braid group and the pure cactus group are fundamental groups of certain spaces with marked points. We refer the reader to [**3, 4, 7**] for further details on this aspect of the cactus group.

1.3. The relationship between the braid and cactus groups. The relationship between the braid group and the cactus group is not completely understood. While the symmetric group is a quotient of both, neither is a quotient of the other. However, there is a homomorphism from the cactus group into the pro-unipotent completion of the braid group (see the proof of Theorem 3.14 in [**6**]). This map is closely related to the unitarization procedure of Drinfel′d to be discussed in Section 6.1.

In the next few sections, we will define categories that are closely related to the braid and cactus groups. We will see that there are some connections between the two. In addition to the unitarization procedure, we will see that braidings satisfy the so-called cactus relation (see Proposition 4.3).

2. Monoidal categories

2.1. Definitions. Recall that for two functors $F, G : \mathcal{C} \to \mathcal{D}$ a *natural transformation* $\varphi : F \to G$ is a collection of morphisms $\varphi_U : F(U) \to G(U)$, $U \in \operatorname{Ob}\mathcal{C}$, such that for all $f \in \operatorname{Hom}_{\mathcal{C}}(U, V)$, we have $\varphi_V \circ F(f) = G(f) \circ \varphi_U : F(U) \to G(V)$. If the maps φ_U are all isomorphisms, we call φ a *natural isomorphism*. We will sometimes refer to the φ_U themselves as *natural isomorphisms* when the functors involved are clear.

DEFINITION 2.1 (Monoidal Category). *A* monoidal category *is a category* $\mathcal{C}$ *equipped with the following:*

(1) *a bifunctor* $\otimes : \mathcal{C} \times \mathcal{C} \to \mathcal{C}$ *called the* tensor product,
(2) *natural isomorphisms (the* associator*)*

$$\alpha_{U,V,W} : (U \otimes V) \otimes W \xrightarrow{\cong} U \otimes (V \otimes W)$$

for all $U, V, W \in \operatorname{Ob}\mathcal{C}$ *satisfying the* pentagon axiom*: for all* $U, V, W, X \in \operatorname{Ob}\mathcal{C}$*, the diagram*

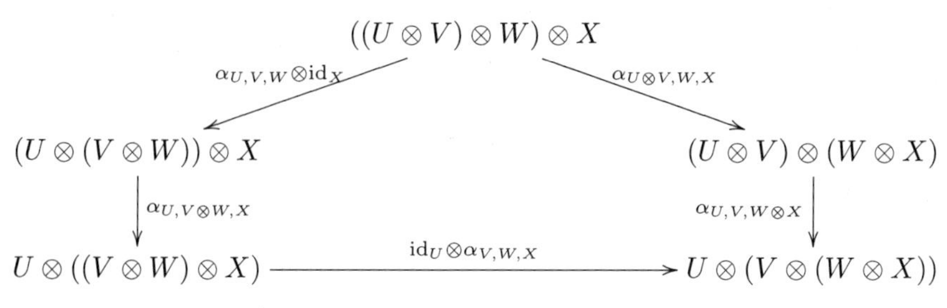

commutes, and

(3) *a* unit object $\mathbf{1} \in \operatorname{Ob}\mathcal{C}$ *and natural isomorphisms*

$$\lambda_V : \mathbf{1} \otimes V \xrightarrow{\cong} V, \quad \rho_V : V \otimes \mathbf{1} \xrightarrow{\cong} V$$

for every $V \in \operatorname{Ob}\mathcal{C}$*, satisfying the* triangle axiom*: for all* $U, V \in \operatorname{Ob}\mathcal{C}$*, the diagram*

$$\begin{array}{ccc} (U \otimes \mathbf{1}) \otimes V & \xrightarrow{\alpha_{U,\mathbf{1},V}} & U \otimes (\mathbf{1} \otimes V) \\ & {\scriptstyle \rho_U \otimes \mathrm{id}_V} \searrow \quad \swarrow {\scriptstyle \mathrm{id}_U \otimes \lambda_V} & \\ & U \otimes V & \end{array}$$

commutes.

A monoidal category is said to be strict *if we can take* $\alpha_{U,V,W}$*,* λ_V *and* ρ_V *to be identity morphisms for all* $U, V, W \in \operatorname{Ob}\mathcal{C}$*. That is, a* strict monoidal category *is one in which*

$$V \otimes \mathbf{1} = V, \quad \mathbf{1} \otimes V = V, \quad (U \otimes V) \otimes W = U \otimes (V \otimes W)$$

for all $U, V, W \in \operatorname{Ob}\mathcal{C}$*.*

The MacLane Coherence Theorem [**18**, §VII.2] states that the pentagon and triangle axioms ensure that any for any two expressions obtained from $V_1 \otimes V_2 \otimes \cdots \otimes V_n$ by inserting $\mathbf{1}$'s and parentheses, all isomorphisms of these two expressions consisting of compositions of α's, λ's and ρ's are equal. This condition is called the *associativity axiom*. In a monoidal category, we can use the natural isomorphisms to identify all expressions of the above type and so we often write multiple tensor products without brackets. In fact, every monoidal category is equivalent to a strict one [**18**, §XI.3].

2.2. Examples. Most of the familiar tensor products yield monoidal categories. For instance, for a commutative ring R, the category of R-modules is a monoidal category. We have the usual tensor product $A \otimes_R B$ of modules A and B. The unit object is R and we have the natural isomorphisms

$$\alpha : A \otimes_R (B \otimes_R C) \cong (A \otimes_R B) \otimes_R C, \quad \alpha(a \otimes_R (b \otimes_R c)) = (a \otimes_R b) \otimes_R c$$
$$\lambda : R \otimes_R A \cong A, \quad \lambda(r \otimes_R a) = ra,$$
$$\rho : A \otimes_R R \cong A, \quad \rho(a \otimes_R r) = ra.$$

In particular, the categories of abelian groups (where $R = \mathbb{Z}$) and vector spaces (where R is a field) are monoidal categories. In a similar fashion, the category of R-algebras is monoidal under the usual tensor product of algebras. For an arbitrary (not necessarily commutative) ring R, the category of R-R bimodules is also monoidal under $\otimes_R$. The categories of sets and topological spaces with the cartesian product are monoidal categories with unit objects 1 (the set with a single element) and $*$ (the single-element topological space) respectively.

3. Braided monoidal categories

3.1. Definitions.

DEFINITION 3.1 (Braided monoidal category). *A* braided monoidal category *(or* braided tensor category*) is a monoidal category $\mathcal{C}$ equipped with natural isomorphisms $\sigma_{U,V} : U \otimes V \to V \otimes U$ for all $U, V \in \mathrm{Ob}\,\mathcal{C}$ satisfying the* hexagon *axiom: for all $U, V, W \in \mathcal{C}$, the diagrams*

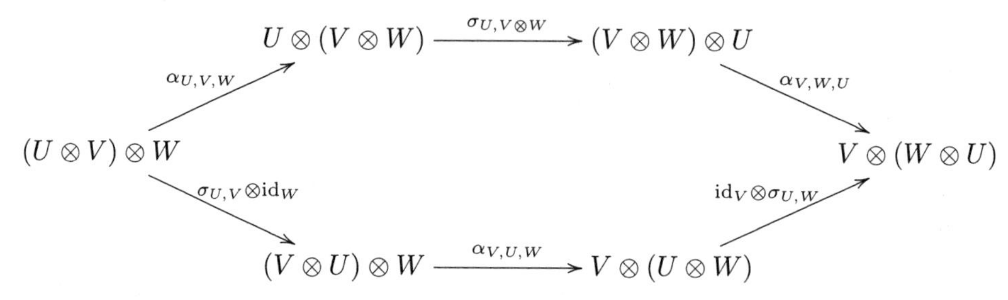

and

$$\begin{array}{ccccc} & U \otimes (V \otimes W) & \xrightarrow{\sigma^{-1}_{V\otimes W,U}} & (V \otimes W) \otimes U & \\ {}^{\alpha_{U,V,W}}\nearrow & & & & \searrow^{\alpha_{V,W,U}} \\ (U \otimes V) \otimes W & & & & V \otimes (W \otimes U) \\ {}_{\sigma^{-1}_{V,U} \otimes \mathrm{id}_W}\searrow & & & & \nearrow_{\mathrm{id}_V \otimes \sigma^{-1}_{W,U}} \\ & (V \otimes U) \otimes W & \xrightarrow{\alpha_{V,U,W}} & V \otimes (U \otimes W) & \end{array}$$

commute. The collection of maps $\sigma_{U,V}$ is called a braiding.

For a braided monoidal category $\mathcal{C}$ and $U, V, W \in \mathrm{Ob}\,\mathcal{C}$, consider the following diagram where we have omitted bracketings and associators (or assumed that $\mathcal{C}$ is strict).

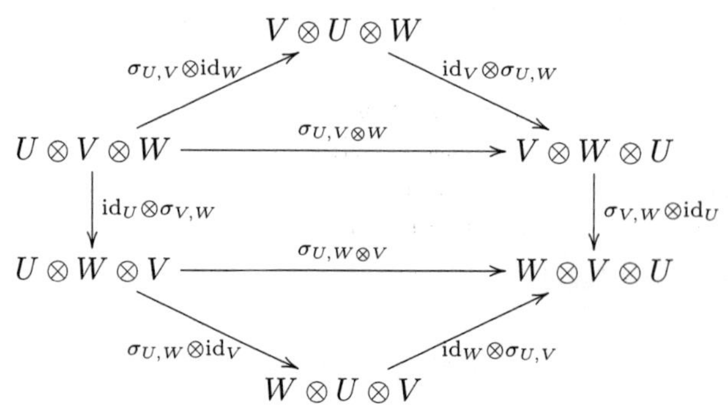

The top and bottom triangles commute by the hexagon axiom and the middle rectangle commutes by the naturality of the braiding (that is, by the fact that it is a natural isomorphism). Therefore, if we write σ_1 for the map $\sigma \otimes \mathrm{id}$ and σ_2 for the map $\mathrm{id} \otimes \sigma$, we have

$$\sigma_1\sigma_2\sigma_1 = \sigma_2\sigma_1\sigma_2,$$

the Yang-Baxter relation for $\mathcal{B}_3$. It follows that in a braided monoidal category, the braid group $\mathcal{B}_n$ acts on n-fold tensor products. That is, if we denote $\mathrm{id}^{\otimes(i-1)} \otimes \sigma \otimes \mathrm{id}^{\otimes(n-i-1)}$ by σ_i, then a composition of such maps depends only on the corresponding element of the braid group.

DEFINITION 3.2 (Symmetric monoidal category). *A* symmetric monoidal category *is a braided monoidal category $\mathcal{C}$ where $\sigma_{V,U} \circ \sigma_{U,V} = \mathrm{id}_{U\otimes V}$ for all $U, V \in \mathrm{Ob}\,\mathcal{C}$.*

In any symmetric monoidal category, the symmetric group S_n acts on n-fold tensor products in the same way that the braid group acts in a braided monoidal category. We note that there is conflicting terminology in the literature. For instance, some authors refer to monoidal categories as tensor categories while others (see, for instance, [**2**]) refer to symmetric monoidal categories as tensor categories and braided monoidal categories as quasitensor categories.

3.2. Examples. Many of the examples in Section 2.2 can in fact be given the structure of a symmetric monoidal category. In particular, the categories of R-modules and R-algebras over a commutative ring R, the category of sets, and the category of topological spaces are all symmetric monoidal categories. In all of these examples, the braiding is given by $\sigma(a \otimes b) = b \otimes a$.

An example that shall be especially important to us is the category of representations of a quantum group. It can be given the structure of a braided monoidal category but is not a symmetric monoidal category.

4. Coboundary monoidal categories

4.1. Definitions. Coboundary monoidal categories are analogues of braided monoidal categories in which the role of the braid group is replaced by the cactus group. As we shall see, they are better suited to the theory of crystals than braided monoidal categories.

DEFINITION 4.1 (Coboundary monoidal category). *A* coboundary monoidal category *is a monoidal category $\mathcal{C}$ together with natural isomorphisms $\sigma^c_{U,V} : U \otimes V \to V \otimes U$ for all $U, V \in \mathrm{Ob}\,\mathcal{C}$ satisfying the following conditions:*

(1) $\sigma^c_{V,U} \circ \sigma^c_{U,V} = \mathrm{id}_{U\otimes V}$, *and*
(2) the cactus relation*: for all $U, V, W \in \mathrm{Ob}\,\mathcal{C}$, the diagram*

$$\begin{array}{ccc} U \otimes V \otimes W & \xrightarrow{\sigma^c_{U,V}\otimes \mathrm{id}_W} & V \otimes U \otimes W \\ \Big\downarrow{\scriptstyle \mathrm{id}_U \otimes \sigma^c_{V,W}} & & \Big\downarrow{\scriptstyle \sigma^c_{V\otimes U, W}} \\ U \otimes W \otimes V & \xrightarrow{\sigma^c_{U,W\otimes V}} & W \otimes V \otimes U \end{array} \tag{4.1}$$

commutes.

The collection of maps $\sigma^c_{U,V}$ is called a cactus commutor.

We will use the term *commutor* for a collection of natural isomorphisms $\sigma_{U,V} : U\otimes V \to V\otimes U$ for all objects $U, V \in \mathrm{Ob}\,\mathcal{C}$ (this is sometimes called a *commutativity constraint*), reserving the term *cactus commutor* for a commutor satisfying the conditions in the definition above.

Suppose $\mathcal{C}$ is a coboundary category and $U_1, \dots, U_n \in \operatorname{Ob}\mathcal{C}$. For $1 \le p < q \le n$, one defines natural isomorphisms

$$\begin{aligned}\sigma^c_{p,q} = (\sigma^c_{p,q})_{U_1,\dots,U_n} &: U_1 \otimes \cdots \otimes U_n \\ &\to U_1 \otimes \cdots \otimes U_{p-1} \otimes U_{p+1} \otimes \cdots \otimes U_q \otimes U_p \otimes U_{q+1} \otimes \cdots \otimes U_n, \\ \sigma^c_{p,q} &:= \mathrm{id}_{U_1 \otimes \cdots \otimes U_{p-1}} \otimes \sigma^c_{U_p, U_{p+1} \otimes \cdots \otimes U_q} \otimes \mathrm{id}_{U_{q+1} \otimes \cdots \otimes U_n}.\end{aligned}$$

For $1 \le p < q \le n$, one then defines natural isomorphisms

$$s_{p,q} : U_1 \otimes \cdots \otimes U_n \to U_1 \otimes U_2 \otimes \cdots \otimes U_{p-1} \otimes U_q \otimes U_{q-1} \otimes \dots U_p \otimes U_{q+1} \otimes U_{q+2} \otimes \cdots \otimes U_n$$

recursively as follows. Define $s_{p,p+1} = \sigma^c_{p,p+1}$ and $s_{p,q} = \sigma^c_{p,q} \circ s_{p+1,q}$ for $q > p+1$. We also set $s_{p,p} = \mathrm{id}$. The following proposition was proved by Henriques and Kamnitzer.

PROPOSITION 4.2 ([**7**, Lemma 3, Lemma 4]). *If $\mathcal{C}$ is a coboundary category and the natural isomorphisms $s_{p,q}$ are defined as above, then*

(1) $s_{p,q} \circ s_{p,q} = \mathrm{id}$,
(2) $s_{p,q} \circ s_{k,l} = s_{k,l} \circ s_{p,q}$ *if $p < q$ and $k < l$ are disjoint, and*
(3) $s_{p,q} \circ s_{k,l} = s_{r,t} \circ s_{p,q}$ *if $p < q$ contains $k < l$, where $r = \hat{s}_{p,q}(l)$ and $t = \hat{s}_{p,q}(k)$.*

Therefore, in a coboundary monoidal category the cactus group J_n acts on n-fold tensor products. This is analogous to the action of the braid group in a braided monoidal category. For this reason, the authors of [**7**] propose the name *cactus category* for a coboundary monoidal category.

PROPOSITION 4.3. *Any braiding satisfies the cactus relation.*

PROOF. Suppose $\mathcal{C}$ is a braided monoidal category with braiding σ and $U, V, W \in \operatorname{Ob}\mathcal{C}$. By the axioms of a braided monoidal category, we have

$$\sigma^{-1}_{V \otimes U, W} = (\mathrm{id}_V \otimes \sigma^{-1}_{U,W})(\sigma^{-1}_{V,W} \otimes \mathrm{id}_U)$$

Taking inverses gives

$$\sigma_{V \otimes U, W} = (\sigma_{V,W} \otimes \mathrm{id}_U)(\mathrm{id}_V \otimes \sigma_{U,W}).$$

Therefore

$$\begin{aligned}(\sigma_{U, W \otimes V})(\mathrm{id}_U \otimes \sigma_{V,W}) &= (\mathrm{id}_W \otimes \sigma_{U,V})(\sigma_{U,W} \otimes \mathrm{id}_V)(\mathrm{id}_U \otimes \sigma_{V,W}) \\ &= (\sigma_{V,W} \otimes \mathrm{id}_U)(\mathrm{id}_V \otimes \sigma_{U,W})(\sigma_{U,V} \otimes \mathrm{id}_W) \\ &= (\sigma_{V \otimes U, W})(\sigma_{U,V} \otimes \mathrm{id}_W).\end{aligned}$$

The first equality uses the definition of a braiding and the second is the Yang-Baxter equation. □

Note that Proposition 4.3 does not imply that every braided monoidal category is a coboundary monoidal category because we require cactus commutors to be involutions whereas braidings, in general, are not. In fact, any commutor that is both a braiding and a cactus commutor is in fact a symmetric commutor (that is, it endows the category in question with the structure of a symmetric monoidal category).

4.2. Examples. The definition of coboundary monoidal categories was first given by Drinfel′d in [**5**]. The name was inspired by the fact that the representation categories of coboundary Hopf algebras are coboundary monoidal categories. Since the cactus group surjects onto the symmetric group, any symmetric monoidal category is a coboundary monoidal category. Our main example of coboundary categories which are not symmetric monoidal categories will be the categories of representations of quantum groups and crystals. Furthermore, we will see that the category of crystals cannot be given the structure of a braided monoidal category. Thus there exist examples of coboundary monoidal categories that are not braided.

5. Quantum groups and crystals

5.1. Quantum groups. Compact groups and semisimple Lie algebras are rigid objects in the sense that they cannot be deformed. However, if one considers the group algebra or universal enveloping algebra instead, a deformation is possible. Such a deformation can be carried out in the category of (noncommutative, non-cocommutative) Hopf algebras. These deformations play an important role in the study of the quantum Yang-Baxter equation and the quantum inverse scattering method. Another benefit is that the structure of the deformations and their representations becomes more rigid and the concepts of canonical bases and crystals emerge.

We introduce here the quantum group, or quantized enveloping algebra, defined by Drinfel′d and Jimbo. For further details we refer the reader to the many books on the subject (e.g. [**2, 8, 17**]). Let $\mathfrak{g}$ be a Kac-Moody algebra with symmetrizable generalized Cartan matrix $A = (a_{ij})_{i,j\in I}$ and symmetrizing matrix $D = \operatorname{diag}(s_i \in \mathbb{Z}_{>0} \mid i \in I)$. Let $P, P^\vee$, and Q_+ be the weight lattice, coweight lattice and positive root lattice respectively. Let $\mathbb{C}_q$ be the field $\mathbb{C}(q^{1/2})$ where q is a formal variable. For $n \in \mathbb{Z}$ and any symbol x, we define

$$[n]_x = \frac{x^n - x^{-n}}{x - x^{-1}}, \quad [0]_x! = 1, \quad [m]_x! = [m]_x[m-1]_x \cdots [1]_x \text{ for } m \in \mathbb{Z}_{>0},$$

$$\begin{bmatrix} k \\ l \end{bmatrix}_x = \frac{[k]_x!}{[l]_x![k-l]_x!} \quad \text{for } k, l \in \mathbb{Z}_{\geq 0}.$$

Definition 5.1 (Quantum group $U_q(\mathfrak{g})$). *The* quantum group *or* quantized enveloping algebra $U_q(\mathfrak{g})$ *is the unital associative algebra over* $\mathbb{C}_q$ *with generators* e_i, f_i $(i \in I)$ *and* q^h $(h \in P^\vee)$ *with defining relations*

(1) $q^0 = 1$, $q^h q^{h'} = q^{h+h'}$ *for* $h, h' \in P^\vee$,
(2) $q^h e_i q^{-h} = q^{\alpha_i(h)} e_i$ *for* $h \in P^\vee$,
(3) $q^h f_i q^{-h} = q^{-\alpha_i(h)} f_i$ *for* $h \in P^\vee$,
(4) $e_i f_j - f_j e_i = \delta_{ij} \dfrac{q^{s_i h_i} - q^{-s_i h_i}}{q^{s_i} - q^{-s_i}}$ *for* $i, j \in I$,
(5) $\sum_{k=0}^{1-a_{ij}} (-1)^k \begin{bmatrix} 1-a_{ij} \\ k \end{bmatrix}_{q^{s_i}} e_i^{1-a_{ij}-k} e_j e_i^k = 0$ *for* $i \neq j$,
(6) $\sum_{k=0}^{1-a_{ij}} (-1)^k \begin{bmatrix} 1-a_{ij} \\ k \end{bmatrix}_{q^{s_i}} f_i^{1-a_{ij}-k} f_j f_i^k = 0$ *for* $i \neq j$.

As $q \to 1$, the defining relations for $U_q(\mathfrak{g})$ approach the usual relations for $\mathfrak{g}$ in the following sense. Taking the "derivative" with respect to q in the second and third relations gives

$$hq^{h-1}e_iq^{-h}+q^he_i\left(-hq^{-h-1}\right)=\alpha_i(h)q^{\alpha_i(h)-1}e_i\xrightarrow{q\to 1}he_i-e_ih=[h,e_i]=\alpha_i(h)e_i,$$

$$hq^{h-1}f_iq^{-h}+q^hf_i\left(-hq^{-h-1}\right)=-\alpha_i(h)q^{-\alpha_i(h)-1}f_i\xrightarrow{q\to 1}hf_i-f_ih=[h,f_i]=-\alpha_i(h)f,$$

Furthermore, if we naively apply L'Hôpital's rule, we have

$$\lim_{q\to 1}\frac{q^{s_ih_i}-q^{-s_ih_i}}{q^{s_i}-q^{-s_i}}=\lim_{q\to 1}\frac{s_ih_iq^{s_ih_i-1}+s_ih_iq^{-s_ih_i-1}}{s_iq^{s_i-1}+s_iq^{-s_i-1}}=\frac{2s_ih_i}{2s_i}=h_i,$$

and so the fourth defining relation of $U_q(\mathfrak{g})$ becomes $[e_i,f_j]=\delta_{ij}h_i$ in the $q\to 1$ limit – called the *classical limit*. Similarly, since we have

$$[n]_{q^{s_i}}\to n,\text{ and }\begin{bmatrix}1-a_{ij}\\ k\end{bmatrix}_{q^{s_i}}\to\binom{1-a_{ij}}{k}\text{ as }q\to\infty,$$

the last two relations (called the *quantum Serre relations*) become the usual Serre relations in the classical limit. Thus, we can think of $U_q(\mathfrak{g})$ as a deformation of $\mathfrak{g}$. For a more rigorous treatment of the classical limit, we refer the reader to [**8**, §3.4].

The algebra $U_q(\mathfrak{g})$ has a Hopf algebra structure given by comultiplication

$$\Delta(q^h)=q^h\otimes q^h,\quad \Delta(e_i)=e_i\otimes q^{s_ih_i}+1\otimes e_i,\quad \Delta(f_i)=f_i\otimes 1+q^{-s_ih_i}\otimes f_i,$$

counit

$$\varepsilon(q^h)=1,\quad \varepsilon(e_i)=\varepsilon(f_i)=0,$$

and antipode

$$\gamma(q^h)=q^{-h},\quad \gamma(e_i)=-e_iq^{-s_ih_i},\quad \gamma(f_i)=-q^{s_ih_i}f_i.$$

There are other choices but we will use the above in what follows.

The representations of $U(\mathfrak{g})$ can be q-deformed to representations of $U_q(\mathfrak{g})$ in such a way that the dimensions of the weight spaces are invariant under the deformation (see [**8, 15**]). The q-deformed notion of a weight space is as follows: for a $U_q(\mathfrak{g})$-module M and $\lambda\in P$, the λ-weight space of M is

$$M^\lambda=\{v\in M\mid q^hv=q^{\lambda(h)}v\ \forall\ h\in P^\vee\}.$$

5.2. Crystal bases. In this section we introduce the theory of crystal bases, which can be thought of as the $q\to\infty$ limit of the representation theory of quantum groups. In this limit, representations are replaced by combinatorial objects called crystal graphs. These objects, which are often much easier to compute with than the representations themselves, can be used to obtain such information as dimensions of weight spaces (characters) and the decomposition of tensor products into sums of irreducible representations. For details, we refer the reader to [**8**]. We note that in [**8**], the limit $q\to 0$ is used. This simply corresponds to a different choice of Hopf algebra structure on $U_q(\mathfrak{g})$. We choose to consider $q\to\infty$ to match the choices of [**2**].

For $n\in\mathbb{Z}_{\geq 0}$ and $i\in I$, define the *divided powers*

$$e_i^{(n)}=e_i^n/[n]_q!,\quad f_i^{(n)}=f_i^n/[n]_q!$$

Let M be an integrable $U_q(\mathfrak{g})$-module and let M^λ be the λ-weight space for $\lambda\in P$. For $i\in I$, any weight vector $u\in M^\lambda$ can be written uniquely in the form

$$u=\sum_{n=0}^{\infty}f_i^{(n)}u_n,\quad u_n\in\ker e_i\cap M^{\lambda+n\alpha_i}.$$

Define the *Kashiwara operators* $\tilde{e}_i, \tilde{f}_i : M \to M$ by

$$\tilde{e}_i u = \sum_{n=1}^{\infty} f_i^{(n-1)} u_n, \quad \tilde{f}_i u = \sum_{n=0}^{\infty} f_i^{(n+1)} u_n.$$

Let A be the integral domain of all rational functions in $\mathbb{C}_q$ that are regular at $q = \infty$. That is, A consists of all rational functions that can be written in the form $g_1(q^{-1/2})/g_2(q^{-1/2})$ for $g_1(q^{-1/2})$ and $g_2(q^{-1/2})$ polynomials in $\mathbb{C}[q^{-1/2}]$ with $g_2(q^{-1/2})|_{q^{-1/2}=0} \neq 0$ (one should think of these as rational functions whose limit exists as $q \to \infty$).

DEFINITION 5.2 (Crystal basis). *A* crystal basis *of a $U_q(\mathfrak{g})$-module M is a pair (L, B) such that*

(1) *L is a free A-submodule of M such that $M = \mathbb{C}_q \otimes_A L$,*
(2) *B is a $\mathbb{C}$-basis of the vector space $L/q^{-1/2}L$ over $\mathbb{C}$,*
(3) *$L = \bigoplus_\lambda L^\lambda$, $B = \bigsqcup_\lambda B^\lambda$ where $L^\lambda = L \cap M^\lambda$, $B^\lambda = B \cap L^\lambda/q^{-1/2}L^\lambda$,*
(4) *$\tilde{e}_i L \subseteq L$, $\tilde{f}_i L \subseteq L$ for all $i \in I$,*
(5) *$\tilde{e}_i B \subseteq B \cup \{0\}$, $\tilde{f}_i B \subseteq B \cup \{0\}$ for all $i \in I$, and*
(6) *for all $b, b' \in B$ and $i \in I$, $\tilde{e}_i b = b'$ if and only if $\tilde{f}_i b' = b$.*

It was shown by Kashiwara [**11**] that all $U_q(\mathfrak{g})$-modules in the category $\mathcal{O}^q_{\text{int}}$ (integrable modules with weight space decompositions and weights lying in a union of sets of the form $\lambda - Q_+$ for $\lambda \in P$) have unique crystal bases (up to isomorphism).

A crystal basis can be represented by a *crystal graph.* The crystal graph corresponding to a crystal basis (L, B) is an edge-colored (by I) directed graph with vertex set B and a i-colored directed edge from b' to b if $\tilde{f}_i b' = b$ (equivalently, if $\tilde{e}_i b = b'$). Crystals can be defined in a more abstract setting where a crystal consists of such a graph along with maps $\text{wt} : B \to P$ and $\varphi_i, \varepsilon_i : B \to \mathbb{Z}_{\geq 0}$ satisfying certain axioms. In this paper, by the *category of $\mathfrak{g}$-crystals* for a symmetrizable Kac-Moody algebra $\mathfrak{g}$, we mean the category consisting of those crystal graphs B such that each connected component of B is isomorphic to some B_λ, the crystal corresponding to the irreducible highest weight $U_q(\mathfrak{g})$-module of highest weight λ, where λ is a dominant integral weight. In this case

$$\text{wt}(b) = \mu \text{ for } b \in B^\mu, \quad \varphi_i(b) = \max\{k \mid \tilde{f}_i^k b \neq 0\}, \quad \varepsilon_i(b) = \max\{k \mid \tilde{e}_i^k \neq 0\}.$$

For the rest of this paper, the word $\mathfrak{g}$-*crystal* means an object in this category.

EXAMPLE 5.3 (Crystal bases of finite-dimensional representations of $U_q(\mathfrak{sl}_2)$). *For $n \in \mathbb{Z}_{\geq 0}$, let V_n be the irreducible $U_q(\mathfrak{sl}_2)$-module of highest weight n. It is a q-deformation of the corresponding $\mathfrak{sl}_2$-module. Let v_n be a highest weight vector of V_n and define*

$$v_{n-2i} = f^{(i)} v_n.$$

Then $\{v_n, v_{n-2}, \dots, v_{-n}\}$ is a basis of V_n and v_j has weight j. Let

$$L = \text{Span}_A\{v_n, v_{n-2}, \dots, v_{-n}\}, \quad \text{and}$$
$$B = \{b_n, b_{n-2}, \dots, b_{-n}\}$$

where b_j is the image of v_j in the quotient $L/q^{-1/2}L$. It is easily checked that (L, B) is a crystal basis of V_n and the corresponding crystal graph is

$$b_n \longrightarrow b_{n-2} \longrightarrow \cdots \longrightarrow b_{-n}$$

(since there is only one simple root for $\mathfrak{sl}_2$, we omit the edge-coloring).

5.3. Tensor products. One of the nicest features of the theory of crystals is the existence of the *tensor product rule* which tells us how to form the crystal corresponding to the tensor product of two representations from the crystals corresponding to the two factors.

THEOREM 5.4 (Tensor product rule [**8**, Theorem 4.4.1], [**2**, Proposition 14.1.14]). *Suppose (L_j, B_j) are crystal bases of $U_q(\mathfrak{g})$-modules M_j $(j = 1, 2)$ in $\mathcal{O}^q_{int}$. For $b \in B_j$ and $i \in I$, let*

$$\varphi_i(b) = \max\{k \mid \tilde{f}_i^k b \neq 0\}, \quad \varepsilon_i(b) = \max\{k \mid \tilde{e}_i^k b \neq 0\}.$$

Let $L = L_1 \otimes_A L_2$ and $B = B_1 \times B_2$. Then (L, B) is a crystal basis of $M_1 \otimes_{\mathbb{C}_q} M_2$, where the action of the Kashiwara operators $\tilde{e}_i$ and $\tilde{f}_i$ are given by

$$\tilde{e}_i(b_1 \otimes b_2) = \begin{cases} \tilde{e}_i b_1 \otimes b_2 & \text{if } \varphi_i(b_1) \geq \varepsilon_i(b_2), \\ b_1 \otimes \tilde{e}_i b_2 & \text{if } \varphi_i(b_1) < \varepsilon_i(b_2), \end{cases}$$

$$\tilde{f}_i(b_1 \otimes b_2) = \begin{cases} \tilde{f}_i b_1 \otimes b_2 & \text{if } \varphi_i(b_1) > \varepsilon_i(b_2), \\ b_1 \otimes \tilde{f}_i b_2 & \text{if } \varphi_i(b_1) \leq \varepsilon_i(b_2). \end{cases}$$

Here we write $b_1 \otimes b_2$ for $(b_1, b_2) \in B_1 \times B_2$ and $b_1 \times 0 = 0 \times b_2 = 0$.

We write $B_1 \otimes B_2$ for the crystal graph $B_1 \times B_2$ of $M_1 \otimes M_2$ with crystal operators defined by the formulas in Theorem 5.4. Note that even though we use a different coproduct than in [**8**], the tensor product rule remains the same as seen in [**2**, Proposition 14.1.14].

5.4. The braiding in the quantum group. The category of representations of $U(\mathfrak{g})$, the universal enveloping algebra of a symmetrizable Kac-Moody algebra $\mathfrak{g}$, is a symmetric monoidal category with braiding given by

$$\sigma_{U,V} : U \otimes V \to V \otimes U, \quad \sigma(u \otimes v) = \mathrm{flip}(u \otimes v) \stackrel{\mathrm{def}}{=} v \otimes u \quad \text{for} \quad u \in U,\ v \in V.$$

However, the analogous map is not a morphism in the category of representations of $U_q(\mathfrak{g})$ and this category is not a symmetric monoidal category. However, it is a braided monoidal category with a braiding constructed as follows. The *R-matrix* is an invertible element in a certain completed tensor product $U_q(\mathfrak{g})\widehat{\otimes}U_q(\mathfrak{g})$ (see [**2, 10**]). It defines a map $U \otimes V \to U \otimes V$ for any representations U and V of $U_q(\mathfrak{g})$. The map given by

$$\sigma_{U,V} : U \otimes V \to V \otimes U, \quad \sigma_{U,V} = \mathrm{flip} \circ R$$

for representations U and V of $U_q(\mathfrak{g})$ is a braiding.

As an example, consider the representation $V_1 \otimes V_1$ of $U_q(\mathfrak{sl}_2)$. In the basis $S_1 = \{v_1 \otimes v_1, v_{-1} \otimes v_1, v_1 \otimes v_{-1}, v_{-1} \otimes v_{-1}\}$, the R-matrix is given by (see [**2**, Example 6.4.12])

$$R = q^{-1/2} \begin{pmatrix} q & 0 & 0 & 0 \\ 0 & 1 & 0 & 0 \\ 0 & q - q^{-1} & 1 & 0 \\ 0 & 0 & 0 & q \end{pmatrix}.$$

Note that $R|_{q=1} = \mathrm{id}$ and so in the classical limit, the braiding becomes the map flip. In the basis S_1, we have

$$(5.1)\quad \mathrm{flip} = \begin{pmatrix} 1 & 0 & 0 & 0 \\ 0 & 0 & 1 & 0 \\ 0 & 1 & 0 & 0 \\ 0 & 0 & 0 & 1 \end{pmatrix}, \quad \text{and so} \quad \mathrm{flip} \circ R = q^{-1/2} \begin{pmatrix} q & 0 & 0 & 0 \\ 0 & q - q^{-1} & 1 & 0 \\ 0 & 1 & 0 & 0 \\ 0 & 0 & 0 & q \end{pmatrix}.$$

Now consider the basis

$$S_2 = \{v_1 \otimes v_1, a, b, v_{-1} \otimes v_{-1}\}, \quad a = v_{-1} \otimes v_1 - q v_1 \otimes v_{-1}, \quad b = v_{-1} \otimes v_1 + q^{-1} v_1 \otimes v_{-1}.$$

Note that

$$ea = fa = 0, \quad f(v_1 \otimes v_1) = b.$$

Thus S_2 is a basis of $V_1 \otimes V_1$ compatible with the decomposition $V_1 \otimes V_1 \cong V_0 \oplus V_2$. In the basis S_2, we have

$$(5.2)\qquad \mathrm{flip} \circ R = \begin{pmatrix} q^{1/2} & 0 & 0 & 0 \\ 0 & -q^{-3/2} & 0 & 0 \\ 0 & 0 & q^{1/2} & 0 \\ 0 & 0 & 0 & q^{1/2} \end{pmatrix}.$$

From this form, we easily see that $\mathrm{flip} \circ R$ is an isomorphism of $U_q(\mathfrak{sl}_2)$-modules $V_1 \otimes V_1 \to V_1 \otimes V_1$. It acts as multiplication by $q^{1/2}$ on the summand V_2 and by $-q^{-3/2}$ on the summand V_0.

Now, from Example 5.3 and the tensor product rule (Theorem 5.4), we see that the crystal basis of $V_1 \otimes V_1$ is given by

$$L = \mathrm{Span}_A\{v_1 \otimes v_1, v_{-1} \otimes v_1, v_1 \otimes v_{-1}, v_{-1} \otimes v_{-1}\}, \quad \text{and}$$
$$B = \{b_1 \otimes b_1, b_{-1} \otimes b_1, b_1 \times b_{-1}, b_{-1} \otimes b_{-1}\}.$$

From the matrix of $\mathrm{flip} \circ R$ in the basis S_1 given in (5.1), we see that it does not preserve the crystal lattice L since it involves positive powers of q. Furthermore, there is no $\mathbb{C}_q$-multiple of $\mathrm{flip} \circ R$ which preserves L and induces an isomorphism of $L/q^{-1/2}L$. To see this, note from (5.1) that in order for $g(q)\,\mathrm{flip} \circ R$, with $g(q) \in \mathbb{C}_q$, to preserve the crystal lattice, we would need $q^{1/2} g(q) \in A$ and thus $q^{-1/2} g(q) \in q^{-1}A \subseteq q^{-1/2}A$. However, we would then have

$$g(q)\,\mathrm{flip} \circ R(v_1 \otimes v_{-1}) = q^{-1/2} g(q) v_{-1} \otimes v_1 \equiv 0 \mod q^{-1/2}L$$

and so $g(q)\,\mathrm{flip} \circ R$ would not induce an isomorphism of $L/q^{-1/2}L$. Therefore, we see that the braiding coming from the R-matrix does not pass to the $q \to \infty$ limit. That is, it does not induce a braiding on the crystal $B_1 \otimes B_1$.

It turns out that the above phenomenon is unavoidable. That is, no braiding on the category of representations of a quantum group passes to the $q \to \infty$ limit. In fact, we have the following even stronger results.

LEMMA 5.5. *The category of $\mathfrak{sl}_2$-crystals cannot be given the structure of a braided monoidal category.*

PROOF. We prove the result by contradiction. Suppose the category of $\mathfrak{sl}_2$-crystals is a braided monoidal category with braiding σ. Consider the crystal B_1. It has crystal graph

$$B_1 \ :\ b_1 \longrightarrow b_{-1}.$$

The crystal graph of the tensor product $B_1 \otimes B_1$ has two connected components:

$$b_1 \otimes b_1 \longrightarrow b_{-1} \otimes b_1 \longrightarrow b_{-1} \otimes b_{-1} \quad \cong B_2$$
$$b_1 \otimes b_{-1} \quad \cong B_0$$

Since σ_{B_1,B_1} is an crystal isomorphism, we see from the above that it must act as the identity. Therefore

$$(\mathrm{id}_{B_1} \otimes \sigma_{B_1,B_1}) \circ (\sigma_{B_1,B_1} \otimes \mathrm{id}_{B_1}) = \mathrm{id}_{B_1 \otimes B_1 \otimes B_1}. \tag{5.3}$$

Now, the graph of the crystal $B_1 \otimes B_2$ has two connected components:

$$b_1 \otimes b_2 \longrightarrow b_{-1} \otimes b_2 \longrightarrow b_{-1} \otimes b_0 \longrightarrow b_{-1} \otimes b_{-2} \quad \cong B_3$$
$$b_1 \otimes b_0 \longrightarrow b_1 \otimes b_{-2} \quad \cong B_1$$

The graph of the crystal $B_2 \otimes B_1$ also has two connected components:

$$b_2 \otimes b_1 \longrightarrow b_0 \otimes b_1 \longrightarrow b_{-2} \otimes b_1 \longrightarrow b_{-2} \otimes b_{-1} \quad \cong B_3$$
$$b_2 \otimes b_{-1} \longrightarrow b_0 \otimes b_{-1} \quad \cong B_1$$

Since σ_{B_1,B_2} is a crystal isomorphism, we must have $\sigma_{B_1,B_2}(b_1 \otimes b_0) = b_2 \otimes b_{-1}$.

Now, consider the inclusion of crystals $j : B_2 \hookrightarrow B_1 \otimes B_1$ given by

$$j(b_2) = b_1 \otimes b_1, \quad j(b_0) = b_{-1} \otimes b_1, \quad j(b_{-2}) = b_{-1} \otimes b_{-1}.$$

By the naturality of the braiding, the following diagram commutes:

$$\begin{array}{ccc} B_1 \otimes B_2 & \xrightarrow{\mathrm{id}_{B_1} \otimes j} & B_1 \otimes B_1 \otimes B_1 \\ \downarrow{\scriptstyle \sigma_{B_1,B_2}} & & \downarrow{\scriptstyle \sigma_{B_1,B_1 \otimes B_1}} \\ B_2 \otimes B_1 & \xrightarrow{j \otimes \mathrm{id}_{B_1}} & B_1 \otimes B_1 \otimes B_1 \end{array}$$

We therefore have

$$\begin{aligned} \sigma_{B_1,B_1 \otimes B_1}(b_1 \otimes b_{-1} \otimes b_1) = \sigma_{B_1,B_1 \otimes B_1} \circ (\mathrm{id}_{B_1} \otimes j)(b_1 \otimes b_0) &= (j \otimes \mathrm{id}_{B_1}) \circ \sigma_{B_1,B_2}(b_1 \otimes b_0) \\ &= (j \otimes \mathrm{id}_{B_1})(b_2 \otimes b_{-1}) = b_1 \otimes b_1 \otimes b_{-1}. \end{aligned}$$

Comparing to (5.3), we see that

$$\sigma_{B_1,B_1 \otimes B_1} \neq (\mathrm{id}_{B_1} \otimes \sigma_{B_1,B_1}) \circ (\sigma_{B_1,B_1} \otimes \mathrm{id}_{B_1}),$$

contradicting the fact that σ is a braiding. □

An alternative proof of Lemma 5.5 was given in [**7**]. We can use the fact that $\mathfrak{g}$-crystals, for $\mathfrak{g}$ a symmetrizable Kac-Moody algebra, can be restricted to $\mathfrak{sl}_2$-crystals to generalize this result.

PROPOSITION 5.6. *For any symmetrizable Kac-Moody algebra $\mathfrak{g}$, the category of $\mathfrak{g}$-crystals cannot be given the structure of a braided monoidal category.*

PROOF. We prove the result by contradiction. Suppose the category of $\mathfrak{g}$-crystals was a braided monoidal category with braiding σ for some symmetrizable Kac-Moody algebra $\mathfrak{g}$. Let α_1 and ω_1 be a simple root and fundamental weight (respectively) corresponding to some vertex in the Dynkin diagram of $\mathfrak{g}$. The restriction of a $\mathfrak{g}$-crystal to the color 1 yields an $\mathfrak{sl}_2$-crystal. More precisely, one forgets the operators $\tilde{e}_i$, $\tilde{f}_i$, φ_i and ε_i for $i \neq 1$ and projects the map wt to the one-dimensional sublattice $\mathbb{Z}\omega_1 \subseteq P$. In general, even if the original $\mathfrak{g}$-crystal was connected (i.e. irreducible), the induced $\mathfrak{sl}_2$-crystal will not be. However, any morphism of $\mathfrak{g}$-crystals induces a morphism of the restricted $\mathfrak{sl}_2$-crystals.

Consider the $\mathfrak{g}$-crystal $B_{k\omega_1}$, $k \geq 1$, corresponding to the irreducible $U_q(\mathfrak{g})$-module of highest weight $k\omega_1$. If we restrict this to an $\mathfrak{sl}_2$-crystal, the connected component of the $\mathfrak{sl}_2$-crystal graph containing the highest weight element $b_{k\omega_1}$ is isomorphic to the $\mathfrak{sl}_2$-crystal B_k. Now, since $b_{\omega_1} \otimes b_{\omega_1}$ is the unique element of $B_{\omega_1} \otimes B_{\omega_1}$ of weight $2\omega_1$, we have

$$\sigma_{B_{\omega_1},B_{\omega_1}}(b_{\omega_1} \otimes b_{\omega_1}) = b_{\omega_1} \otimes b_{\omega_1}.$$

The connected $\mathfrak{sl}_2$-subcrystal containing the element $b_{\omega_1} \otimes b_{\omega_1}$ is

$$b_{\omega_1} \otimes b_{\omega_1} \longrightarrow \tilde{f}_1 b_{\omega_1} \otimes b_{\omega_1} \longrightarrow \tilde{f}_1 b_{\omega_1} \otimes \tilde{f}_1 b_{\omega_1} \quad \cong B_2 \tag{5.4}$$

(as in the proof of Lemma 5.5). Thus, since $\sigma_{B_{\omega_1} \otimes B_{\omega_1}}$ is a morphism of $\mathfrak{sl}_2$-crystals, we must have

$$\sigma_{B_{\omega_1} \otimes B_{\omega_1}}(\tilde{f}_1 b_{\omega_1} \otimes b_{\omega_1}) = \tilde{f}_1 b_{\omega_1} \otimes b_{\omega_1}.$$

Now, the only element of $B_{\omega_1} \otimes B_{\omega_1}$ of weight $2\omega_1 - \alpha_1$ not contained in the connected $\mathfrak{sl}_2$-subcrystal mentioned above is $b_{\omega_1} \otimes \tilde{f}_1 b_{\omega_1}$. Therefore, we must also have

$$\sigma_{B_{\omega_1},B_{\omega_1}}(b_{\omega_1} \otimes \tilde{f}_1 b_{\omega_1}) = b_{\omega_1} \otimes \tilde{f}_1 b_{\omega_1}.$$

Thus,

$$\begin{aligned}(\mathrm{id}_{B_{\omega_1}} \otimes \sigma_{B_{\omega_1},B_{\omega_1}}) \circ (\sigma_{B_{\omega_1},B_{\omega_1}} \otimes \mathrm{id}_{B_{\omega_1}})(b_{\omega_1} \otimes \tilde{f}_1 b_{\omega_1} \otimes b_{\omega_1}) \\ = (\mathrm{id}_{B_{\omega_1}} \otimes \sigma_{B_{\omega_1},B_{\omega_1}})(b_{\omega_1} \otimes \tilde{f}_1 b_{\omega_1} \otimes b_{\omega_1}) = b_{\omega_1} \otimes \tilde{f}_1 b_{\omega_1} \otimes b_{\omega_1}.\end{aligned} \tag{5.5}$$

Now, the connected $\mathfrak{sl}_2$-subcrystal of $B_{\omega_1} \otimes B_{2\omega_1}$ containing the element $b_{\omega_1} \otimes b_{2\omega_1}$ is

$$b_{\omega_1} \otimes b_{2\omega_1} \longrightarrow \tilde{f}_1 b_{\omega_1} \otimes b_{2\omega_1} \longrightarrow \tilde{f}_1 b_{\omega_1} \otimes \tilde{f}_1 b_{2\omega_1} \longrightarrow \tilde{f}_1 b_{\omega_1} \otimes \tilde{f}_1^2 b_{2\omega_1} \quad \cong B_3,$$

and the connected $\mathfrak{sl}_2$-subcrystal containing the element $b_{\omega_1} \otimes \tilde{f}_1 b_{2\omega_1}$ is

$$b_{\omega_1} \otimes \tilde{f}_1 b_{2\omega_1} \longrightarrow b_{\omega_1} \otimes \tilde{f}_1^2 b_{2\omega_1} \quad \cong B_1$$

(as in the proof of Lemma 5.5). Similarly, we have the following $\mathfrak{sl}_2$-subcrystals of $B_{2\omega_1} \otimes B_{\omega_1}$:

$$b_{2\omega_1} \otimes b_{\omega_1} \longrightarrow \tilde{f}_1 b_{2\omega_1} \otimes b_{\omega_1} \longrightarrow \tilde{f}_1^2 b_{2\omega_1} \otimes b_{\omega_1} \longrightarrow \tilde{f}_1^2 b_{2\omega_1} \otimes \tilde{f}_1 b_{\omega_1} \quad \cong B_3,$$

$$b_{2\omega_1} \otimes \tilde{f}_1 b_{\omega_1} \longrightarrow \tilde{f}_1 b_{2\omega_1} \otimes \tilde{f}_1 b_{\omega_1} \quad \cong B_1.$$

Now, since $b_{\omega_1} \otimes b_{2\omega_1}$ and $b_{2\omega_1} \otimes b_{\omega_1}$ are the unique elements of $B_{\omega_1} \otimes B_{2\omega_1}$ and $B_{2\omega_1} \otimes B_{\omega_1}$ (respectively) of weight $3\omega_1$, we must have

$$\sigma_{B_{\omega_1},B_{2\omega_1}}(b_{\omega_1} \otimes b_{2\omega_1}) = b_{2\omega_1} \otimes b_{\omega_1}.$$

Also, since $b_{\omega_1} \otimes \tilde{f}_1 b_{2\omega_1}$ and $b_{2\omega_1} \otimes \tilde{f}_1 b_{\omega_1}$ are the only elements of $B_{\omega_1} \otimes B_{2\omega_1}$ and $B_{2\omega_1} \otimes B_{\omega_1}$ (respectively) of weight $3\omega_1 - \alpha_1$ not contained in the connected $\mathfrak{sl}_2$-subcrystal containing $b_{\omega_1} \otimes b_{2\omega_1}$ and $b_{2\omega_1} \otimes b_{\omega_1}$ (respectively), we must have

$$\sigma_{B_{\omega_1},B_{2\omega_1}}(b_{\omega_1} \otimes \tilde{f}_1 b_{2\omega_1}) = b_{2\omega_1} \otimes \tilde{f}_1 b_{\omega_1}.$$

Now, consider the inclusion of $\mathfrak{g}$-crystals $j : B_{2\omega_1} \hookrightarrow B_{\omega_1} \otimes B_{\omega_1}$ determined by $j(b_{2\omega_1}) = b_{\omega_1} \otimes b_{\omega_1}$. Restricting to the connected $\mathfrak{sl}_2$-crystals containing the elements $b_{2\omega_1}$ and $b_{\omega_1} \otimes b_{\omega_1}$, we see from (5.4) that

$$j(\tilde{f}_1 b_{2\omega_1}) = \tilde{f}_1 b_{\omega_1} \otimes b_{\omega_1}, \quad j(\tilde{f}_1^2 b_{2\omega_1}) = \tilde{f}_1 b_{\omega_1} \otimes \tilde{f}_1 b_{\omega_1}.$$

By the naturality of the braiding σ, the following diagram is commutative:

$$\begin{array}{ccc} B_{\omega_1} \otimes B_{2\omega_1} & \xrightarrow{\mathrm{id}_{B_{\omega_1}} \otimes j} & B_{\omega_1} \otimes B_{\omega_1} \otimes B_{\omega_1} \\ \downarrow \sigma_{B_{\omega_1}, B_{2\omega_1}} & & \downarrow \sigma_{B_{\omega_1}, B_{\omega_1} \otimes B_{\omega_1}} \\ B_{2\omega_1} \otimes B_{\omega_1} & \xrightarrow{j \otimes \mathrm{id}_{B_{\omega_1}}} & B_{\omega_1} \otimes B_{\omega_1} \otimes B_{\omega_1} \end{array}$$

Therefore

$$\begin{aligned} \sigma_{B_{\omega_1}, B_{\omega_1} \otimes B_{\omega_1}}(b_{\omega_1} \otimes \tilde{f}_1 b_{\omega_1} \otimes b_{\omega_1}) &= \sigma_{B_{\omega_1}, B_{\omega_1} \otimes B_{\omega_1}} \circ (\mathrm{id}_{B_{\omega_1}} \otimes j)(b_{\omega_1} \otimes \tilde{f}_1 b_{2\omega_1}) \\ &= (j \otimes \mathrm{id}_{B_{\omega_1}}) \circ \sigma_{B_{\omega_1}, B_{2\omega_1}}(b_{\omega_1} \otimes \tilde{f}_1 b_{2\omega_1}) \\ &= (j \otimes \mathrm{id}_{B_{\omega_1}})(b_{2\omega_1} \otimes \tilde{f}_1 b_{\omega_1}) \\ &= b_{\omega_1} \otimes b_{\omega_1} \otimes \tilde{f}_1 b_{\omega_1}. \end{aligned}$$

Comparing this to (5.5) we see that

$$\sigma_{B_{\omega_1}, B_{\omega_1} \otimes B_{\omega_1}} \neq (\mathrm{id}_{B_{\omega_1}} \otimes \sigma_{B_{\omega_1}, B_{\omega_1}}) \circ (\sigma_{B_{\omega_1}, B_{\omega_1}} \otimes \mathrm{id}_{B_{\omega_1}}).$$

This contradicts the fact that σ is a braiding. □

6. Crystals and coboundary categories

In this section, we discuss how the categories of $U_q(\mathfrak{g})$-modules and $\mathfrak{g}$-crystals can be given the structure of a coboundary category. For the case of $\mathfrak{g}$-crystals, we mention several different constructions and note the relationship between them.

6.1. Drinfel′d's unitarization. In [**5**], Drinfel′d defined the *unitarized R-matrix*

$$\bar{R} = R(R^{\mathrm{op}} R)^{-1/2},$$

where $R^{\mathrm{op}} = \mathrm{flip}(R)$ (as an operator on $M_1 \otimes M_2$, R^{op} acts as $\mathrm{flip} \circ R \circ \mathrm{flip}$) and the square root is taken with respect to a certain filtration on the completed tensor product $U_q(\mathfrak{g}) \hat{\otimes} U_q(\mathfrak{g})$. He then showed that $\mathrm{flip} \circ \bar{R}$ is a cactus commutor and so endows the category of $U_q(\mathfrak{g})$-modules with the structure of a coboundary category (see the comment after the proof of Proposition 3.3 in [**5**]). That is, it satisfies the conditions of Definition 4.1.

Consider the representation $V_1 \otimes V_1$ of $U_q(\mathfrak{sl}_2)$. In Section 5.4, we described the action of the R-matrix on this representation in two different bases S_1 and S_2. It follows from (5.2) that in the basis S_2,

$$R^{\mathrm{op}} R = (\mathrm{flip} \circ R)^2 = \begin{pmatrix} q & 0 & 0 & 0 \\ 0 & q^{-3} & 0 & 0 \\ 0 & 0 & q & 0 \\ 0 & 0 & 0 & q \end{pmatrix}.$$

Therefore, we can take the (inverse of the) square root

$$(R^{\mathrm{op}} R)^{-1/2} = q^{-1/2} \begin{pmatrix} 1 & 0 & 0 & 0 \\ 0 & q^2 & 0 & 0 \\ 0 & 0 & 1 & 0 \\ 0 & 0 & 0 & 1 \end{pmatrix}.$$

In the basis S_1, we then have

$$(R^{\mathrm{op}}R)^{-1/2} = q^{-1/2}\begin{pmatrix}1 & 0 & 0 & 0\\ 0 & \frac{2q^2}{1+q^2} & \frac{q-q^3}{1+q^2} & 0\\ 0 & \frac{q-q^3}{1+q^2} & \frac{1+q^4}{1+q^2} & 0\\ 0 & 0 & 0 & 1\end{pmatrix},$$

$$\bar{R} = R(R^{\mathrm{op}}R)^{-1/2} = \begin{pmatrix}1 & 0 & 0 & 0\\ 0 & \frac{2q}{1+q^2} & \frac{1-q^2}{1+q^2} & 0\\ 0 & \frac{q^2-1}{1+q^2} & \frac{2q}{1+q^2} & 0\\ 0 & 0 & 0 & 1\end{pmatrix}.$$

Therefore, in the basis S_1,

$$\text{(6.1)}\quad \mathrm{flip}\circ R = q^{-1/2}\begin{pmatrix}q & 0 & 0 & 0\\ 0 & q-q^{-1} & 1 & 0\\ 0 & 1 & 0 & 0\\ 0 & 0 & 0 & q\end{pmatrix},\qquad \mathrm{flip}\circ\bar{R} = \begin{pmatrix}1 & 0 & 0 & 0\\ 0 & \frac{q^2-1}{1+q^2} & \frac{2q}{1+q^2} & 0\\ 0 & \frac{2q}{1+q^2} & \frac{1-q^2}{1+q^2} & 0\\ 0 & 0 & 0 & 1\end{pmatrix}.$$

And in the basis S_2,

$$\text{(6.2)}\quad \mathrm{flip}\circ R = \begin{pmatrix}q^{1/2} & 0 & 0 & 0\\ 0 & -q^{-3/2} & 0 & 0\\ 0 & 0 & q^{1/2} & 0\\ 0 & 0 & 0 & q^{1/2}\end{pmatrix},\qquad \mathrm{flip}\circ\bar{R} = \begin{pmatrix}1 & 0 & 0 & 0\\ 0 & -1 & 0 & 0\\ 0 & 0 & 1 & 0\\ 0 & 0 & 0 & 1\end{pmatrix}.$$

In the above, we have recalled the computation of $\mathrm{flip}\circ R$ from Section 5.4 for the purposes of comparison.

We note two important properties of $\mathrm{flip}\circ\bar{R}$. First of all, we see from (6.1) that the matrix coefficients in the basis S_1 of $\mathrm{flip}\circ\bar{R}$ lie in A, the ring of rational functions in $\mathbb{C}_q$ that are regular at $q=\infty$. Thus, $\mathrm{flip}\circ\bar{R}$ preserves the crystal lattice of $V_1\otimes V_1$. In the $q\to\infty$ limit (more precisely, when passing to the quotient $L/q^{-1/2}L$), we have (in the basis S_1)

$$\mathrm{flip}\circ\bar{R} = \begin{pmatrix}1 & 0 & 0 & 0\\ 0 & 1 & 0 & 0\\ 0 & 0 & -1 & 0\\ 0 & 0 & 0 & 1\end{pmatrix} \quad \mathrm{mod}\ q^{-1/2}L.$$

Thus, $\mathrm{flip}\circ\bar{R}$ passes to the $q\to\infty$ limit and, up to signs, induces an involution on the crystal $B_1\otimes B_1$ (see Section 6.4 for details). As noted in Section 5.4, the same is not true of $\mathrm{flip}\circ R$.

The second important property of $\mathrm{flip}\circ\bar{R}$ is that it is a cactus commutor. We see immediately from (6.2) that $(\mathrm{flip}\circ\bar{R})^2 = \mathrm{id}$. A straightforward (if somewhat lengthy) computation shows that $\mathrm{flip}\circ\bar{R}$ also satisfies the cactus relation (see [**5**, §3] for the proof in a more general setting).

The unitarized R-matrix has shown up in several different places. In [**1**] it arose naturally in the development of the theory of braided symmetric and exterior algebras. The reason for this is that if one wants to have interesting symmetric or exterior algebras, one needs an operator with eigenvalues of positive or negative one. Notice from (6.2) that while $\mathrm{flip}\circ R$ does not have this property, the operator $\mathrm{flip}\circ\bar{R}$ does. Essentially, in the basis S_2, the matrix for $\mathrm{flip}\circ\bar{R}$ is obtained from the matrix for flip $\circ R$ by setting $q = 1$. Note that this does not imply that flip $\circ \bar{R}$ is an operator in the classical limit, merely that its matrix coefficients in a certain basis do not involve powers of q. The $q \to \infty$ limit of the unitarized R-matrix also appeared independently in the study of cactus commutors for crystals. We discuss this in the next two subsections.

6.2. The crystal commutor using the Schützenberger involution. Let $\mathfrak{g}$ be a simple complex Lie algebra and let I denote the set of vertices of the Dynkin graph of $\mathfrak{g}$. If w_0 is the long element in the Weyl group of $\mathfrak{g}$, let $\theta : I \to I$ be the involution such that $\alpha_{\theta(i)} = -w_0 \cdot \alpha_i$. Define a crystal $\overline{B_\lambda}$ with underlying set $\{\bar{b} \mid b \in B_\lambda\}$ and

$$\tilde{e}_i \cdot \bar{b} = \overline{\tilde{f}_{\theta(i)} \cdot b}, \quad \tilde{f}_i \cdot \bar{b} = \overline{\tilde{e}_{\theta(i)} \cdot b}, \quad \mathrm{wt}(\bar{b}) = w_0 \cdot \mathrm{wt}(b).$$

There is a crystal isomorphism $\overline{B_\lambda} \cong B_\lambda$. We compose this isomorphism with the map of sets $B_\lambda \to \overline{B_\lambda}$ given by $b \mapsto \bar{b}$ and denote the resulting map by $\xi = \xi_{B_\lambda} : B_\lambda \to B_\lambda$. We call the map ξ the *Schützenberger involution.* When $\mathfrak{g} = \mathfrak{gl}_n$, there is a realization of B_λ using tableaux. In this realization, ξ is the usual Schützenberger involution on tableaux (see [**14**]).

For an arbitrary $\mathfrak{g}$-crystal B, write $B = \bigoplus_{i=1}^k B_{\lambda_i}$. This is a decomposition of B into connected components. Then define $\xi_B : B \to B$ by $\xi_B = \bigoplus_{i=1}^k \xi_{B_{\lambda_i}}$. That is, we apply $\xi_{B_{\lambda_i}}$ to each connected component B_{λ_i}.

For crystals A and B, define

$$\sigma^S : A \otimes B \to B \otimes A, \quad \sigma^S(a \otimes b) = \xi_{B\otimes A}(\xi_B(b) \otimes \xi_A(a)).$$

THEOREM 6.1 ([**7**, Proposition 3, Theorem 3]). *We have*

(1) $\sigma^S_{B,A} \circ \sigma^S_{A,B} = \mathrm{id}$, *and*
(2) σ^S *satisfies the cactus relation* (4.1).

In other words, σ^S endows the category of $\mathfrak{g}$-crystals with the structure of a coboundary category.

6.3. The crystal commutor using the Kashiwara involution. Let $\mathfrak{g}$ be a symmetrizable Kac-Moody algebra and let B_∞ be the $\mathfrak{g}$-crystal corresponding to the lower half $U_q^-(\mathfrak{g})$ of the associated quantized universal enveloping algebra. Let $* : U_q(\mathfrak{g}) \to U_q(\mathfrak{g})$ be the $\mathbb{C}_q$-linear anti-automorphism given by

$$e_i^* = e_i, \quad f_i^* = f_i, \quad (q^h)^* = q^{-h}.$$

The map $*$ sends $U_q^-(\mathfrak{g})$ to $U_q^-(\mathfrak{g})$ and induces a map $* : B_\infty \to B_\infty$ (see [**12**, §8.3]). We call the map $*$ the *Kashiwara involution.*

Let B_λ be the $\mathfrak{g}$-crystal corresponding to the irreducible highest weight $U_q(\mathfrak{g})$-module of highest weight λ and let b_λ be its highest weight element. For two integral dominant weights λ and μ, there is an inclusion of crystals $B_{\lambda+\mu} \hookrightarrow B_\lambda \otimes B_\mu$ sending $b_{\lambda+\mu}$ to $b_\lambda \otimes b_\mu$. It follows from the tensor product rule that the image of this inclusion contains all elements of the form $b \otimes b_\mu$ for $b \in B_\lambda$. Thus we define a map

$$\iota_\lambda^{\lambda+\mu} : B_\lambda \to B_{\lambda+\mu}$$

which sends $b \in B_\lambda$ to the inverse image of $b \otimes b_\mu$ under the inclusion $B_{\lambda+\mu} \hookrightarrow B_\lambda \otimes B_\mu$. While this map is not a morphism of crystals, it is $\tilde{e}_i$-equivariant for all i and takes b_λ to $b_{\lambda+\mu}$.

The maps $\iota_\lambda^{\lambda+\mu}$ make the family of crystals B_λ into a directed system and the crystal B_∞ can be viewed as the limit of this system. We have $\tilde{e}_i$-equivariant maps $\iota_\lambda^\infty : B_\lambda \to B_\infty$ which we will simply denote by ι^∞ when it will cause no confusion. Define $\varepsilon^* : B_\infty \to P_+$ by

$$\varepsilon^*(b) = \min\{\lambda \mid b \in \iota^\infty(B_\lambda)\}$$

where we put the usual order on P_+, the positive weight lattice of $\mathfrak{g}$, given by $\lambda \geq \mu$ if and only if $\lambda - \mu \in Q_+$. Recall that we also have the map $\varepsilon : B_\infty \to P_+$ given by $\varepsilon(b)(h_i) = \varepsilon_i(b)$. Then by [**12**, Proposition 8.2], the Kashiwara involution preserves weights and satisfies

$$\varepsilon^*(b) = \varepsilon(b^*). \tag{6.3}$$

Consider the crystal $B_\lambda \otimes B_\mu$. Since $\varphi(b) = \varepsilon(b) + \mathrm{wt}(b)$ for all $b \in B_\lambda$, we have that $\varphi(b_\lambda) = \mathrm{wt}(b_\lambda) = \lambda$. It follows from the tensor product rule for crystals that the highest weight elements of $B_\lambda \otimes B_\mu$ are those elements of the form $b_\lambda \otimes b$ for $b \in B_\mu$ with $\varepsilon(b) \leq \lambda$. Thus $\varepsilon^*(b^*) = \varepsilon(b) \leq \lambda$ and so, by the definition of ε^*, we have $b^* \in \iota^\infty(B_\lambda)$. So we can consider b^* as an element of B_λ. Furthermore, $\varepsilon(b^*) = \varepsilon^*(b) \leq \mu = \varphi(b_\mu)$ since $b \in B_\mu$. Thus $b_\mu \otimes b^*$ is a highest weight element of $B_\mu \otimes B_\lambda$. Since $B_\lambda \otimes B_\mu \cong B_\mu \otimes B_\lambda$ as crystals, we can make the following definition.

Definition 6.2 ([**9**, §3]). *Let $\sigma^c_{B_\lambda,B_\mu} : B_\lambda \otimes B_\mu \xrightarrow{\cong} B_\mu \otimes B_\lambda$ be the crystal isomorphism given uniquely by $\sigma^c_{B_\lambda,B_\mu}(b_\lambda \otimes b) = b_\mu \otimes b^*$ for $b_\lambda \otimes b$ a highest weight element of $B_\lambda \otimes B_\mu$.*

Theorem 6.3 ([**9**, Theorem 3.1]). *For $\mathfrak{g}$ a simple complex Lie algebra, $\sigma^S = \sigma^c$ and so σ^c satisfies the cactus relation.*

We call σ^c the *crystal commutor.* Note that Theorem 6.3 only implies that it is a cactus commutor for $\mathfrak{g}$ of finite type.

6.4. The relationship between the various commutors. We have described three ways of constructing commutors in the categories of $U_q(\mathfrak{g})$-modules or $\mathfrak{g}$-crystals. The three definitions are closely related. In [**7**], Henriques and Kamnitzer defined a cactus commutor on the category of finite-dimensional $U_q(\mathfrak{g})$-modules when $\mathfrak{g}$ is of finite type, using an analogue of the Schützenberger involution on $U_q(\mathfrak{g})$. This definition of the commutor involves some choices of normalization. In [**10**], Kamnitzer and Tingley showed that Drinfel′d's commutor coming from the unitarized R-matrix corresponds to the commutor coming from the Schützenberger involution up to normalization. From there it follows that Drinfel′d's commutor preserves crystal lattices and acts on crystal bases as the crystal commutor, up to signs. The precise statement is the following.

Proposition 6.4 ([**10**, Theorem 9.2]). *Suppose (L_j, B_j) are crystal bases of two finite-dimensional representations V_j, $j = 1, 2$, of $U_q(\mathfrak{g})$ for a simple complex Lie algebra $\mathfrak{g}$. Let $\sigma^D_{V_1,V_2}$ be the isomorphism $V_1 \otimes V_2 \cong V_2 \otimes V_1$ given by* $\mathrm{flip} \circ \bar{R}$. *Then*

$$\sigma^D_{V_1,V_2}(L_1 \otimes L_2) = L_2 \otimes L_1$$

and thus $\sigma^D_{L_1 \otimes L_2}$ induces a map

$$\sigma^{D \mod q^{-1/2}(L_1\otimes L_2)}_{V_1 \otimes V_2} : (L_1 \otimes L_2)/q^{-1/2}(L_1 \otimes L_2) \to (L_2 \otimes L_1)/q^{-1/2}(L_2 \otimes L_1).$$

For all $b_j \in B_j$, $j = 1, 2$,

$$\sigma^{D \mod q^{-1/2}(L_1 \otimes L_2)}_{V_1 \otimes V_2}(b_1 \otimes b_2) = (-1)^{\langle \lambda + \mu - \nu, \rho^\vee \rangle} \sigma^c_{B_1, B_2}(b_1 \otimes b_2)$$

where λ, μ *and* ν *are the highest weights of the connected components of* B_1, B_2 *and* $B_1 \otimes B_2$ *containing* b_1, b_2 *and* $b_1 \otimes b_2$ *respectively,* ρ *is half the sum of the positive roots of* $\mathfrak{g}$ *and* $\langle \cdot, \cdot \rangle$ *denotes the pairing between the Cartan subalgebra* $\mathfrak{h} \subset \mathfrak{g}$ *and its dual* $\mathfrak{h}^*$.

We thus have essentially two definitions of the crystal commutor. The first, using the Schützenberger involution (and coinciding with Drinfel′d's commutor in the crystal limit) only applies to $\mathfrak{g}$ of finite type but with this definition, it is apparent that the commutor satisfies the cactus relation. The second definition, using the Kashiwara involution, applies to $\mathfrak{g}$ of arbitrary type but it is not easy to see that it satisfies the cactus relation. In the next section, we will explain how a geometric interpretation of this commutor using quiver varieties allows one to prove that this is indeed the case.

7. A geometric realization of the crystal commutor

In this section we describe a geometric realization of the crystal commutor defined in Section 6.3 in the language of quiver varieties. This realization yields new insight into the coboundary structure and equips us with new geometric tools. Using these tools, one is able to show that the category of $\mathfrak{g}$-crystals for an arbitrary symmetrizable Kac-Moody algebra $\mathfrak{g}$ can be given the structure of a coboundary category. This extends the previously known result, which held for $\mathfrak{g}$ of finite type.

7.1. Quiver varieties. Lusztig [**16**], Nakajima [**20, 21, 22**] and Malkin [**19**] have introduced varieties associated to quivers (directed graphs) built from the Dynkin graph of a Kac-Moody algebra $\mathfrak{g}$ with symmetric Cartan matrix. These varieties yield geometric realizations of the quantum group $U_q(\mathfrak{g})$, the representations of $\mathfrak{g}$, and tensor products of these representations in the homology (or category of perverse sheaves) of such varieties. In addition, Kashiwara and Saito [**13, 23**], Nakajima [**22**] and Malkin [**19**] have used quiver varieties to give a geometric realization of the crystals of these objects. Namely, they defined geometric operators on the sets of irreducible components of quiver varieties, endowing these sets with the structure of crystals. In the current paper, we will focus on the $\mathfrak{sl}_2$ case of these varieties for simplicity. In this case, the quiver varieties are closely related to grassmannians and flag varieties.

In this section, all vector spaces will be complex. Fix integers $w \geq 0$ and $n \geq 1$, and $\mathbf{w} = (w_i)_{i=1}^n \in (\mathbb{Z}_{\geq 0})^n$ such that $\sum_{i=1}^n w_i = w$. Let W be a w-dimensional vector space and let

$$0 = W_0 \subseteq W_1 \subseteq \cdots \subseteq W_n = W, \quad \dim W_i / W_{i-1} = w_i \text{ for } 1 \leq i \leq n,$$

be an n-step partial flag in W. Define the *tensor product quiver variety*

$$\mathfrak{T}(\mathbf{w}) = \{(U, t) \mid U \subseteq W,\ t \in \operatorname{End} W,\ t(W_i) \subseteq W_{i-1}\ \forall i,\ \operatorname{im} t \subseteq U \subseteq \ker t\}.$$

We use the notation $\mathfrak{T}(\mathbf{w})$ since, up to isomorphism, this variety depends only on the dimensions of the subspaces W_i, $0 \leq i \leq n$. We have

$$\mathfrak{T}(\mathbf{w}) = \bigsqcup_{u=0}^{w} \mathfrak{T}(u, \mathbf{w}), \quad \text{where} \quad \mathfrak{T}(u, \mathbf{w}) = \{(U, t) \in \mathfrak{T}(\mathbf{w}) \mid \dim U = u\}.$$

Let $B(u, \mathbf{w})$ denote the set of irreducible components of $\mathfrak{T}(u, \mathbf{w})$ and set $B(\mathbf{w}) = \bigsqcup_u B(u, \mathbf{w})$.

Define

$$\mathrm{wt} : B(\mathbf{w}) \to P, \quad \mathrm{wt}(X) = w - 2u \text{ for } X \in B(u, \mathbf{w}),$$
$$\varepsilon : \mathfrak{T}(\mathbf{w}) \to \mathbb{Z}_{\geq 0}, \quad \varepsilon(U, t) = \dim U / \operatorname{im} t,$$
$$\varphi : \mathfrak{T}(\mathbf{w}) \to \mathbb{Z}_{\geq 0}, \quad \varphi(U, t) = \dim \ker t / U = \varepsilon(U, t) + w - 2 \dim U.$$

For $k \in \mathbb{Z}_{\geq 0}$, let

$$\mathfrak{T}(u, \mathbf{w})_k = \{(U, t) \in \mathfrak{T}(u, \mathbf{w}) \mid \varepsilon(U, t) = k\},$$

and for $X \in B(u, \mathbf{w})$, define $\varepsilon(X) = \varepsilon(U, t)$ and $\varphi(X) = \varepsilon(U, t)$ for a generic point (U, t) of X. Let

$$B(u, \mathbf{w})_k = \{X \in B(u, \mathbf{w}) \mid \varepsilon(X) = k\}, \quad B(\mathbf{w})_k = \bigsqcup_u B(u, \mathbf{w})_k.$$

The map

$$\mathfrak{T}(u, \mathbf{w})_k \to \mathfrak{T}(u - k, \mathbf{w})_0, \quad (U, t) \mapsto (\operatorname{im} t, t)$$

is a Grassmann bundle and thus induces an isomorphism

$$B(u, \mathbf{w})_k \cong B(u - k, \mathbf{w})_0. \tag{7.1}$$

We then define crystal operators on $B(\mathbf{w})$ as follows. Suppose $X' \in B(u - k, \mathbf{w})_0$ corresponds to $X \in B(u, \mathbf{w})_k$ under the isomorphism (7.1). Define

$$\tilde{f}^k : B(u - k, \mathbf{w})_0 \to B(u, \mathbf{w})_k, \quad \tilde{f}^k(X') = X,$$
$$\tilde{e}^k : B(u, \mathbf{w})_k \to B(u - k, \mathbf{w})_0, \quad \tilde{e}^k(X) = X'.$$

For $k > 0$, we then define $\tilde{e}_i : B(\mathbf{w}) \to B(\mathbf{w})$ by

$$\tilde{e} : B(u, \mathbf{w})_k \xrightarrow{\tilde{e}^k} B(u - k, \mathbf{w})_0 \xrightarrow{\tilde{f}^{k-1}} B(u - 1; \mathbf{w})_{k-1},$$

and set $\tilde{e}_i(X) = 0$ for $X \in B(u, \mathbf{w})_0$. For $k > 2u - w$, define

$$\tilde{f} : B(u, \mathbf{w})_k \xrightarrow{\tilde{e}^k} B(u - k, \mathbf{w})_0 \xrightarrow{\tilde{f}^{k+1}} B(u + 1, \mathbf{w})_{k+1},$$

and set $\tilde{f}(X) = 0$ for $X \in B(u, \mathbf{w})_k$ with $k \leq 2u - w$. The maps $\tilde{e}^k$ and $\tilde{f}^k$ defined above can be considered as the kth powers of $\tilde{e}$ and $\tilde{f}$ respectively.

THEOREM 7.1 ([**22**, §7]). *The operators $\varepsilon, \varphi, \mathrm{wt}, \tilde{e}$, and $\tilde{f}$ endow the set $B(\mathbf{w})$ with the structure of an $\mathfrak{sl}_2$-crystal and $B(\mathbf{w}) \cong B_{w_1} \otimes \cdots \otimes B_{w_n}$ as $\mathfrak{sl}_2$-crystals.*

We let $\phi : B(\mathbf{w}) \cong B_{w_1} \otimes \cdots \otimes B_{w_n}$ denote the isomorphism of Theorem 7.1.

7.2. The geometric realization of the crystal commutor. Fix a hermitian form on W. Let $t^\dagger$ denote the hermitian adjoint of $t \in \operatorname{End} W$ and let $S^\perp$ denote the orthogonal complement to a subspace $S \subseteq W$. If we let $\hat{W}_i = W_{n-i}^\perp$ for $0 \leq i \leq n$, and $\hat{\mathbf{w}} = (\hat{w}_i)_{i=1}^n$ where

$$\hat{w}_i = \dim \hat{W}_i / \hat{W}_{i-1} = \dim W_{n-i}^\perp / W_{n-i+1}^\perp = w_{n-i+1},$$

then

$$\mathfrak{T}(\hat{\mathbf{w}}) = \{(U, t) \mid U \subseteq W,\ t \in \operatorname{End} W,\ t(\hat{W}_i) \subseteq \hat{W}_{i-1}\ \forall i,\ \operatorname{im} t \subseteq U \subseteq \ker t\}.$$

Note that $\varepsilon(U, t) = 0$ if and only if $U = \operatorname{im} t$. Also, for $t \in \operatorname{End} W$,

$$t(W_i) \subseteq W_{i-1} \Rightarrow t^\dagger(\hat{W}_{n-i+1}) \subseteq \hat{W}_{n-i}.$$

Therefore,

$$(\operatorname{im} t, t) \in \mathfrak{T}(\mathbf{w}) \iff (\operatorname{im} t^\dagger, t^\dagger) \in \mathfrak{T}(\hat{\mathbf{w}}),$$

and the map $(\operatorname{im} t, t) \mapsto (\operatorname{im} t^\dagger, t^\dagger)$ induces isomorphisms

$$\mathfrak{T}(u, \mathbf{w})_0 \cong \mathfrak{T}(u, \hat{\mathbf{w}})_0, \quad B(u, \mathbf{w})_0 \cong B(u, \hat{\mathbf{w}})_0.$$

We denote the isomorphism $B(u, \mathbf{w})_0 \cong B(u, \hat{\mathbf{w}})_0$ by $X \mapsto X^\dagger$ for $X \in B(u, \mathbf{w})_0$. Since the elements of $B(\mathbf{w})_0$ are precisely the highest weight elements of the crystal $B(\mathbf{w})$, a commutor is uniquely determined by its action on these elements.

THEOREM 7.2 ([**24**, §4.2]). (1) *If $n = 2$ and $X \in B(\mathbf{w})_0$, we have*

$$\phi^{-1} \circ \sigma^c_{B_{w_1}, B_{w_2}} \circ \phi(X) = X^\dagger,$$

and thus the map $X \mapsto X^\dagger$ corresponds to the crystal commutor on highest weight elements.

(2) *If $n = 3$ and $X \in B(\mathbf{w})_0$,*

$$\phi^{-1} \circ \left(\sigma^c_{B_{w_1}, B_{w_3} \otimes B_{w_2}} \circ \left(\mathrm{id}_{B_{w_1}} \otimes \sigma^c_{B_{w_2} \otimes B_{w_3}}\right)\right) \circ \phi(X) = X^\dagger$$
$$= \phi^{-1} \circ \left(\sigma^c_{B_{w_2} \otimes B_{w_1}, B_{w_3}} \circ \left(\sigma^c_{B_{w_1}, B_{w_2}} \otimes \mathrm{id}_{B_{w_3}}\right)\right) \circ \phi(X),$$

and thus the crystal commutor satisfies the cactus relation.

One advantage of the geometric interpretation of the crystal commutor defined here is that it extends to any symmetrizable Kac-Moody algebra $\mathfrak{g}$. In particular, if $\mathfrak{g}$ has symmetric Cartan matrix, then there exists a tensor product quiver variety whose irreducible components can be given the structure of a tensor product crystal. There then exists a map $X \mapsto X^\dagger$, which generalizes the map defined above. One can show that, in the case of two factors, this map corresponds to the crystal commutor. For three factors, the compositions $\sigma^c_{B_{\lambda_1}, B_{\lambda_3} \otimes B_{\lambda_2}} \circ \left(\mathrm{id}_{B_{\lambda_1}} \otimes \sigma^c_{B_{\lambda_2} \otimes B_{\lambda_3}}\right)$ and $\sigma^c_{B_{\lambda_2} \otimes B_{\lambda_1}, B_{\lambda_3}} \circ \left(\sigma^c_{B_{\lambda_1}, B_{\lambda_2}} \otimes \mathrm{id}_{B_{\lambda_3}}\right)$ both correspond (on highest weight elements) to the map $X \mapsto X^\dagger$ and are therefore equal. Thus the commutor satisfies the cactus relation. When $\mathfrak{g}$ is symmetrizable but with non-symmetric Cartan matrix, one can use a well-known folding argument to obtain the same result from the symmetric case. We therefore have the following theorem.

THEOREM 7.3 ([**24**, Theorem 6.4]). *For a symmetrizable Kac-Moody algebra $\mathfrak{g}$, the category of $\mathfrak{g}$-crystals is a coboundary monoidal category with cactus commutor σ^c.*

This generalizes the previously known result for $\mathfrak{g}$ of finite type. We refer the reader to [**24**] for details.

References

[1] A. Berenstein and S. Zwicknagl. Braided symmetric and exterior algebras. *Trans. Amer. Math. Soc.*, 360(7):3429–3472, 2008.

[2] V. Chari and A. Pressley. *A guide to quantum groups*. Cambridge University Press, Cambridge, 1995. Corrected reprint of the 1994 original.

[3] M. Davis, T. Januszkiewicz, and R. Scott. Fundamental groups of blow-ups. *Adv. Math.*, 177(1):115–179, 2003.

[4] S. L. Devadoss. Tessellations of moduli spaces and the mosaic operad. In *Homotopy invariant algebraic structures (Baltimore, MD, 1998)*, volume 239 of *Contemp. Math.*, pages 91–114. Amer. Math. Soc., Providence, RI, 1999.

[5] V. G. Drinfel′d. Quasi-Hopf algebras. *Leningrad Math. J.*, 1(6):1419–1457, 1990.
[6] P. Etingof, A. Henriques, J. Kamnitzer, and E. Rains. The cohomology ring of the real locus of the moduli space of stable curves of genus 0 with marked points. To appear in Ann. of Math. (2), available at arXiv:math/0507514v2.
[7] A. Henriques and J. Kamnitzer. Crystals and coboundary categories. *Duke Math. J.*, 132(2):191–216, 2006.
[8] J. Hong and S.-J. Kang. *Introduction to quantum groups and crystal bases*, volume 42 of *Graduate Studies in Mathematics*. American Mathematical Society, Providence, RI, 2002.
[9] J. Kamnitzer and P. Tingley. A definition of the crystal commutor using Kashiwara's involution. To appear in J. Algebraic Combin., available at arXiv:math.QA/0610952v2.
[10] J. Kamnizter and P. Tingley. The crystal commutor and Drinfeld's unitarized R-matrix. To appear in J. Algebraic Combin., available at arXiv:0707.2248v2.
[11] M. Kashiwara. On crystal bases of the Q-analogue of universal enveloping algebras. *Duke Math. J.*, 63(2):465–516, 1991.
[12] M. Kashiwara. On crystal bases. In *Representations of groups (Banff, AB, 1994)*, volume 16 of *CMS Conf. Proc.*, pages 155–197. Amer. Math. Soc., Providence, RI, 1995.
[13] M. Kashiwara and Y. Saito. Geometric construction of crystal bases. *Duke Math. J.*, 89(1):9–36, 1997.
[14] A. Lascoux, B. Leclerc, and J.-Y. Thibon. Crystal graphs and q-analogues of weight multiplicities for the root system A_n. *Lett. Math. Phys.*, 35(4):359–374, 1995.
[15] G. Lusztig. Quantum deformations of certain simple modules over enveloping algebras. *Adv. in Math.*, 70(2):237–249, 1988.
[16] G. Lusztig. Quivers, perverse sheaves, and quantized enveloping algebras. *J. Amer. Math. Soc.*, 4(2):365–421, 1991.
[17] G. Lusztig. *Introduction to quantum groups*, volume 110 of *Progress in Mathematics*. Birkhäuser Boston Inc., Boston, MA, 1993.
[18] S. Mac Lane. *Categories for the working mathematician*, volume 5 of *Graduate Texts in Mathematics*. Springer-Verlag, New York, second edition, 1998.
[19] A. Malkin. Tensor product varieties and crystals: the ADE case. *Duke Math. J.*, 116(3):477–524, 2003.
[20] H. Nakajima. Instantons on ALE spaces, quiver varieties, and Kac-Moody algebras. *Duke Math. J.*, 76(2):365–416, 1994.
[21] H. Nakajima. Quiver varieties and Kac-Moody algebras. *Duke Math. J.*, 91(3):515–560, 1998.
[22] H. Nakajima. Quiver varieties and tensor products. *Invent. Math.*, 146(2):399–449, 2001.
[23] Y. Saito. Crystal bases and quiver varieties. *Math. Ann.*, 324(4):675–688, 2002.
[24] A. Savage. Crystals, quiver varieties and coboundary categories for Kac-Moody algebras. arXiv:0802.4083.

UNIVERSITY OF OTTAWA, OTTAWA, ONTARIO, CANADA
E-mail address: alistair.savage@uottawa.ca

Contemporary Mathematics
Volume **483**, 2009

Bases for direct powers

Phill Schultz

ABSTRACT. A power basis of an infinite direct power of a ring is an independent spanning set. I describe a class of rings for which power bases can be classified. There are connections with column finite matrices over a ring and with slender rings.

1. Introduction and Notation

Let R be a unital ring and J a set. The *J–power of R* is the left R–module R^J of functions from J to R with addition and scaling defined pointwise. Elements of R^J are called *J–vectors over R*. The J–vector $\mathbf{a} \in R^J$ whose value at $j \in J$ is a_j is denoted (a_j), and a_j is called the *j–component* of $\mathbf{a}$. It is clear that sums of finitely many J–vectors are defined, but certain infinite sums are also possible. Let $B = \{\mathbf{b}_i : i \in I\}$ be a family of J–vectors over R, where the j–component of $\mathbf{b}_i$ is denoted b_{ij}. B is called *summable* [**GT06**, Definition 1.4.9] if for each $j \in J$, only finitely many b_{ij} are non–zero. If B is summable then $\sum_{i \in I} \mathbf{b}_i$ is defined to be the J–vector whose j–component for all $j \in J$ is $\sum_{i \in I} b_{ij}$, a finite sum in R. Note that if $B = \{\mathbf{b}_i : i \in I\}$ is summable and $\mathbf{r} = (r_i : i \in I) \in R^I$, then $\mathbf{r}B = \{r_i \mathbf{b}_i : i \in I\}$ is also summable.

The *standard power basis* E of R^J is the summable set of J–vectors $\{\mathbf{e}_i : i \in J\}$ for which for all $i \in J$, $e_{ij} = \delta_{ij}$, the Kronecker delta function on J. Thus for all J–vectors (r_i), the J–vector $\sum_{i \in J} r_i \mathbf{e}_i$ is defined and equals (r_i). This set E is an example of a power basis defined as follows:

DEFINITION 1.1. Let B be an infinite family in R^J. B is a *power basis* if

[Sum] B is summable.

[Span] For all $\mathbf{a} \in R^J$ and for all $\mathbf{b} \in B$, there exist $r_{\mathbf{b}} \in R$ such that $\mathbf{a} = \sum_{\mathbf{b} \in B} r_{\mathbf{b}} \mathbf{b}$

[Ind] If $\sum_{\mathbf{b} \in B} r_{\mathbf{b}} \mathbf{b} = 0$, then each $r_{\mathbf{b}} = 0$.

Note that [Span] and [Ind] are well defined precisely when B satisfies [Sum], and just as in Linear Algebra, [Span] and [Ind] can be summarised by stating that each vector in R^J has a unique representation as an element of $\prod_{\mathbf{b} \in B} \langle \mathbf{b} \rangle$. We say

---

2000 *Mathematics Subject Classification.* Primary 16S10, Secondary 16D10.
Key words and phrases. Direct products, bases.

that $\mathbf{a} = \sum_{\mathbf{b}\in B} r_{\mathbf{b}}\mathbf{b}$ is the *decomposition of* $\mathbf{a}$ *with respect to the power basis* B. For example, the J–vector $\mathbf{a}$ has decomposition $\sum_{\mathbf{e}\in \mathrm{E}} a_{\mathbf{e}}\mathbf{e}$ with respect to E.

The question addressed in this paper is the construction of other power bases for R^J. The motivating example is the ring $\mathbb{F}[[x]]$ of power series over a field $\mathbb{F}$, considered as an $\mathbb{F}$ vector space. Of course, there we have the advantage that $\mathbb{F}[[x]]$ is a commutative graded ring, which makes it easy to construct power bases. The standard basis $\{x^i : i \in \mathbb{N}\}$ is a power basis for $\mathbb{F}[[x]]$ and in [**CS07**] E. F. Cornelius Jr. and the author described a power basis $\{\rho_i : i \in \mathbb{N}\}$, based on a sequence $(s_i : i \in \mathbb{N})$ of distinct elements of $\mathbb{F}$, by $\rho_0 = 1$ and $\rho_{i+1} = \rho_i(x - s_i)$. The point of that construction was to prove a power series version of the Lagrange Interpolation Theorem, namely, if $\mathbf{a} = (a_i : i \in \mathbb{N})$ is any sequence in $\mathbb{F}^{\mathbb{N}}$, then there is a unique power series $\ell_{\mathbf{a}}(x)$ in $\mathbb{F}[[x]]$ such that $\mathbf{a} = (\ell_{\mathbf{a}}(i) : i \in \mathbb{N})$.

When both R and J are infinite, dealing with R^J inevitably leads to problems of a foundational nature. In order to avoid set theoretical complications, I shall henceforth assume the axioms of ZFC, together with the *generalized continuum hypothesis* in the form:

[GCH] for all sets A, B and C with $|A| > 1$, $|A^B| = |A^C|$ implies $|B| = |C|$.

An obvious observation can be made: by [Sum], [Scan] and [GCH], if B is a power basis for R^J, then $|B| = |J|$. Hence we can index each power basis by J and when we need to well–order a power basis, by the least ordinal of cardinality $|J|$.

Let B be a power basis for R^J and $\mathbf{a} = \sum_{\mathbf{b}\in B} r_{\mathbf{b}}\mathbf{b} \in R^J$. The *support* of $\mathbf{a}$ with respect to B, $\operatorname{supp}_B(\mathbf{a}) = \{\mathbf{b} \in B : r_{\mathbf{b}} \neq 0\}$. If B is a power basis of R^J, by [Ind], B generates a free submodule, denoted $R_B^{(J)} = \{\mathbf{a} \in R^J : \operatorname{supp}_B(\mathbf{a}) \text{ is finite}\}$. By the remarks above, the free submodules of R^J generated by different power bases are isomorphic, and by convention we denote that generated by E simply by $R^{(J)}$.

2. Relations between power bases

Let B be a power basis for R^J and $\alpha \in \operatorname{Aut}(R^J)$, the automorphism group of the module R^J. It does not follow that $B\alpha = \{\mathbf{b}\alpha : \mathbf{b} \in B\}$ is a power basis. For example, let $\mathbb{F}$ be a field and λ an infinite ordinal. Then $\mathbb{F}^\lambda$ is an $\mathbb{F}$–space of dimension 2^λ and the standard power basis E is a vector space basis of $\mathbb{F}^{(\lambda)}$ which can be extended to a vector space basis $\overline{\mathrm{E}}$ of $\mathbb{F}^\lambda$. For a fixed $\mathbf{e}_0 \in \mathrm{E}$, let $B = \{\mathbf{e}_0, \mathbf{e}_0 + \mathbf{b} : \mathbf{e}_0 \neq \mathbf{b} \in \overline{\mathrm{E}}\}$. Then B is also a vector space basis for $\mathbb{F}^\lambda$ and hence $\mathbb{F}^\lambda$ has an automorphism α mapping $\mathbf{e}_0$ onto $\mathbf{e}_0$ and $\mathbf{b}$ onto $\mathbf{e}_0 + \mathbf{b}$ for all $\mathbf{e}_0 \neq \mathbf{b} \in \overline{\mathrm{E}}$. But B is not a power basis for $\mathbb{F}^\lambda$ since it is not summable.

However, we do have a sort of converse result:

PROPOSITION 2.1. *Let* $B = \{\mathbf{b}_i : i \in J\}$ *and* $C = \{\mathbf{c}_i : i \in J\}$ *be power bases for* R^J. *Then there exists a unique* $\alpha \in \operatorname{Aut}(R^J)$ *for which* $\mathbf{b}_i\alpha = \mathbf{c}_i$ *for all* $i \in J$.

Furthermore, for all $\mathbf{a} = \sum_{i<\lambda} r_i\mathbf{b}_i \in R^J$, $\mathbf{a}\alpha = \sum_{i<\lambda} r_i\mathbf{c}_i$.

PROOF. Define α by the latter equation. By [Sum], [Ind] and [Span], α is a well–defined bijection. By the definition of addition and scaling, α is the unique automorphism of R^J satisfying $\mathbf{b}_i\alpha = \mathbf{c}_i$ for all $i < \lambda$. □

We say that R^J is *suitable* if for all $\alpha \in \operatorname{Aut}(R^J)$, $\mathrm{E}\alpha = \{\mathbf{e}\alpha : \mathbf{e} \in \mathrm{E}\}$ is summable. The importance of suitability lies in the following theorem:

THEOREM 2.2. *Let* R^J *be suitable and let* $\alpha \in \operatorname{Aut}(R^J)$. *Then:*

(1) *For all* $\sum_{\mathbf{e}\in \mathrm{E}} r_{\mathbf{e}}\mathbf{e} \in R^J$, $\left(\sum_{\mathbf{e}\in \mathrm{E}} r_{\mathbf{e}}\mathbf{e}\right)\alpha = \sum_{\mathbf{e}\in \mathrm{E}} r_{\mathbf{e}}(\mathbf{e}\alpha)$

(2) $\mathrm{E}\alpha$ *is a power basis for* R^J.

(3) $\mathrm{Aut}(R^J)$ *permutes the set of power bases of* R^J.

Proof. (1) For each $j \in J$, only finitely many $r_{\mathbf{e}}(\mathbf{e}\alpha)$ have non-zero j-component, so the sum $\sum r_{\mathbf{e}}(\mathbf{e}\alpha)$ makes sense. I show that for each $j \in J$, the j components of each side of the equation are identical. For each $j \in J$, let $F_j = \{\mathbf{e} \in \mathrm{E} : (\mathbf{e}\alpha)_j \neq 0\}$. Then F_j is finite and for all $j \in J$,

$$\sum_{\mathbf{e}\in \mathrm{E}} r_{\mathbf{e}}(\mathbf{e}\alpha)_j = \sum_{\mathbf{e}\in F_j} r_{\mathbf{e}}(\mathbf{e}\alpha)_j = \Big(\big(\sum_{\mathbf{e}\in F_j} r_{\mathbf{e}}\mathbf{e}\big)\alpha\Big)_j$$

Hence $\sum_{\mathbf{e}\in \mathrm{E}} r_{\mathbf{e}}(\mathbf{e}\alpha) = \left(\sum_{\mathbf{e}\in \mathrm{E}} r_{\mathbf{e}}\mathbf{e}\right)\alpha$.

(2) Since R^J is suitable, $\mathrm{E}\alpha$ is summable. For all $\mathbf{b} \in R^J$, $\mathbf{b}\alpha^{-1}$ can be expressed as $\sum_{\mathbf{e}\in \mathrm{E}} r_{\mathbf{e}}\mathbf{e}$, so by (1) again, $\mathbf{b} = \sum_{\mathbf{e}\in \mathrm{E}} r_{\mathbf{e}}(\mathbf{e}\alpha)$. Thus $\mathrm{E}\alpha$ satisfies [Span]. Furthermore, if $\sum_{\mathbf{e}\in \mathrm{E}} s_{\mathbf{e}}(\mathbf{e}\alpha) = 0$, then $\sum_{\mathbf{e}\in \mathrm{E}} s_{\mathbf{e}}\mathbf{e} = 0$ so all $s_{\mathbf{e}} = 0$. Thus $\mathrm{E}\alpha$ satisfies also [Ind] so forms a power basis.

(3) Let B be a power basis and let $\alpha \in \mathrm{Aut}(R^J)$. By Proposition 2.1, there exists $\beta \in \mathrm{Aut}(R^J)$ such that $B = \mathrm{E}\beta$. By (2), $B\alpha = \mathrm{E}\beta\alpha$ is a power basis. Conversely, Proposition 2.1 shows that if B and C are power bases, then there exists $\alpha \in \mathrm{Aut}(R^J)$ such that $B\alpha = C$. □

Theorem 2.2 can also be stated in the language of matrix theory. Let $\mathrm{CFM}(J, R)$ be the set of $|J| \times |J|$ matrices over the ring R in which each column contains only finitely many non-zero entries. It is routine to verify that $\mathrm{CFM}(J, R)$ is a ring under pointwise addition and standard convolutional matrix multiplication, using column finiteness in an essential way. Denote by $\mathrm{CFGL}(J, R)$ the unit group of $\mathrm{CFM}(J, R)$, that is, the group of column finite invertible matrices.

Note that the map $r \mapsto$ right multiplication by r is a ring isomorphism of R onto $\mathrm{End}(R)$. For all $i, j \in J$, let ι_i be the embedding of the ith component in R^J and let π_j be the projection of R^J onto the jth component. Let $\alpha \in \mathrm{Aut}(R^J)$. Then $\iota_i\alpha\pi_j \in \mathrm{End}(R)$ is right multiplication by an element $m_{ij} \in R$. Let $M(\alpha)$ be the matrix (m_{ij}).

Theorem 2.3. *Let* R^J *be suitable. The mapping* $\alpha \mapsto M(\alpha)$ *is a group isomorphism of* $\mathrm{Aut}(R^J)$ *onto* $\mathrm{CFGL}(J, R)$.

Proof. Since for each $j \in J$, only finitely many $m_{ij} \neq 0$, $M(\alpha) \in \mathrm{CFM}(J, R)$. With the usual operations of addition and multiplication of matrices, the mapping $\alpha \mapsto M(\alpha)$ is a group monomorphism into $\mathrm{CFGL}(J, R)$. To see that it is a surjection, let $N \in \mathrm{CFGL}(J, R)$, and define $\alpha \in \mathrm{Aut}(R^J)$ by $\alpha : \mathbf{e}_i \mapsto$ the ith row of N, considered as an element of R^J. Then it is routine to check that $N = M(\alpha)$. □

Corollary 2.4. *Let* R^J *be suitable.*

(1) *For all* $\alpha \in \mathrm{Aut}(R^J)$, α *is realised by right multiplication of* J*-vectors by* $M(\alpha)$.

(2) *The group* $\mathrm{Aut}(R^J)$ *is isomorphic by the map* $\alpha \mapsto M(\alpha)$ *to the group* $\mathrm{CFGL}(J, R)$ *of invertible matrices in* $\mathrm{CFM}(J, R)$. □

We can now characterize power bases of suitable powers R^J.

Theorem 2.5. *Let* R^J *be suitable and let* $B \subset R^J$ *be a summable set. The following are equivalent:*

(1) *B is a power basis.*
(2) $B = \mathrm{E}\alpha$ *for some* $\alpha \in \mathrm{Aut}(R^J)$.
(3) *The matrix M whose rows are the elements of B (in any order) is an element of* $\mathrm{CFGL}(J, R)$.

PROOF. (1) $\Rightarrow$ (2) This is a consequence of Proposition 2.1.

(2) $\Rightarrow$ (3) Since B satisfies [Sum], $M \in \mathrm{CFM}(J, R)$ and since B satisfies [Span] and [Ind], M is invertible.

(3) $\Rightarrow$ (1) Since $M \in \mathrm{CFGL}(J, R)$, its rows form a summable set B in R^J. Since M is invertible, B satisfies [Span] and [Ind]. □

Note that by an appropriate change of wording, E in Theorem 2.5 (2) could be replaced by any power basis of R^J.

3. Slenderness

We have characterized the power bases of suitable R^J, but have not found the structure of unital rings R for which R^J is suitable. To do so, we now consider a stronger property for which not just automorphisms but endomorphisms preserve summability of the standard basis. We say that R^J is *very suitable* if for all $\nu \in \mathrm{End}(R^J)$, $\mathrm{E}\nu = \{\mathbf{e}\nu : \mathbf{e} \in \mathrm{E}\}$ is summable.

This means that very suitable rings satisfy the crucial property (1) of Theorem 2.2, with endomorphisms in place of automorphisms:

PROPOSITION 3.1. *Let R^J be very suitable and let $\nu \in \mathrm{End}(R^J)$. Then for all* $\sum_{\mathbf{e}\in\mathrm{E}} r_{\mathbf{e}}\mathbf{e} \in R^J$, $\left(\sum_{\mathbf{e}\in\mathrm{E}} r_{\mathbf{e}}\mathbf{e}\right)\nu = \sum_{\mathbf{e}\in\mathrm{E}} r_{\mathbf{e}}(\mathbf{e}\nu)$. □

The proof is omitted, since it is the same, *mutatis mutandis*, as the proof of Theorem 2.2 (1).

A matrix version of Proposition 3.1 is also valid. With the notation of Theorem 2.3, summability implies that for all $\nu \in \mathrm{End}(R^J)$, the matrix $M(\nu) = (n_{ij})$, where $\iota_i\nu\pi_j$ is right multiplication by n_{ij}, is in $\mathrm{CFM}(J, R)$, and the proofs of the following proposition and corollary mimic closely those of Theorem 2.3 and Corollary 2.4.

PROPOSITION 3.2. *Let R^J be very suitable. The mapping $\nu \mapsto M(\nu)$ is a ring isomorphism of* $\mathrm{End}(R^J)$ *onto* $\mathrm{CFM}(J, R)$. □

COROLLARY 3.3. *Let R^J be very suitable.*
(1) *For all $\nu \in \mathrm{End}(R^J)$, ν is realised by right multiplication of J–vectors by $M(\nu)$.*
(2) *The ring* $\mathrm{End}(R^J)$ *is isomorphic by the map $\nu \mapsto M(\nu)$ to the matrix ring* $\mathrm{CFM}(J, R)$. □

The reason for this specialization is that very suitable rings are well known in other contexts. Recall [**GT06**, Definition 1.3.5] that a *slender* ring is one for which every module homomorphism $f : R^\omega \to R$ maps all but finitely many components to zero. It follows immediately that R is slender if and only if the natural module homomorphism $\theta : \mathrm{Hom}_R(R^\omega, R) \to \mathrm{Hom}_R(R, R)^{(\omega)}$ is an isomorphism. What is less evident, but nonetheless true is that ω may be replaced by any non–measurable ordinal:

PROPOSITION 3.4. [**GT06**, Corollary 1.4.14] *Let λ be any non–measurable ordinal. The following are equivalent:*

(1) *R is slender.*
(2) *Every module endomorphism $f : R^\lambda \to R$ maps all but finitely many components to zero.*
(3) *The natural map $\theta : \mathrm{Hom}_R(R^\lambda, R) \to \mathrm{Hom}_R(R, R)^{(\lambda)}$ is an isomorphism.* □

THEOREM 3.5. *R^J is very suitable if and only if R is slender and $|J|$ is non-measurable.*

PROOF. ($\Rightarrow$) Suppose R^J is very suitable, but not slender. Let $f \in \mathrm{Hom}(R^J, R)$ be non-zero on infinitely many components $j \in J$. Define $\nu \in \mathrm{End}(R^J)$ by $\mathbf{e}\nu = \mathbf{e}f\iota_j$. Then $\{\mathbf{e}\nu : \mathbf{e} \in \mathrm{E}\}$ is not summable, contradicting suitability.

In [**F73**, Remark, p 161] Fuchs shows that if $|J|$ is measurable, then there is an epimorphism of $\mathbb{Z}^J$ onto $\mathbb{Z}$ which maps $\mathbb{Z}^{(J)}$ to zero. The same proof works with $\mathbb{Z}$ replaced by any unital ring. This implies by Theorem 2.2 that if $|J|$ is measurable, then R^J is not suitable for any unital ring R.

($\Leftarrow$) If R^J is not suitable, let $j \in J$ and $\nu \in \mathrm{End}(R^J)$ be such that $\mathbf{e}_i\nu\pi_j \neq 0$ for infinitely many $i \in J$. Then $\nu\pi_j \in \mathrm{Hom}(R^J, R)$ is non–zero on infinitely many components, so R is not slender. □

Theorem 3.5 and Corollary 3.3 have an amusing consequence, pointed out by my colleague E. F. Cornelius Jr.:

COROLLARY 3.6. *The power bases of the Baer–Specker group $\mathbb{Z}^\omega$* [**F73**, Section 94] *coincide with the rows of matrices in* $\mathrm{CFGL}(\omega, \mathbb{Z})$. □

The problem now arises of characterizing the slender rings among all unital rings. The concept of slenderness was introduced in 1954 by Łoś [**L54**] in the context of abelian groups. The definition was extended to modules over arbitrary unital rings in 1973 by Lady [**L73**]. There has been an extensive literature on the topic since then, the main results being presented in [**GT06**, Sections 1.4 and 1.5]. It turns out that almost all the results on the structure of slender rings use topological methods which are relevant only for commutative integral domains, and definitive results are only available for Dedekind domains.

Theorem 3.5 allows us to use the methods of this paper to obtain a new characterization of arbitrary slender rings. If $N \leq M$ are R–modules, we say that $\mathrm{End}(M)$ is *determined by* N if whenever f and $g \in \mathrm{End}(M)$ satisfy $f|_N = g|_N$ then $f = g$. Note that if R^J is a very suitable ring with power basis B then B is a basis for the free module $R_B^{(J)}$ and by Proposition 3.1, every $\nu \in \mathrm{End}(R_B^J)$ is determined by $R^{(J)}$ and every $\nu \in \mathrm{End}(R_B^{(J)})$ extends uniquely to $\mathrm{End}(R^J)$. Hence it is of interest to consider powers R^J with the two properties:

(1) Every $\nu \in \mathrm{Hom}(R^{(J)}, R^J)$ extends to $\mathrm{End}(R^J)$
(2) Every $\nu \in \mathrm{End}(R^J)$ is determined by $R^{(J)}$.

PROPOSITION 3.7. *Let R be a unital ring and J a set. Then R^J is very suitable if and only if every $\nu \in \mathrm{End}(R^{(J)})$ extends to $\mathrm{End}(R^J)$ and for every $\mu \in \mathrm{End}(R^{(J)})$ there is a unique $\nu \in \mathrm{End}(R^J)$ whose restriction to $R^{(J)}$ is μ.*

PROOF. If R^J is very suitable, then by Corollary 3.3, the ring $\mathrm{CFM}(J, R)$ represents both $\mathrm{End}(R^J)$ and $\mathrm{End}(R^{(J)})$ in both cases by right multiplication. Hence every $\nu \in \mathrm{End}(R^{(J)})$ extends to $\mathrm{End}(R^J)$ and such an extension is unique.

Conversely, since $R^{(J)}$ is a free module, $\text{End}(R^{(J)})$ is realised by $\text{CFM}(J, R)$, which also acts by right multiplication on R^J. Since every element of $\text{End}(R^{(J)})$ extends, every $\mu \in \text{End}(R^J)$ is such a multiplication, and since extensions are unique, the action is faithful. Hence for all $\nu \in \text{End}(R^J)$, $\left(\sum_{\mathbf{e}\in \mathbf{E}} r_{\mathbf{e}}\mathbf{e}\right)\nu = \sum_{\mathbf{e}\in \mathbf{E}} r_{\mathbf{e}}(\mathbf{e}\nu)$, so R^J is very suitable, □

Proposition 3.7 provides, in view of Theorem 3.5, a new algebraic characterization of slender rings. We use the natural isomorphisms $\text{End}(R^J) \cong \text{Hom}(R^J, R)^J$ and $\text{Ext}(R^J, R^J) \cong \text{Ext}(R^J, R)^J$ [**R70**, Theorems 2.4 and 7.11] to describe this alternative characterization.

Denote by R_J the *reduced product* $R^J/R^{(J)}$.

LEMMA 3.8. *Let R be a ring and J a set. Then*

(1) *Every $\nu \in \text{Hom}(R^J, R)$ is determined by $R^{(J)}$ if and only if* $\text{Hom}(R_J, R) = 0$
(2) *Every $\nu \in \text{Hom}(R^{(J)}, R)$ can be extended to* $\text{Hom}(R^J, R)$ *if and only if the natural map* $\text{Ext}(R_J, R) \to \text{Ext}(R^J, R)$ *is monic.*

PROOF. The natural short exact sequence of modules $R^{(J)} \rightarrowtail R^J \twoheadrightarrow R_J$ induces an exact sequence of abelian groups

$$\begin{aligned}\text{Hom}(R_J, R) \rightarrowtail \text{Hom}(R^J, R) \overset{res}{\to} \text{Hom}(R^{(J)}, R)\\ \to \text{Ext}(R_J, R) \twoheadrightarrow \text{Ext}(R^J, R)\end{aligned}$$

where res is the restriction map and the last map is epic since $R^{(J)}$ is projective.

Thus the restriction map is monic if and only if $\text{Hom}(R_J, R) = 0$ and epic if and only if the natural map $\text{Ext}(R_J, R) \to \text{Ext}(R^J, R)$ is monic. □

COROLLARY 3.9. *A unital ring R is slender if and only if*

(1) $\text{Hom}(R_\omega, R) = 0$
(2) *The natural map* $\text{Ext}(R_\omega, R) \to \text{Ext}(R^\omega, R)$ *is monic.* □

This characterization my be compared with some of those in the literature:

- [**GT06**, Theorem 1.4.23] If R is commutative, then R is slender if and only if $\text{Hom}(R_\omega, R) = 0$ and R is not complete in any non–discrete Haussdorff linear topology.
- [**GT06**, Theorem 1.5.15] If R is a Dedekind domain with quotient field Q, then R is slender if and only if R is not a field or a complete discrete valuation ring and $\text{Ext}(Q, R) \neq 0$.

To conclude, I should note that it remains unsettled whether every suitable ring is very suitable. While I believe this is unlikely, I have not been able to construct a counter–example.

References

CS07. E. F. Cornelius Jr. and P. Schultz, *Root bases of polynomials over integral domains*, Contributions to Module Theorey, de Gruyter, 2007, 1–14

F73. L. Fuchs, *Infinite Abelian Groups*, Vol. 2, Academic Press, 1973

GT06. R. Göbel and J. Trlifaj, *Approximations and Endomorphism Algebras of Modules*, De Gruyter Expositions in mathematics 41, W. de Gruyter, 2006

L54. J. Łoś, *On the complete direct sum of countable abelian groups*, Publ. Math. Debrecen, **3**, (1954), 269–272

L73. E. L. Lady, *Slender rings and modules*, Pcific J. Math. **49**, (1973), 397–406

R70. J. J. Rotman, *Notes on Homological Algebra*, Van Nostrand Reinhold mathematical Studies 26, 1970

School of Mathematics and Statistics, The University of Western Australia, Nedlands, Australia 6009

E-mail address: schultz@maths.uwa.edu.au

Contemporary Mathematics
Volume **483**, 2009

STRUCTURE AND REPRESENTATIONS OF JORDAN ALGEBRAS ARISING FROM INTERMOLECULAR RECOMBINATION

SERGEI R. SVERCHKOV

ABSTRACT. In this work we investigate the structure and representations of Jordan algebras arising from intermolecular recombination. It is proved that the variety of all these algebras is special. The basis and multiplication table are constructed for the free algebra of this variety. It is also shown that all the identities satisfying the operation of intermolecular recombination are consequences of only one identity of degree 4.

1. Introduction. The commutative algebra J over the field F is the *algebra arising from intermolecular recombination*, if it satisfies the identity

$$(x^2 \cdot y) \cdot z + 2((x \cdot z) \cdot x) \cdot y - 2(x^2 \cdot z) \cdot y = x^2 \cdot (zy). \tag{1}$$

Assuming that $z = x$ in the identity (1) we get the identity $(x^2 \cdot y) \cdot x = x^2 \cdot (x \cdot y)$. Consequently, all the algebras arising from intermolecular recombination are Jordan algebras. We will denote by IR-algebras (intermolecular recombination), the algebras arising from intermolecular recombination. Let IR denote the variety of all IR-algebras. The IR-algebras were introduced by M. Bremner in the work [1] and algebraically formalized IR-operations, operations of intermolecular recombination.

In the general theory of DNA computing (see G. Păun, G. Rozenberg and A. Salomaa [2], L. Landweber and L. Kari [3]) the IR-operation has the form

$$u_1xv_1 + u_2xv_2 \Rightarrow u_1xv_2 + u_2xv_1,$$

where u_1, u_2, v_1, v_2 are words over some alphabet S. Formalization of IR-operations by M. Bremner [1] can be defined as follows. Let consider the free

1991 *Mathematics Subject Classification.* 17C50.

F-module J, generated by the set $C = A \times B$, where $A = \{a_i,\, i \in I_1\}$, $B = \{b_i,\, i \in I_2\}$ are some finite or countable sets. Let turn F-module J into F-algebra by defining the operation ¡¡$\rhd$¿¿ (splicing) on basis elements $a_i b_j$ $(i \in I_1,\ j \in I_2)$ following the rule

$$a_i b_j \rhd a_k b_l = a_i b_l,$$

and extend it linearly to all of the F-module J. It is easy to check that this operation is associative. M. Bremner [1] notes that the operation of intermolecular recombination

$$a_i b_j \cdot a_k b_l = \frac{1}{2}(a_i b_l + a_k b_j)$$

is a symmetrized product on the algebra $(J, \rhd, +)$. And in fact,

$$a_i b_j \circ a_k b_l = \frac{1}{2}(a_i b_j \rhd a_k b_l + a_k b_l \rhd a_i b_j) = \frac{1}{2}(a_i b_l + a_k b_j) = a_i b_j \cdot a_k b_l.$$

So, the algebra J of intermolecular recombination is a special Jordan algebra, i.e.

$$(J, \circ, +) = (J, \rhd, +)^{(+)}.$$

In the work [1] it is proved that the algebra J satisfies the identity (1) and all the identities of degree ≤ 6 of this algebra are consequences of the identity (1). In the same work we get the question if all the IR-algebras are special Jordan algebras. In this work we investigate the structure and representations of IR-algebras. We prove that all the identities of the algebra J are consequences of the identity (1) and that the variety of IR-algebras is special.

1.1. Standard IR-algebras. Definitions and notations. All algebras in this work are considered over the field F of characteristic 0, so the defining identities of varieties are linearized. We will use right-handed bracketing in non-associative words. Standard definitions and notations can be found in [4].

Denote $a_{ij} = a_i b_j$, where $i, j \in \mathbb{N}$. Then the associative splicing operation defines the associative algebra C with the basis a_{ij}, $i, j \in \mathbb{N}$, and the following multiplication table:

$$a_{ij} a_{kl} = a_{il}. \tag{2}$$

In cases when $|A| = |B| = \infty$; $|A| = n$, $|B| = m$; $|A| = n$, $|B| = \infty$, the corresponding associative algebras with the multiplication table (2) will be denoted correspondingly C_∞; $C_{n,m}$; C_n.

We will call C_∞; $C_{n,m}$; C_n *the standard algebras of splicing*, or *standard S-algebras* for short. We'll also define the Jordan algebras $J_\infty = C_\infty^{(+)}$; $J_{n,m} = C_{n,m}^{(+)}$; $J_n = C_n^{(+)}$ and call them *standard algebras of intermolecular recombination*, or *standard IR-algebras* for short. It is clear that the standard IR-algebras have the basis a_{ij}, $i, j \in \mathbb{N}$ $(1 \leq i \leq n,\ 1 \leq j \leq m$ for $J_{n,m}$ and $1 \leq i \leq n$, $j \in \mathbb{N}$ for $J_n)$, and the multiplication table

$$a_{ij} \cdot a_{kl} = \frac{1}{2}(a_{il} + a_{kj}). \tag{3}$$

For multiplication of basis elements in standard IR-algebras it is convenient to use the following correlations (see [1]):

$$(x \cdot y) \cdot z = \frac{1}{2}(x \cdot z + y \cdot z), \quad (x \cdot y) \cdot (z \cdot t) = \frac{1}{4}(x \cdot z + x \cdot t + y \cdot z + y \cdot t). \quad (4)$$

In fact,

$$\begin{aligned}(a_{i_1 i_2} \cdot a_{j_1 j_2}) a_{k_1 k_2} &= \frac{1}{2}(a_{i_1 j_2} + a_{j_1 i_2}) \cdot a_{k_1 k_2} \\ &= \frac{1}{4}(a_{i_1 k_2} + a_{k_1 j_2} + a_{j_1 k_2} + a_{k_1 i_2}) = \frac{1}{2}(a_{i_1 i_2} \cdot a_{k_1 k_2} + a_{j_1 j_2} \cdot a_{k_1 k_2});\end{aligned}$$

$$\begin{aligned}&(a_{i_1 i_2} \cdot a_{j_1 j_2}) \cdot (a_{k_1 k_2} \cdot a_{l_1 l_2}) = \frac{1}{4}(a_{i_1 j_2} + a_{j_1 i_2}) \cdot (a_{k_1 l_2} + a_{l_1 k_2}) \\ &= \frac{1}{8}(a_{i_1 l_2} + a_{i_1 k_2} + a_{k_1 j_2} + a_{l_1 j_2} + a_{j_1 l_2} + a_{j_1 k_2} + a_{k_1 i_2} + a_{l_1 i_2}) \\ &\qquad = \frac{1}{4}(a_{i_1 i_2} \cdot a_{k_1 k_2} + a_{i_1 i_2} \cdot a_{l_1 l_2} + a_{j_1 j_2} \cdot a_{k_1 k_2} + a_{j_1 j_2} \cdot a_{l_1 l_2}).\end{aligned}$$

Let's note that the correlations (4) are not valid for arbitrary elements. For example,

$$\begin{aligned}((x - y) \cdot z) \cdot t &= (x \cdot z - y \cdot z) \cdot t \underset{(4)}{=} \frac{1}{2}(x \cdot t - y \cdot t) \underset{(3)}{=} \frac{1}{2}(a_{13} - a_{23}) \\ &\neq \frac{1}{2}(x - y) \cdot t + \frac{1}{2} z \cdot t = \frac{1}{2} x \cdot t - \frac{1}{2} y \cdot t + \frac{1}{2} z \cdot t = \frac{1}{2} a_{13} - \frac{1}{2} a_{23} + \frac{1}{2} a_{33}.\end{aligned}$$

for $x = a_{11}$, $y = a_{22}$, $z = t = a_{33}$. Using the correlation (4) it is easy to check that in standard IR-algebras the identity (1) is valid.

1.2. General results. Let's review the main results of this work. The multiplication table (3) shows that the standard IR-algebras are algebras with genetic realization (see [5]). Among the algebras with genetic realization the class of Bernstein algebras holds an important position.

Let's remind that the Bernstein algebra over the field F is a commutative algebra J, with a non-zero algebra homomorphism $\omega : J \to F$, satisfying the identity

$$x^2 \cdot x^2 = \omega(x)^2 x^2.$$

These algebras were introduced by P. Holgate [9], It is known [5] that the algebra J can be represented as $J = Fe \oplus N$, where $N = \operatorname{Ker} \omega$, e is idempotent and $n^2 \cdot n^2 = 0$ for all $n \in \mathbb{N}$. If $\operatorname{ch}(F) \neq 2$, then $N = U \oplus Z$, where $U = \{u \in N \mid e \cdot u = \frac{1}{2}u\}$, $Z = \{z \in N \mid e \cdot z = 0\}$. On the algebra N the following Bernstein graduation is defined:

$$U^2 \subseteq Z, \quad Z^2 \subseteq U, \quad U \cdot Z \subseteq U.$$

A Bernstein algebra is called Jordan, if it also satisfies the Jordan identity

$$(x^2 \cdot y) \cdot x = x^2 \cdot (y \cdot x).$$

The Jordan Bernstein algebras were first introduced by P. Holgate [10], who proved that the genetic algebras for the simple Mendel inheritance are special Jordan algebras. Later this result was generalized by A. Wörs-Busekros [11]. It was shown that finite-dimensional Bernstein algebras with zero multiplication in N are special Jordan algebras. Also in the paper [11] the necessary and sufficient conditions for a Bernstein algebra to be Jordan were obtained: $Z^2 = 0$ and $N = U \oplus Z$ is nil-index 3 algebra. Jordan Bernstein algebras play an important role in the theory of Bernstein algebras (see [5, 12, 13]).

Definition 1. Bernstein algebra $B = Fe \oplus U \oplus Z$ is called *annihilator algebra* if Z coincides with annihilator of the algebra B, i.e. $Z = \operatorname{Ann}(B)$.

It is easy to note that in the annihilator algebras $Z^2 = N^3 = 0$. That is why all annihilator algebras are Jordan Bernstein algebras.

In the Section 2 of the present work we prove that all standard IR-algebras are Bernstein algebras (Theorem 1) and furthermore, the class of standard IR-algebras coincides with the class of annihilator Bernstein algebras of a special type (Theorem 2).

We will denote by $F(c_i,\, i \in I)$ the free F-module, generated by the elements $c_i,\, i \in I$.

The Section 2 describes the annihilator of a standard IR-algebra J_∞. It is found that

$$\operatorname{Ann}(J_\infty) = F((a_{11} - a_{ij})^2,\ i > 1,\ j > 1)$$

and the following isomorphism of the modules takes place:

$$J_\infty / \operatorname{Ann}(J_\infty) \simeq F(a_{ij},\ i = 1 \text{ or } j = 1),$$

where $\operatorname{Ann}(J_\infty / \operatorname{Ann}(J_\infty)) = 0$ (Lemma 1). Let's adjoin a formal unit 1 to the algebra $J_\infty / \operatorname{Ann}(J_\infty)$. It is shown that the algebra $J^\sharp = F \cdot 1 + J_\infty / \operatorname{Ann}(J_\infty)$ is a Jordan algebra of symmetrical bilinear form over F (Lemma 3).

The Section 3 of this work proves the speciality of the variety of all IR-algebras (Theorem 3).

Let $B_\infty = F \cdot 1 + V$ denote the Jordan algebra of non-degenerate symmetric bilinear form $f : V \times V \to F$, where V is an infinite dimentional vector space over F. By the determining identities of the variety $\operatorname{Var}(B_\infty)$ (see the results by S. Vasilovsky [7]), it is shown that the variety IR is a proper subvariety of $\operatorname{Var}(B_\infty)$, i.e.

$$IR \subsetneq \operatorname{Var}(B_\infty).$$

In view of the results [8] the variety $\operatorname{Var}(B_\infty)$ is special. That is why any commutative algebra satisfying the identity (1) is a special Jordan algebra. Let's note that the class of Jordan Bernstein algebras is not special (see [6]).

In the Section 4 we study the free IR-algebras.

Definition 2. Associative algebra A is called *splicing alsebra* or *S-algebra*, if it satisfies the identity

$$x[y, z]t = 0. \tag{5}$$

It is easy to note that all standard splicing algebras are S-algebras. Let S denote the variety of all S-algebras. Let $S[X]$ be a free algebra in the variety S

with set of free generators $X = \{x_1, \ldots, x_n, \ldots\}$. The Theorem 4 finds the basis of the identities of a standard splicing algebra C_∞. It is found that $S = \mathrm{Var}(C_\infty)$, i.e. all the identities of a standard splicing algebra C_∞ follow from the identity (5).

Let $IR[X]$ denote a free algebra in the variety IR with generating set X. It is shown that $S[X]$ is an associative envelope algebra for $IR[X]$. In the Lemmas 5, 6, 7 the basis and multiplication tables are constructed for the free algebras $S[X]$ and $IR[X]$.

The Theorem 5 proves that all the identities of a standard IR-algebra J follow from the identity (1), i.e.

$$IR = \mathrm{Var}(J_\infty).$$

In the Section 5 we describe the annihilator of a free algebra $IR[X]$. and prove that

$$\mathrm{Ann}(IR[X]/\,\mathrm{Ann}(IR[X])) = 0.$$

The Theorem 6 proves that the following isomorphism of F-modules takes place:

$$IR[X] \simeq F[X] \oplus D,$$

where $F[X]$ is a free associative-commutative algebra with generating set $X = \{x_1, \ldots, x_n, \ldots\}$, $D = D(IR[X])$ is an associator ideal of the algebra $IR[X]$. Furthermore, $D = D_0 \oplus D_1$, where $D_1 = \mathrm{Ann}(IR[X])$, $D_0^2 \subset D_1$ and $D^3 = 0$. From this we conclude that $D = M(IR[X])$, where $M(IR[X])$ is the McCrimmon radical of $IR[X]$.

Section 6 is devoted to the study of the basis of the identities of standard IR-algebras J_n and $J_{n,m}$. It is proved that all the variety of IR-algebras is generated by a minimal nontrivial standard algebra $J_{1,2}$. This algebra has the basis $a = a_{11}$, $b = a_{12}$ and the multiplication table

$$a^2 = a, \quad b^2 = b, \quad a \cdot b = \frac{1}{2}(a + b).$$

Theorem 7. *The variety* $\mathrm{Var}(J_{1,1}) = K$ *is the variety of associative commutative algebras and has the determining identity* $xD_{y,z} = 0$, *the variety* $\mathrm{Var}(J_{1,2}) = \mathrm{Var}(J_{n,m}) = \mathrm{Var}(J_\infty) = IR$, $(n, m) \neq (1, 1)$, *and has the determining identity*(1).

Consequently, To prove the speciality of the variety IR in the Section 3, we used two complicated results: the description of the identities of variety $\mathrm{Var}(B_\infty)$ [7] and the speciality of the variety $\mathrm{Var}(B_\infty)$ [8]. The basis and the multiplication table of free algebras $S[X]$ and $IR[X]$ allow us to prove the specialty of the variety IR by means of a rather simple method. It is found that the variety IR possesses the following property: $IR[X] = HS[X]$, i.e. the algebra $IR[X]$ coincides with a Jordan algebra of symmetric elements of associative envelope $S[X]$ under the standard involution. We will call the varieties of Jordan algebras satisfying this property the reflective varieties.

In the Section 7 we will prove the following theorem:

Theorem 8. *Any reflexive variety of Jordan algebras is special.*

2. Annihilator Bernstein algebras. In this Section we will prove that the class of standard IR-algebras coincides with the class of annihilator Bernstein's algebras of a special type.

2.1. Annihilator of the standard IR-algebras. Let

$$b_{ij} = a_{11} + a_{ij} - 2a_{11} \cdot a_{ij} = (a_{11} - a_{ij})^2 = a_{11} + a_{ij} - a_{1j} - a_{i1} \text{ for all } i, j > 1.$$

First observe that $b_{ij} \in \operatorname{Ann}(J_\infty)$ for all $i, j > 1$. Indeed, for any basis element $a = a_{kl}$ of the algebra J_∞ we have

$$b_{ij} \cdot a = (a_{11} + a_{ij} - 2a_{11} \cdot a_{ij}) \cdot a \underset{(4)}{=} a_{11} \cdot a + a_{ij} \cdot a - a_{11} \cdot a - a_{ij} \cdot a = 0.$$

Further, for any $a = \sum\limits_{k,l} \alpha_{kl} a_{kl}$ from J_∞ we'll have

$$b_{ij} \cdot \sum_{k,l} \alpha_{kl} a_{kl} = \sum_{k,l} \alpha_{kl} (b_{ij} a_{kl}) = 0.$$

It is easy to check that the set $\{b_{ij},\ i > 1,\ j > 1\}$ is linearly independent over F. By definition of b_{ij}, we have

$$\sum_{i,j>1} \alpha_{ij} b_{ij} = 0 \Rightarrow \sum_{i,j>1} \alpha_{ij} (a_{11} + a_{ij} - a_{1j} - a_{i1}) = 0 \Rightarrow \alpha_{ij} = 0 \text{ for all } i > 1,\ j > 1.$$

We will denote by $L(c_i;\, i \in I)$ the F -module generated by the elements c_i, $i \in I$, i.e.

$$L(c_i;\, i \in I) = \{\sum_i \alpha_i c_i \mid \alpha_i \in F\}.$$

We will call $L(c_i;\, i \in I)$ a linear envelope of the set $\{c_i;\, i \in I\}$.

By the above, we have

$$L(b_{ij};\, i > 1,\ j > 1) = F(b_{ij};\, i > 1,\ j > 1).$$

Let $I = F(b_{ij};\, i > 1,\ j > 1)$.

Lemma 1. *The following isomorphism of the F-modules takes place:*

$$J_\infty / \operatorname{Ann}(J_\infty) \simeq F(a_{ij};\ i = 1 \text{ or } j = 1),$$

where $\operatorname{Ann}(J_\infty) = I$ *and* $\operatorname{Ann}(J_\infty / \operatorname{Ann}(J_\infty)) = 0$.

Proof. We first prove the isomorphism

$$J_\infty / I \simeq F(a_{ij};\ i = 1 \text{ or } j = 1).$$

Let's consider an arbitrary basis element a_{ij}, $i > 1$, $j > 1$. By the definition of b_{ij}, $i > 1$, $j > 1$, we have

$$a_{ij} = b_{ij} - a_{11} + a_{1j} + a_{i1},$$

hence

$$a_{ij} \in L(a_{11}, a_{1j}, a_{i1}) + I.$$

Consequently,

$$J_\infty/I = L(a_{ij};\ i = 1 \text{ or } j = 1).$$

We now prove that the elements a_{11}, a_{1i}, a_{j1}, where $i > 1$ and $j > 1$, are linearly independent in J_∞/I. Let's suggest the contrary, then in the algebra J_∞ we'll have

$$0 \neq a = \alpha_{11}a_{11} + \sum_{i=2}^{n} \beta_{1i}a_{1i} + \sum_{i=2}^{n} \gamma_{i1}a_{i1} \in I \subseteq \mathrm{Ann}(J).$$

Then

$$a \cdot a_{11} = \alpha_{11}a_{11} + \frac{1}{2}\left(\sum_{i=2}^{n} \beta_{1i}(a_{11} + a_{1i})\right) + \frac{1}{2}\left(\sum_{i=2}^{n} \gamma_{i1}(a_{11} + a_{i1})\right) = 0.$$

Consequently, $\beta_{1i} = 0$ and $\gamma_{i1} = 0$ for all $i \geq 2$. Then $a = \alpha_{11}a_{11}$, but a_{11} is an idempotent, so $\alpha_{11} = 0$. The obtained contradiction proves that

$$J_\infty/I \simeq F(a_{ij};\ i = 1 \text{ or } j = 1).$$

Let now $a \in \mathrm{Ann}(J)$. Then

$$a = \alpha_{11}a_{11} + \sum_{i=2}^{n} \beta_{1i}a_{1i} + \sum_{i=2}^{n} \gamma_{i1}a_{i1} + v,$$

where $v \in I$. As $a \in \mathrm{Ann}(J_\infty)$, $v \in I \subseteq \mathrm{Ann}(J_\infty)$, then $a \cdot a_{11} = 0$. But in this case $\alpha_{11} = 0$, $\beta_{1i} = \gamma_{i1} = 0$, $i \geq 2$. So, $a = v \in I$ and $\mathrm{Ann}(J_\infty) = I$. Let now $a \in \mathrm{Ann}(J_\infty/I)$, then $a \cdot a_{11} = 0$. Similar arguments apply to this case, we'll get $a = 0$, i.e. $\mathrm{Ann}(J_\infty/I) = 0$. This proves the lemma.

Let's now construct a structure of Bernstein Jordan algebra on the algebra J_∞. To do this, we'll introduce the following notations:

$$e = a_{11};\quad e_{1i} = a_{11} - a_{1i},\ i > 1;\quad e_{i1} = a_{11} - a_{i1},\ i > 1;$$
$$U = F(e_{1i}, e_{i1}; i > 1);\quad Z = \mathrm{Ann}(J_\infty) = F(b_{ij};\ i > 1,\ j > 1).$$

Lemma 2. *In the algebra J_∞ the following relations are valid:*
(1) $J_\infty = Fe \oplus U \oplus Z$ *— the direct sum of the F-modules;*
(2) $e^2 = e$, $e \cdot u = \frac{1}{2}u$, $u \in U$, *and* $e \cdot z = 0$, $z \in Z$*;*
(3) $U^2 \subseteq Z$*;*
(4) $N^3 = 0$, *where* $N = U \oplus Z$*.*

Proof. (1) follows from the Lemma 1.
(2) Calculating with the multiplication table (3) we obtain:

$$e^2 = a_{11}^2 = a_{11} = e,\quad e \cdot Z = 0,$$

$$e \cdot e_{1i} = (a_{11} - a_{1i}) \cdot a_{11} = a_{11} - \frac{1}{2}a_{1i} - \frac{1}{2}a_{11} = \frac{1}{2}(a_{11} - a_{1i}) = \frac{1}{2}e_{1i},$$

$$e \cdot e_{i1} = (a_{11} - a_{i1}) \cdot a_{11} = a_{11} - \frac{1}{2}a_{i1} - \frac{1}{2}a_{11} = \frac{1}{2}(a_{11} - a_{i1}) = \frac{1}{2}e_{i1}.$$

Let now

$$u = \sum_{i>1} \alpha_{1i} e_{1i} + \sum_{i>1} \beta_{i1} e_{i1},$$

then

$$e \cdot u = \sum_{i>1} \alpha_{1i}(e_{1i} \cdot e) + \sum_{i>1} \beta_{i1}(e_{i1} \cdot e) = \frac{1}{2}u.$$

(3) According to the distributivity of the multiplication it is sufficient to check the statement 3 on the basis elements U. We have

$$\begin{aligned} e_{1i} \cdot e_{1j} &= (a_{11} - a_{1i}) \cdot (a_{11} - a_{1j}) \\ &= a_{11} - \frac{1}{2}a_{1j} - \frac{1}{2}a_{11} - \frac{1}{2}a_{11} - \frac{1}{2}a_{1i} + \frac{1}{2}a_{1j} + \frac{1}{2}a_{1i} = 0 \quad \text{for all } i > 1,\ j > 1. \end{aligned}$$

Analogously, $e_{i1} \cdot e_{j1} = 0$, $i, j > 1$. Let's now prove that

$$e_{i1} \cdot e_{1j} \in \operatorname{Ann}(J_\infty) \quad \text{for all } i, j > 1.$$

Indeed,

$$\begin{aligned} e_{i1} \cdot e_{1j} &= (a_{11} - a_{i1}) \cdot (a_{11} - a_{1j}) = a_{11} - \frac{1}{2}a_{1j} - \frac{1}{2}a_{11} - \frac{1}{2}a_{i1} - \frac{1}{2}a_{11} + \frac{1}{2}a_{ij} + \frac{1}{2}a_{11} \\ &= \frac{1}{2}(a_{11} + a_{ij} - a_{1j} - a_{i1}) = \frac{1}{2}(a_{11} + a_{ij} - 2a_{11} \cdot a_{ij}) = \frac{1}{2}b_{ij} \in \operatorname{Ann}(J). \end{aligned}$$

Hence, $U^2 \subseteq Z$.

(4) We have the following sequence of inclusions:

$$N^3 = (U \oplus Z)^2 \cdot N \subseteq U^2 \cdot N \subseteq Z \cdot N = 0.$$

The lemma is proved.

Let's adjoin a formal unit 1 to the algebra $J_\infty / \operatorname{Ann}(J_\infty)$. Let $J^\sharp$ denote the obtained algebra as

$$J^\sharp = F \cdot 1 + J_\infty / \operatorname{Ann}(J_\infty)$$

Lemma 3. *The algebra $J^\sharp$ is a Jordan algebra of symmetric bilinear form over F.*

Proof. Let $e_{11} = (2e - 1)$. From the Lemma 1 we conclude that

$$J^\sharp = F \cdot 1 + F(e_{11}, e_{1i}, e_{j1};\ i, j > 1).$$

It follows from the proof of the Lemma 2 that

$$e_{1i}^2 = (2e-1)^2 = 4e - 4e + 1 = 1; \quad e_{1i} \cdot e_{1j} = e_{i1} \cdot e_{j1} = e_{i1} \cdot e_{1j} = 0 \text{ for } i, j > 1.$$

Consequently, $\{e_{11}, e_{1i}, e_{j1};\ i, j > 1\}$ is a standard basis of a Jordan algebra of symmetric bilinear form over F. This proves the Lemma.

Observe that the bilinear form defined in the Lemma 3 is degenerate.

Theorem 1. *Standard IR-algebras J_∞, $J_{n,m}$, and J_n are Bernstein Jordan algebras with the following (U,Z)-graduation:*

1. $J_\infty = Fe \oplus U \oplus Z$, *where* $U = F(e_{1i}, e_{j1};\, i,j > 1)$, $Z = \operatorname{Ann}(J) = F(b_{ij};\, i,j > 1)$, $e = a_{11}$.
2. $J_{n,m} = Fe \oplus U_{(n,m)} \oplus Z_{(n,m)}$, *where* $U_{(n,m)} = F(e_{1j}, e_{i1};\, 1 < j \le m,\, 1 < i \le n)$, $Z_{(n,m)} = \operatorname{Ann}(U_{(n,m)}) = F(b_{ij};\, 1 < i \le n,\, 1 < j \le m)$, $e = a_{11}$.
3. $J_n = F_e \oplus U_{(n)} \oplus Z_{(n)}$, *where* $U_{(n)} = F(e_{1j}, e_{i1};\, 1 < i \le n,\, j > 1)$, $Z_n = F(b_{ij};\, 1 < i \le n,\, j > 1)$, $e = a_{11}$, *and nontrivial homomorphism* $\omega : J_\infty(J_{n,m}, J_n) \to F$, *which is defined by:*

$$\omega(x) = \alpha \quad \forall x = \alpha e + u + z \in J,$$

where $\alpha \in F$, $u \in U(U_{(n,m)}, U_{(n)})$, $z \in Z(Z_{(n,m)}, Z_{(n)})$.

Proof. It follows from the Lemma 2 that the introduced graduations are Bernstein graduations (see [6]) and $Z^2 = 0$ and $N^3 = 0$. In view of the proposition 3.1 [6], the algebras J_∞, $J_{n,m}$, and J_n are Bernstein Jordan algebras. The theorem is proved.

2.2. Annihilator algebras of the type (V,W). Let's note that $Z = \operatorname{Ann}(J)$ for all Bernstein algebras $J = Fe + U + Z$, defined in the Theorem 1.

In accordance with the definition 1 in the Section 1.2, such algebras are called annihilator algebras. Our next goal is to construct a free algebra in the class of annihilator algebras. Let $I = (X)_J$ denote the ideal of the algebra J, generated by $X \subseteq J$. Let $BJ = BJ[X;Y]$ — a free (U,Z)-graded Bernstein algebra (see [6]) from U-generators $X = \{x_1, \ldots, x_n, \ldots\}$ and Z-generators $Y = \{y_1, \ldots, y_n, \ldots\}$. It is easy to see that the algebra

$$J = F \cdot e + BJ[X;Y]/I,$$

where $I = (w \cdot u \mid w \in Z,\, u \in BJ[X;Y])_{BJ}$, is a free annihilator algebra.

We will denote by $\operatorname{Ann} BJ[X;Y]$, the nucleus of the Bernstein algebra J, i.e.

$$\operatorname{Ann} BJ[X;Y] = BJ[X;Y]/I.$$

Let now $X = V \cup W$, where $V = \{v_1, \ldots, v_n, \ldots\}$, $W = \{w_1, \ldots, w_n, \ldots\}$ — some sets and $Y = \varnothing$.

Definition 3. The annihilator algebra

$$J = Fe + \operatorname{Ann} BJ[X;\varnothing]/I,$$

where $I = (v_i \cdot v_j, w_i \cdot w_j;\, i,j \in \mathbb{N})_{\operatorname{Ann} BJ[X;\varnothing]}$, is called the *annihilator algebra of the type* (V,W).

Proposition 1. *The annihilator algebra of the type (V,W) has the basis*

$$\{e, v_i, w_j, v_i w_j;\, i,j \in \mathbb{N}\}$$

and the following multiplication table:

$$\begin{gathered} e^2 = e, \quad e \cdot u_i = \frac{1}{2} u_i, \quad e \cdot w_j = \frac{1}{2} w_j, \\ v_i \cdot w_j = w_i \cdot w_j = 0, \quad v_i \cdot w_j = v_i \cdot w_j, \\ v_i \cdot w_j \in \mathbb{Z} = \operatorname{Ann}(J) \quad \textit{for all } i,j \in \mathbb{N}. \end{gathered} \tag{6}$$

Proof. In view of the properties of free (U, Z)-graded Bernstein algebra (see [6]) we have

$$Y = \varnothing \Rightarrow U^2 = Z.$$

But $Z = \mathrm{Ann}(J)$ since $U = F(v_i, w_j;\ i, j \in \mathbb{N})$. Therefore $Z = F(u_i w_j;\ i, j \in \mathbb{N})$. The proposition is proved.

Theorem 2. *The class of standard IR-algebras coincides with the class of annihilator algebras of the type (V, W). An annihilator algebra of the type (V, W) is isomorphic to J_∞, if $|V| = |W| = \infty$, to J_n, if $|V| = n$, $|W| = \infty$, to $J_{n,m}$, if $|V| = n$, $|W| = m$.*

Proof. We first consider a standard IR-algebra J_∞. It follows from the Theorem 1, that

$$J = Fe \oplus U \oplus Z,$$

where $U = F(e_{1i}, e_{j1};\ i, j > 1)$, $Z = \mathrm{Ann}(J) = F(b_{ij};\ i > 1,\ j > 1),\ e = a_{11}$. Let us introduce the following notations:

$$v_i = e_{1i}, \quad w_i = e_{i1},\ i > 1; \quad V = \{v_i;\ i > 1\}, \quad W = \{w_i; i > 1\}.$$

Let's prove that the algebra J_∞ is an annihilator algebra of the type (V, W). From the proof of the Lemma 2 it follows that

$$v_i \cdot v_j = w_i \cdot w_j = 0 \quad \text{for } i, j > 1; \quad v_i \cdot w_j = e_{1i} \cdot e_{j1} = \frac{1}{2} b_{ij} \in Z.$$

We conclude from the Lemma 1 that the set $\{v_i \cdot w_j = \frac{1}{2} b_{ij},\ i, j > 1\}$ is linearly independent over F. Hence, the algebra J_∞ has the basis and the multiplication table (6). Consequently, it is isomorphic to an annihilator algebra of the type (v, w), where $|v| = |w| = \infty$.

Conversely let us consider an annihilator algebra B of the type (V, W) with a basis and a multiplication table (6). Let

$$a_{11} = e; \quad a_{1i} = e - v_{i-1},\ i \geq 2;$$

$$a_{j1} = e - w_{j-1},\ j \geq 2; \quad a_{ij} = 2v_{i-1}w_{j-1} - v_{i-1} - w_{j-1} + e,\ i, j \geq 2.$$

At first it is necessary to note that $B = F(a_{ij};\ i, j \in \mathbb{N})$, i.e. $\{a_{ij},\ i, j \in \mathbb{N}\}$, is the basis of the algebra B. Now we will found a multiplication table in this basis:

$$a_{11}^2 = a_{11}; \quad a_{11} \cdot a_{1i} = e - \frac{1}{2} v_{i-1} = \frac{1}{2} a_{11} + \frac{1}{2} a_{1i};$$

$$a_{11} \cdot a_{j1} = e - \frac{1}{2} w_{j-1} = \frac{1}{2} a_{11} + \frac{1}{2} a_{j1};$$

$$a_{11} \cdot a_{ij} = -\frac{1}{2} v_{i-1} - \frac{1}{2} w_{j-1} + e = \frac{1}{2} a_{1j} + \frac{1}{2} a_{i1}, \quad i, j \geq 2;$$

$$a_{1i} \cdot a_{1j} = (e - v_{i-1}) \cdot (e - v_{j-1}) = e - \frac{1}{2} v_{i-1} - \frac{1}{2} v_{j-1} = \frac{1}{2} a_{1i} + \frac{1}{2} a_{1j}, \quad i, j \geq 2.$$

Analogously,

$$a_{j1} \cdot a_{i1} = \frac{1}{2}a_{i1} + \frac{1}{2}a_{j1}, \quad i, j \geq 2;$$

$$a_{ij} \cdot a_{kl} = (-v_{i-1} - w_{j-1} + e) \cdot (-v_{k-1} - w_{l-1} + e) = v_{i-1}w_{l-1} + v_{k-1}w_{j-1} \\ - \frac{1}{2}v_{i-1} - \frac{1}{2}w_{j-1} - \frac{1}{2}v_{k-1} - \frac{1}{2}w_{l-1} + e = \frac{1}{2}(a_{il} + a_{kj}); \quad i, j, k, l \geq 2.$$

Therefore, the algebra B has the multiplication table (3) in the basis a_{ij}, $i, j \in \mathbb{N}$. Hence, it is isomorphic to a standard IR-algebra J_∞. The same proof works for the cases $|V| = n$, $|W| = m$ and $|V| = n$, $|W| = \infty$ This proves the theorem.

3. The specialty of the variety IR. In this Section we will prove that the variety IR is a proper subvariety of the variety of special algebras $\mathrm{Var}(B_\infty)$ [8].

We will use the standard notations [4]:

$$xRy = x \cdot y, \quad D_{x,y} = R_xR_y - R_yR_x, \quad U_{x,y} = R_xR_y + R_yR_x - R_{x \cdot y}.$$

Lemma 4. *In the variety IR the following identities are valid:*

$$zU_{x,x}R_y = zR_xU_{x,y}; \tag{7}$$

$$2xD_{x,z} \cdot y = yD_{x^2,z}; \tag{8}$$

$$xD_{y,z}R_t + xD_{z,t}R_y + xD_{t,y}R_z = 0; \tag{9}$$

$$3xD_{y,z} \cdot t = tD_{x \cdot y,z} + tD_{y,x \cdot z}; \tag{10}$$

$$xD_{y,z}U_{t,t} = 0; \tag{11}$$

$$((x \cdot z)U_{y,t} + (y \cdot t)U_{x,z} - (x \cdot t)U_{y,z} - (y \cdot z)U_{x,t}) \cdot u = 0; \tag{12}$$

$$zU_{x^2,y} = zR_xU_{x,y}; \tag{13}$$

$$[x, y]^2 \cdot z = 0, \quad \textit{where } [x, y]^2 = 2yD_{y,x^2} - 4yD_{y,x} \cdot x; \tag{14}$$

$$zR_{x_1} \dots R_{x_n}U_{y,y} = zR_{x_{6(1)}} \dots R_{x_{6(n)}}U_{y,y}, \quad \textit{for any } \sigma \in S_n. \tag{15}$$

Furthermore, the identities (1), (7), (8) *are equivalent in the variety of commutative algebras.*

Proof. (1) $\Rightarrow$ (7). We have

$$2(zU_{x,x}R_y - zR_xU_{x \cdot y}) = 2(2z \cdot x \cdot x \cdot y - z \cdot x^2 \cdot y - z \cdot x \cdot x \cdot y - z \cdot x \cdot y \cdot x + (z \cdot x) \cdot (x \cdot y)) \\ \underset{(J)}{=} 4z \cdot x \cdot x \cdot y - 2z \cdot x^2 \cdot y + 2(z \cdot x) \cdot (x \cdot y) - 2z \cdot x \cdot x \cdot y + x^2 \cdot y \cdot z - x^2 \cdot (y \cdot z) - 2(x \cdot z) \cdot (x \cdot y) \\ = 2z \cdot x \cdot x \cdot y - 2z \cdot x^2 \cdot y + x^2 \cdot y \cdot z - x^2 \cdot (y \cdot z) \underset{(1)}{=} 0,$$

where (J) is a Jordan identity

$$2z \cdot x \cdot y \cdot x + x^2 \cdot y \cdot z = 2(z \cdot x) \cdot (x \cdot y) + x^2 \cdot (y \cdot z).$$

(7) $\Rightarrow$ (1). In view of the above proved it is enough to check that (7) $\Rightarrow$ (J). Substituting $y = x$ into (7), we have

$$2z \cdot x \cdot x \cdot x - (z \cdot x^2) \cdot x = 2z \cdot x \cdot x \cdot x - (z \cdot x) \cdot x^2,$$

i.e. $(z \cdot x^2) \cdot x = (z \cdot x) \cdot x^2$. Therefore, the identities (1) and (7) are equivalent for commutative algebras.

(1) $\Rightarrow$ (8). It is easy to notice that the identity (8) is a D-operator representation of the identity (1):

$$x^2 \cdot y \cdot z - x^2 \cdot (z \cdot y) + 2x \cdot z \cdot x \cdot y - 2x^2 \cdot z \cdot y = 2xD_{z,x} \cdot y + yD_{x^2,z} = 0.$$

(1) $\Rightarrow$ (9). Rewriting the left part of (9) in the terms of U-operator, we have:

$$xU_{y,z} - yU_{x,z} = x \cdot y \cdot z + x \cdot z \cdot y - (y \cdot z) \cdot x - (x \cdot y) \cdot z - (y \cdot z) \cdot x + (x \cdot z) \cdot y = 2zD_{x,y},$$

Consequently,

$$2zD_{x \cdot y} = xU_{y,z} - yU_{x,z}. \tag{16}$$

Hence,

$$\begin{aligned} 4xD_{y,z}R_t &\underset{(16)}{=} 2(yU_{z,x} - yU_{y,x})R_t \underset{(7)}{=} ((y \cdot z)U_{x,t} \\ &+ (x \cdot y)U_{z,t} - (z \cdot y)U_{x,t} - (z \cdot x)U_{y,t}) = (x \cdot y)U_{z,t} - (x \cdot z)U_{y,t}. \end{aligned}$$

Writing $\sigma f(y,z,t) = f(y,z,t) + f(z,t,y) + f(t,y,z)$ gives

$$4\sigma(xD_{y,z}R_t) = \sigma(x \cdot y)U_{z,t} - \sigma(x \cdot z)U_{y,t} = \sigma(x \cdot y)U_{z,t} - \sigma(x \cdot y)U_{t,z} = 0.$$

(1) $\Rightarrow$ (10). We have

$$\begin{aligned} 3xD_{y,z} \cdot t &= (2xD_{y,z} - yD_{z,x} - zD_{x,y}) \cdot t \\ &= (xD_{y,z} + yD_{x,z} + xD_{y,z} + zD_{y,x}) \cdot t \underset{(8)}{=} tD_{x \cdot y,z} + tD_{y,z \cdot x}. \end{aligned}$$

(1) $\Rightarrow$ (11). By (8), (10)

$$\begin{aligned} 3xD_{y,z} \cdot t^2 &\underset{(10)}{=} t^2 D_{x \cdot y \cdot z} + t^2 D_{y,x \cdot z} = -(x \cdot y)D_{z,t^2} - zD_{t^2,x \cdot y} - yD_{x \cdot z,t^2} - (x \cdot z)D_{t^2,y} \\ &\underset{(8)}{=} -2tD_{z,t}R_{(x \cdot y)} - 2tD_{t,x \cdot y}R_z - 2tD_{x \cdot z,t}R_y - 2tD_{t,y}R_{(x \cdot z)} \\ &\underset{(9)}{=} 2tD_{x \cdot y \cdot z} + tD_{y,x \cdot z}R_t \underset{(10)}{=} 6xD_{y,z} \cdot t \cdot t. \end{aligned}$$

This gives $xD_{y,z}U_{t,t} = 0$.

(1) $\Rightarrow$ (12). From (7), (11) we obtain

$$\begin{aligned} &2((x \cdot z)U_{y,t} + (y \cdot t)U_{x,z} - (x \cdot t)U_{y,z} - (y \cdot z)U_{x,t})R_u \underset{(7)}{=} ((x \cdot z \cdot y)U_{t,u} \\ &+ (x \cdot z \cdot t)U_{y,u} + (y \cdot t \cdot x)U_{z,u} + (y \cdot t \cdot z)U_{x,u} - (x \cdot t \cdot y)U_{z,u} - (x \cdot t \cdot z)U_{y,u} \\ &-(y \cdot z \cdot x)U_{t,u} - (y \cdot z \cdot t)U_{x,u} = (zD_{x,y}U_{t,u} + xD_{z,t}U_{y,u} + tD_{y,x}U_{z,y} + yD_{t,z}U_{x,u}) \underset{(11)}{=} 0. \end{aligned}$$

(1) ⇒ (13). Since IR is a Jordan variety, the known Jordan identity is valid in IR (e.g., see [4]):

$$zU_{x,x}R_y + zU_{x^2,y} = 2zR_xU_{x,y}.$$

From this,

$$zU_{x^2,y} = 2zR_xU_{x,y} - zU_{x,x}R_y \underset{(7)}{=} zR_xU_{x,y}.$$

(1) ⇒ (14). We have

$$\begin{aligned}[x,y]^2 = 2yD_{y,x^2} - 4yD_{y,x}\cdot x &\underset{(16)}{=} yU_{x^2,y} - x^2U_{y,y} \\ - 2yU_{x,y}R_x + 2xU_{y,y}R_x &\underset{(13),(7)}{=} (x\cdot y)U_{x,y} \\ - x^2U_{y,y} - (x\cdot y)U_{x,y} - y^2U_{x,x} + 2(x\cdot y)U_{x,y} &= 2(x\cdot y)U_{x,y} - x^2U_{y,y} - y^2U_{x,x}.\end{aligned}$$

Therefore,

$$[x,y]^2\cdot u = (2(x\cdot y)U_{x,y} - x^2U_{y,y} - y^2U_{x,x})\cdot u \underset{(12),z=y,t=x}{=} 0.$$

(1) ⇒ (15). It suffices to show that:

$$zD_{x_1,x_2}R_{y_1}\dots R_{y_n}U_{y,y} = 0,$$

where the operators $R_{y_1},\dots,R_{y_n}$ can be missing. We induct on n. The base of induction is the identity (11), when $R_{y_1},\dots,R_{y_n}$ are missing. Now

$$\begin{aligned}3zD_{x_1,x_2}R_{y_1}\dots R_{y_n}U_{y,y} &\underset{(10)}{=} y_1D_{z\cdot x_1,x_2}R_{y_2}\dots R_{y_n}U_{y,y} \\ &\quad + y_1U_{x_1,z\cdot x_2}R_{y_2}\dots R_{y_n}U_{y,y} = 0\end{aligned}$$

by induction. This proves the lemma.

Theorem 3. *The variety IR is a proper subvariety* $\mathrm{Var}(B_\infty)$, *i.e.*

$$IR \subsetneq \mathrm{Var}(B_\infty).$$

The variety IR is special, i.e. any commutative algebra satisfying the identity (1) *is a special Jordan algebra.*

Proof. By S. Vasilovsky's results [7], the variety $\mathrm{Var}(B_\infty)$ is defined by the following identities:

$$zD_{[x,y]^2,t} = 0, \quad \sigma(x^2D_{y,z}R_t - 2xD_{y,z}R_tR_x) = 0,$$

It is easily seen that the first identity is the consequence of the identity (14), and the second is the consequence of the identity (9). Therefore,

$$IR \subseteq \mathrm{Var}(B_\infty).$$

It is obvious that (14) is not valid in B_∞. Hence,

$$IR \neq \mathrm{Var}(B_\infty).$$

By the Theorem 3.1 [8], $\mathrm{Var}(B_\infty)$ is a special variety. Therefore the variety IR is special. This proves the theorem.

4. Free$S$- and IR-algebras. We will denote by $IR[X]$ the free algebra in the variety IR of generating set $X = \{x_1, \dots, x_n, \dots\}$. In this Section we will construct a basis and a multiplication table of the free algebra $IR[X]$ and will prove that the variety of all IR-algebras is generated by a standard IR-algebra J_∞, i.e. $IR = \mathrm{Var}(J_\infty)$.

4.1. The variety of the algebras of splicing. Let $\mathrm{Ass}[X]$, $SJ[X]$ be free associative, free special Jordan algebras. Let's remind (Definition 2) that the associative algebra A is called *S-algebra* if it satisfies the identity (5)

$$x[y, z]t = 0.$$

It is easy to check that the standard algebras of splicing are S-algebras. By (2),

$$a_{i_1j_1}a_{i_2j_2}a_{i_3j_3}a_{i_4j_4} \underset{(2)}{=} a_{i_1j_4},$$

hence,

$$a_{i_1j_1}[a_{i_2j_2}, a_{i_3j_3}]a_{i_4j_4} = 0.$$

Let us denote by S the variety of all S-algebras and let $S[X]$ be a free algebra in this variety. Let's construct a basis and multiplication table for a free algebra S. We will define an ordering operator $\langle\ \rangle : \mathrm{Ass}[X] \to SJ[X]$ by a rule: If $u = x_{s_1} \dots x_{s_m}$ is a monomial from $\mathrm{Ass}[X]$ then

$$\langle u \rangle = x_{i_1} \cdot \dots \cdot x_{i_m} \in SJ[X],$$

where $i_1 \leq \dots \leq i_m$ and the set $\{s_1, \dots, s_m\}$ and $\{i_1, \dots, i_m\}$ coincide with respect to repetitors of all the symbols. Then we will extend the ordering operator on the algebra $\mathrm{Ass}[X]$ by linearity: if $f = \sum \alpha_i \cdot u_i$, where $\alpha_i \in F$, u_i are monomials, then $\langle f \rangle = \sum \alpha_i \langle u_i \rangle$. For example, $\langle x_1x_2x_1 + 3x_3x_1 \rangle = x_1 \cdot x_i \cdot x_2 + 3x_1 \cdot x_3$.

By definition, the operation of multiplication of the elements of the algebra $\mathrm{Ass}[X]$, and consequently of the algebra $SJ[X]$ within the brackets $\langle\ \rangle$ is associative-commutative. Therefore, for any $v_1, \dots, v_n \in \mathrm{Ass}[X]$ and $\sigma \in S_n$ we have

$$\langle v_1 \cdot \dots \cdot v_n \rangle = \langle v_1 v_2 \dots v_n \rangle = \langle v_{6(1)} \dots v_{6(n)} \rangle.$$

Let's consider the algebra $A[X]$ with the basis

$$B = \{x_i,\ x_ix_j = x_i\langle 1 \rangle x_j,\ x_i\langle u \rangle x_j\},$$

where x_i, $\langle u \rangle$ runs over all ordered monomials of $SJ[X]$, 1 — a formal unit; with the multiplication table:

$$x_i \triangleright x_j = x_i\langle 1 \rangle x_j;$$

$$x_i \triangleright x_j\langle u \rangle x_k = x_i\langle u \rangle x_j \triangleright x_k = x_i\langle u x_j \rangle x_k;$$

$$x_i\langle u \rangle x_j \triangleright x_k\langle v \rangle x_l = x_i\langle u v x_k x_j \rangle x_l,$$

where $i, j, k, l \in \mathbb{N}$.

Lemma 5. *The algebras* $S[X]$ *and* $A[X]$ *are isomorphic.*

Proof. It follows from the definition of the multiplication and ordering operator in the algebra $A[X]$ that the algebra $A[X]$ is an S-algebra. Consequently, the identical mapping $\tau : X \to X$ has a unique extension to canonical homomorphism $\tau : S[X] \to A[X]$. From (5) we conclude that

$$S[X] = L(B).$$

It is clear that the images of the basis elements from B under homomorphism τ are linearly independent in $A[X]$. Hence,

$$S[X] = F(B),$$

Consequently, the algebras $S[X]$ and $A[X]$ are isomorphic. This proves the lemma.

Theorem 4. $S = \mathrm{Var}(C_\infty)$, *i.e. all the identities of the standard algebra of splicing* C_∞ *follow from the identity* (5).

Proof. The algebra C_∞ is an S-algebra, hence $\mathrm{Var}(C_\infty) \subseteq S$. Let's prove the reverse inclusion.

Let a homogenous polynomial

$$f = \sum_{i,j} \alpha_{ij} x_i \langle u_{ij} \rangle x_j \in S[X],$$

where $\alpha_{ij} \in F$, be an identity on C_∞. Consider the mapping $\varphi : x_i \to a_{ii}$ an and extend it up to the homomorphism $\varphi : S[X] \to C_\infty$. Such extension exists due to the fact that $S[X]$ is a free algebra in the variety S. Then,

$$\varphi(f) = \sum_{i,j} \alpha_{ij} a_{ii} \langle \varphi(u_{ij}) \rangle a_{jj} = \sum_{i,j} \alpha_{ij} a_{ij} = 0.$$

Hence, $\alpha_{ij} = 0$ for all i, j. Consequently, $f = 0$ in the algebra $S[X]$ and f is the consequence of the identity (5). This proves the theorem.

4.2. Basis and multiplication table of the algebra $IR[X]$. We will denote by $D = D(IR[X])$ the associator ideal of the algebra $IR[X]$, i.e. the ideal generated by all Jordan associators $aD_{b,c}$, where $a, b, c \in IR[X]$.

Proposition 2. *In the algebra* $IR[X]$ *the following relations are valid:*

$$D = L(aD_{b,c}, \text{ where } a, b, c \in IR[X]); \tag{17}$$

$$uU_{x,y} = \langle u \rangle U_{x,y}; \tag{18}$$

$$\langle u \rangle U_{x,y} R_z = \frac{1}{2} \langle ux \rangle U_{y,z} + \frac{1}{2} \langle uy \rangle U_{x,z}; \tag{19}$$

$$\langle u \rangle U_{x,y} \cdot \langle v \rangle U_{z,t} = \frac{1}{4} (\langle uvyz \rangle U_{x,t} + \langle uvyt \rangle U_{x,z} + \langle uvxt \rangle U_{y,z} + \langle uvxz \rangle U_{y,t}), \tag{20}$$

where $u, v, x, y, z, t \in IR[X]$ and u, v can be formal units.

Proof. The relation (17) follows immediately from the identity (10). It is evident that $u = \langle u\rangle + d$, where $d \in D$. Hence,

$$uU_{x,y} = \langle u\rangle U_{x,y} + dU_{x,y} \underset{(17),(11)}{=} \langle u\rangle U_{x,y}.$$

We have

$$\langle u\rangle U_{x,y}R_z \underset{(7)}{=} \frac{1}{2}\langle u\rangle R_x U_{y,z} + \frac{1}{2}\langle u\rangle R_y U_{x,z} \underset{(18)}{=} \frac{1}{2}\langle ux\rangle U_{y,z} + \frac{1}{2}\langle uy\rangle U_{x,z}.$$

Analogously,

$$\begin{aligned}
4\langle u\rangle U_{x,y}\cdot\langle v\rangle U_{z,t} &\underset{(7),(18)}{=} 2\langle ux\rangle U_{y,\langle v\rangle}U_{z,t} + 2\langle uy\rangle U_{x,\langle v\rangle}U_{z,t}\\
&\underset{(13),(18)}{=} \langle uxvz\rangle U_{y,t} + \langle uxt\rangle U_{y,\langle v\rangle\cdot z} + \langle uxvt\rangle U_{y,z}\\
&+ \langle uxz\rangle U_{y,\langle v\rangle\cdot t} - \langle uxv\rangle U_{y,z\cdot t} - \langle uxzt\rangle U_{y,v} + \langle yuvz\rangle U_{x,t}\\
&\langle uyt\rangle U_{x,\langle v\rangle\cdot z} + \langle uyvt\rangle U_{x,z} + \langle uyz\rangle U_{x,\langle t\rangle\cdot t}\\
&- \langle uyv\rangle U_{x,z\cdot t} - \langle uyzt\rangle U_{x,v}\\
&\underset{(13),(18)}{=} \langle uvyz\rangle U_{x,t} + \langle uvyt\rangle U_{x,z} + \langle uvxt\rangle U_{y,z} + \langle uvxz\rangle U_{y,t}.
\end{aligned}$$

This proves the proposition.

Let's consider a subset $B \subseteq IR[X]$ of the following type:

$$B = \{x_i,\ x_i\cdot x_j = \langle 1\rangle U_{x_i,x_j},\ \langle u\rangle U_{x_i,x_j}\},$$

where $x_i, x_j \in X$, $\langle u\rangle$ runs over all ordered monomials of $SJ[X]$, 1 — a formal unit. In view of the relations (19), (20) the linear envelope $L(B)$ is a subalgebra of $IR[X]$, i.e. it is closed under multiplication.

Lemma 6. *The set B is the basis of the algebra $IR[X]$.*

Proof. We first show that $IR[X] = L(B)$. Let's consider an arbitrary homogenous polynomial $w \in IR[X]$. We will prove by induction on $\deg(w)$ that $w \in L(B)$. If $\deg(w) \le 3$, then $w \in L(B)$ by definition of B. Let's assume that $\deg(w) = n$, $n \ge 4$, and all homogenous monomials of the length $< n$ belong to $L(B)$.

It is well known that the algebra of multiplication $R(SJ[X])$ is generated by the set of operators

$$\{R_{x_i}, U_{x_i,x_j};\ \text{where } x_i, x_j \in X\}.$$

Hence,

$$w \in L(uU_{x_i,x_j}, vR_{x_i}),$$

where $x_i, x_j \in X$, u, v are homogenous monomials of $IR[X]$ of degree $n-2$ and $n-1$ correspondingly. It follows from the relation (18) that

$$uU_{x_i,x_j} = \langle u\rangle U_{x_i,x_j} \in L(B).$$

By induction assumption

$$vR_{x_i} = \left(\sum_{j,k} \alpha_{jk}\langle v_{jk}\rangle U_{x_i,x_k}\right)R_{x_i} = \sum_{j,k} \alpha_{jk}\langle v_{jk}\rangle U_{x_i,x_k}R_{x_i},$$

where $\alpha_{jk} \in F$. We have

$$\langle v_{jk}\rangle U_{x_j,x_k}R_{x_i} \underset{(19)}{=} \frac{1}{2}\langle v_{jk}x_j\rangle U_{x_k,x_i} + \frac{1}{2}\langle v_{jk}x_k\rangle U_{x_j,x_i} \in L(B).$$

Consequently, $w \in L(B)$ and $IR[X] = L(B)$. Let's now prove that the set B is linearly independent over F, i.e. that $IR[X] = F(B)$. Let's suppose that

$$f = \sum_{i,j} \alpha_{ij}\langle u_{ij}\rangle U_{x_i,x_j} = 0$$

in algebra $IR[X]$, where f is a homogenous polynomial and $\alpha_{ij} = \alpha_{ji} \in F$.

Consider the mapping $\varphi : x_i \to a_{ij}$ and extend it to homomorphism $\varphi : IR[X] \to J_\infty$. Such extension exists due to the fact that J_∞ is a IR-algebra. Then in the algebra J_∞ we have:

$$\begin{aligned}\varphi(f) = \sum_{i,j} \alpha_{ij}\varphi(\langle u_{ij}\rangle)U_{a_{ii},a_{jj}} &\underset{(4)}{=} \frac{1}{2}\sum_{i,j} \alpha_{ij}(\varphi\langle u_{ij}\rangle) \cdot a_{jj} \\ + a_{ii}\cdot a_{jj} + \varphi(\langle u_{ij}\rangle)\cdot a_{ii} &+ a_{ii}\cdot a_{jj} - \varphi(\langle u_{ij}\rangle)\cdot a_{ii} - \varphi(\langle u_{ij}\rangle)\cdot a_{jj}) \\ &= \sum_{i,j} \alpha_{ij}(a_{ii}\cdot a_{jj}) = \frac{1}{2}\sum_{i,j} \alpha_{ij}(a_{ij} + a_{ji}) = 0.\end{aligned}$$

Consequently,, $\alpha_{ij} = 0$ for all i, j and $f = 0$. So, it follows that $IR[X] = F(B)$. The lemma is proved.

Note that the multiplication table on the basis B of the algebra $IR[X]$ is defined by the relations (19), (20) in case when $x, y, z, t \in X$ which follows from the Lemma 6.

4.3. Associative envelope for $IR[X]$ and basis of the identities of the algebra J_∞. On the algebra $S[X]$ it is defined a standard involution *, which is set on the basis words in accordance with the following rule:

$$x_i^* = x_i, \quad (x_i x_j)^* = x_j x_i, \quad (x_i\langle u\rangle x_j)^* = x_j\langle u\rangle x_i,$$

where $x_i, x_j \in X$ and u runs over all ordered monomials of $SJ[X]$ and linearly extends to the whole algebra $S[X]$.

Let $HS[X]$ be a Jordan algebra $H(S[X], *)$ of symmetrical elements of the algebra $S[X]$ relatively to $*$. And let $JS[X]$ be a Jordan subalgebra $S[X]^{(+)}$, generated by the set X.

Proposition 3. $HS[X] = JS[X]$.

Proof. It is obvious that $JS[X] \subseteq HS[X]$. Let's prove the reverse inclusion. Consider an arbitrary homogenous element $f \in HS[X]$. Then

$$f = \sum_{i,j} \alpha_{ij} x_i \langle u_{ij} \rangle x_j \quad \text{and} \quad f^* = f,$$

where $\alpha_{ij} \in F$. It follows that

$$\sum_{i,j} \alpha_{ij} x_i \langle u_{ij} \rangle x_j = \sum_{i,j} \alpha_{ij} x_j \langle u_{ij} \rangle x_i.$$

Consequently, $\alpha_{ij} = \alpha_{ji}$ for all i, j. Finally,

$$f = \frac{1}{2}(f + f^*) = \frac{1}{2} \sum_{i,j} \alpha_{ij} (x_i \langle u_{ij} \rangle x_j + x_j \langle u_{ij} \rangle x_i) = \sum_{i,j} \alpha_{ij} \langle u_{ij} \rangle U_{x_i, x_j} \in JS[X].$$

This proves the proposition.

Lemma 7. *The algebra* $S[X]$ *is an associative envelope for the algebra* $IR[X]$.

Proof. It suffices to prove that the algebras $IR[X]$ and $JS[X]$ are isomorphic. Let us consider the set

$$B = \{x_i, \ x_i \cdot x_j = \langle 1 \rangle U_{x_i, x_j}, \ \langle u \rangle U_{x_i, x_j}\},$$

where $x_i, x_j \in X$, and u runs over all ordered monomials of $SJ[X]$. It follows from the Lemma 6 and the proof of the Proposition 3 that B is a basis of $JS[X]$. Let us find the multiplication table in this basis:

$$x_i \circ x_j = \langle 1 \rangle U_{x_i, x_j};$$

$$\begin{aligned} x_i \circ \langle u \rangle U_{x_k, x_l} &= \frac{1}{2} x_i \circ (x_k \langle u \rangle x_l + x_l \langle u \rangle x_k) = \frac{1}{4} (x_i \langle x_k u \rangle x_l \\ &+ x_k \langle u x_l \rangle x_i + x_i \langle x_l u \rangle x_k + x_l \langle u x_k \rangle x_i) = \frac{1}{2} \langle u x_k \rangle U_{x_i, x_l} + \frac{1}{2} \langle u x_l \rangle U_{x_k, x_l}; \end{aligned}$$

analogously,

$$\begin{aligned} &\langle u \rangle U_{x_i, x_j} \circ \langle v \rangle U_{x_k, x_l} \\ &= \frac{1}{4} (\langle uvx_i x_k \rangle U_{x_j, x_l} + \langle uvx_i x_l \rangle U_{x_j, x_k} + \langle uvx_j x_k \rangle U_{x_i, x_l} + \langle uvx_j x_l \rangle U_{x_i, x_k}). \end{aligned}$$

Thus, the basis and the multiplication tables of the algebras $IR[X]$ and $JS[X]$ coincide, what proves the lemma.

Theorem 5. $IR = \mathrm{Var}(J_\infty)$, *i.e. all the identities of a standard IR-algebra J_∞ follow from the identity* (1).

Proof. We have $\mathrm{Var}(J_\infty) \subseteq IR$. Let's prove a reverse inclusion. Let the homogenous polynomial

$$f = \sum_{i,j} \alpha_{ij} \langle u_{ij} \rangle U_{x_i,x_j},$$

where $\alpha_{ji} = \alpha_{ij} \in F$, be an identity on the algebra J_∞. Let's consider the homomorphism $\varphi : IR[X] \to J_\infty$, defined in the Lemma 6. Then

$$\varphi(f) = \frac{1}{2} \sum_{i,j} \alpha_{ij}(a_{ij} + a_{ji}) = 0.$$

Hence, $\alpha_{ij} = 0$ for all i, j and $f = 0$ in the algebra $IR[X]$. Consequently, the determining identities of the algebras $IR[X]$ and J_∞ coincide. This proves the theorem.

5. Annihilator of the free algebra $IR[X]$. In this section we will describe the generators of the $Ann(IR[X])$. The examples of non-zero elements of $\mathrm{Ann}(IR[X])$ were found in the Lemma 4:

$$(x \cdot z)U_{y,t} + (y \cdot t)U_{x,z} - (x \cdot t)U_{y,z} - (y \cdot z)U_{x,t} \underset{(12)}{\in} \mathrm{Ann}(IR[X]);$$

$$yD_{y,x^2} - 2yD_{y,x} \cdot x \underset{(13)}{\in} \mathrm{Ann}(IR[X]),$$

where $x = x_1$, $y = x_2$, $z = x_3$, $t = x_4$. Let

$$n = n(x,y,z,t,u) = \langle uzt \rangle U_{x,y} + \langle uxy \rangle U_{z,t} - \langle uzy \rangle U_{x,t} - \langle uxt \rangle U_{z,y},$$

where $x, y, z, t, u \in IR[X]$ and u can be a formal unit. Let's check that $n \in Ann(IR[X])$. Indeed,

$$\begin{aligned} 2n \cdot v &= \langle uzt \rangle U_{x,y}R_v + \langle uxy \rangle U_{z,t}R_v - \langle uzy \rangle U_{x,t}R_v - \langle uxt \rangle U_{z,y}R_v \\ &\underset{(19)}{=} \langle uztx \rangle U_{y,v} + \langle uzty \rangle U_{x,v} + \langle uxyz \rangle U_{t,v} + \langle uxyt \rangle U_{z,v} \\ &\quad - \langle uzyx \rangle U_{t,v} - \langle uzyt \rangle U_{x,v} - \langle uxtz \rangle U_{y,v} - \langle uxty \rangle U_{z,v} = 0. \end{aligned}$$

Lemma 8. *In the algebra $IR[X]$ the following relations are valid:*

1. $\mathrm{Ann}(IR[X]) = L(n(x,y,z,t,u))$, *where $x, y, z, t \in X$, and u runs over all ordered monomials of $SJ[X]$, including a formal* 1.
2. $\mathrm{Ann}(IR[X]/\mathrm{Ann}(IR[X])) = 0$.

Proof. 1. According to the above remark, we have

$$L = L(n(x,y,z,t,u)) \subseteq \mathrm{Ann}(IR[X]).$$

Let's now prove the reverse inclusion. Consider an arbitrary non-zero homogenous element $f \in \mathrm{Ann}(IR[X])$. It is obvious that $\deg(f) \geq 4$. Let's decompose it by the basis of $IR[X]$:

$$f = f(x_1, \ldots, x_n) = \sum_{i,j=1}^{n} \alpha_{ij} \langle u_{ij} \rangle U_{x_i,x_j},$$

where $\alpha_{ij} = \alpha_{ji} \in F$, $\deg(u_{ij}) \geq 2$. Renumerating, if required, the generators $x_1, \ldots, x_n$ in $f = f(x_1, \ldots x_n)$, we will have

$$\deg_{x_1}(f) \geq \deg_{x_2}(f) \geq \cdots \geq \deg_{x_n}(f) \geq 1.$$

If $\deg_{x_1}(f) \geq 2$, then due to the definition of the elements n we have

$$\langle ux_1^2\rangle U_{x_i,x_j} + \langle ux_ix_j\rangle U_{x_1,x_1} - \langle ux_1x_i\rangle U_{x_j,x_1} - \langle ux_1x_j\rangle U_{x_i,x_1} \in L.$$

Consequently,

$$f = \sum_{i=1}^{n} \alpha_{1i}\langle u_i\rangle U_{x_1,x_i} + u,$$

where $\alpha_{1i} \in F$, $u \in L$. Then

$$2f \cdot x_1 \underset{(7)}{=} 2\alpha_{11}\langle u_1x_1\rangle U_{x_1,x_1} + \sum_{i=2}^{n} \alpha_{1i}(\langle u_ix_1\rangle U_{x_1,x_i} + \langle u_ix_i\rangle U_{x_1,x_1}) = 0.$$

It follows from the view of the basis words of the algebra $IR[X]$ that $\alpha_{1i} = 0$ for $2 \leq i \leq n$, and hence $\alpha_{11} = 0$. Consequently, $f = u \in L$. If $\deg_{x_1}(f) = 1$, then $f(x_1, \ldots, x_n)$ is a multilinear polynomial. We have

$$\langle ux_1x_2\rangle U_{x_i,x_j} + \langle ux_ix_j\rangle U_{x_1,x_2} - \langle ux_ix_2\rangle U_{x_1,x_j} - \langle ux_1x_j\rangle U_{x_2,x_j} \in L,$$

$$\langle ux_1x_3\rangle U_{x_2,x_4} + \langle ux_2x_4\rangle U_{x_1,x_3} - \langle ux_2x_3\rangle U_{x_1,x_4} - \langle ux_1x_4\rangle U_{x_2,x_3} \in L.$$

Therefore,

$$f = \sum_{i=2}^{n} \alpha_i\langle u_{1i}\rangle U_{x_1,x_i} + \beta\langle v\rangle U_{x_2,x_3} + u,$$

where $\alpha_i, \beta \in F$, $u \in L$. Then

$$2f \cdot x_{n+1} = \sum_{i=2}^{n} \alpha_i(\langle u_{1i}x_1\rangle U_{x_i,x_{n+1}} + \langle u_{1i}x_i\rangle U_{x_i,x_{n+1}}) + \beta\langle vx_2\rangle U_{x_3,x_{n+1}} + \beta\langle vx_3\rangle U_{x_2,x_{n+1}} = 0.$$

It follows from the view of the basic words of $IR[X]$ that $\alpha_i = 0$ at $2 \leq i \leq n$, hence, $\beta = 0$, too. Consequently, $f = u \in L$.

2. Let us consider an arbitrary non-zero homogenous element

$$f = f(x_1, \ldots, x_n) \in IR[X]/\operatorname{Ann}(IR[X]).$$

From what has already been proved, it follows that $\deg(f) \geq 4$. Let

$$f \in \operatorname{Ann}(IR[X]/\operatorname{Ann}(IR[X])).$$

Then we have $f \cdot u \cdot v = 0$ for any $u, v \in IR[X]$. Proceeding analogously as above, we come to two cases: either

1. $f = \sum_{i=1}^{n} \alpha_{1i}\langle u_i\rangle U_{x_1,x_i}$, where $\alpha_{1i} \in F$. But in this case $f \cdot x_1 \cdot x_1 = 0$ and $\alpha_{1i} = 0$ for $2 \leq i \leq n$, and hence $\alpha_{11} = 0$.

or

2. $f = \sum_{i=2}^{n} \alpha_i\langle u_{1i}\rangle U_{x_1,x_i} + \beta\langle v\rangle U_{x_2,x_3}$, where $\alpha_i, \beta \in F$. But in this case $f \cdot x_{n+1} \cdot x_{n+2} = 0$ and $\alpha_i = 0$ for $2 \leq i \leq n$, and hence $\beta = 0$, too, what makes a contradiction. Consequently, $\operatorname{Ann}(IR[X]/\operatorname{Ann}(IR[X])) = 0$. The Lemma is proved.

Corollary. $\mathrm{Ann}(IR[X]) \subseteq D = D(IR[X])$.

Proof. It follows from the Lemma 8 that it is sufficient to prove that $\forall x, y, z, t, u \in IR[X] \quad n = n(x, y, z, t, u) \in D$. We have

$$n = \langle uzt\rangle U_{x,y} - \langle uzy\rangle U_{x,t} + \langle uxy\rangle U_{z,t} - \langle uxt\rangle U_{z,y} \underset{(18)}{=} \langle uz\rangle(R_t U_{x,y} - R_y U_{x,t}) + \langle ux\rangle(R_y U_{z,t} - R_t U_{z,y}).$$

But

$$4xD_{y,z}R_t \underset{(16)}{=} 2(yU_{z,x} - zU_{y,z})R_t \underset{(7)}{=} x(R_y U_{z,t} - R_z U_{y,t}),$$

thus, $n \in D$. This proves the corollary.

Let us choose a basis $\langle u\rangle$ in the free associative commutative algebra $F[X]$ of generation $X = \{x_1, \dots, x_n, \dots\}$, where $\langle u\rangle$ runs over all ordered monomials of $SJ[X]$. Then $F[X] = F(\langle u\rangle)$.

Define $D_1 = \mathrm{Ann}(IR[X])$. By the corollary, $D_1 \subseteq D$. Let D_0 be a direct complement of F-module D_1 in D, then $D = D_0 \oplus D_1$.

Theorem 6. *The following isomorphism of F-modules takes place:*

$$IR[X] \simeq F[X] \oplus D,$$

where $D = D_0 \oplus D_1$, $D_0^2 \subseteq D_1$, $D_1 = \mathrm{Ann}(IR[X])$ and $D^3 = 0$.

Proof. It is obvious that $u = \langle u\rangle + d$ where a monomial $u \in F[X]$ and $d \in D$. Therefore, $IR[X] \simeq F[X] + D$. If $\langle w\rangle \in F[X] \cup D$, then it is clear that $\langle\langle u\rangle\rangle = \langle u\rangle = 0$. Hence, $IR[X] = F[X] \oplus D$. Let's verify that $D_0^2 \subseteq D_1$. By (17) it is sufficient to prove that $\forall a, b, c, x, y, z \in IR[X] \quad aD_{b,c} \cdot xD_{y,z} \in D_1$. We have

$$3aD_{b,c} \cdot xD_{y,z} = 3aD_{b,c}D_{y,z} \cdot x + 3(aD_{b,c} \cdot x)D_{y,z} \underset{(10)}{=} 3aD_{b,c}D_{y,z} \cdot x + xD_{a\cdot b,c}D_{y,z} + xD_{b,c\cdot a}D_{y,z}.$$

On the other hand,

$$2aD_{b,c}D_{y,z} \underset{(16)}{=} yU_{(aDb,c),z} - zU_{(aDb,c),y} \underset{(13)}{=} \frac{1}{2}(\langle yab\rangle U_{c,z} + \langle yc\rangle U_{a\cdot b,z} - \langle yac\rangle U_{b,z} - \langle yb\rangle U_{a\cdot c,z} - \langle zab\rangle U_{c,y} + \langle zc\rangle U_{a\cdot b,y} + \langle zac\rangle U_{b,y} - \langle zb\rangle U_{a\cdot c,y})$$
$$\underset{(13)}{=} \frac{1}{2}(\langle ayb\rangle U_{c,z} + \langle azc\rangle U_{y,b} - \langle ayc\rangle U_{b,z} - \langle azc\rangle U_{b,y}) + \frac{1}{4}(\langle yca\rangle U_{b,z} - \langle yba\rangle U_{c,z} - \langle zca\rangle U_{b,y} + \langle zba\rangle U_{c,y}) \underset{(20)}{=} \mathrm{Ann}(IR[X]).$$

Hence, $3aD_{b,c} \cdot xD_{y,z} \in D_1$. Further, $D^3 = (D_0 \oplus D_1)^3 \subseteq D_0^3 \subseteq D_1 \cdot D_0 = 0$. The theorem is proved.

Corollary. $D = M(IR[X])$, *where $M(IR[X])$ is a MacCrimmon radical of $IR[X]$.*

Proof. By the Theorem 6, $\forall d \in D$, $z \in IR[X]$, $zU_{d,d} \in \mathrm{Ann}(IR[X]) \subseteq \mathbb{Z}(IR[X])$, where $\mathbb{Z}(IR[X])$ is an ideal of $IR[X]$ generated by all absolute zero divisors. That is why $D \subseteq M(IR[X])$. But $IR[X]/D = F[X]$, hence $D = M(IR[X])$. The corollary is proved.

6. Basis of the identities of the algebras J_n and $J_{n,m}$. Let's find the basis of the algebra $J_{1,2}$. This algebra has the basis $a = a_{11}$, $b = a_{12}$ an and the following multiplication table:

$$a^2 = a, \quad b^2 = b, \quad a \cdot b = \frac{1}{2}(a+b).$$

Lemma 9. $\mathrm{Var}(J_\infty) = \mathrm{Var}(J_{1,2})$.

Proof. Since $J_{1,2}$ is a subalgebra of J_∞, then $\mathrm{Var}(J_{1,2}) \subseteq \mathrm{Var}(J_\infty)$. Let's suppose that $\mathrm{Var}(J_{1,2}) \neq \mathrm{Var}(J_\infty)$. Then there exists a nonzero homogenous multilinear polynomial $f = f(x_1, \ldots, x_n) \in IR[X]$, which is an identity on $J_{1,2}$. It is evident that $\deg f \geq 3$. Let's decompose it over the basis of the algebra $IR[X]$:

$$f = \sum_{1 \leq i < j \leq n} \alpha_{ij} \langle u_{ij} \rangle U_{x_i, x_j},$$

where $\alpha_{ij} \in F$. Reindexing if necessary the generators of f we can assume that $\alpha_{1,2} \neq 0$. Let $\bar{f} = f|_{x_1=v_1,\ldots,x_n=v_n}$ is the value of polynomial f in the algebra $J_{1,2}$ for $x_1 = v_1, \ldots, x_n = v_n$, where $v_i \in J_{1,2}$. Let $x_1 = a_{12}$, $x_2 = \cdots = x_n = a_{11}$. Then

$$\bar{f} = \sum_{i=2}^{n} \alpha_{1i} \langle \bar{u}_{1i} \rangle U_{a_{12},a_{11}} + \sum_{2 \leq i < j \leq n} \alpha_{ij} \langle \bar{u}_{ij} \rangle U_{a_{11},a_{11}}$$

$$\underset{(4)}{=} \sum_{i=2}^{n} \alpha_{1i}(a_{12}a_{11}) + \sum_{2 \leq i < j \leq n} \alpha_{ij} a_{11} = \frac{1}{2} \sum_{i=2}^{n} \alpha_{1i}(a_{11}+a_{12}) + \sum_{2 \leq i < j \leq n} \alpha_{ij} a_{11} = 0.$$

Consequently,

$$\sum_{i=2}^{n} \alpha_{1i} = 0.$$

Similarly,

$$\sum_{i=1, j \neq i}^{n} \alpha_{ij} = 0.$$

Let $x_1 = a_{11} + 2a_{12}$, $x_2 = a_1 + 3a_{12}$, $x_3 = \cdots = x_n = a_{11}$. Then

$$\bar{f} = \alpha_{12} \langle \bar{u}_{12} \rangle U_{(a_{11}+2a_{12}),(a_{11}+3a_{12})} + \sum_{i=3}^{n} \alpha_{1i} \langle \bar{u}_{1i} \rangle U_{(a_{11}+2a_{12}),a_{11}}$$
$$+ \sum_{i=3}^{n} \alpha_{2i} \langle \bar{u}_{2i} \rangle U_{(a_{11}+3a_{12}),a_{11}} + \sum_{3 \leq i < j \leq n} \alpha_{ij} \langle \bar{u}_{ij} \rangle U_{a_{11},a_{11}} = 0.$$

By (4):

$$\langle \bar{u}_{12} \rangle U_{(a_{11}+2a_{12}),(a_{11}+3a_{12})} = (a_{11} + 2a_{12}) \cdot (a_{11} + 3a_{12})$$
$$= a_{11} + \frac{3}{2}(a_{12} + a_{11}) + (a_{11} + a_{12}) + 6a_{12} = \frac{7}{2}a_{11} + \frac{17}{2}a_{12};$$

$$\langle \bar{u}_{1i}\rangle U_{(a_{11}+2a_{12}),a_{11}} = (a_{11}+2a_{12})\cdot a_{11} = 2a_{11}+a_{12};$$

$$\langle \bar{u}_{2i}\rangle U_{(a_{11}+3a_{12}),a_{11}} = (a_{11}+3a_{12})\cdot a_{11} = a_{11}+\frac{3}{2}(a_{11}+a_{12}) = \frac{5}{2}a_{11}+\frac{3}{2}a_{12};$$

$$\langle \bar{u}_{ij}\rangle U_{a_{11},a_{11}} = a_{11}\cdot a_{11} = a_{11}.$$

Consequently,

$$\alpha_{12}\left(\frac{7}{2}a_{11}+\frac{17}{2}a_{12}\right)+\sum_{i=3}^{n}\alpha_{1i}(2a_{11}+a_{12})$$
$$+\sum_{i=3}^{n}\alpha_{2i}\left(\frac{5}{2}a_{11}+\frac{3}{2}a_{12}\right)+\sum_{3\le i<j\le n}\alpha_{ij}a_{11}=0.$$

Hence

$$\alpha_{12}(7a_{11}+17a_{12})+\sum_{i=3}^{n}\alpha_{1i}(4a_{11}+2a_{12})+\sum_{i=3}^{n}\alpha_{2i}(5a_{11}+3a_{12})+\sum_{3\le i<j\le n}2\alpha_{ij}a_{11}=0.$$

Since $\sum_{i=2}^{n}\alpha_{1i}=0$ and $\sum_{1\le i\le n, i\ne 2}\alpha_{2i}=0$, we have

$$\alpha_{12}(3a_{11}+15a_{12})+\sum_{i=3}^{n}\alpha_{2i}(5a_{11}+3a_{12})+\sum_{3\le i<j\le n}2\alpha_{ij}a_{11}=0.$$

and

$$12\alpha_{12}a_{12}+\sum_{i=3}^{n}\alpha_{2i}(2a_{11})+\sum_{3\le i<j\le n}2\alpha_{ij}a_{11}=0.$$

Consequently,, $\alpha_{12}=0$. This contradicts our assumption. This proves the lemma.

Theorem 7. *The variety* $\mathrm{Var}(J_{1,1})=K$ *is the variety of associative commutative algebras and has the determining identity* $xD_{y,z}=0$*, the variety* $\mathrm{Var}(J_{1,2})=\mathrm{Var}(J_{n,m})=\mathrm{Var}(J_\infty)=IR$*,* $(n,m)\ne(1,1)$*, and has the determining identity*(1).

Proof. It is obvious that $\mathrm{Var}(J_{1,1})=K$. From the chain of inclusion

$$J_{1,2}\subseteq J_{n,m}\subseteq J_\infty$$

we get

$$\mathrm{Var}(J_{1,2})\subseteq \mathrm{Var}(J_{n,m})\subseteq \mathrm{Var}(J_\infty).$$

From the Lemma 9 and the Theorem 5 it follows that

$$\mathrm{Var}(J_{1,2})=\mathrm{Var}(J_{n,m})=\mathrm{Var}(J_\infty)=IR.$$

The theorem is proved.

7. Reflexive varieties of Jordan algebras. The proof of the specialty of variety IR in the Section 3 required application of two rather complex results: description of identities of the variety $\mathrm{Var}(B_\infty)$ [7] and the specialty of the variety

$\mathrm{Var}(B_\infty)$ [8]. The basis and the multiplication table of the algebras $S[X]$ $IR[X]$ (Section 4) allow us to easily prove the specialty of the variety IR.

Let $\mathfrak{M}$ be some homogenous variety of Jordan algebras. Let a free algebra $\mathfrak{M}[X]$ in $\mathfrak{M}$ be a special Jordan algebra and $A[X]$ be some associative envelope algebra for $\mathfrak{M}[X]$.

There exists a natural involution $*$ on $A[X]$ which acts on the monomials by $(x_1, \ldots, x_n)^* = x_n \ldots x_1$ and linearly extends over the whole algebra $A[X]$. We will denote by $HA[X]$ a Jordan algebra of symmetric elements of $A[X]$ in regard to $*$. It is obvious that $\mathfrak{M}[X] \subseteq HA[X]$.

Definition. The variety $\mathfrak{M}$ is called *reflexive*, if $\mathfrak{M}[X] = HA[X]$ for some associative envelope algebra $A[X]$ of $\mathfrak{M}[X]$.

Theorem 8. *Any reflexive variety of Jordan algebras is special.*

Proof. It suffices to prove that all homomorphic images of $\mathfrak{M}[X]$ are special Jordan algebras. Let's consider the algebra $J \simeq \mathfrak{M}[X]/I$ and let's prove that J is a special algebra. In accordance with the Cohn lemma [8] it is sufficient to show that

$$\widehat{I} = (I)_{A[X]} \cap \mathfrak{M}[X] = I,$$

where $(I)_{A[X]}$ is the ideal of the algebra $A[X]$, generated by the set I.

Let $f \in \widehat{I}$. Then $f = \sum_i \alpha_i a_i u_i b_i$, where a_i, b_i are the monomials of $A[X]$ or formal units, and $u_i \in I$. Since $f \in \mathfrak{M}[X]$, then $f^* = f$. Consequently, in view of reflexivity of the $\mathfrak{M}$ we have

$$f = \frac{1}{2} \sum_i \alpha_i (a_i u_i b_i + b^* u_i a_i^*) = \sum_i h_i(u_i, X),$$

where $h_i(u_i, X)$ are Jordan polynomials of u_i and X. Therefore, $f \in I$, $\widehat{I} = I$. This proves the theorem.

Corollary. *The variety IR is reflexive and, consequently, it is special.*

Proof. The reflexivity of IR is proved in the Proposition 3, and the corollary is proved.

It would be interesting to describe the reflexive varieties of Jordan algebras.

Acknowledgements. I thank professor I. Hentzel for bringing my interest in investigation of use of Jordan algebras in mathematical genetics and professor M. Bremner for setting an interesting question on specialty of Jordan algebras arising from intermolecular recombination.

References

1.. Bremner M., *Jordan algebras arising from inter molecular recombination*, SIGSAM Bull. **39** (2005), no. 4, 106–117.
2. Păun G., Rozenberg G, Salomaa A., *DNA computung: new computing paradigms*, Springer-Verl., Berlin, 1998.
3. Landweber L., Kari L., *The evolution of cellular computing: nature's solution to a computational problem*, Biosystems **52** (1999), 3–13.
4. Zhevlakov K. A., Slinko A. M., Shestakov I. P., and Shirshov A. I., *Rings that are nearly associative*, Nauka, Moscow, 1978. (in Russian).

5. Wörz-Busekros A., *Algebras in genetics*, Springer-Verl., New York, 1980 (Lecture notes in biomath; N 36).
6. Concepción López-Díaz M., Shestakov I. P., Sverchkov S. R., *On speciality of Bernstein Jordan algebras*, Comm. Algebra **28** (2000), no. 9, 4375–4387.
7. Vasilovskiy S., *Basis of the identity of a Jordan algebra of bilinear form over the infinite field, Investigation in the theory of rings and algebras,*, Sib. Adv. Math. **1** (1991), no. 4, 142–185.
8. Sverchkov S. R., *Varieties of special algebras*, Comm. Algebra **16** (1988), no. 9, 1877–1919.
9. Holgate P., *Genetic algebras satisfying Bernstein's stationary principle*, J. London Math. Soc. **2** (1975), no. 9, 613–623.
10. Holgate P., *Jordan algebras arising in population genetics*, Proc. Edinburgh Math. Soc. **15** (1967), no. 2, 291–294.
11. Wörs-Busekros A., *Bernstein algebras*, Arch. Math. **48** (1987), 388–398.
12. Lubich U., *Bernstein Algebras*, Uspehi Mat. Nauk **32** (1977. (in Russian)), no. 6, 261–262.
13. Walcher S., *Bernstein algebras which are Jordan algebras*, Arch. Math. **50** (1988), 218–222.

Sergei R. Sverchkov
Department of Algebra & Logic
Novosibirsk State University
2 Pirogov Str., 630090 Novosibirsk, Russia
E-mail address: sverchkovSR@yandex.ru